Cover Photomicrograph

The photomicrograph on the cover shows synthetic diamonds which were grown on a silicon substrate by microwave assisted chemical vapor deposition. A more detailed description of this photomicrograph and other synthetic diamond films is provided in the article, "Microscopic Evaluation of Synthetic Diamond Films" by Richard H. Lee on page17.

Microstructural Science
Volume 22

Metallographic Techniques and the Characterization of Composites, Stainless Steels, and Other Engineering Materials

Proceedings of the Twenty-Seventh Annual Technical Meeting of the International Metallographic Society

Edited by

M.T. Shehata, Technical Meeting Chairman
Canada Centre for Mineral and Energy Technology

T.R. Leduc, General Chairman
Gerton Consulting

I. LeMay, General Co-Chairman
Canada Centre for Mineral and Energy Technology

M.R. Louthan, Jr., Series Editor
Savannah River Technology Center

International Metallographic Society, Inc.
an affiliate of ASM International®

ASM INTERNATIONAL®
MATERIALS PARK, OHIO USA

The Materials Information Society

First printing, June 1995

This book is a collective effort involving hundreds of technical specialists. It brings together a wealth of information from world-wide sources to help scientists, engineers, and technicians solve current and long-range problems.

Comments, criticisms, and suggestions are invited, and should be forwarded to ASM International.

Library of Congress Cataloging Card Number: 95-76678
ISBN: 0-87170-537-0
SAN: 204-7586

ASM International®
Materials Park, OH 44073-0002

Printed in the United States of America

Preface

Volume 22 of Microstructural Science contains selected manuscripts based on presentations made to the Technical Meeting of the 27th Annual Convention of the International Metallographic Society. The Technical Meeting meeting was in held Montreal, Canada on July 26-27, 1994. These proceedings, prepared from camera ready manuscripts, summarize recent work of authors and coauthors representing eight countries.

The Technical Meeting began with the ASM-IMS Distinguished Lecture in Metallography, "Control of Recrystallization in Aluminum Alloys" by Dr. T. H. Sanders, Jr. from the Georgia Institute of Technology. Dr. Underwood's lecture provided an excellent keynote review for the conference by emphasizing the importance of second phase particles to the recrystallization behavior of precipitation hardenable aluminum alloys.

Technical Sessions were chaired by U. Harnisch, C. Bagnall, D. W. Stevens, R. H. Lee, J. D. Boyd, J. Lo, G. F. VanderVoort, C. Forget, W. R. Kanne, Jr., M. French, W. E. White, G. Johnson, G. Blann, and M. Braunovic. The arrangement of the papers in the proceedings does not correspond to the order of presentations during the Technical Sessions, partially because every presentation is not included in the proceedings.

The success of the Technical Meeting was primarily due to the presentations and to the discussions that the presentations inspired. However, valuable contributions were also provided through the support of the IMS Board of Directors and IMS Corporate Members. The proceedings of the Technical Meeting represents only a small fraction of the results of a most successful Annual Convention.

The real work required to prepare the proceedings was accomplished under the direction of Veronica Flint of ASM International. The editors are most grateful to her for assuming this challenging role. Veronica has continually provided the guidance and direction necessary to publish Microstructural Science on a timely basis. Many thanks for her outstanding contributions.

M. T. Shehata, T. R. Leduc, I. LeMay and M. R. Louthan, Jr.

CONTRIBUTORS

Asensio, J.
Universidad de Oviedo
Oviedo, Spain

Belley, M.
Alcan International Ltd.
Jonquière, Québec, Canada

G.A. Blann
Buehler Ltd.
Lake Bluff IL

Blum, W.
University Erlangen-Nürnberg
Erlangen, Germany

Bousfield, B.
Buehler Europe Ltd.
Coventry, England

Bowker, J.T.
CANMET/MTL
Ottawa, Ontario, Canada

Boutin, J.
Alcan International Ltd.
Jonquière, Québec, Canada

Boyd, J.D.
Queen's University
Kingston, Ontario, Canada

Burger, G.
Alcan International Ltd.
Kingston, Ontario, Canada

Carpenter, G.
MTL/CANMET
Ottawa, Ontario, Canada

Casault, B.R.
MTL/CANMET
Ottawa, Ontario, Canada

Cheung, P.
Tito Silveira Engenharia e Consultaria Ltda.
Rio de Janeiro, Brazil

Clark, E.A.
Westinghouse Savannah River Co.
Aiken SC

Covino, B.S., Jr.
U.S. Bureau of Mines
Albany OR

Cramer, S.D.
U.S. Bureau of Mines
Albany OR

Czajkowski, C.
Brookhaven National Lab
Upton NY

da Silva, M.H.P.
COPPE/UFRJ
Rio de Janeiro, Brazil

da Silveira, T.L.
Tito Silveira Engenharia e Consultaria Ltda.
Rio de Janeiro, Brazil

de Almeida, L.H.
COPPE/UFRJ
Rio de Janeiro, Brazil

de Almeida Soares, G.C.
COPPE/UFRJ
Rio de Janeiro, Brazil

Danials, R.
University of Oklahoma
Norman OK

Dascoli, M.
Riverhead High School
Riverhead NY

Diaz, D.J.
Buehler Ltd.
Lake Bluff IL

Drew, R.A.L.
McGill University
Montréal, Québec, Canada

Dufour, H.
Alcan International Ltd.
Jonquière, Québec, Canada

Dunning, J.S.
U.S. Bureau of Mines
Albany OR

Edwards, E.
Rensselaer Polytechnic Institute
Troy NY

Eisele, P.
Lumonics, Inc.
Kanata, Ontario, Canada

Eissner, G
Max-Planck-Institut für Metallforschung
Stuttgart, Germany

Ellis, F.V.
Tordonato Energy Consultants, Inc.
Chattanooga TN

Entezarian, M.
McGill University
Montréal, Québec, Canada

Feng, C
U.S. Army R & D
Picatinny Arsenal NJ

Fowler, J.P.
Carleton University
Ottawa, Ontario, Canada

Gendron, A.
Canadian Liquid Air Ltd.
Montréal, Québec, Canada

Habib, K.
Kuwait Institute for Scientific Research
Safat, Kuwait

Hassenzahl, W.V. (ret'd)
Lawrence Livermore National Lab
Livermore CA

Herrera, J.M.
The University of Texas
El Paso TX

Hetzner, D.W.
Timken Research
Canton OH

Hull, D.A.
Engineering Material Research
Toronto, Ontario, Canada

Immarigeon, J-P.
NRC Institute for Aerospace Research
Ottawa, Ontario, Canada

Imrich, K.J.
Westinghouse Savannah River Co.
Aiken SC

Innus, A.B.
Alcan Smelters and Chemical Ltd.
Jonquière, Québec, Canada

Iyer, N.C.
Westinghouse Savannah River Co.
Aiken SC

Katz, O.M.
Westinghouse Electric Corp.
West Mifflin PA

Konopleva, E.V.
Institute Solid State Physics
Chernogolovka, Russia

Korn, D.
Max-Planck-Institut für Metallforschung
Stuttgart, Germany

Krishnadev, M.R.
Laval University
Québec, Canada

Leader, D.R.
Westinghouse Savannah River Co.
Aiken SC

Lebleu, M.
Alcan International Ltd.
Jonquière, Québec, Canada

Lee, R.H.
Argonne National Lab
Argonne IL

Leithner, K.A.
Buehler Ltd.
Lake Bluff IL

Le May, I.
Metallurgical Consulting Services Ltd.
Saksatoon, Canada

Lin, Y.C.
Tordonato Energy Consultants, Inc.
Chattanooga TN

Lloyd, D.J.
Alcan International Ltd.
Kingston, Ontario, Canada

Lo, J.
MTL/CANMET
Ottawa, Ontario, Canada

Lockley, A.J.
AECL Research
Chalk River, Ontario, Canada

Louisnathan. S.J.
Corning Inc. R, D & E Div.
Corning NY

Louthan, M.R., Jr.
Westinghouse Savannah River Co.
Aiken SC

Luo, A.
Institute of Magnesium Technology, Inc.
Sainte-Foy, Québec, Canada

Magrani, S.J.G.
COPPE/UFRJ
Rio de Janeiro, Brazil

Maldonado, J.G.
The University of Texas
El Paso TX

Malis, T.
MTL/CANMET
Ottawa, Ontario, Canada

McQueen, H.J.
Concordia University
Montréal, Québec, Canada

Miller, I.
Lumonics Inc.
Kanata, Ontario, Canada

Mirtchi, A.
Alcan Intrnational Ltd.
Jonquière, Québec, Canada

Montgomery, J.
Lumonics Inc.
Kanata, Ontario, Canada

Morrison, J.
Defence Research Establishment
Victoria, B.C., Canada

Mueller, H.J.
American Dental Association
Chicago IL

Murr, L.E.
The University of Texas
El Paso TX

Nelson, J.A.
Buehler Ltd.
Lake Bluff IL

Niou, C-S.
The University of Texas
El Paso TX

Pappu, S
The University of Texas
El Paso TX

Pekguleryuz, M.
Institute of Magnesium Technology, Inc.
Sainte-Foy, Québec, Canada

Pero-Sanz, J.A.
Universidad Politécnica de Madrid
Madrid, Spain

Pilkey, A.K.
Carleton University
Ottawa, Ontario, Canada

Putatunda, S.K.
Wayne State University
Detroit MI

Quinones, S.
The University of Texas
El Paso TX

Raymond, Y.
Alcan International Ltd.
Jonquière, Québec, Canada

Réne, R.
MIL Davie, Inc.
Lévis, Québec, Canada

Ribeiro, A.F.
COPPE/UFRJ
Rio de Janeiro, Brazil

Rivas, J.M.
The University of Texas
El Paso TX

Romero, R.J.
The University of Texas
El Paso TX

Saenz, N.T.
Pacific Northwest Lab
Richland WA

Schaeffer, J.
Detroit Edison
Detroit MI

Sengupta, A.
Wayne State University
Detroit MI

Shehata, M.
MTL/CANMET
Ottawa, Ontario, Canada

Simmons, J.W.
U.S. Bureau of Mines
Albany OR

Stafford, S.W.
The University of Texas
El Paso TX

Stevens, D.W.
Lawrence Livermore National Lab
Livermore CA

Stone, R.
Buehler Ltd.
Lake Bluff IL

Tao, S.
Queen's University
Kingston, Ontario, Canada

Townley, N.
Queen's University
Kingston, Ontario, Canada

Trillo, E.A.
The University of Texas
El Paso TX

Tsantrizos, P.
Pegasus Refractory Materials Inc.
Montréal, Québec, Canada

Vaidya, V.
Canadian Liquid Air, Ltd.
Montréal, Québec, Canada

Vega, L.
The University of Texas
El Paso TX

Verdeja, J.I.
Imoversodad de Pvoedp
Oviedo, Spain

Warburton, K.
Brookhaven National Lab
Upton NY

West, S.L.
Westinghouse Savannah River Co.
Aiken SC

Williams, G.
MTL/CANMET
Ottawa, Ontario, Canada

Worswick, M.J.
Carleton University
Ottawa, Ontario, Canada

Xia, X.
Concordia University
Montréal, Québec, Canada

Zhai, Z.
Defence Research Establishment
Victoria, B.C., Canada

Zhang, W.L.
Laval University
Québec, Canada

CONTENTS

Metallographic Techniques

Composites and Polymers

Light Metals

Fatigue, Fracture and Failure Analysis

Chapter 1

Metallographic Applications

METALLOGRAPHY OF HYPERVELOCITY IMPACT CRATER FORMATION IN ALUMINUM AND COPPER

Stella Quinones, J. M. Rivas, and L. E. Murr[1]

ABSTRACT

In this preliminary study optical and transmission electron microscopy has been utilized to observe and compare the microstructures associated with hypervelocity (>5km/s) impact craters in an OFHC copper target and an 1100 aluminum target. In the regions proximate to the crater wall, the microstructure exhibits dynamic recrystallization in both targets. Beyond this recrystallized zone, the grain structure exhibits heavy distortion which declines with distance from the crater wall. Microhardness measurements along the impact axis for the copper revealed a low value within the recrystallized zone with increasing hardness with increasing distance from the crater wall. A region of maximum hardness extended well beyond the point where recognizable grain distortion ceased, and in copper this region contained microbands coincident with {111}. For aluminum, the corresponding zone was characterized by dislocation cells whose size increased with increasing distance from the crater wall.

INTRODUCTION

Studies of ballistic and hypervelocity penetration, especially laboratory simulations using high velocity guns, have been a popular

[1]Department of Metallurgical and Materials Engineering, The University of Texas at El Paso, El Paso, TX 79968

means of investigating cratering in spacecraft materials for more than 40 years (1-4). It is remarkable to realize that while considerable data regarding parametric relationships between impacting particle diameter (dp) density (ρ_p), velocity (u_o), and the target crater geometry (depth of penetration, p; crater diameter, D_c) and target density, ρ_t, etc. have been accumulated, there is not a single, systematic metallurgical analysis of crater-related deformation phenomena in metallic targets. While shock effects are normally associated with hypervelocity impact, most commonly through the Bernoulli condition: $P(\sigma) = 0.5\ \rho_p\rho_t/(\rho_p^{1/2} + \rho_t^{1/2})^2\ u_o^2$, there have never been any attempts to observe these effects in connection with residual crater microstructures, or to differentiate them from plastic deformation which most surely accompanies the crater formation.

In this study, we have observed the metallographic details of crater-related deformation in OFHC copper and 1100 aluminum targets using optical and transmission electron microscopy.

EXPERIMENTAL DETAILS

Using optical metallography and transmission electron microscopy techniques, archived impact craters in OFHC copper and 1100 aluminum targets have been examined in this preliminary work. For the 1.3 cm thick copper target, an 1100 aluminum impact sphere 3.17 mm in diameter impacted the OFHC copper target at 6.73 km/s. This produced a crater 1.1 cm in diameter (D_c) and a penetration depth, P, of 5.5 mm. The corresponding Bernoulli pressure was 25.4 GPa. The 1100 aluminum target crater measured 1.4 cm in diameter and had a penetration depth of 7 mm produced by a soda-lime glass (ρ = 2.2 g/cm^3) sphere 3.1 mm in diameter at 5.85 km/s. The corresponding Bernoulli pressure was 10.4 GPa. The crater samples were cut in half and the half-sections ground, polished, and etched to reveal the residual microstructures. Figure 1 illustrates the corresponding crater half sections. Transmission electron microscopy techniques developed previously for micrometeoroid craters in LDEF samples (5) were employed in systematically punching TEM discs from specific locations in relation to the crater wall as illustrated in Fig. 2. In addition, Vickers microhardness measurements were made along the crater impact axes below the crater wall bottom using a 100 gf load.

RESULTS AND DISCUSSION

Figure 3 illustrates the typical microstructures and microstructure zones below the OFHC copper crater bottom to consist of a dynamically recrystallized region next to the crater wall which extends into a highly deformed region characterized by heavy grain distortion; indicated by Zone I. Zone I in Fig. 3 is followed by a

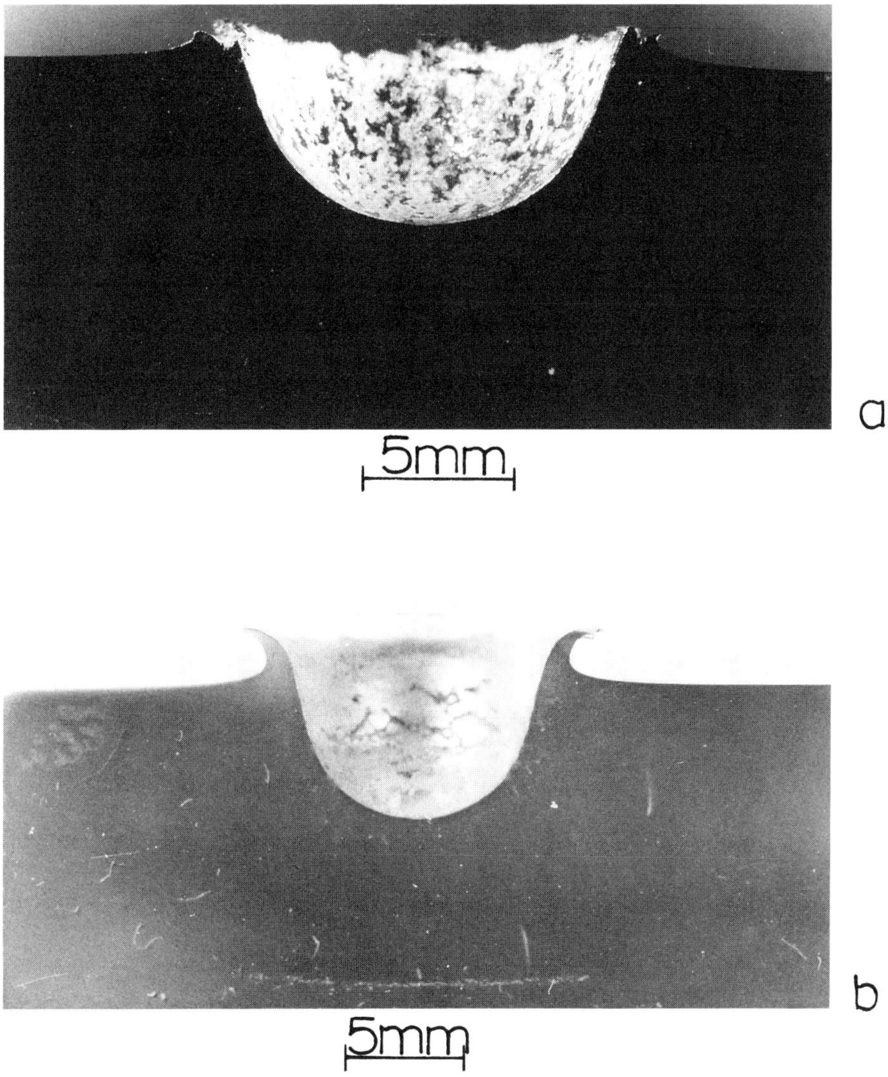

Figure 1. Polished and etched crater half sections in OFHC copper (a) and 1100 aluminum (b).

transition zone which includes grain distortion and other linear microstructure features within the grains. These linear features extend into Zone II which is characterized by relatively undistorted grains. These features are illustrated in more detail in the magnified views of the prominent zone microstructures along with the selected microhardness values at locations marked in Fig. 3.

The microstructural details typical of the crater-related deformation zones indicated in Fig. 3 are illustrated in Fig. 4 and compared with the microstructure for the undeformed (unimpacted) plate in Fig. 4(d). The heavily deformed or distorted grain regions in Zone I and the transition zone (Tr) (Fig. 3) exhibit varying degrees of dislocation cells as shown in Fig. 4(a). The cell sizes generally increase with distance from the crater which corresponds to a decrease in the effective strain. However, cell sizes decrease near the recrystallized zone, then increase in the grain distorted region and then decrease again. However, the cell size is often very small and dislocation density large in the linear microstructure regions shown in Fig. 4(b) and (c) which are recognized as microbands with dense dislocation walls. These microbands are aligned with the trace of {111} and their density declines with distance from the crater wall. Beyond these microbands in Zone II the microstructure is characterized by dislocation cells as in Fig. 4(a) which increase in size with distance from the crater. Figure 4(d) shows large dislocation cells in the original OFHC copper plate as a consequence of roll fabrication.

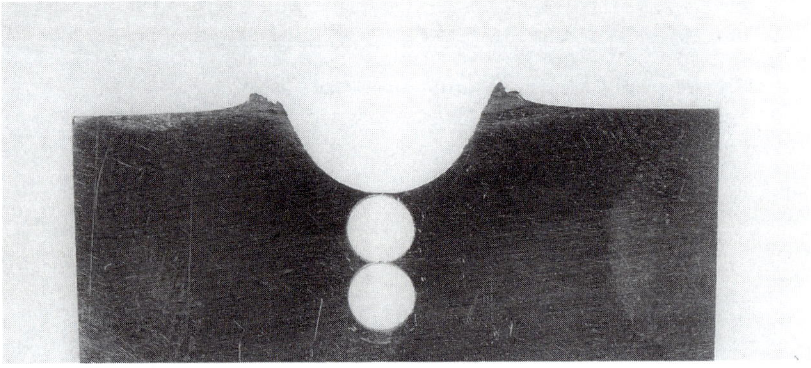

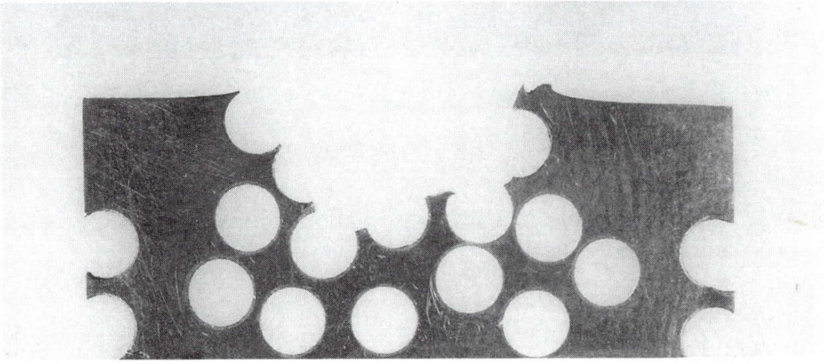

Figure 2. Examples of TEM disc punching from slices cut from the half section shown in Fig. 1(a) for OFHC copper. The initial 0.2 mm thick slices were ground and polished to roughly 40 μm thickness prior to 3 mm disc punching.

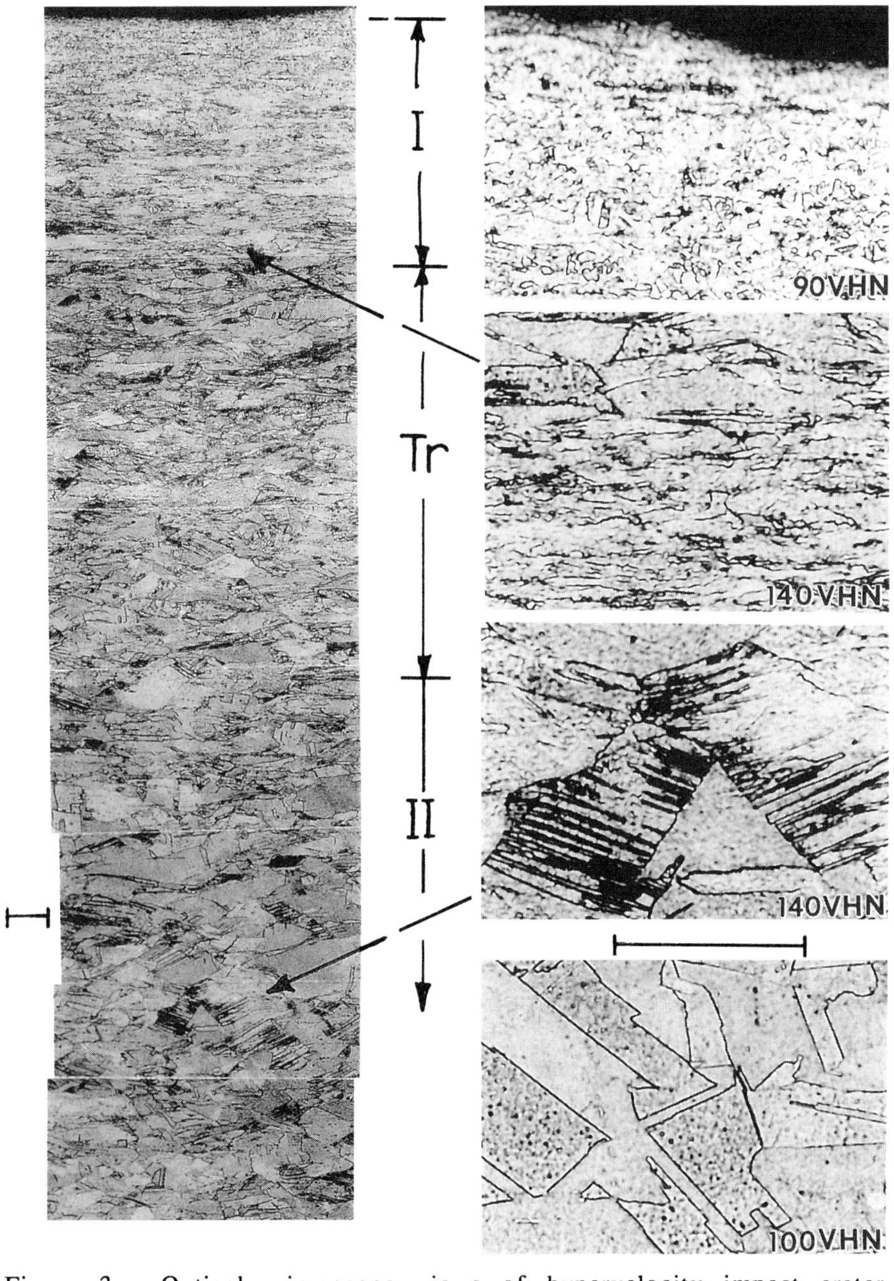

Figure 3. Optical microscope views of hypervelocity impact crater-related microstructures in OFHC copper. The composite to the left corresponds to the impact axis direction below the crater bottom. Progressive zones are noted, and the right-hand figures show magnified views of characteristic microstructures. The lower right micrograph illustrates the base (undeformed) microstructure for the plate for comparison. Markers are 50 μm.

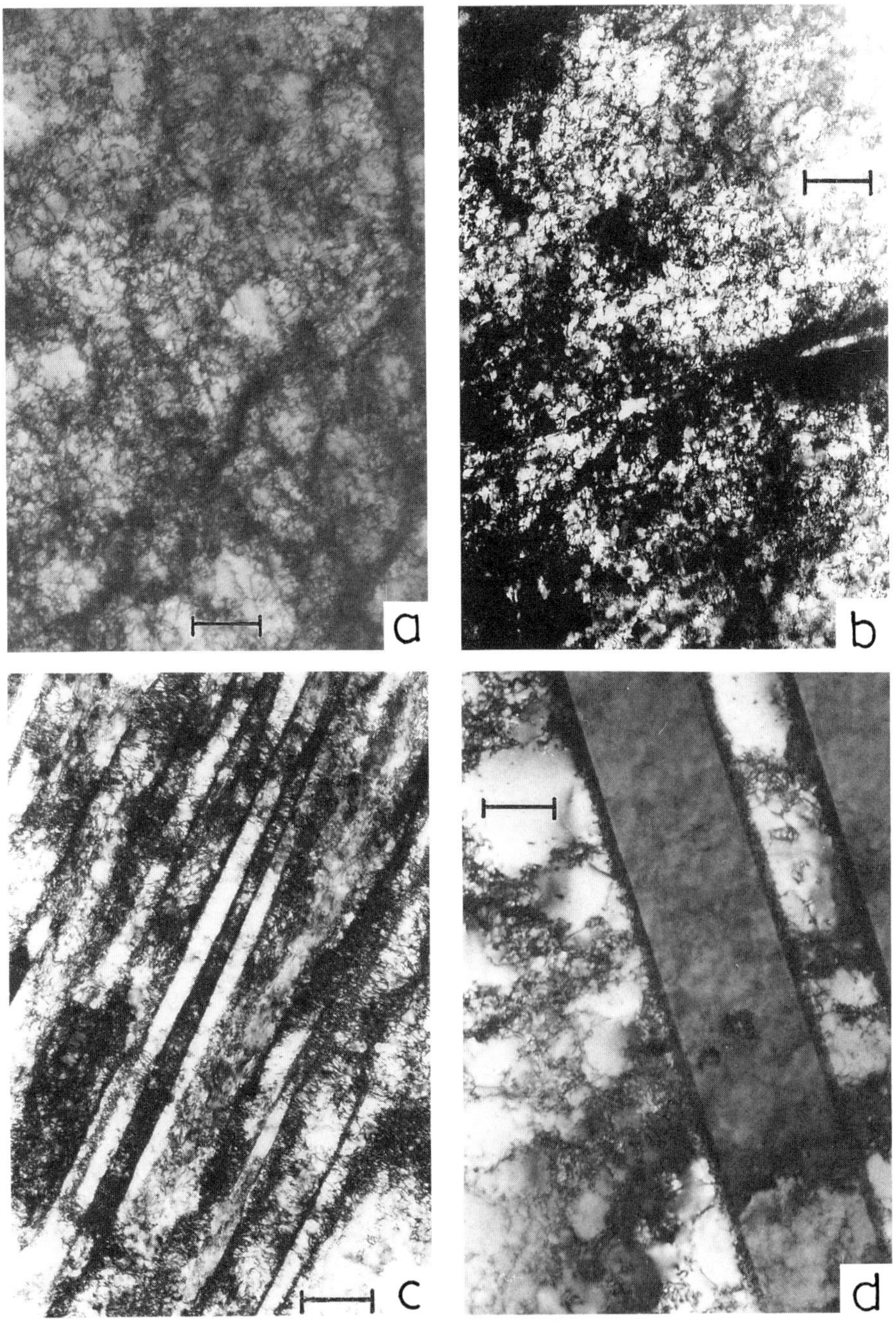

Figure 4. Comparison of prominent microstructural zone features associated with impact crater in OFHC copper (Fig. 3). (a) Dislocation cells near the Zone I/Tr region. (b) Dislocation microbands beginning. (c) Dense microbands. (d) Undeformed plate microstructure characterized by large dislocation cells. Markers are 0.5 μm.

The 1100 aluminum crater microstructure bore some resemblance to that for copper although the deformation zone sizes were different (as were the grain sizes) and there were no linear, microband features. However, recrystallization dominated the microstructure near the crater wall bottom and deformed and distorted grain regions exhibited dislocation cells which varied in size with effective strain similar to copper. However dislocation cells persisted well beyond the region of distorted cells probably as a consequence of the shock pulse. Figures 5 and 6 illustrate the 1100 aluminum microstructures which are similar to those for copper (Figs. 3 and 4) except for the absence of microbands as in Fig. 4(c). If the microbands are related in some way to the shock pressure, it may be that their absence is a consequence of the lower Bernoulli pressure in the 1100 aluminum (10.4 GPa) in contrast to that in the OFHC copper (25.4 GPa). While deformation twinning might be expected in the copper crater microstructure because the critical twinning pressure is around 20 GPa (3), the corresponding pressure for aluminum is around 40 GPa (6). This is somewhat consistent with the variations in stacking fault free energy between copper and aluminum; 70 mJ/m^2 and 160 mJ/m^2 respectively (7).

Except for the microbands in the regions below the copper impact crater (Figs. 3 and 4(c)), the microstructures and microstructure evolution for both the copper and aluminum impact craters are consistent with those expected for shock wave effects and plastic deformation effects (6) as well as the principles of low energy dislocation structures (LEDS) (8, 9).

It is particularly interesting to note in Fig. 3 that the region of maximum residual hardness below the crater wall is associated with Zone II copper crater microstructures, where the grains are not recognizably distorted, but heavily deformed with microbands, heavy dislocation density, and microband clusters. It is obvious that the deformation zones represent plastic wave effects and shock wave effects in Zones I and II respectively (with reference to Fig. 3) and that these deformation modes are mixed in the transition (Tr) region in Fig. 3. Figure 7 is an attempt to represent the prominent impact-related parameters involved in the deformation of the target below the crater: strain representing the plastic wave profile (ε), pressure representing the shock wave profile ($P(\sigma)$), the residual microhardness representing the flow characteristics of the residual microstructure along the impact axis (H); and the temperature profiles corresponding to the plastic and shock waves: T(a) representing the adiabatic heating in the plastic regime which is related to the product ($\varepsilon \cdot \dot{\varepsilon}$), where $\dot{\varepsilon}$ is the plastic strain rate; and T(s), representing the thermal pulse trailing the shock front, and proportional to the instantaneous peak shock pressure. Only the microhardness data shown plotted in Fig. 7 represents actual experimental data. The other curves are representations of trends

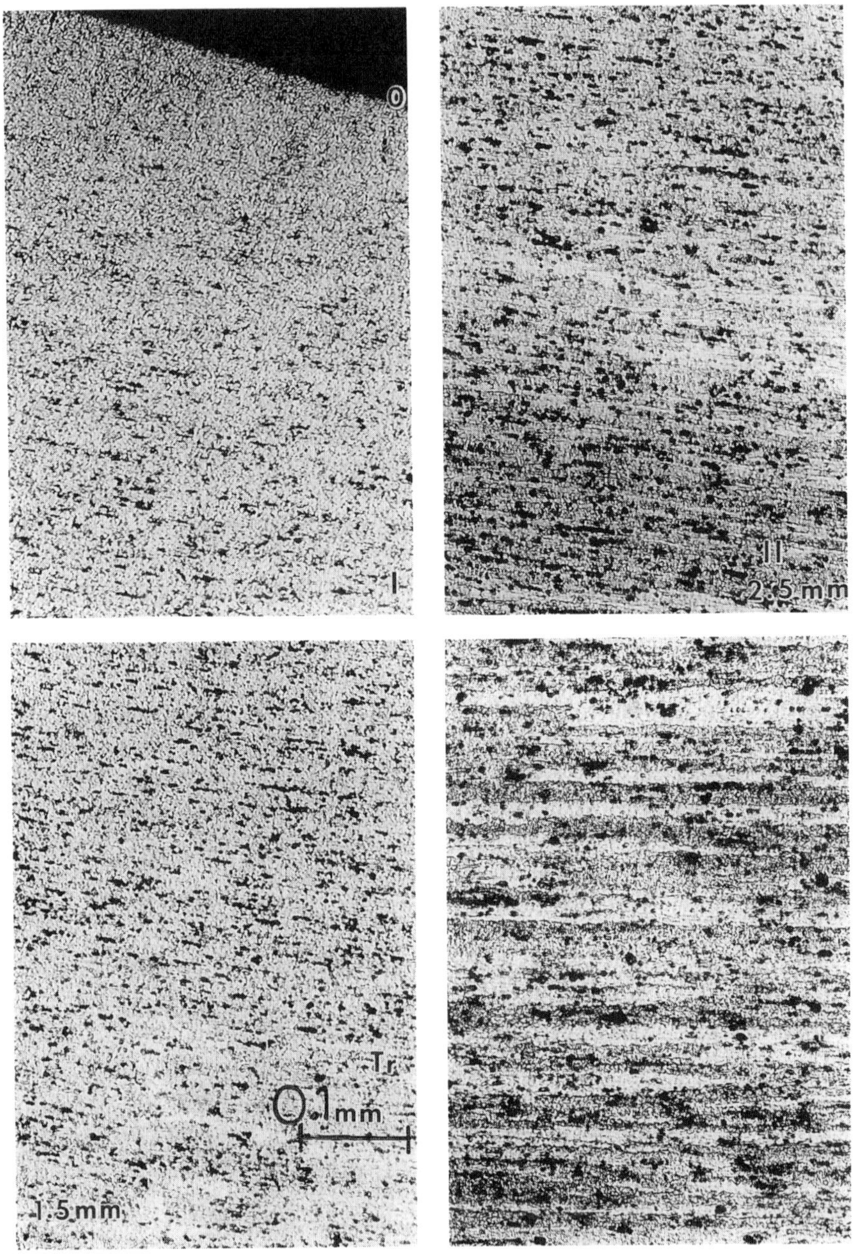

Figure 5. Optical microscope views of hypervelocity impact crater-related microstructure in 1100 aluminum for impact velocity of 5.16 km/s. The sequence of micrographs represent distances from the crater bottom along the impact axis as noted in mm. The lower right micrograph is representative of the target plate grain structure away from the crater region. The three microstructures shown here correspond roughly to the three zones (I, Tr, II) denoted for copper in Fig. 3. Markers are 50 μm.

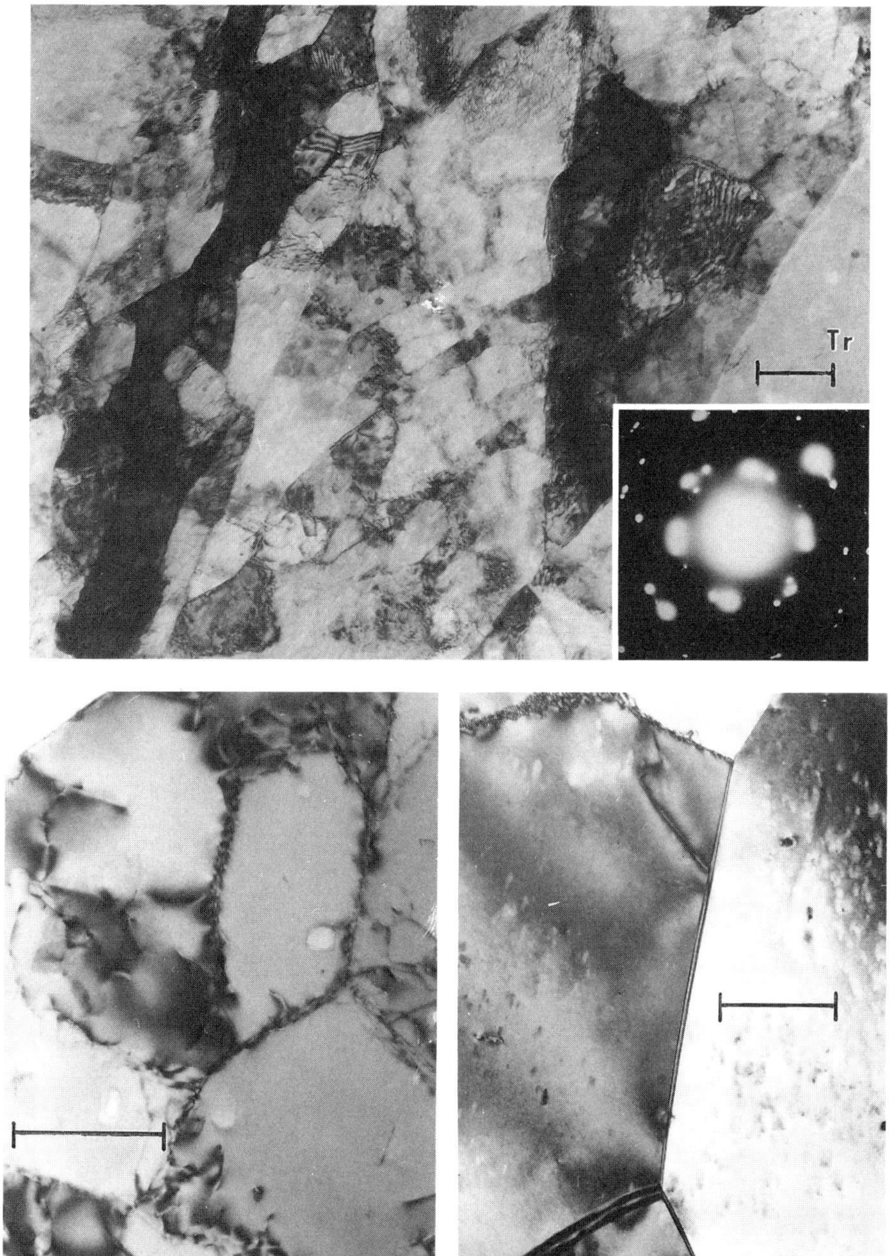

Figure 6. TEM images corresponding to the regions and zones illustrated in Fig. 5, but for the slightly larger crater shown in Fig. 1(b) corresponding to an impact velocity of 5.85 km/s. The lower right image corresponds to the undeformed plate microstructure for comparison. Markers are 0.5 μm.

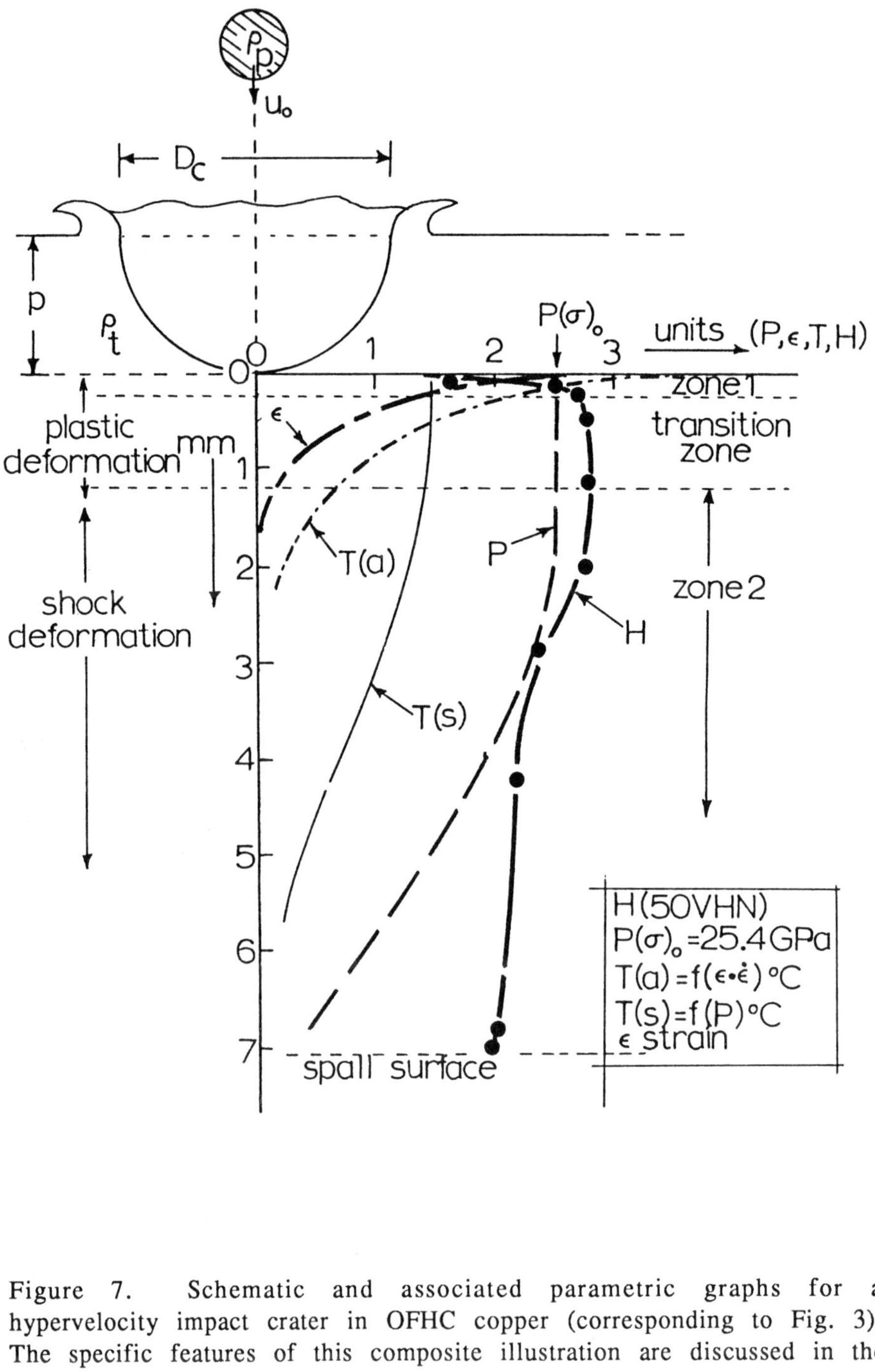

Figure 7. Schematic and associated parametric graphs for a hypervelocity impact crater in OFHC copper (corresponding to Fig. 3). The specific features of this composite illustration are discussed in the text.

which are expected along the impact axis, referenced to the crater bottom at zero. The pressure P(σ) on the crater bottom reference axis is calculated from the Bernoulli condition, and the temperatures, and strain are assumed to be maximum on this axis.

It is observed that as the microband density declines, the microhardness declines, and the maximum microhardness is associated with the region of maximum microband density as illustrated in Fig. 4(c). While the maximum shock pressure exceeds the critical twinning pressure for copper (~20 GPa), it may be that attenuation reduces this pressure below the critical twinning pressure in the region where strain and temperature effects are minimized. It is certainly apparent that the microstructures below the crater wall are quite different within Zones I and II which correspond to plastic and shock deformation respectively. Since microband clusters which occupy entire grains as illustrated in Fig. 4(c) have not been observed in previous shock or other modes of deformation of copper, it is not a trivial matter to describe the reasons or mechanism for their formation as a consequence of shock front propagation from the impact-forming crater. Howcver, Thuillier and Rauch (10) have very recently demonstrated similar microband clusters in mild steel during cross loading. Here, as in the microstructures observed in Fig. 4(c), the microbands were ascribed to deformation localization and were essentially planar structures associated with a single crystallographic slip system. The intergranular density of the microbands in mild steel (10) was observed to increase with increasing macroscopic shear strain. In contrast, the intergranular density of the microbands in the region below the impact crater wall in copper in this study is observed to vary in a kind of Gaussian distribution which is mapped out by the microhardness data in Zone II in Fig. 7. This is probably due to parametric interactions in the region closer to the crater bottom and shock wave attenuation in the region farthest from the crater wall where the microband density and the microhardness decline.

Future studies involving higher shock pressures well in excess of the minimum (critical) twinning pressure for copper (>20 GPa) may shed some additional light on this feature, and special studies for target alloys where the critical twinning pressure is much lower may help to clarify this issue. It may also be possible to map out the microstructures and associated microhardnesses for a large area beyond the crater walls for impact craters developed for a range of impact projectile velocities and densities to produce a wide range of corresponding Bernoulli pressures.

CONCLUSIONS

The residual microstructures associated with a hypervelocity impact crater in an OFHC copper plate consisted of a narrow zone of dynamically recrystallized grains containing few if any annealing or growth twins which evolved into a zone of severe grain distortion which declined with increasing distance from the crater wall bottom. This zone evolved into a plateau region of maximum hardness which extended roughly 6 mm from the crater wall, and then began to decline to the base hardness of the undeformed plate. The corresponding TEM microstructures for these three zones (recrystallized and heavily distorted, a transition zone from severe grain distortions to unrecognizable grain distortions, and a zone exhibiting microstructures (microbands) within undistorted grains) exhibited recrystallized grains, varying dislocation cell sizes and cell structures, and a zone of heavy microbands and microband clusters coincident with {111} planes. In the aluminum, the recrystallized microstructure was followed by a continuous zone of heavy dislocation density and small dislocation cells which increased slightly over a plateau region and then increased in cell size at large distances (>5 mm) from the crater wall. The aluminum sample did not exhibit the microbands observed in the copper sample.

The zones containing the recrystallized grains distorted grains, and small dislocation cell sizes are characteristic of plastic deformation while the deformation microstructures (microbands in copper and dislocation cells in aluminum) well beyond these regions of recognizable grain distortions are characterized by shock deformation. The relative dimensions of these deformation zones are apparently impact pressure-related and more extensive studies may begin to develop deformation-zone maps which can be matched with corresponding microhardness maps.

ACKNOWLEDGEMENTS

This research was supported by a NASA-Johnson Space Center Grant (NAG-9-481) and a NASA Graduate Student Research Training Grant (NGT-70343) (S. Quinones). We are grateful to Dr. Fred Hörz (NASA-Johnson) for providing archived impact crater specimens examined in this work.

REFERENCES

1. R. J. Naumann, "The Near-Earth Meteoroid Environment", NASA TN D-3717, 1966.
2. B. G. Cour-Palais, "Space Vehicle Meteoroid Shielding Design", Proc. Comet Halley Micrometeoroid Hazard Workshop", ESA SP-153, pp. 85-92, 1979.
3. D. R. Christman, *AIAA Journal, 4(10),* 1872 (1966).

4. J. F. Lundeberg, P. H. Stern, and R. J. Bristow, "Meteoroid Protection for Spacecraft Structures", NASA CR 915, Washington, D.C., 1968.
5. J. M. Rivas, et al., *Scripta Metall. et Materialia, 27,* 919(1992).
6. L. E. Murr, Chap. 37 in *Shock Wave and High-Strain-Rate Phenomena in Metals,* M. A. Meyers and L. E. Murr, Eds., Plenum Press, New York, 1981, p. 607.
7. L. E. Murr, *Interfacial Phenomena in Metals,* Addison-Wesley, 1975, Reprinted by Tech Books, Herndon, VA, 1991.
8. D. Kuhlmann-Wilsdorf, *Phys. Stat. Sol. (a), 104,* 121(1987).
9. D. Kuhlmann-Wilsdorf, *Mater. Sci. Engng. A113,* 1(1989).

MICROSCOPIC EVALUATION OF SYNTHETIC DIAMOND FILMS*

Richard H. Lee[1]

Abstract

This paper summarizes how we microscopically examine and analyze synthetic diamond coatings that have been nucleated and grown on various substrates. Several methods of enhancing nucleation and growth on silicon, silicon carbide, and refractory metals were evaluated. Changes in experimental conditions resulted in many differing crystal morphologies, which we characterized by optical and scanning electron microscopy and Raman spectroscopy. Special procedures were developed for examining diamond crystals and films, and some unusual growth forms were discovered.

---

*Work supported by the Tribology Program, Office of Transportation Materials, U.S. Department of Energy, under Contract W-31-109-ENG-38.

[1]Energy Technology Division, Argonne National Laboratory, Argonne, IL 60439 U.S.A.

INTRODUCTION

Synthetic diamond and diamondlike carbon (DLC) coatings are becoming important in applications where diamond properties are useful. At Argonne National Laboratory, we are interested in coatings that are hard and wear-resistant; exhibit excellent heat conductivity; and are corrosion-resistant and show low-friction. For prosthesis applications, corrosion resistance and low friction are probably the most important properties.

We began our research by making diamond and hard carbon coatings under various conditions and then examining them to determine their structure and properties. We used a scanning electron microscope (SEM), laser Raman spectroscopy, and microhardness testing to evaluate the coatings. Electrical resistivity was also measured for some of the coatings. Several investigators have used these techniques and others to understand the effects of gas chemistry, temperature, microwave energy, and substrate preparation on the nucleation and growth of synthetic-diamond films [1-3].

Most of our efforts were concentrated on evaluating diamond films by examining structure with the SEM, which we routinely use to evaluate new batches of coatings. The characteristics we look for with the SEM are general morphology and uniformity, thickness, crystallite size, roughness, and resistivity (implied by charging behavior).

We investigated experimental conditions to find the best parameters for the desired coating morphology and found microscope operating parameters that allow us to successfully examine these high-resistivity materials. We also tested the films for adhesion and bonding to the substrates [4].

EXPERIMENTAL PROCEDURES

Initially we worked on silicon wafer substrates that had been polished for semiconductor industry (100) orientation. This material is one of the standard substrates used by many researchers, and carbon will usually nucleate and grow on it easily. Later, we experimented with other substrates of more practical interest, such as silicon carbide and silicon nitride from Kyocera ceramics. Figure 1 shows a synthetic diamond coating with typical morphology. The spaces between the diamond crystals are the result of low density nucleation. Figure 2 shows the same material as Fig. 1 but polished on a vibratory polisher and with a very fine diamond abrasive before coating. This pretreatment, which results in many fine scratches on the surface, increases nucleation sites.

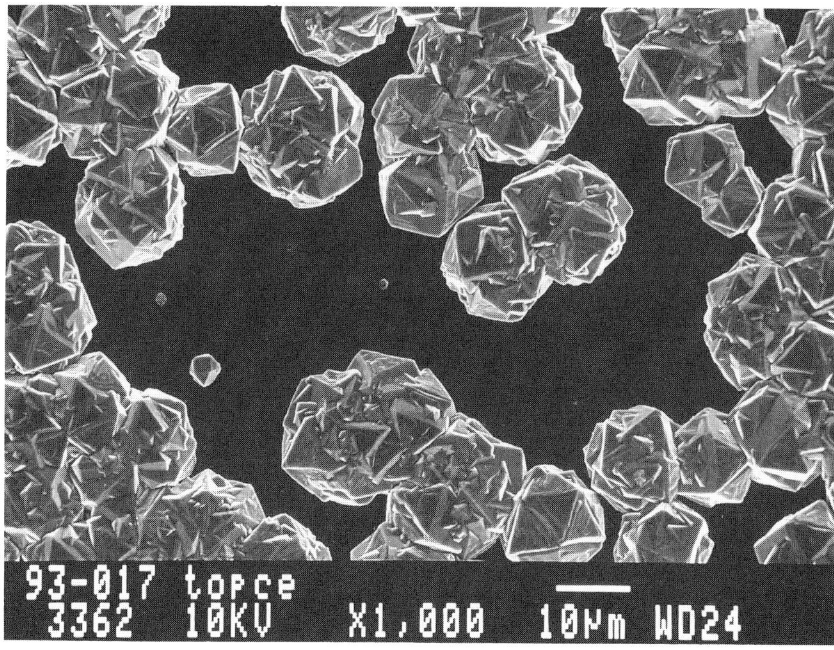

Figure 1. Synthetic diamond coating with typical morphology.

We have tried three pretreatments for samples that were to be diamond-coated. Mechanical polishing with a cloth-covered wheel and fine diamond abrasive is much better than no pretreatment at all. Figure 3 shows a silicon substrate that was mechanically polished and coated with synthetic diamond at 1273 K. Some abrasive particles remain on the surface and act as nucleation sites. Ultrasonic agitation in a fine diamond abrasive slurry has been tried also, with limited success.

Vibratory polishing on a hard synthetic cloth and fine (0.1 micron) diamond abrasive has best achieved complete coverage and diamond coatings with smaller grain size. The plasma gas chemistry during coating in the Astex microwave assisted chemical vapor deposition system also has a strong effect on the morphology and composition of

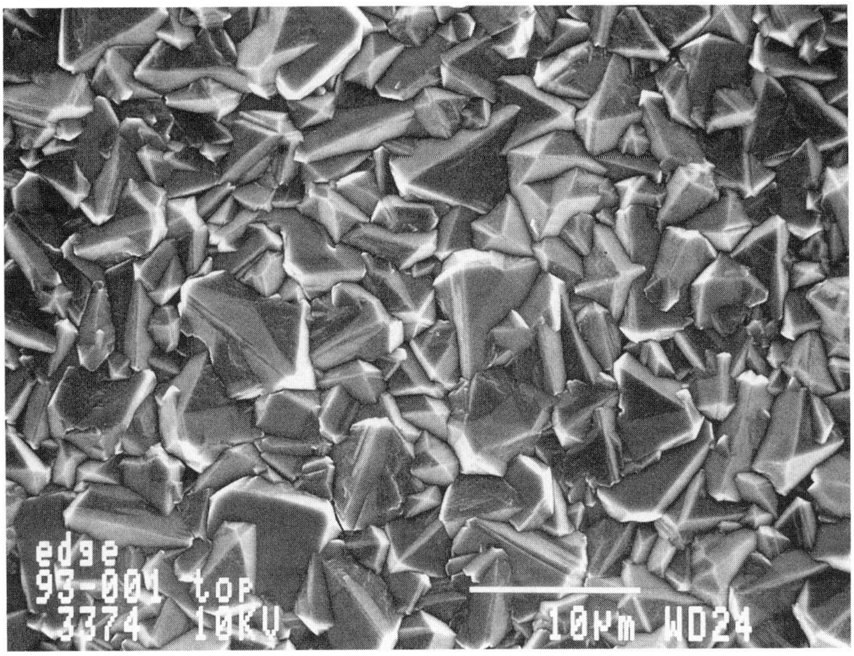

Figure 2. Dense diamond coating; same material as that shown in Fig. 1, but polished on a vibratory polisher and with very fine diamond abrasive before coating.

the coatings. We have tried varying the methane concentration in a pure hydrogen atmosphere from 1% to 4.5% and also tried various partial pressures. Figure 4 shows a coating with very fine-grained crystals that results when growth starts out at a high temperature (1273 K), which is later reduced to a slow growth conditions.

Fracture cross-sections are useful to examine the structure of the coating and to measure its thickness. This is done with silicon by simply scribing the back of the substrate with a sharp diamond scribe and rapidly bending the piece. Both the brittle silicon and the thin coating usually break at the scratch and yield a fairly clean fracture. Sometimes, the coating will also show delamination from the substrate. Figure 5 shows a typical coarse-grained diamond coating on silicon. Notice the sharp facets on the coating surface on the left side of the photo.

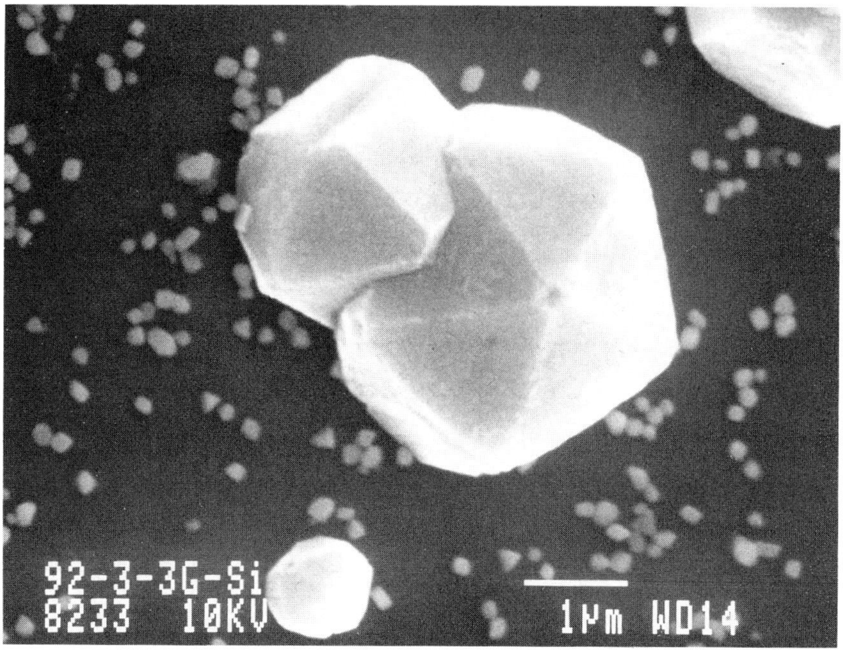

Figure 3. Silicon substrate, mechanically polished and then diamond-coated.

Figure 6 shows a fracture cross section of a "smooth" diamond coating on silicon carbide. The surface on the right side of the figure shows a finer texture and a relatively smooth finish. The surface appears to "glow" because of the excitation of the penetrating electron beam.

RESULTS

We have identified the parameters that control the structure and properties of coatings and provide complete coverage and a fine (submicron) crystal size. These parameters give a much lower friction and wear of mating sliding surfaces. Some test results have recently

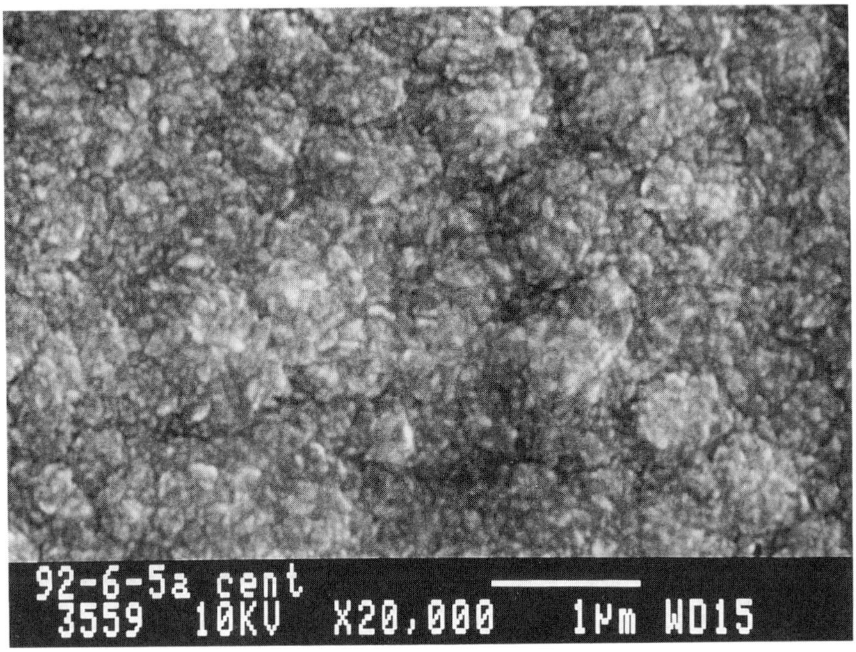

Figure 4. Fine-grained diamond coating on silicon.

been reported elsewhere [5,6]. Figure 7 shows a diamond coating on silicon carbide that is better than most, mainly because of the small crystal size of the coating. Figure 8 shows the coating with the best (low) friction and wear and with better coverage of the substrate surface.

Examination of synthetic diamond coatings in the SEM is best accomplished under the following conditions:

Low accelerating voltage - 5 kV to 10 kV
Close working distance - 15 to 25 mm
Low emission current - 40 to 60 microamps
Small aperture in the column.

By examining the coated surfaces at the periphery, one may find unusual growth forms with the SEM. The fine details revealed in some of our samples indicate that some kind of "whisker" forms under certain conditions. Figure 9 shows some fine filamentary growth from the surface on the edge of a niobium sample. Addition to the SEM of a

Figure 5. Cross section of typical coarse-grained diamond coating on silicon.

computer that allows images to be captured and enhanced has proved very useful for observing such phenomena.

Finally, the SEM has revealed wear behavior and mechanisms from examination of wear-tested surfaces. Figure 10 shows a sample of diamond-coated silicon carbide that has had another uncoated SiC sample sliding across it for many hours. The fine-grained diamond coating is evident in the center of the photo where a slight depression has prevented any contact with the sliding piece.

Surrounding the depression is a smooth surface where worn SiC debris has coated the diamond by a smearing action. This can be verified by taking a backscatter electron image of the same area and looking for atomic-number contrast. Diamond, being essentially carbon, will appear darker than adjacent silicon carbide, which has a higher average atomic number and backscatters more. This technique is also useful for examining free-standing diamond coatings that have

Figure 6. Cross section of "smooth" diamond coating on silicon carbide.

been removed from silicon by etching away the substrate with acid. Diffusion or reaction with the substrate can be seen with backscatter imaging.

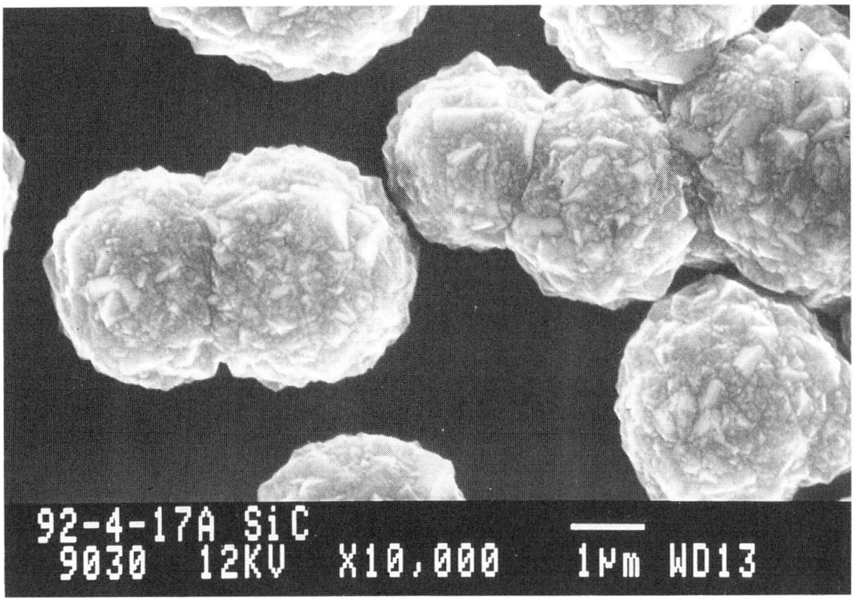

Figure 7. Diamond coating on silicon carbide, showing good coverage.

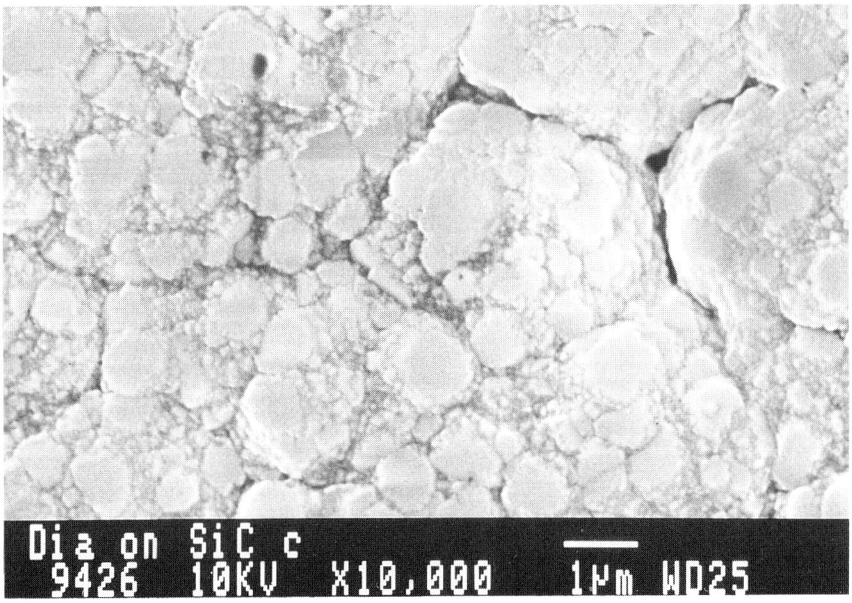

Figure 8. Excellent coverage of silicon carbide by diamond coating.

Figure 9. Diamond "whiskers" on the surface of niobium, in profile.

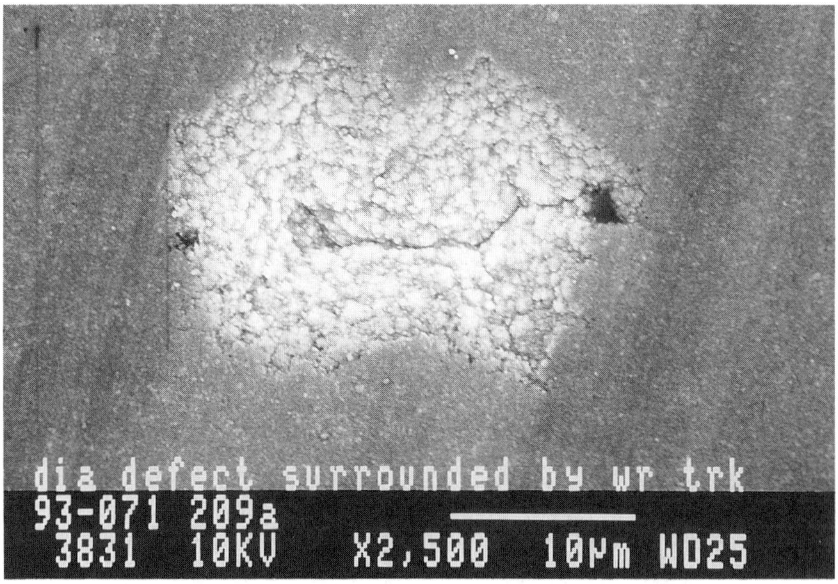

Figure 10. Diamond-coated silicon carbide after sliding against silicon carbide.

REFERENCES

1) F. Nichols and D. K. Moores, Workshop on Diamond and Diamond-Like Carbon Films for the Transportation Industry, Argonne National Laboratory CONF-920262, Argonne, IL (Feb. 4-5, 1992).
2) H. C. Shih, C. P. Sung, Y. S. Tang, and J. G. Chen, "Microstructure and Characterization of Diamond Films Grown on Various Substrates," Surface Coatings Technol., 52, 105, 1992.
3) J. C. Angus, Y. Want, and M. Sunkara, "Metastable Growth of Diamond and Diamondlike Phases," Ann. Rev. Mater. Sci., 21, 221, 1992.
4) R. Erck and J. F. Dierks, "Semiannual Progress Report DOE-OTM Tribology Program, ANL/OTM-93/2, p. 99, Jan. 1994.
5) A. Erdemir, F. A. Nichols, R. Wei, and P. Wilbur, "Formation and Self-lubricating Mechanism of Transfer Layers on Surfaces Sliding Against Diamond-like Carbon," 4th European Conference on Diamond, Diamond-like, and Related Materials, Albufeira, Algarve, Portugal, Sept. 20-24, 1993.
6) M. Languell, M. A. George, J. J. Wert, and J. L. Davidson, "Morphological Response of Diamond Films in Dry Sliding Contact," J. Met., V46, 66, 1994.

THE MICROSTRUCTURAL ANALYSIS

OF METAL FASTENERS

James A. Nelson*

ABSTRACT

Compliance with PL-101-592, the Fastener Quality Act, requires the Microscopical analysis of samples that have been prepared according to the guidelines of ASTM E 3-80. The analysis of such specimens to comply with the law encompasses Flaw Detection and Verification, Qualitative Analysis, and Quantitative Measurements. Some specific analysis items are: flaws below the pitch line, grain size and grain flow, retained austenite, decarburization, and case hardness. To certify fasteners to be in compliance with the law requires that testing be performed by a laboratory that has been accredited by a recognized authority.

INTRODUCTION

The purpose of this paper is to describe the metallographic tests that might be required for the various grade marked fastener specifications. These tests are listed in Fig. 1 and are arranged under three separate categories, Flaw Detection and Verification, Qualitative Analysis and Quantitative Analysis. However, it should be noted that not all tests apply to all fastener types and grades but each individual specification states the requirements for that grade. Some specifications require multiple tests while others require only two or three.

* BUEHLER Ltd., Lake Bluff, IL 60044

Fig. 1: Metallographic Tests for Grade Marked Fasteners

Flaw Detection Verification	Qualitative Analysis	Quantitative Analysis
identification	phases	microhardness
location	grain flow	case depth
extent	decarburization	micro dimensions
mode	retained austenite	% retained austenite

To fulfill the requirements of the specification, most fastener specifications require that the threaded section of a fastener including an additional length of the unthreaded shank equal to 1/2 the diameter of the fastener must be prepared in the longitudinal direction to evaluate the product for thread discontinuities. (Fig. 2)

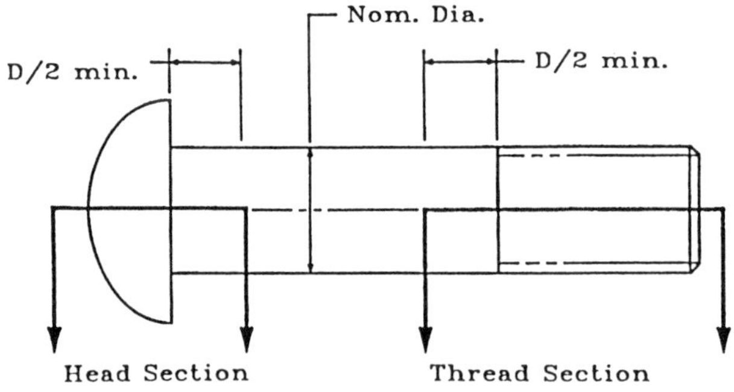

Fig. 2: Sampling plan for fasteners

FLAW DETECTION AND VERIFICATION

The longitudinal section previously described provides a view of the thread profile and any defects that are present and their location and extent. This analysis is performed in the as-polished condition because etching can mask fine cracks or laps. However, it is advisable to etch the samples lightly followed by an abbreviated repeat of the final polish to remove any residual smear. Most specifications for grade-marked fasteners illustrate the areas of the thread profile that are most critical. Generally, discontinuities above the pitch line are considered less critical than those below the pitch line. The pitch line is a horizontal line that dissects the vertical dimension of the thread profile as shown in Fig. 3a. Figures 3b and 3c, show actual defects that can occur relative to the pitch line.

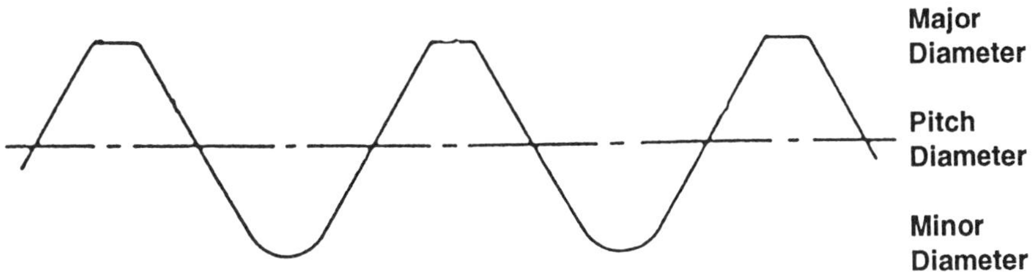

Fig. 3a: Location of the pitch line

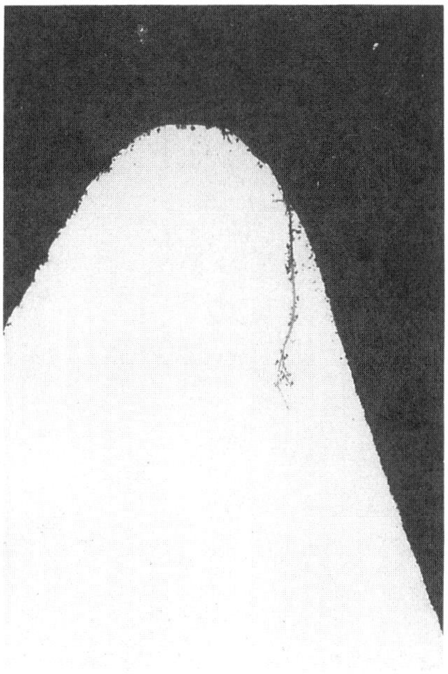

Fig. 3b: Thread crest lap with oxide

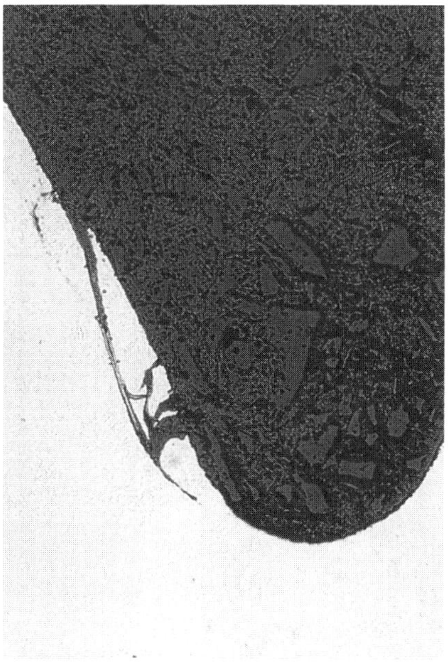

Fig. 3c: Lap below the pitch line

Conditions for both photomicrographs: Mag. 100X Etchant: none

In the above illustrations, Fig. 3a is a diagram showing the critical pitch line that is used as a guide to determine what discontinutie are allowable. Fig. 3b shows an underfilled thread crest condition that would be accepted in many cases except where specifically disallowed.by stricter standards. Fig. 3c is a lap in the thread root that is well below the pitch line and therefore unacceptable in all grade-marked fastener specifications. Cracks and

Another useful application of metallographic cross-sectioning is to determine the extent of a flaw indication that has been revealed by visual inspection or nondestructive testing. In some specifications, surface discontinuities are allowed to a specified maximum depth. An as-polished cross-section provides the opportunity to accurately measure the depth of such indications using a measuring micrometer on any microscope. This will be discussed more fully later under quantitative analysis.

Fig. 4a is a macrophoto showing the extent of a crack in the critical root area of a grade 5 fastener. Fig. 4b is a polished transverse section showing the limited extent of the flaw. In the etched condition, this section provides an opportunity to determine the probable mode of propagation, i.e. whether it is intergranular, transgranular or a combination of both modes. With this information plus other pertinent historical data concerning the manufacturing and heat treatment of the lot, a valid judgment can be made.

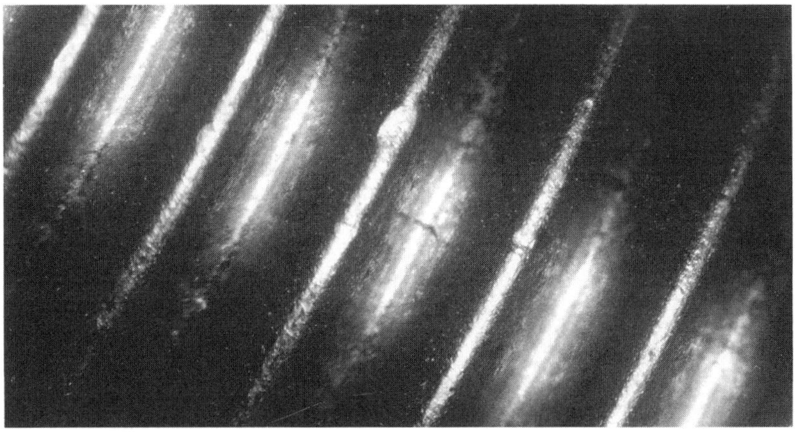

Fig. 4a: Macrophoto of a surface indication in the threaded section Mag.: 5X

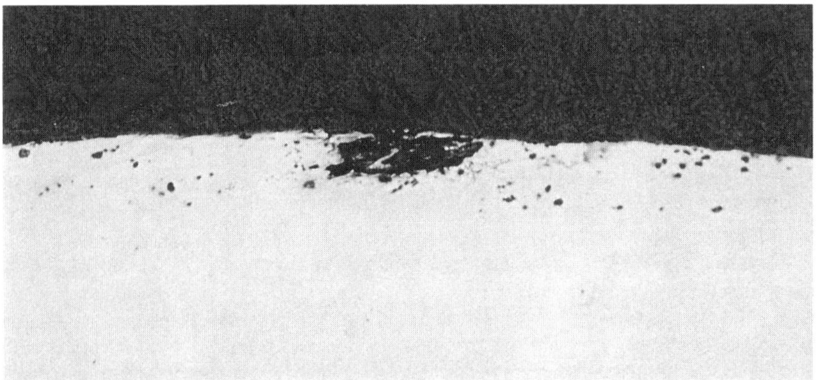

Fig. 4b: Transverse section showing the extent of the above defect Mag.: 100X

Fig. 4b: Transverse section showing the extent of the above defect Mag.: 100X

QUALITATIVE ANALYSIS

Flow lines are visible patterns of grain displacement which occur in the head and thread areas of fasteners. They are the direct result of the head forming and thread rolling operations. Although they seldom present a problem, many specifications require fastener test samples be either a macro etched or micro etched to evaluation the flow lines in specific fasteners. This is done to verify that the grain flow is not interrupted by the sharp radius where the head and root meet or in the high deformation root area of the threads.

A macro etch is a strong acid solution such as equal parts water and hydrochloric acid heated to 140°F. Fig. 5a shows the flow lines in cold headed bolt. The original surface was ground through the 240 grit abrasive.

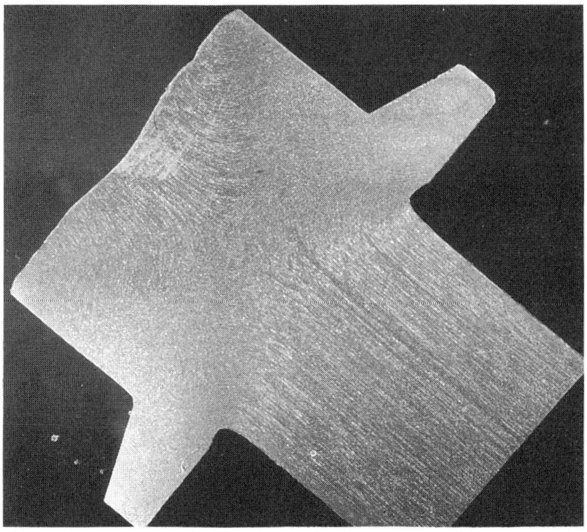

Fig. 5a: (left): Flow lines in a cold headed carbon steel bolt after macro etching

Mag. 2 X

Etchant: equal parts of hydrochloric acid and water at 140 °F (60° C)

A micro etch is a milder etchant such as 2% nital that is applied at room temper- ature to a fully polished sample.

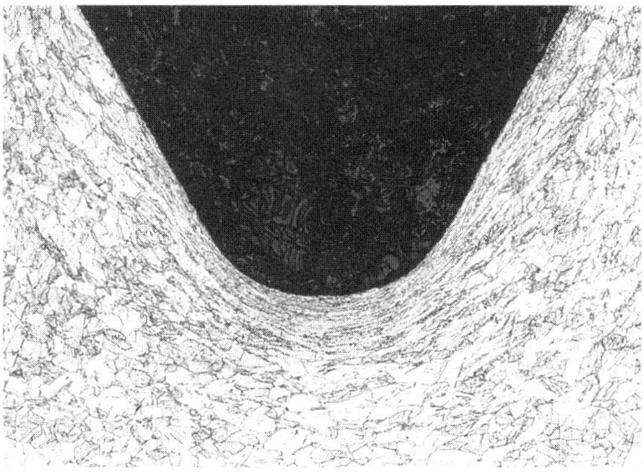

Fig. 5b (left): Flow lines resulting from thread rolling a carbon steel

Mag:: 200X

Etchant: 2% nital

To meet the require- ments of a fastener specifi- cation, it may be necessary to verify that a specific micro-has been achieved. The specifications may also require that certain microstructures such as decarburization not be present or restrict the amount that is allowed. Although well trained metallographers may be able to identify various microstructural constituents, these often critical determinations are best performed by professional metallurgists.

The microstructure of suitably prepared fastener specimen is revealed by micro etching the polished sample in a chemical solution (etchant) that has been selected according to the chemical composition of the fastener alloy. Micro etchants for low or medium carbon steels will differ from those intended for higher alloy steels, stainless steels and non-ferrous alloys such as titanium.

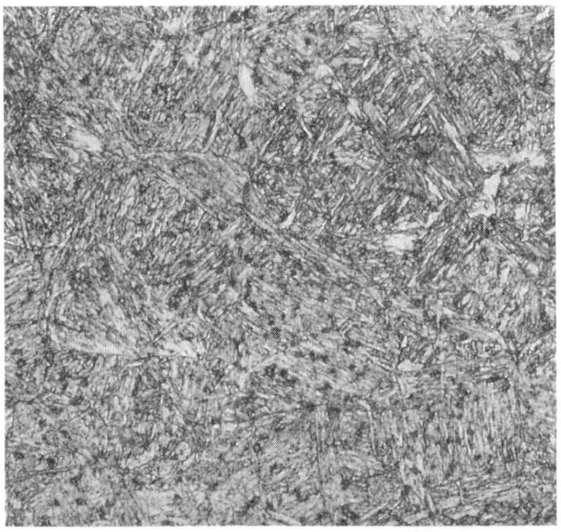

Fig. 6a (left) Martensitic in through-hardened steel fastener.
Mag.: 1000X
Etchant: 2% nital

However, if a fastener lot is heat treated in a furnace whose atmosphere reacts with carbon, the material would be decarburized. as illustrated in Fig. 6b This lowers the carbon content at the surface to a depth that depends on the time at temperature and the carbon content of the atmosphere. In general the lower the carbon in the atmosphere the greater severity of the decarburization.

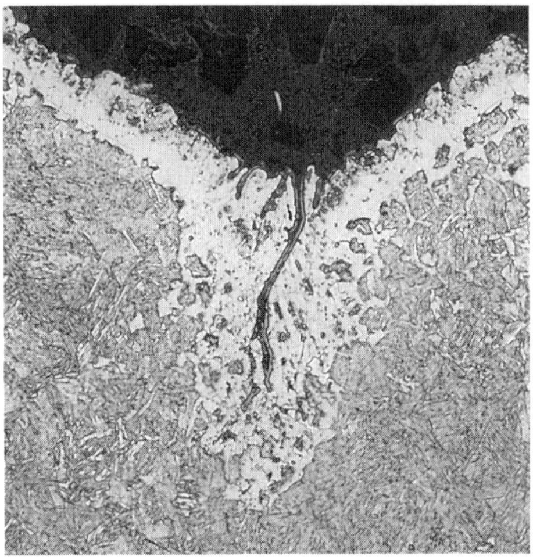

Figure 6b: Total decarb in a case hardened steel fastener
Mag.: 400 X
Etchant: 2% nital

Total "decarb" is the total loss of carbon from the surface of the fastener causing the microstructure to be become ferrite, a very soft microstructure that is an ideal site for the initiation of cracks and corrosion that could lead to failure. Total decarb is easily recognized by the obvious white areas with characteristic ferrite grain boundaries.

Partial decarb, however, is less obvious and is best determined by microhardness tests. Case hardened fasteners are formed from lower carbon steels that are heat-:treated in a carburizing atmosphere to increase the carbon content at the surface so that when they are quenched, the surface will be hardened without significantly increasing the hardness and reducing the impact resistance of the *core.* The best way to evaluate the effectiveness of case hardening is microhardness measurements. If you rely on a visual qualitative measurement of case depth, as shown in Fig. 8, you will be measuring *Total Case*, the distance from the edge to the point at which the microstructure becomes core structure. Effective case, on the other hand, is the distance measured from the edge (surface) of the fastener that meets the specified hardness level. The ASM International Metal's Handbook defines effective hardness as a case having a minimum hardness of 50 HRC or higher. The relevant specification, however should be the final authority. This can only be determined by microhardness measurements. Figure 8 illustrates the difference between *Total Case* and *Effective Case.*

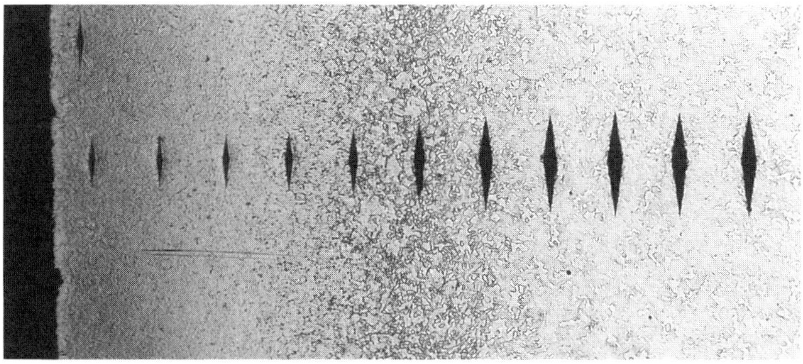

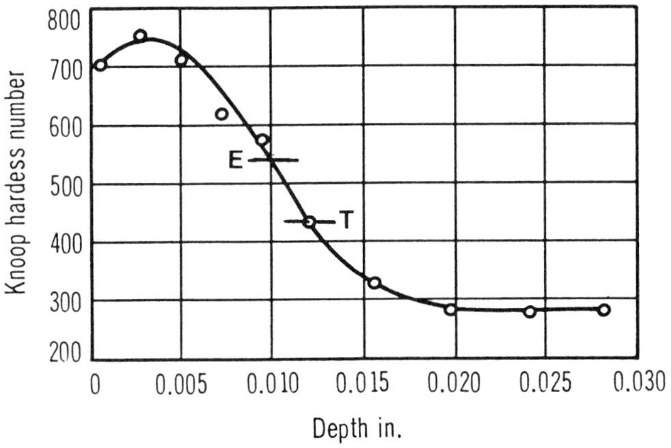

Fig. 7 (above):
Hardened case
Mag.: 50 X
Etchant: 2% nital

Fig. 8 (left):
Total case Vs
effective case

E = Effective case

T = Total case

QUANTITATIVE ANALYSIS

Linear Measurements are the simplest and most economical method of measuring the thickness or length of a feature by obtaining a direct linear measurement using a Digital Filar Micrometer. This is an accessory that consists of an encoder assembly that is inserted into one of the microscope eyepiece tubes. The Encoder consists of a stationary and a movable cross-hair which defines the distance to be measured. The encoder translates this distance into an electrical signal that is sent to a Digital Display Device that displays the measurement reading in either English or Metric units Each objective lens of the microscope that ill be used to make measurements must be calibrated using a stage micrometer. The calibration factor for the particular lens being used is entered into the Display device before measurements are made, so that the values that appear in the display are actual values.

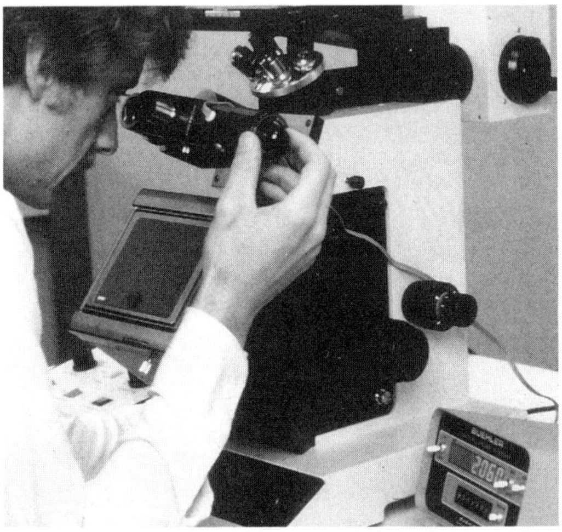

Fig 9 (right): Digital Filar Micrometer in use.

Microhardness: measurements are essential to determine the effective case of case hardened fasteners and to positively identify and quantify partial decarb. In addition, microhardness may be used to help identify heat-treated microstructures. Transformation microstructures such as martensite and bainite are sometime difficult to positively identify and hardness values can be help make the identification more positive.

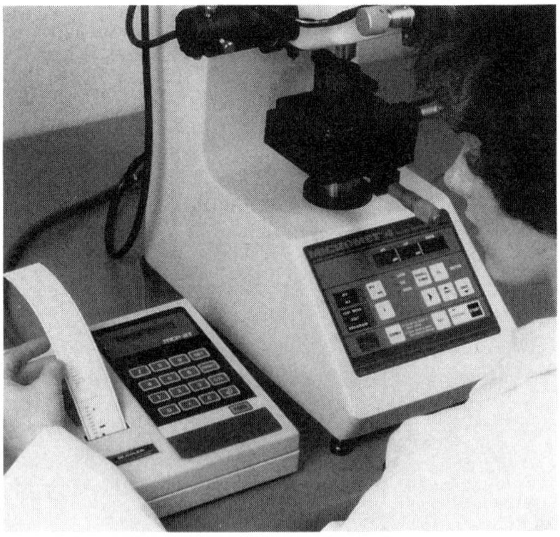

Fig. 10 (right): Microhardness tester and data printer

Retained austenite is a condition that results from incomplete transformation of austenite to martensite during quenching . However, when identification is more difficult, light load microhardness indentations carefully placed in the suspicious area may indicate that a lower hardness such as bainite has been produced. Visual estimates of % retained austenite are generally inaccurate and therefore not reliable. A fastener manufacturer who must rely on such quantitative measurements is advised to consider automatic image analysis, a technique to be discussed later.

The crystalline grain size of materials has a significant affect on their physical properties. There are two distinct types of grain sizes that may be seen in carbon steel alloys: ferritic and austenitic. Ferritic grain size primarily affects the formability of the metal and ferritic grains are prominent in lower carbon steels that have not been through hardened. The application of a 2-5% nital etch will reveal distinct ferrite grains. In this condition, grain flow is easily seen.

Austenitic grain size primarily affects heat-treatability and the tensile strength of ferrous materials.

Fig. 11 (right): A-286 alloy austenitic grain structure

Mag.: 200X
Etchant: HCl-H_2O_2 (3%)

Prior austenitic grains are always present in ferrous alloys but are sometimes difficult to reveal because the austenite has transformed to other phase depending on the heat-treatment. Special etchants and or heat-treatments used to reveal and measure prior austenetic grain size in steels are described in ASTM E-112. One specific requirement that appears in certain fastener specifications is that the grain size may not differ more than 1 grain size in the shank of the fastener. Methods of measuring grain size include comparison wall charts and reticules that fit into the light path of the microscope, linear and circular intercept methods, clearly described in ASTM E-112. and automatic Image Analysis.

Automatic Image Analyzers are video based computer driven analysis systems that capture and process a microscope image so that specific quantitative analyses may be performed accurately and rapidly. This technique for performing quantitative measurements has been mentioned several times previously in this paper. This methodology is gaining wide usage throughout the materials community because stricter specifications are slow and tedious and are not as reliable as automated analysis.

Figure 12 shows one such unit that may be used to perform a variety of fastener related measurements such as totally automated inclusion analysis, layer measurements including depth of total decarburization, plating thicknesses, % retained austenite, and in some applications, grain size measurements.

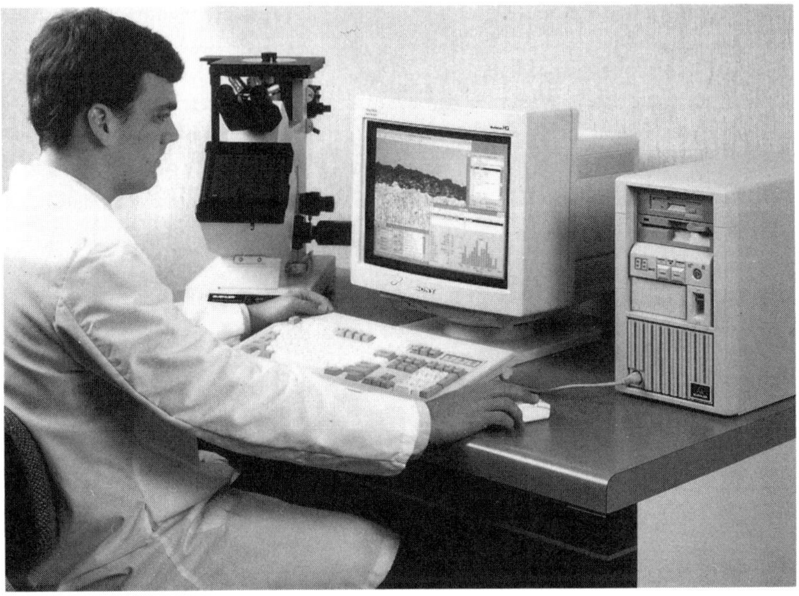

Fig.12: An image analyzer used to perform various quantitative measurements

SUMMARY

Fastener specifications require that various metallographic tests be performed to comply with the provisions of PL 101- 592, the Fastener Quality Act. This law primarily impacts grade marked fasteners and all who handle them in the market place. Fastener distributors are not affected unless they change the original product in any way through heat-treatment, plating or any other altering process/. They are, however responsible to maintain the original documents of all lots they handle.

Fasteners may only be certified by laboratories that gave been accredited by agencies designated by the US Department of the Interior. The performance of these tests must comply with the provisions of the Fastener Quality Act through it's auditing requirements. It must be remembered that specimen preparation must produce specimens that are flat, free of surface deformation and preparation produced artifacts and are carefully etched to reveal the microstructure after the as-polished evaluations have been made. In addition to the normal regular maintenance of the equipment, analytical devices such as microscopes, linear measuring devices and microhardness testers must be in calibrated according to the provisions of the law.

METALLOGRAPHIC PREPARATION OF Zr-2.5Nb PRESSURE TUBE MATERIAL FOR EXAMINATION OF INCLUSIONS

A.J. Lockley[1]

ABSTRACT

The traditional final polish of Zr-2.5Nb alloy comprises an attack polish that contains a 0.05 μm alumina or fly-ash slurry with dilute hydrofluoric acid. This polish preferentially etches the material adjacent to the inclusions and distorts or removes the inclusions. A final polish has been developed that uses a caustic alumina slurry to produce a chemical-mechanical polish that keeps the inclusions intact. This preparation is reproducible, suitable for automation, and retains smaller inclusions.

INTRODUCTION

The metallographic examination of inclusions in Zr-2.5Nb is required to determine their effect on the mechanical properties of CANDU[2] pressure tubes.

[1]AECL Research, Chalk River, Ontario

[2]CANada Deuterium Uranium, registered trademark of Atomic Energy of Canada Limited

Traditionally, the final polish used for zirconium alloys comprises a chemical-mechanical polish, referred to as an attack polish, that contains hydrofluoric acid (HF) as the aggressive etching agent, mixed with a fine abrasive of either 0.05 µm alumina or fly-ash, appropriately proportioned to provide the required balance of abrasion and etching. Samples prepared using a final polish comprising dilute HF and fly-ash have allowed inclusions to be deteriorated or removed. A similar method using alumina [1] has been used and has provided improved retention of the inclusions, but this technique requires personal skill to ensure consistency. Neither method is suitable for automation.

Superior inclusion retention is obtained using colloidal silica during the final polish. This technique is suitable for automated polishing, but is not appropriate when surface mapping is required to determine the presence of silicon, as the colloidal silica becomes imbedded in the surface. An alternate metallographic preparation has been developed that reveals inclusions with more geometric detail, retains smaller particles, does not contaminate the surface with silica, and is suitable for automation.

EXPERIMENTAL PROCEDURES

The cold-worked Zr-2.5Nb materials used in the development of the caustic polishing procedure were taken from production CANDU pressure tubes. Material containing inclusions was sectioned and mounted to present the plane normal to the transverse direction with respect to the pressure tube. The samples were mounted in Bakelite using a 3.175 cm diameter mounting press, and were ground through a standard progression of grits of 240, 320, 400, and 600 using an aliphatic hydrocarbon (Varsol)[3] as a lubricant. The samples were rough polished using a slurry comprising 1.0 µm alumina, water and hydrogen peroxide (H_2O_2), and were charged onto a napless chemotextile cloth. A polishing time of one to two minutes was usually sufficient to remove the 600 grit scratches.

Three formulations of slurries were used for the final polish. The first slurry comprised 0.05 µm alumina, water and H_2O_2, being similar in composition to the slurry used for the rough polish. The two reformulations added potassium hydroxide (KOH) to increase the alkalinity of the slurry. The final reformulation included aqueous ammonium hydroxide (NH_4OH). Two types of polishing cloths were used: a chemically resistant cloth and a synthetic velvet cloth.

The procedure was adapted for use on an automated polishing apparatus. The final polish could be consistently produced in about two minutes using an applied load of 4.5 kg per sample on a synthetic velvet cloth. The polish was considered complete when scratches could not be seen in bright field illumination. The contrast

[3]Registered Tradename of Imperial Oil Canada

between the inclusions and the zirconium matrix was increased by the deposition of an anodized layer [2].

For the purpose of comparison, a sample was first prepared using the HF and fly-ash attack polish and then prepared using the caustic polish. The size distribution and population of the inclusions for each preparation were quantified using an image analyzer. About 100 observation fields were selected at random for each prepared surface; each field covered an area of 2500 μm^2.

The caustic prepared sample was prepared for chemical analysis. Areas around inclusions to be analysed were marked, using a microhardness indenter, and photographed before the anodized layer was removed using the caustic slurry. The sample was extracted from the Bakelite, ultrasonically cleaned in methanol and chemically analysed using a scanning auger microscope (SAM). Three elemental maps were made, for phosphorous, silicon, and carbon. The elemental maps were compared with the optical photomicrographs, to allow the inclusions to be identified and correlated.

RESULTS & DISCUSSION

Initially, a final polish comprising 0.05 μm alumina, water, and H_2O_2 was tried, but polishing times extended beyond ten minutes because the surface contained scratches that resulted from continued polishing. The alkalinity of the slurry was increased to a pH of 9.0 using KOH, which noticeably reduced the number of scratches, but not enough to eliminate their appearance in bright field illumination. In an attempt to further reduce the number of scratches, the amount of KOH was increased, which raised the alkalinity to a pH of 11.0. This did not appear to reduce the number of scratches, but it did induce surface pitting. The slurry was optimized when KOH was used to establish a pH of 9.0, followed by the addition of NH_4OH to increase the alkalinity to a pH of 11.0.

The best polishes were achieved when the optimized slurry was used with the synthetic velvet cloth. The chemically resistant cloth produced scratches on the surface regardless of the slurry's alkalinity or the duration of the polishing time. For automation, a final polish was achieved in two minutes when the synthetic velvet cloth was used. The slurry was charged to the cloth at 30-second intervals, to keep the surface wet, and an applied load of 4.5 kg per specimen was used.

In bright field illumination a properly polished surface was generally featureless. The only distinguishable features were small and dark angular objects, which interference contrast illumination revealed to be gaps within the inclusions. Interference contrast illumination also revealed a surface of fine scratches, with inclusions appearing as flat unscratched areas standing slightly proud of the scratched background (Figure 1).

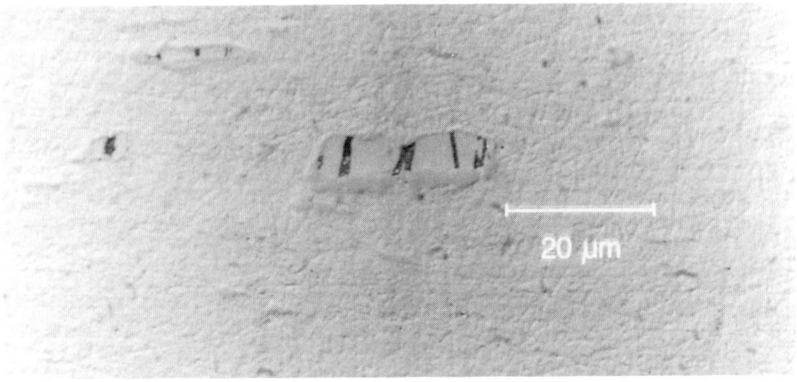

Figure 1. The caustic polished surface viewed using interference contrast illumination.

Figure 2. The same surface as in Figure 1, anodized and viewed in bright field illumination.

The anodized layer presented the primary inclusions as pale blue and bright yellow, and the (αZr) and the (βZr) phases in the background Zr-2.5Nb matrix became visible (Figure 2). The cracks that occurred within the inclusions were observed without any deterioration in detail. Even inclusions perched at the edges of large cracks or voids retained their edges. The inclusions that produced a pale blue anodized film consisted of zirconium and carbon, while inclusions that produced a yellow anodized film consisted of either phosphorous and zirconium, silicon and zirconium, or a combination of phosphorous and silicon with zirconium (Figure 3a). There was no surface contamination from silica, which allowed surface mapping for silicon without confounding (Figure 3b). Although colloidal silica provides an outstanding final polish, the silicon imbeds in the surface confounding SAM examinations for fine silicon particles.

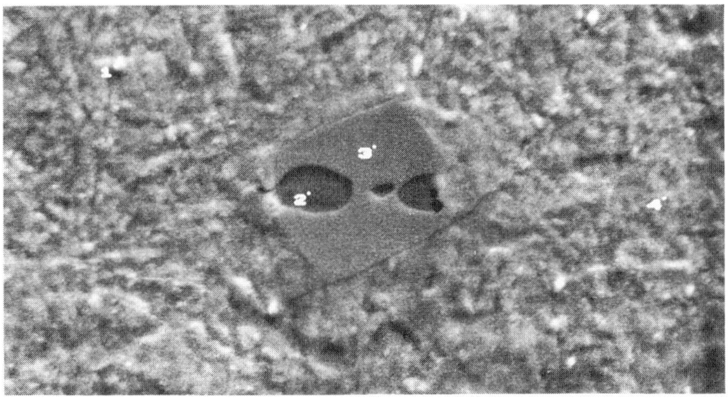

a

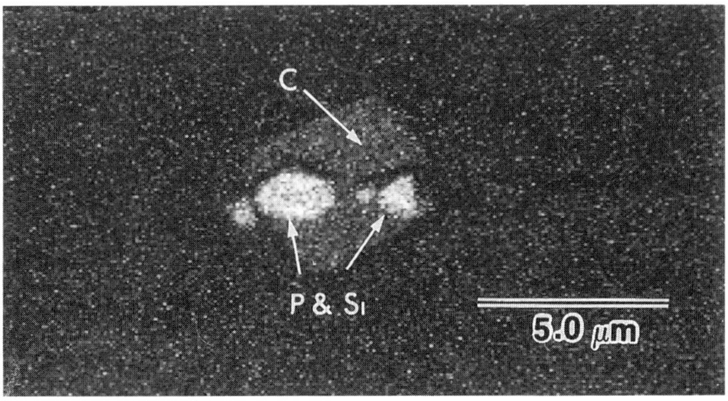

b

Figure 3. Secondary electron image of an inclusion (a) and an elemental map showing carbon, and silicon and phosphorous, as indicated (b).

The sample prepared using the standard acid-attack polish showed preferential etching at the interface between the inclusions and the Zr-2.5Nb matrix and at the cracks within the inclusions. The edges of the inclusions appear to be more rounded, and the image is not as sharp and clear as the surface prepared using the caustic preparation. The caustic polish does not induce preferential etching at the interface between the Zr-2.5Nb, and the inclusion geometric detail is maintained at the edges of the inclusion.

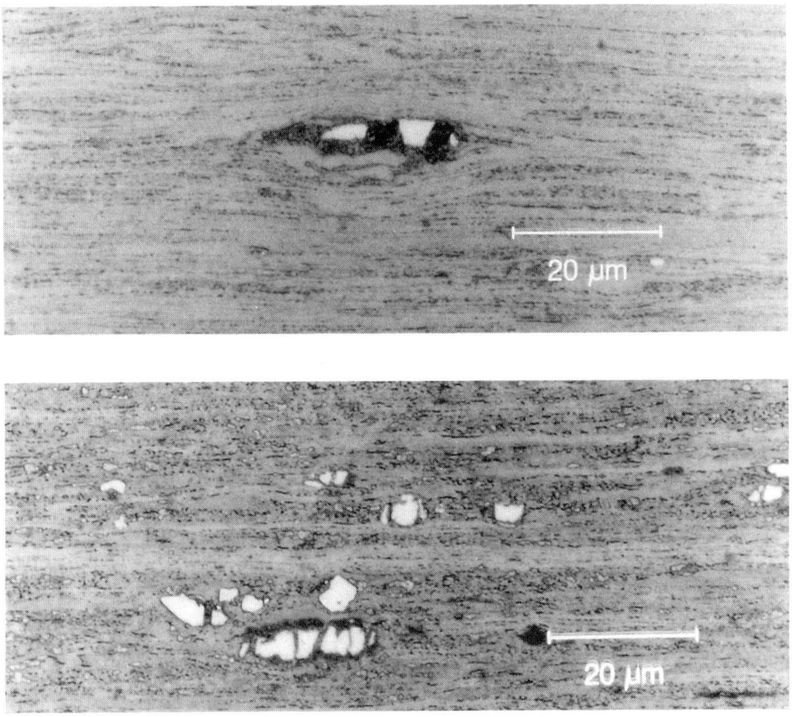

Figure 4. Inclusions prepared using acid polish (top) and caustic polish (bottom).

A quantitative analysis that compared the effect of the acid and caustic preparations on the size distribution and population of inclusions revealed that the caustic preparation exhibited 37% more particles than the acid preparation. Figure 5a shows the normalized distributions of inclusion size for both preparations. The distributions are similar (which is expected, as the analysis was conducted on the same sample), except that the acid preparation has 8% fewer inclusions in the <0.1 μm^2 size range. Figure 5b shows the normalized distributions of the number of particles observed per field for both preparations; the distributions are different. The surface prepared using the caustic preparation had a higher frequency of fields that contained more inclusions and a lower frequency of fields that contained no inclusions. It seems likely that the preferential etching at the interface between the inclusion and the Zr-2.5Nb interface, associated with an acid-attack polish, undermines small inclusions, allowing them to detach from the Zr-2.5Nb matrix. Conversely, the caustic preparation does not cause preferential etching; therefore, no undermining of the inclusion takes place and the presence of small inclusions is maintained.

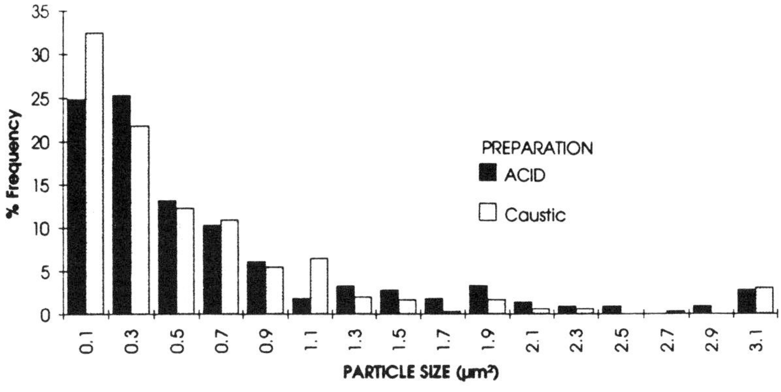

Figure 5a. Normalized distributions of inclusion size for both preparations.

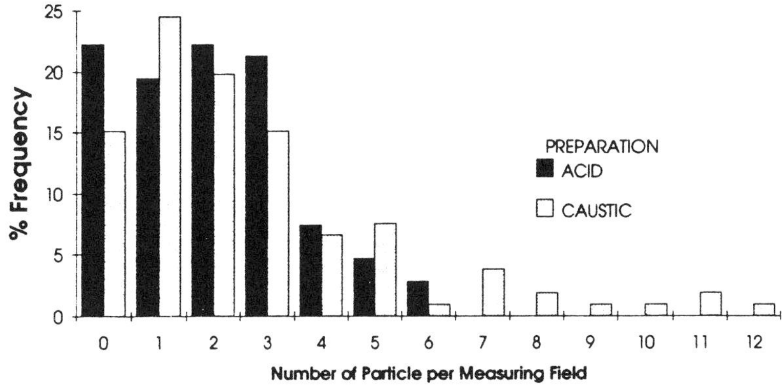

Figure 5b. Normalized distributions of the number of particles per field.

CONCLUSIONS

A new technique for preparing Zr-2.5Nb material that uses a caustic final polish has been developed. The advantages of the technique are that:

(a) the caustic slurry does not induce preferential etching at the interface between the Zr-2.5Nb matrix and the inclusions, or at cracks within the inclusions, and is well suited to the surface examination of small inclusions (less than 0.2 μm^2),
(b) no silica, which can confound SAM mapping for silicon, is used,
(c) the surface structure and the inclusions appear to be sharper, due to the lack of

matrix removal adjacent to the inclusions, and

(d) it has been adapted for use with an automatic polishing apparatus, making this procedure expedient. An optimal final polish was achieved in two minutes using a synthetic velvet cloth charged with optimized caustic slurry at intervals appropriate to keep the surface wet, using an applied load of 4.5 kg per specimen.

ACKNOWLEDGMENTS

I would like to thank CANDU Owners Group for financing the development of this technique, and J.F. Watters for the SAM analysis.

REFERENCES

1) P.E. Danielson, Zirconium and Hafnium and Their Alloys, Metals Handbook, Ninth Edition, Vol 9, p 497, 1986.

2) M.L. Picklesimer, Anodizing for Controlled Microstructural Contrast by Color, **The Microscope**, Vol 15, 1967.

ZINC CONTAMINATION FROM BRASS UPON HEAT TREATING A SUPERCONDUCTING MAGNET

Donald W. Stevens* and William V. Hassenzahl**

ABSTRACT

Theoretical calculations predicted that zinc outgassing from brass spacers during a planned heat treatment would likely damage a lab-scale superconducting magnet. This specter was reinforced by a simulated heat treatment, the samples of which were analyzed by gravimetry, metallography, and microprobe chemical analysis. It was found that zinc escaping from the brass could diffuse 80 µm into copper electrical conductors and degrade their conductivity. To avoid this, steel was temporarily substituted for the brass during the heat treatment process.

INTRODUCTION

As part of the effort to develop superconducting magnets for the International Thermonuclear Experimental Reactor (ITER), the Conductor Performance Qualification (CPQ) test coil shown in Fig. 1 was wound and assembled at the Lawrence Livermore National Laboratory (LLNL).[1] This was a wind-and-react magnet coil where a ductile, unreacted conductor is wound into a coil and later thermally

---

*L-437, Environmental Technologies, Lawrence Livermore National Laboratory, Livermore, CA 94550.
**Retired from Lawrence Livermore National Laboratory.

Fig. 1. CPQ coil. The coil cover is open to expose the outermost layer of insulated conductor (CICC), C, and three brass spacers, B.

reacted to form its internal, brittle, superconducting phase, Nb_3Sn. The conductor, known as a cable-in-conduit-conductor (CICC), along with its unreacted wire made by the internal tin process,[2,3,4] had been manufactured by Intermagnetics General Corporation (IGC) of Guilderland, NY. The unreacted wire is largely stabilizer copper on the outside with a core of tin rods, copper, and niobium filaments in intimate contact. The reaction treatment, depicted in Fig. 2, involves heating the entire CPQ coil to 200 and 340°C to interdiffuse the tin and copper to form bronze in the core, then raising the temperature to 660 and 725°C to react the Nb and Sn (in the bronze) to form superconducting Nb_3Sn. The resulting wire structure is shown in the light optical micrograph of a polished cross section, Fig. 3. The Cu stabilizer, which serves as an emergency shunt for current should the

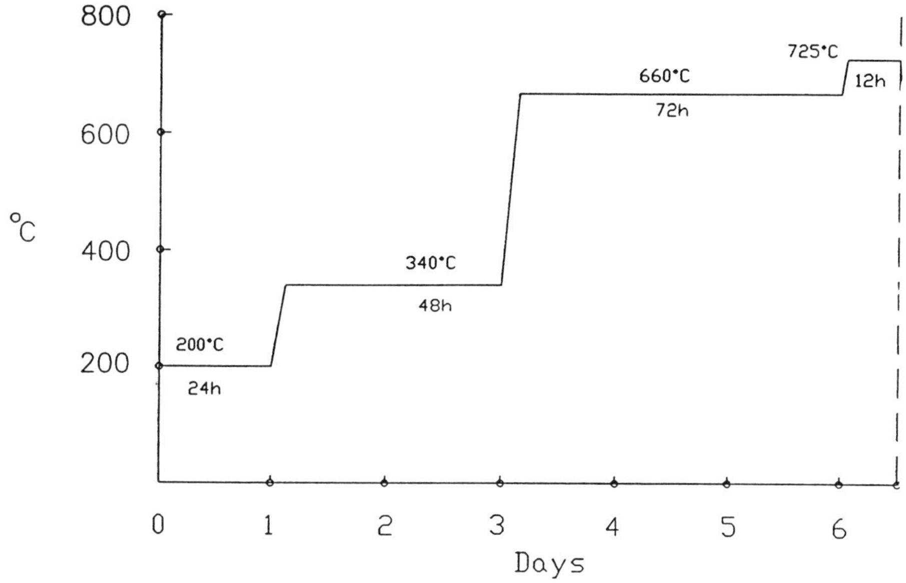

Fig. 2. Reaction heat treatment to form Nb_3Sn.

superconductor becomes temporarily resistive, is protected against alloying with core metals by diffusion barrier layers of Ta and Nb.

Four different diameters of CICC were mandrel-wound to form the CPQ coil, and shortages of some of these imposed gaps in the cylindrical windings. To fill these gaps, commercial Cu-34 wt% Zn brass plates were machined, curved on a mandrel, cut, and filed to fit snugly. Brass was selected for its ease of machining and forming. In total, the spacer plates averaged 10mm (0.4 in.) in thickness, weighed 20 kg (43 lb.) and had about 0.7 m^2 (1000 $in.^2$) of surface area. Three such spacers are seen in Fig. 1, along with the outermost winding of the CICC cable (insulated with Nextel ceramic cloth). Fortunately, it turned out that 80-90% of all the brass spacers could be accessed without disturbing the CICC (brittle after the heat treatment reaction) by removing the three outermost ones and those underneath them, as was done after this investigation showed the advisability of substituting steel for the brass.

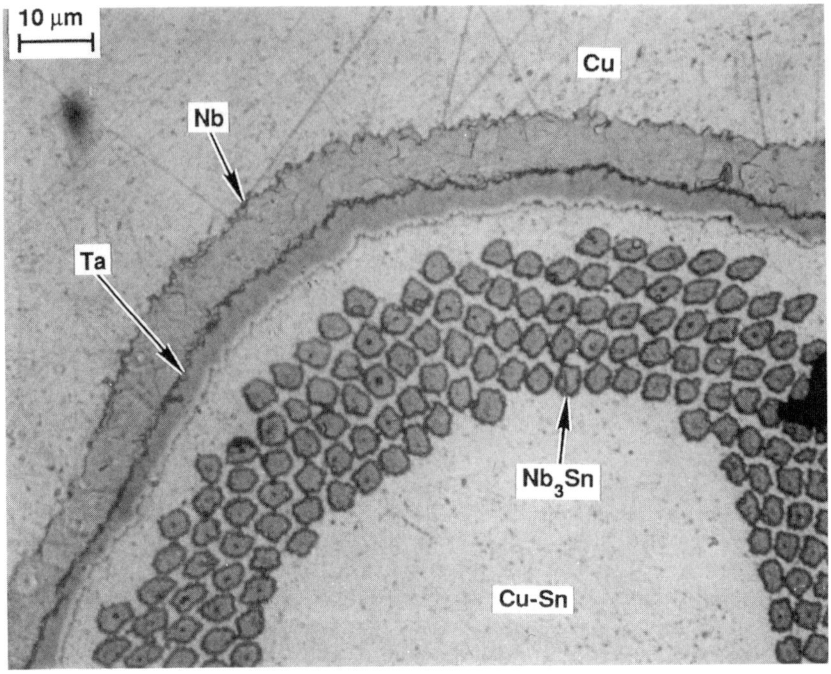

Fig. 3. Superconducting wire within the conductor, CICC. Shown is a cross-section, light-optical, micrograph of reacted (heat treated) wire with phases present after heat treatment.

THEORETICAL ZINC TRANSFER

In planning to do the reaction heat treatment, the prospect of high vapor pressure of zinc from the brass lead to the calculation of vapor pressures of zinc and copper above the brass at according to

$$P_i = P_i^o N_i \qquad (1)$$

where

i = Cu or Zn

P_i = vapor pressure of i above the alloy,

P_i^o = vapor pressure of pure metal, i,

N_i = mole fraction of i in the alloy.

Published values[5] of P_i^o at 725°C are

$$P_{Cu}^o = 10^{-8} \text{ Torr (mm Hg)}$$

$$P_{Zn}^o = 10^{2} \text{ Torr.}$$

The mole fractions at the outset are calculated to be

$N_{Cu} = 0.67$

$N_{Zn} = 0.33.$

Substituting the above values into Eq. 1 yields the following vapor pressures near the brass in the furnace at the outset of the heat treatment:

$P_{Cu} = 6.7 \times 10^{-9}$ Torr

$P_{Zn} = 33$ Torr.

It is seen that the vapor pressure of zinc is high at the outset, amounting to 1/20 atm or 5% of the gas in the argon-rich atmosphere of the furnace. Even with expected decreases in its surface concentration (and thus, vapor pressure) with time, this raised the specter of a significant quantity of zinc vapor (from the 0.7 m^2 surface area of brass) condensing onto, and then diffusing into copper joints and leads (surface area <0.7 m^2). This would transform low-resistivity copper to high-resistivity Cu-Zn alloy (brass) in the diffusion zone.

An estimate of this diffusion penetration was calculated by assuming that a continuous film of Zn was immediately condensed and maintained on the Cu joints and leads, then diffused inward to a depth given by[6]

$$x = 2(Dt)^{1/2} \tag{2}$$

The interdiffusion coefficient is relatively insensitive to composition in the range experienced here for the contaminated Cu (N_{Zn}<10%) and is[7]

$D = 10^{-9} cm^2/s.$

Diffusion time at 725°C was

$t \approx 24$ h.

This value is a combination of 12 h at 725°C plus 12 h estimated for the equivalent amount of 725°C-diffusion occurring over the longer times (a few days) at 660°C and less, as per Fig. 2.

Substituting the above values into Eq. 2 yields

$$x \approx 200\ \mu m.$$

This is clearly a deep zone of alloying which might increase the resistivity of copper substantially. To interrogate this possibility further, the heat treatment of the CPQ coil in a dynamic atmosphere of argon gas was simulated with brass and copper samples in a laboratory furnace, as described below.

EXPERIMENTAL ZINC TRANSFER

The simulated heat treat arrangement is shown in Fig. 4. It

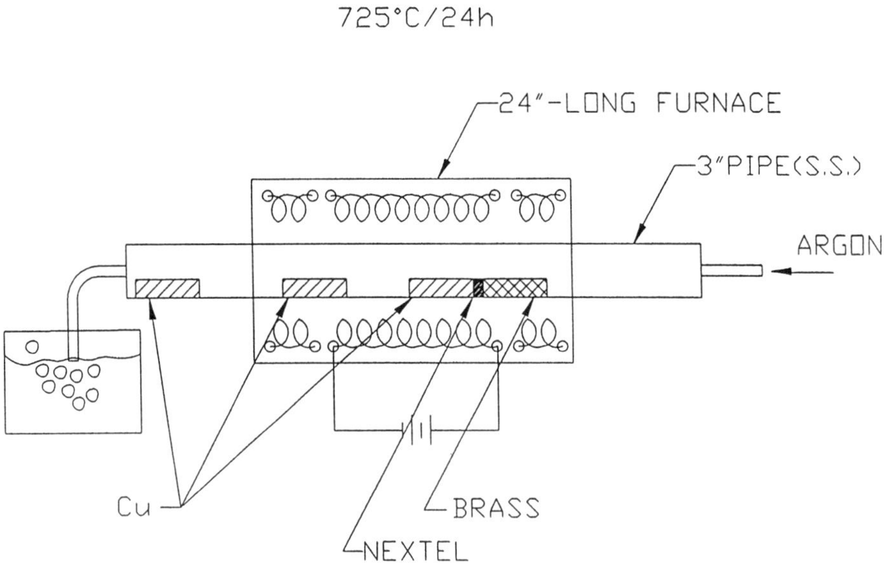

Fig. 4. Furnace arrangement to simulate heat treating the CPQ coil. The brass (Zn source) is upstream of the three copper samples in the flowing argon atmosphere.

featured a 24 in.-long, three-zone, horizontal tube furnace with a modest flow of argon which exits through an ethylene glycol bubble chamber. One brass (the same alloy as the spacers) and three copper cylindrical samples, 1 in. dia. X 3 in. long, were arranged with the brass upstream to facilitate the transfer of zinc vapor to copper. The Nextel ceramic insulator cloth (porous) was added to simulate its use in the coil. Two of the copper cylinders in the furnace were at 725±5°C. One of these was adjacent to the brass and the other began 6 in. from the downstream end of the brass. The copper cylinder in the cool zone outside the furnace was 24 in. from the brass and was about 100°C.

The samples were weighed to the nearest 0.01 g before and after the 725°C/24 h heat treatment. Cross section metallography by light optics was done to observe structural changes. A JEOL 833 analytical electron microscope was employed for SEM imaging and to obtain electron microprobe analyses (also known as wavelength dispersive X-ray spectroscopy) of the zinc distribution in metallographic cross sections. Zinc quantification was within ±0.2 wt% and was based on the MAGIC program of ZAF and instrument corrections, using the brass and copper t=0 (before heat treatment) samples as *in situ* standards.

Changes in the color and weight of the samples upon heat treatment are given in Table I. These were consequences of zinc transfer from the

Table I. Color and weight changes upon heat treatment

Mat-erial	Position	T, °C	Color before	Color after	Weight change, g
Brass	upstream	725	yellow	dull yellow	-2.29
Cu	adjacent	725	copper	yellow	+0.30
Cu	6" downstream	725	copper	mostly yellow	+0.20
Cu	24" downstream	100	copper	copper	<0.01

brass to the copper. The brass sample lost over 2 g of mass and developed a dull luster. This dullness is shown in Fig. 5 to be a consequence of 20-μm-deep grain boundary cavitation upon zinc outgassing. The copper sample adjacent to the brass picked up 13% of the mass lost by the brass, and turned a brassy yellow color in the process of its case converting from copper to a copper-zinc (brass) alloy. Its metallographic cross-section is seen in Fig. 6 where a) was etched in ammonium persulfate and repolished, then re-etched to reveal the crystallographic boundaries shown in b). Both show an 80 μm-deep zone of zinc diffusion penetration delineated by the interface, I. It is seen in b) that zinc alloying transformed the larger-grained,

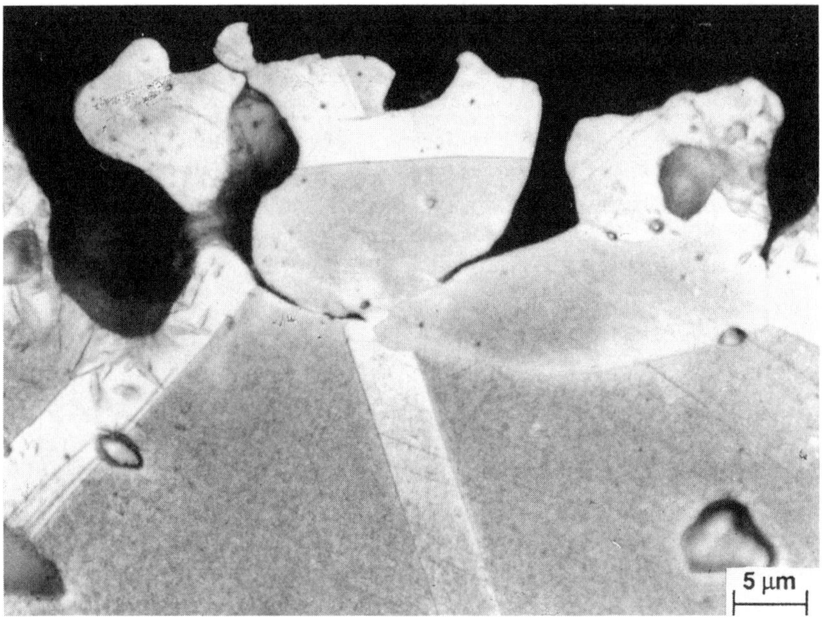

Fig. 5. Brass structure after heat treatment. SEM micrograph of cross section, etched in ammonium persulfate. Grain boundary cavitation, 20 μm deep, is a result of zinc outgassing.

heavily twinned copper into smaller-grained, twin-free brass. This observation from Fig. 6 is substantiated by the 80 μm-deep zone of zinc diffusion revealed in the electron microprobe results of Fig. 7, which also shows zinc loss from the brass sample.

Concerning the two copper samples further downstream, the one closer to the brass and still in the hot zone gained 9% of the zinc lost by the brass and, accordingly, did not turn as yellow as its upstream cousin. The final and furthest sample from the brass was cold and picked up only <0.5% weight, and remained copper-colored. These color- and weight-change gradients reflect downstream losses of zinc vapor by dilution in the argon carrier gas and absorption/condensation onto other surfaces (furnace tube, etc.).

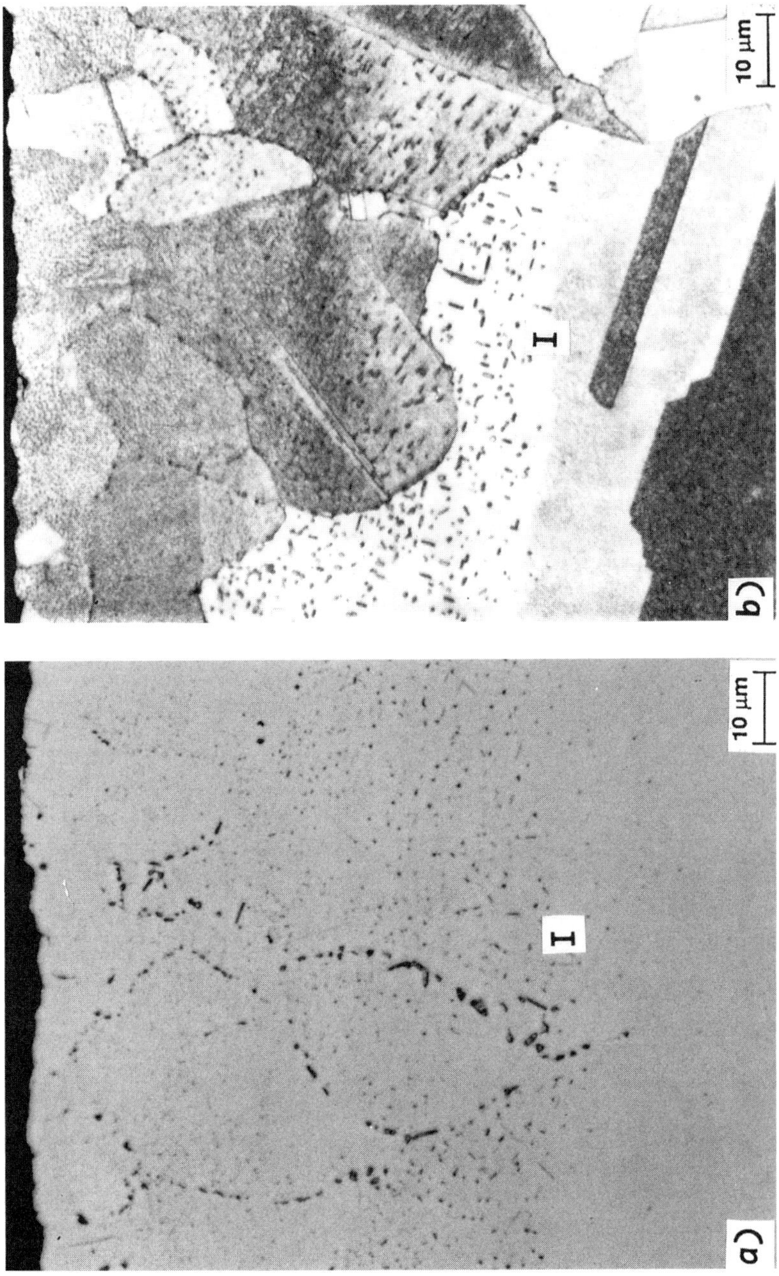

Fig. 6. Copper structure after heat treatment. See text for discussion.

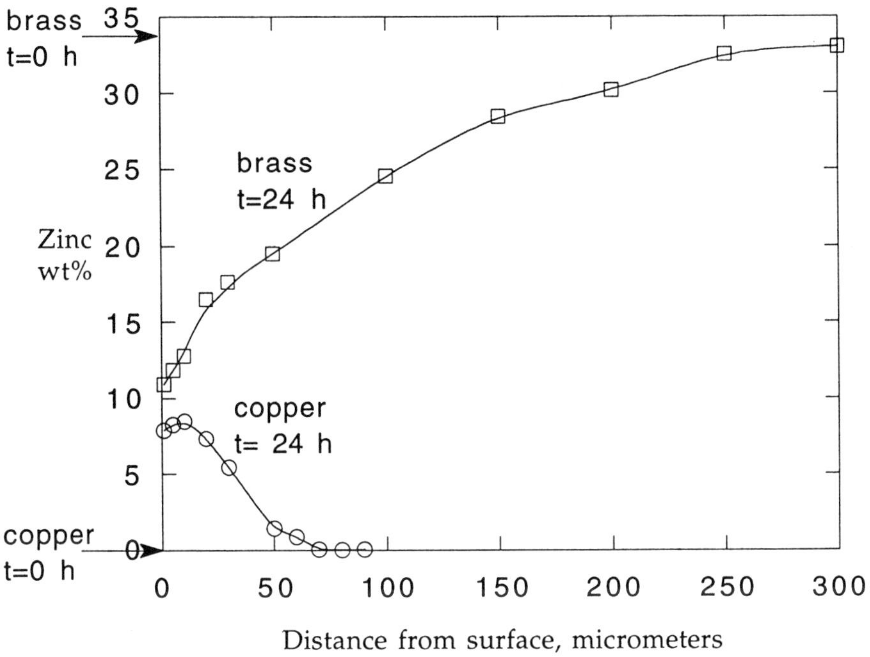

Fig. 7. Distribution of zinc in brass and copper. Data from electron microprobe analysis of metallographically polished cross sections.

DISCUSSION

Zinc outgassing from brass, its condensation onto nearby copper, and its diffusion penetration of 80 μm into the copper, have been experimentally demonstrated. The theoretical prediction based on vapor pressure and diffusion kinetics is reasonably close at 200 μm. The smaller penetration obtained by experiment is likely due to the vapor flux of zinc being insufficient to maintain a continuous zinc layer everywhere on the surface of copper, especially in the latter stages when zinc depletion from the brass surface (as seen in Fig. 7) diminished its effective vapor pressure according to Eq. 1.

As a result of this work, plain carbon steel spacers were fabricated and substituted for their brass counterparts during the heat treatment process. These are to be replaced after heat treatment if the ferromagnetism of the steel interferes with coil testing.

ACKNOWLEDGMENTS

The authors are grateful for the capable assistance of the following persons.

James Bowman	Heat treating and sample preparation
Aaron Horner	Heat treating, sample preparation, and spacer plate fabrication
Robert Jenkins	Spacer plate fabrication
Donald McCoy	Analytical electron microscopy
Robert Kershaw	Light optical metallography and cross-section sample preparation

This work was performed under the auspices of the U. S. Department of Energy by Lawrence Livermore National Laboratory under contract No. W-7405-Eng-48.

REFERENCES

[1]S. Shen et al, "Design and Fabrication of the CPQ Coil", 11th International Conference on Magnet Technology (MT11). T. Sekiguchi and S. Shimamoto, eds, 2 (Elsevier, New York, 1990) 1318-1329.

[2]D.B. Smathers, "A15 Superconductors", *Metals Handbook*, Tenth ed., **2,** Properties and Selection of Materials (ASM International, Materials Park, OH, 1990)1060.

[3]G.M. Ozeryansky and E. Gregory, **IEEE Trans. on Magnetics, 27**(2),1755 (Mar. 1991).

[4]E. Gregory, G.M. Ozeryansky and B.A. Zeitlin, "Improvement of the Structure and Properties of Internal Tin Nb_3Sn Conductors", *Adv. Cryo. Eng. (Mater.), **38**,* F.R. Fickett and R.P. Reed, Eds. (Plenum Press, New York, 1992) 579-586.

[5]S. Margrave, ed., *The Characterization of High Temperature Vapors* (J. Wiley, New York, 1967).

[6]P.G. Shewmon, *Diffusion in Solids* (McGraw-Hill, New York, 1963).

[7]W. Jost, *Diffusion in Solids, Liquids, Gases* (Academic Press, New York, 1960) 232.

CONTRIBUTION OF THE QUANTITATIVE METALLOGRAPHY TO THE STUDY OF THE THERMAL REFINEMENT IN SOME MICROALLOYED STEELS.

Asensio J.[1], Verdeja J.I.[1], and Pero-Sanz J.A.[2].

ABSTRACT

In the present work the effect of a normalizing heat treatment at laboratory scale on different controlled rolled HSLA steels is analysed. The base chemistry of the steels used was similar, the only difference being the type and amount of microalloying element added. This experimental material was surveyed in two conditions: as rolled (V, Nb and V+Nb controlled rolled steels) and industrially normalized after rolling (Nb steel).

Quantitative metallographic determinations of the pearlite volume fractions and of the grain size were carried out on the as-received samples. Preliminary observations suggest that a post-normalizing treatment given to a Nb controlled rolled steel provides a further refinement of the ferrite. The metallographic observations on V and Nb+V controlled rolled steels after a laboratory normalizing heat treatment suggests that V favours the formation of inhomogeneous ferrite-pearlite characteristics, and a more pronounced banded structure.

---

[1] Departamento de Ciencia de los Materiales e Ingeniería Metalúrgica.Escuela Técnica Superior de Ingenieros de Minas. Universidad de Oviedo. Oviedo (Asturias). SPAIN.

[2] Departamento de Ingeniería de los Materiales. Escuela Técnica Superior de Ingenieros de Minas. Universidad Politécnica de Madrid. Madrid. SPAIN.

INTRODUCTION

The criterion for selection of steels for pipelines, pressure vessels, plate for shipbuilding, offshore platforms, etc., has evolved during almost fifty years towards paying more attention to the yield stress σ_y - as a factor of design, than to the tensile strength σ_m. On the other hand there is a need for an answer to two requirements in apparent controversy: higher yield stress (giving name to the HSS steels) and better toughness at low temperatures ("artic" qualities) obtained with small ferritic grain size and low carbon content, which also improves weldability.

The interest shown among users in steels with very fine ferritic grain size seems reasonable as it appears to be the only parameter which increases the yield stress and the toughness at once, and also results in favourable structures for welding.

Four different steels have been examined with less than 0.2% carbon, 1.5% manganese aproximately and niobium and/or vanadium -as microalloying elements- with the usual contents in the HSLA steels. The reasons for selecting those compositions are explained below.

The effect of manganese on the mechanical properties is to raise the yield stress of the steel by solid solution strengthening in the ferrite without impairing the toughness level; it also allows formation of smaller ferrite grains because it lowers the transformation temperature A_3 -as a consequence of this the yield stress and toughness are increased at once- allowing for lower carbon contents in the steels. The present tendency for the manganese content in ferritic-pearlitic HSLA steels with small grain size is to fix it around 1.5% .Higher contents than this will increase the hardenability too much.

A complementary grain refinement -almost always much superior to the one obtained with Mn- is obtained with small additions of elements, which precipitate as small intermetalic compounds in the γ phase- inhibiting the grain growth at high temperatures and consequently giving much smaller ferrite grains. The first steels of this type were obtained using aluminium. This element removes the free nitrogen by forming aluminium nitride. A similar effect occurs with elements like vanadium, niobium or titanium. The result of this is that it is possible to obtain steels with no Cottrell ageing behaviour [1].

Unlike steels containing aluminium only, the steels containing V, Nb, Ti do not require the presence of nitrogen for the formation of intermetallic compounds which restrain the mobility of γ grain boundaries [2]. The existence of carbon in solid solution allows for existence of carbide precipitates playing a similar role on the grain boundary migration. The beneficial effect of the small additions of these elements (microalloyed steels) is not only limited to the arrest of the grain growth in a normalizing heat treatment but it also affects in an important way the size and shape of the austenite during controlled rolling.

In the present investigation different microalloyed steels with a similar base chemistry (C, Mn, S, P, ...) were used. A first series of steels with only one microalloying element, vanadium or niobium, with the usual contents for grain refinement were used. Complementary to these, another composition containing V and Nb simultaneously in the range of the steel alloyed with only one element has been used.

EXPERIMENTAL TECHNIQUES AND RESULTS

The chemical compositions of the four steels used in the present research work are listed in Table 1. The steels referred to as A, B and C were in the «as rolled» condition; steel D was industrially normalized after «controlled rolling».

Table 1 Chemical composition in weight % of the steels studied.

Steel	%C	%Mn	%Si	%Nb	%V	$\%Al_{total}$	%P	%S
A	0.18	1.50	0.40	0.02	0.064	0.043	0.029	0.012
B	0.14	1.46	0.24	----	0.046	0.028	0.025	0.008
C	0.18	1.49	0.45	0.02	------	0.030	0.020	0.015
D	0.16	1.52	0.41	0.02	------	0.033	0.022	0.013

As an orientating value, previous to the experimental determination, the critical temperature A_{c3} and the eutectoid temperature A_e have been calculated using the formulas given by Andrews [7]. For the determination of the carbon content which corresponds to the eutectoid transformation C_p in each steel, it has been assumed that the maximum solubility of carbon in the austenite A_{cm} is similar to that of a binary steel. The actual numerical value for C_p was determined by the intersection of the horizontal line A_e (calculated value) with the curve for binary steels $A_{cm} = \psi(\%C)$. The results are presented in Table 2.

Table 2 Critical temperatures and expected volume fraction of ferrite.

Steels	A_{c3}	A_e	C_p (%)	$\rho_{pearlite}$ (g/cm^3)	Vv_t (α) %
A	1138 K 865 ºC	995 K 722 ºC	0.75	7.78	76
B	1107 K 834 ºC	991 K 718 ºC	0.74	7.7	81
C	1098 K 825 ºC	997 K 724 ºC	0.76	7.78	76
D	1102 K 829 ºC	995 K 722 ºC	0.756	7.782	78

In the same table the densities of the proeutectoid pearlite, $\rho_{pearlite}$ for each steel has been indicated . This value was calculated from the theoretical weight proportion of ferrite and cementite which corresponds to each of the steels, and the densities of ferrite and cementite (assuming a constant value in all the cases of 7.86 and 7.66 g/cm^3 respectively). The theoretical value of the primary ferrite volume fraction was also calculated from: $Vv_t(\alpha)$ = 100 - Vv_t(pearlite). For the steels considered the theoretical value remained between 76% and 81%.

In order to determine the actual ferrite volume fractions $Vv(\alpha)$, the volume fraction of pearlite was measured using an automatic image analyzer (OPTOMAX V) in 250 observation fields, with a relative error of 5% or lower. The fields were selected randomly at the appropriate magnification in a microstructure with a marked banded orientation of the pearlite. The results are listed in Table 3, together with the measured volume fractions of ferrite $Vv(\alpha)$. The values were very similar and remained in the range of 72% to 79%, which in average was slightly lower than the theoretical value. This seems reasonable as the actual microstructures are homogeneous.

Table 3 Quantitative metallographic determinations and mechanical parameters.

Steel	Vv (α) % measured	L_3 (α) (μm)	φ (α) (μm)	$\Delta\sigma_y$ (MPa)	Δ_ϕ ITT (ºC)
A	77	8.4	12.6	189	- 125
B	78	11.0	16.5	166	- 110
C	72	10.6	16.0	169	- 111
D	78	8.3	12.5	191	- 126

The alpha grain diameters, $\phi(\alpha)$, were determined by direct measurement on micrographs at 500X on two test lines of 220 mm in length with a random orientation in the observation field. Preliminary determinations of the grain size using the image analyzer revealed misleading results as the grain boundary delineation was not the same in all the areas, therefore including more than two grains enclosed in the same grain boundary. During the grain counting it was taking as 1/2 every time that the test line intersected in less than two points a grain. In Table 4 the number of grains counted in each case has been quoted to obtain the listed standard deviations in $\Phi(\alpha)$.

Table 4 Total number of grains counted to reach the standard deviation

Steel	Number of grains counted	Standard Deviation (%) in $\Phi(\alpha)$
A	222	12
B	293	11
C	300	10
D	309	10

The mean linear intercept of each grain, L_3 (α), in Table 3 was determined as a function of the volume fraction of ferrite as:

$$\overline{L_3}(\alpha) = L_T \cdot V_v(\alpha) \qquad (1)$$

For the determination of the grain volumetric diameter as a function of the mean linear intercept $L_3(\alpha)$, it was considered that ϕ =1.5 L_3, following the recommendation given for the case of equiaxed structures in ASTM E112-82.

The structures of the steels A, B, C and D are shown in Figs. 1, 2, 3 and 4, respectively.

The effect on the yield stress σ_y of the grain diameter could be evaluated by means of the corresponding term in the *Hall-Petch* equation; that is: $\Delta\sigma_y = 1.5\ L_3^{-1/2}$ (MPa). This does not include other important factors like solid solution strengthening and the precipitation hardening. The calculated values are listed in Table 3.

It is a well known fact that the grain refinement improves toughness at low temperatures. A rough estimate of the impact transition temperature ITT (ºC) based on the grain size and ignoring the unfavourable effect of the pearlite, structural strengthening, etc. can be obteined from: $\Delta ITT\ (ºC) = -11.5\ L_3^{-1/2}$.

Steels C and D are of similar compositions, only microalloyed with Nb, but they differ in their thermomechanical history during the industrial processing after controlled rolling: steel C corresponds to the «as rolled» condition and steel D was rolled and subsequently normalized. The observed differences between the ferritic grain sizes suggest the possibility of a complementary grain refinement with a normalizing after controlled rolling despite of the formation of a more pronounced banded structure (see Figs. 3 and 4).

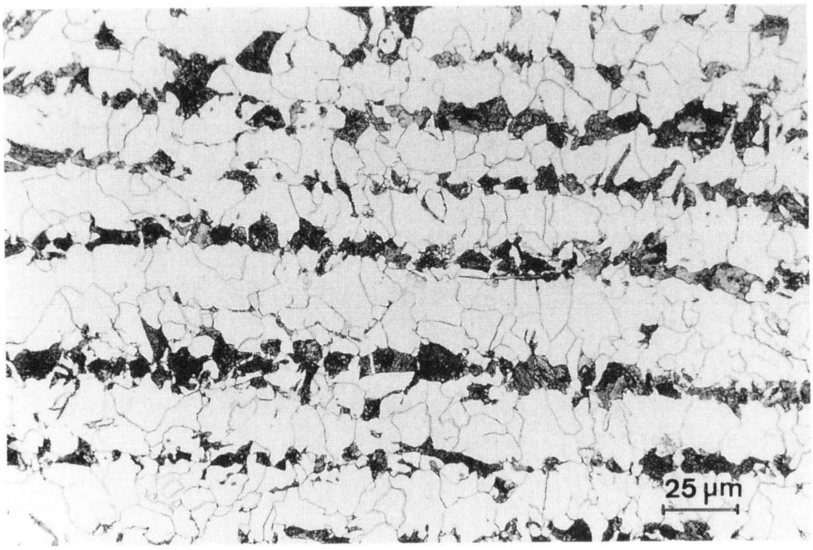

Figure 1.- Transverse section parallel to the rolling direction (R.D.) in steel A. (400X).

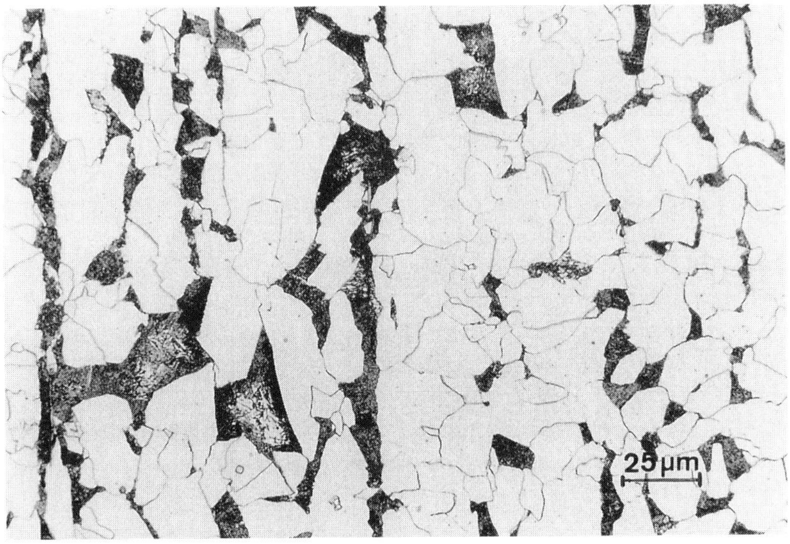

Figure 2.- Transverse section parallel to the rolling direction in steel B (400X)

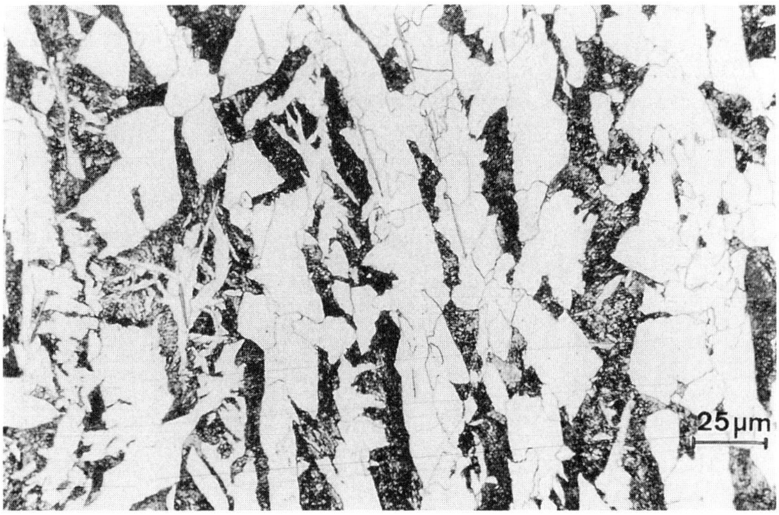

Figure 3.- Transverse section parallel to the rolling direction (R.D.) in steel C. (400X).

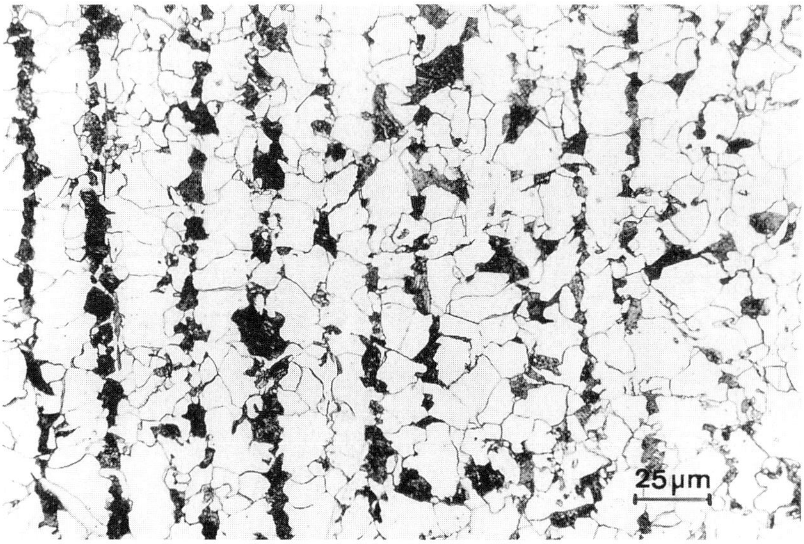

Figure 4.- Transverse section parallel to the rolling direction in steel D (400X)

In order to quantify the degree of improvement of a «thermal refining treatment» on a «thermomechanical refined structure» obtained by controlled rolling and to asses the relevance of V or Nb acting separately, a normalizing heat treatment was carried out in the laboratory at 940ºC for 25 minutes on the V-steel (steel B) and the Nb-steel (steel C). The results were obtained in a similar way as described for Table 3, are listed in Table 5.

Table 5 Quantitative metallographic determinations and expected mechanical behaviour in the laboratory normalized samples.

Steel	V_v (α)% measured	L_3 (α) (μm)	$\phi(\alpha)$ (μm)	$\Delta\sigma_y$ (MPa)	Δ ITT (ºC)
B_N	77	8.5	12.8	188	- 124
C_N	72	6.7	10.0	213	- 141

The structures in Figures 5 and 6 correspond to the steel B and C in the normalized condition, B_N and C_N respectively. The reduction detected in the grain diameter after normalizing was greater in the Nb steel than in the V steel: ϕ_{CN}:ϕ_C =0.63 compared to ϕ_{BN}:ϕ_B =0.77. It seems reasonable that the solution temperatures for niobium carbide is higher than that for vanadium carbides and hence the pinning effect on the mobility of the grain boundaries at the normalizing temperatures is greater for the niobium-containing steel.

It has been detected that during the heating-up process and the holding periods during the normalizing the vanadium intermetallic compounds were dissolved away rapidly giving abnormal austenitic grain growth and an inhomogeneous grain structure on the transformation. This is in sharp contrast with the observed structure in the Nb steel. A possible reason for this could be a lower solubility of the niobium carbides capable of restraining grain growth at temperatures above A_3 and thus giving a more uniform α structure **[3,4]**.

In the case of the Nb+V microalloyed steel the solution of V during normalizing appears to give structural inhomogeneities after transformation at room temperature **[5]**. Figure 7 corresponds to the steel A in the normalized condition (A_N).

In order to verify the effect of V on the inhomogeneity of the γ grains, a series of heat treatments at the normalizing temperatures followed by quenching in water was carried out to observe the austenite structure. This technique allows revelations of the existing structures at high temperatures (above A_{r3}). The micrographs in figures 8 to 10 correspond to the vanadium steel (steel B), the niobium steel (steel C) and the vanadium-niobium steel (steel A). It is observed that those steels containing V present a «duplex» gamma grain structure, probably due to the displacement of the grain boundaries in the absence of vanadium carbide precipitates. In contrast, it was found the γ grains in the Nb steel were more uniform in size and smaller $\phi(\gamma)$ = 22 μm . The same mesurements done for the V and the Nb+V steels yield 41 μm and 29 μm for the grain diameters, respectively.

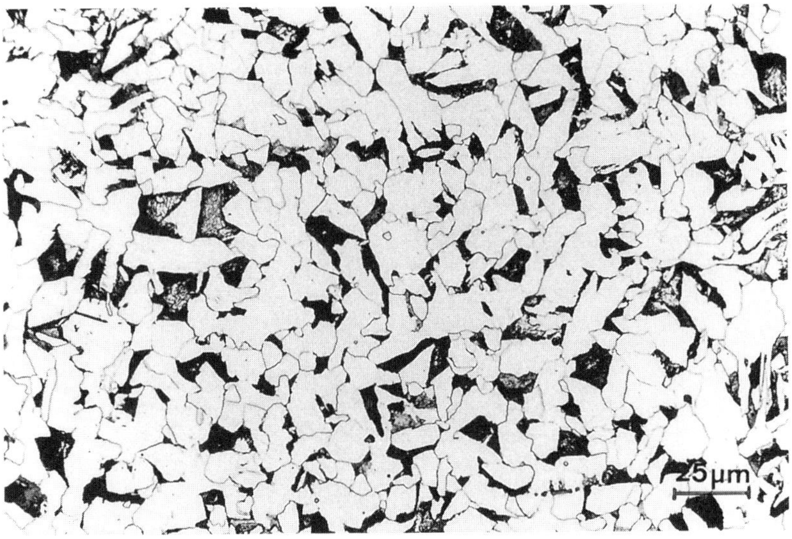

Figure 5.- Steel B as normalized (400X)

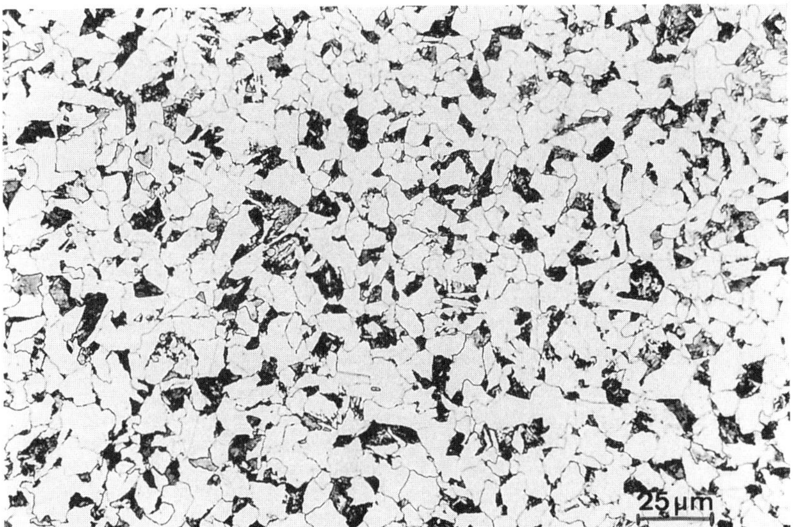

Figure 6.- Steel C as normalized (400X)

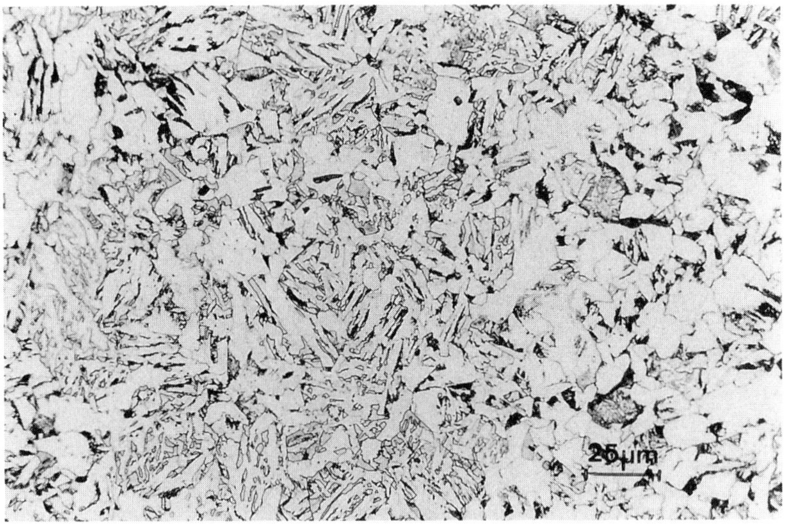

Figure 7.- Steel A as normalized (400X)

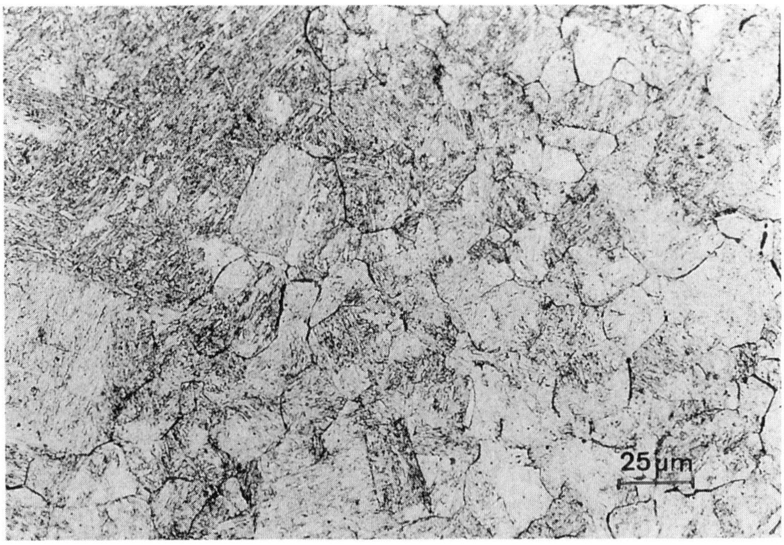

Figure 8.- Steel B as quenched (400X)

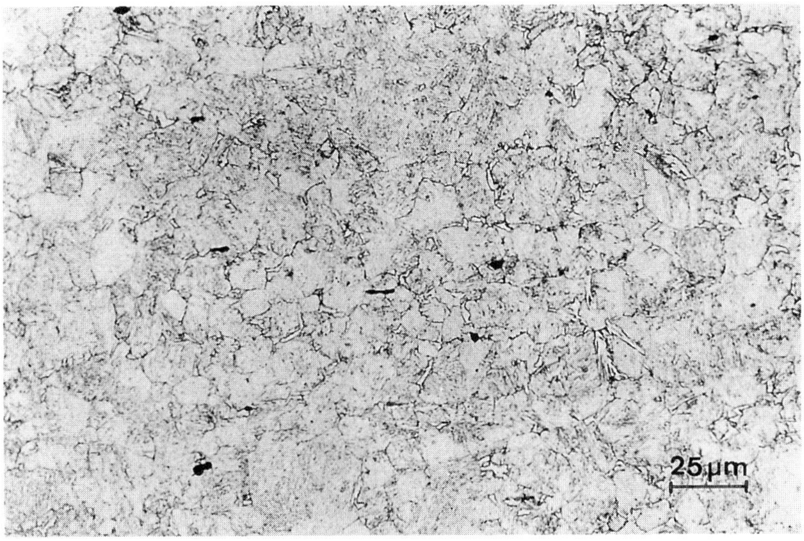

Figure 9.- Steel C as quenched (400X)

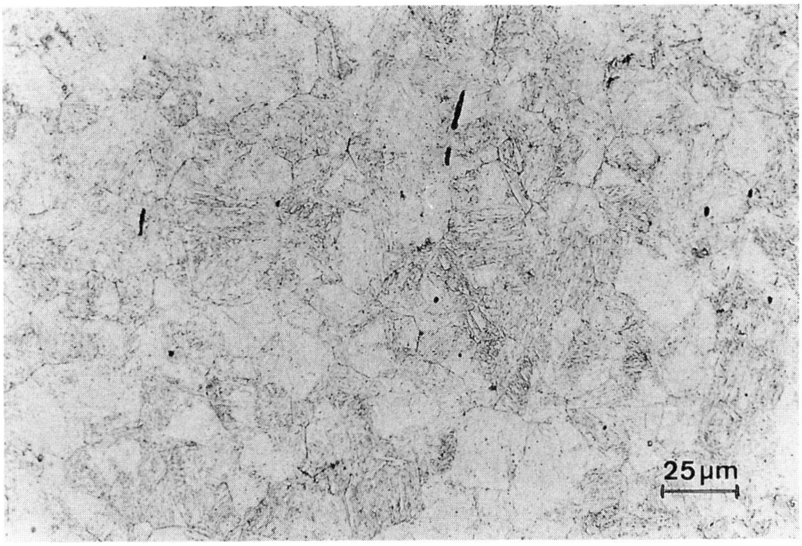

Figure 10.- Steel A as quenched (400X).

CONCLUSIONS

Thermomechanical refinement.

The presence of small quantities of niobium and/or vanadium in a HSLA steel of 0.15-0.20% of carbon and 1.5% manganese gives a much finer ferritic grain size when it is controlled rolled, than in a C-Mn steel after conventional hot rolling.

The thermomechanical refinement obtained in C-1.5%Mn steels with only one microalloying element gives for the degree of refinement for the ferrite structure in the case of the 0.02%Nb steel (steel C) than for the 0.05%V steel (steel B). In the case of the 0.02%Nb-0.064%V (steel A) the degree of refinement was better than in the previous steels ($\phi_{Nb+V} : \phi_{Nb} = 0.79$).

Thermal Refinement Complementary to Controlled Rolling.

A further refinement of a controlled rolling structure is possible with a normalizing heat treatment.The ferrite grain size observed in the hot rolled and normalized samples was smaller than in the only hot rolled samples. Besides it seems as if the banded structure is less pronounced after a post-normalizing.

The ratio of the ferrite grain diameter in the Nb steel (hot rolled and normalized, C_N) and a plain C-Mn steel of say 7 ASTM grain size for the ferrite is 0.25 aproximately. The increment in the yield stress due to this factor could be estimated as more than 100 MPa, which also contributes to the structural strenghthening.

It seems from the observations on vanadium and vanadium-niobium steels, that vanadium favours the formation of inhomogeneous ferritic-pearlitic structures perhaps owing to the coarsening of vanadium carbides during the normalizing heat treatment which ultimately favours an abnormal grain growth of the austenite [6].

It must not be concluded from the above that a heat treatment of normalizing a vanadium containing steels will always be detrimental to the mechanical properties. vanadium in solid solution at the normalizing temperatures will improve the mechanical strength of the steel because of its precipitation as V carbide in α phase which on cooling gives structural strenghthening. The effect of Nb on the precipitation strengthening in our steel composition is probably meaningless as Nb remains as a precipitate which partially coarsens during the heat treatment.

ACKNOWLEDGMENTS

The authors want to thank the FICYT for the funding of the project, ENSIDESA for the experimental material and Dr. Agustín Suárez, Mr. J.M. Martínez and J.O. García for their valuable contribution and help.

REFERENCES

1) F.B. Pickering, "Physical Metallurgy and the Design of Steels". **Applied Science Publishers** 1983.
2) M.M.A. Bepari, "Effects of precipitates on ferrite grain size in 0.15%C steels". **Materials Science and Technology**. Vol.5 (13), January 1989.
3) H.-G. Hillenbrand, C. Passer, and P. Schwaab, "Quantitative Investigation of the Precipitation State of Microalloyed Steels, and its Connection with the Mechanical Properties". **Practica Metallurgica 26**, (17) 1989.
4) R.C. Cochrane, B. Mintz, and J.L. Ward. "Influence of prior microstructure on normalising response of C-Mn-Al-Nb steels". **Materials Science and Technology**, Vol.5, (20), January 1989.
5) T.M. Hoogendoorn, and M.J. Spanraft, "Quantifying the Effect of Microalloying Elements on Structures during Processing" **Microalloying 75 Conference Proceedings**, (75), 1975.
6) T. Gladman *et al.* "Structure-Property Relationships in High-Strength Microalloyed Steels" **MicroAlloying 75 Conference Proceedings. October 1-3, 1975. Washington, D.C., USA.** (32) 1975.
7) Andrews. "Empirical Formulae for the Calculation of some Transformation Temperatures". **Journal Iron and Steel Inst.** (721), 1965.

DEFORMATION MICROSTRUCTURES IN TANTALUM EXPLOSIVELY FORMED PENETRATORS

C-S. Niou[1], L.E. Murr[1], C. Feng[2] and S. Pappu[1]

ABSTRACT

A systematic optical and transmission electron microscope investigation of recovered tantalum explosively formed penetrators (EFP's) has been conducted. Variations of straining range from essentially zero to in excess of 300% in various EFP locations. Corresponding observations depict an evolution of microstructures involving dislocation development, dislocation cells which elongate to form sub-boundaries or related low-energy dislocation structures (LEDS) which resemble what are popularly called dynamic recovery or recrystallization structures. Dark-field TEM is used to differentiate re-oriented cell block (CB) regions and examine the microstructural evolution in terms of misorientation angles or grain surface variations observable in the selected-area electron diffraction patterns. The microstructural features of interfacial evolution are also critically examined in the context of dislocation cell walls, dense dislocation cell walls, and high-angle grain boundaries or so-called geometrically necessary boundaries characteristic of recrystallized regions in the most heavily deformed regions of the EFP. Comparisons are made with dynamic recovery and recrystallization microstructures in a recovered, tantalum shaped charge slug.

---

[1]Department of Metallurgical and Materials Engineering, The University of Texas at El Paso, El Paso, TX 79968 and [2]Metallic Materials Branch, Armaments Tech., U.S. Army Armament Research, Development, and Engineering Center, Picatinny Arsenal, NJ 07806

INTRODUCTION

Murr, et al (1) have recently described residual microstructures in tantalum shaped charge components to be characterized primarily by dynamic recrystallization where new, initially strain free grains are formed in a continuous (or discontinuous) deformation process involving strains which accumulate to 900 percent at strain rates in excess of about 10^6 s^{-1}. While the shaped charge jet appears to be nearly completely dynamically recrystallized, the slug cross-section consists of an outer zone characteristic of dynamic recovery, a transition region of mixed recovery and recrystallization, and a central region which is dynamically recrystallized.

In more recent investigations of residual microstructures in tantalum explosively formed penetrators (EFP's), regions of severe deformation exhibited primarily dynamic recovery microstructures, with some evidence for dynamic recrystallization (2). Here we present a more detailed investigation of tantalum EFP's in which we look critically at the variations in, and evolution of, dislocation boundaries and walls, block wall structures or geometrically necessary boundaries (3, 4), and their corresponding misorientation angles (or crystallographic variations across such boundary features), and intermixed (recrystallized) regions using transmission electron microscopy.

OBSERVATIONS AND COMPARISON OF EFP MICROSTRUCTURES

Because the starting grain size in the tantalum EFP preform consisted of 75 μm equiaxed grains, it might be expected that the initial, explosively driven shock wave would impart at least a basic shock-induced defect structure at $\varepsilon \sim 0$ (5). As the preform is dynamically shaped in flight (Fig. 1(a)), strains vary irregularly, corresponding to concentric zones on the original preform. These features are also illustrated for several recovered EFP's in Fig. 1(b) and (c) and examples of typical, residual microstructural regimes denoted 1 to 3 for the accompanying optical micrographs.

Transmission electron microscope (TEM) observations corresponding to the microstructural regimes illustrated in Fig. 1 (for recovered tantalum EFP's) have demonstrated a range of dislocation cell

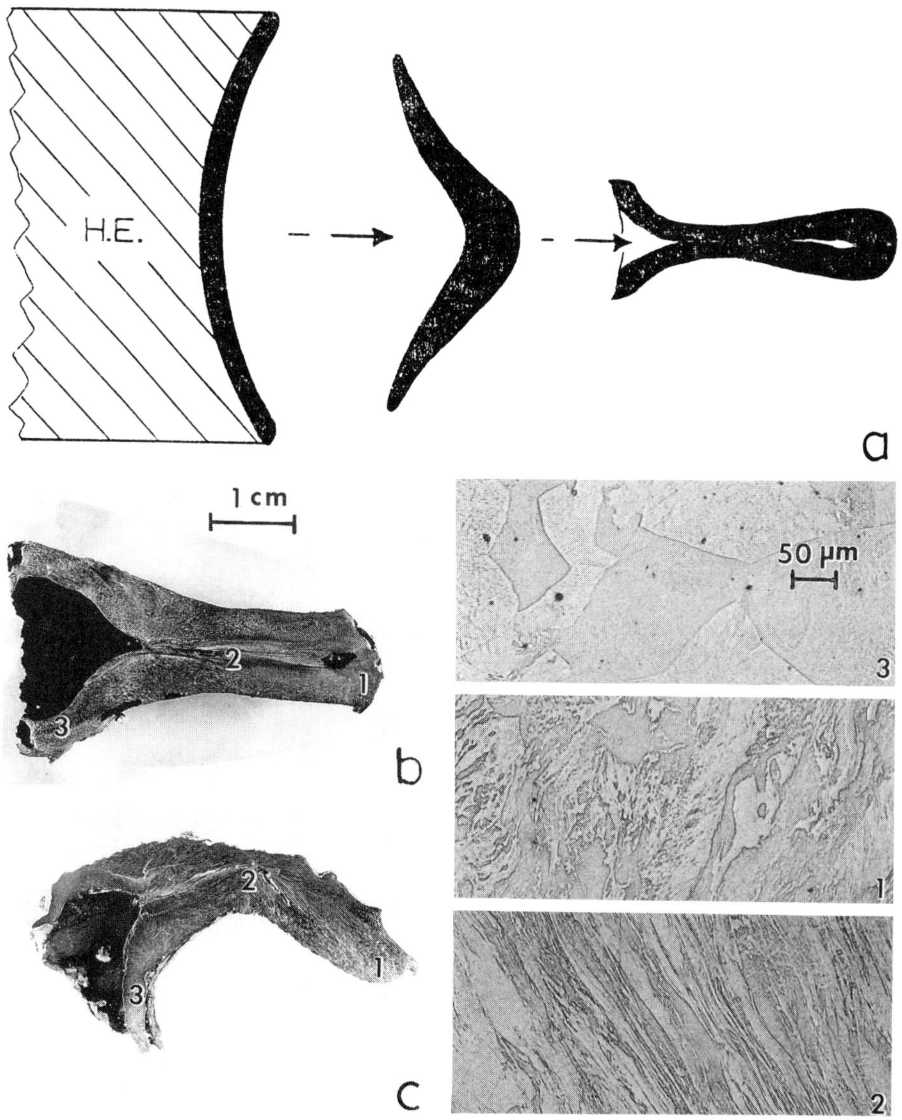

Figure 1. Schematic and actual views of EFP development and residual structures and microstructures. (a) Schematic view of EFP development (self-formation). (b) and (c) show recovered tantalum EFP's cut in half. The optical metallographic views marked 1, 2, 3 show distinct microstructures variously located as illustrated generally for (b) and (c).

and cell wall structures, geometrically necessary boundaries, and grain refined areas having different, neighboring blocks. Figure 2 shows for comparison a relatively unaltered preform microstructure (region 1 in Fig. 1) and an intermixed region of dense dislocation cell structures and sub-boundaries near the head of the recovered EFP in Fig. 1(b). The selected-area electron diffraction pattern inserted into Fig. 2(b) is indicative of simple misorientations (Θ) ranging from less than 1° to 2°.

Figure 3 shows for comparison the microstructures typical of the most heavily strained regions within the EFP, corresponding to regions illustrated typically by 3 in Fig. 1. These represent an intermixing of very fine sub-grains and dislocation cells (including elongated cells) with various cell wall structures (3, 4). (Fig. 3(a)) and (Fig. 3(b)). The selected-area electron diffraction pattern inserted into Fig. 3(a) shows average misorientations for the elongated cell boundaries between 2 and 5°. In Fig. 3(b), the small, equiaxed, sub-grains are characterized by geometrically necessary boundaries (composed of dense dislocation walls) and regular grain boundaries. Where neighboring grain surface orientations are the same, misorientations are observed to be greater than 12°, while most grains are truly "recrystallized" or refined, possessing different grain surface orientations, but with no common crystallographic directions which could be used to rationalize some characteristic misorientation angle (6). These features, unique to dynamic recrystallization in the classical sense, (7, 8) are illustrated in the comparative sequences of bright and dark-field TEM images, and corresponding selected-area electron (micro-) diffraction patterns for specific grains, shown in Figs. 4 and 5.

It is of interest to compare the variations of residual microstructures in the recovered EFP (slug), as illustrated in Figs. 2 to 5, with the variations of residual microstructures in the recovered slug for a detonated, tantalum shaped charge. The shaped charge slug is characterized by a region of elongated grains which extend from the surface of the slug into a central region exhibiting a transition from these elongated sub-structures to small, recrystallized, equiaxed grains in a small central core region. Figure 6 shows the elongated dislocation cell-type microstructures where the cell wall/sub-boundary misorientations average 2 to 5° and their transition to dynamically recrystallized grains having large misorientations (> 12°) or different crystal (grain) surface orientations. These features are

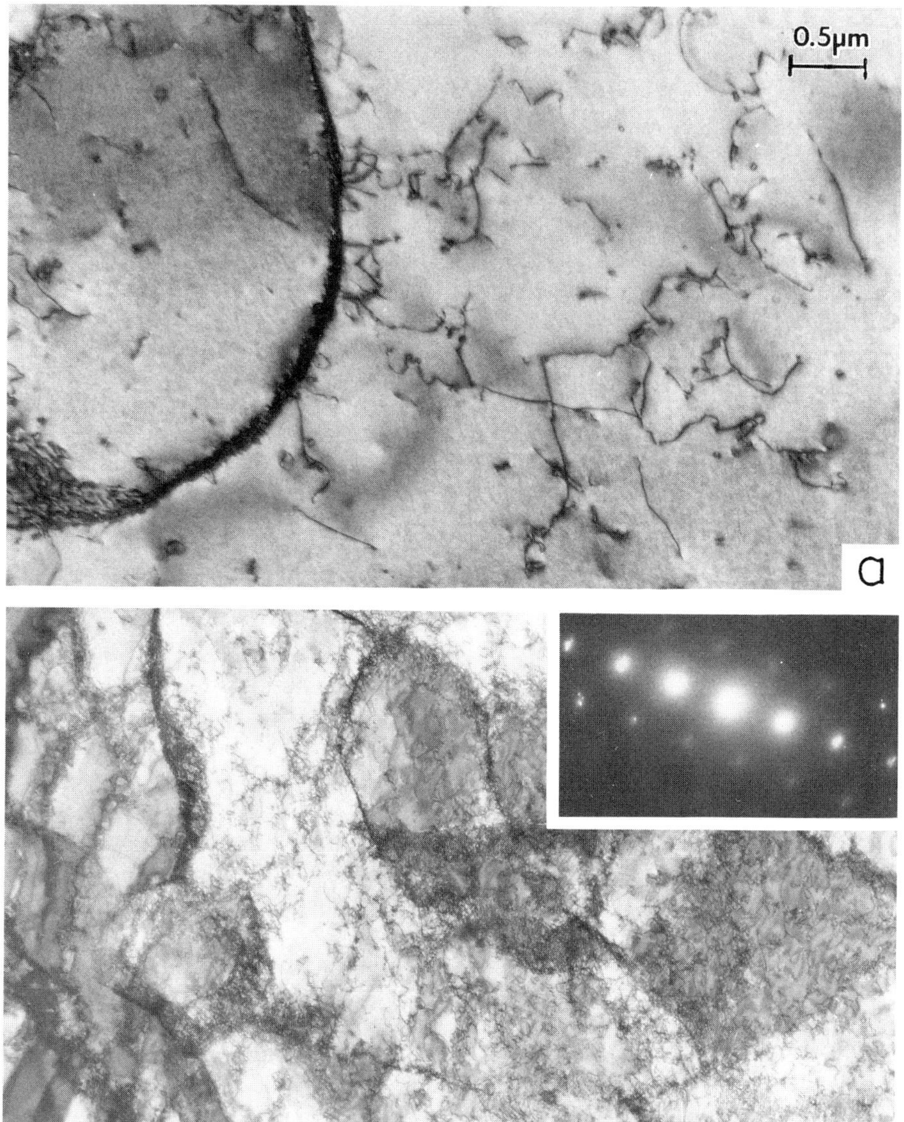

Figure 2. Bright-field TEM images of tantalum EFP microstructures. (a) Relatively undeformed or lightly deformed regions typical of 1 in Fig. 1. (b) Dislocation cells and cell wall structures showing small misorientation features (SAD insert) typical of many EFP regions.

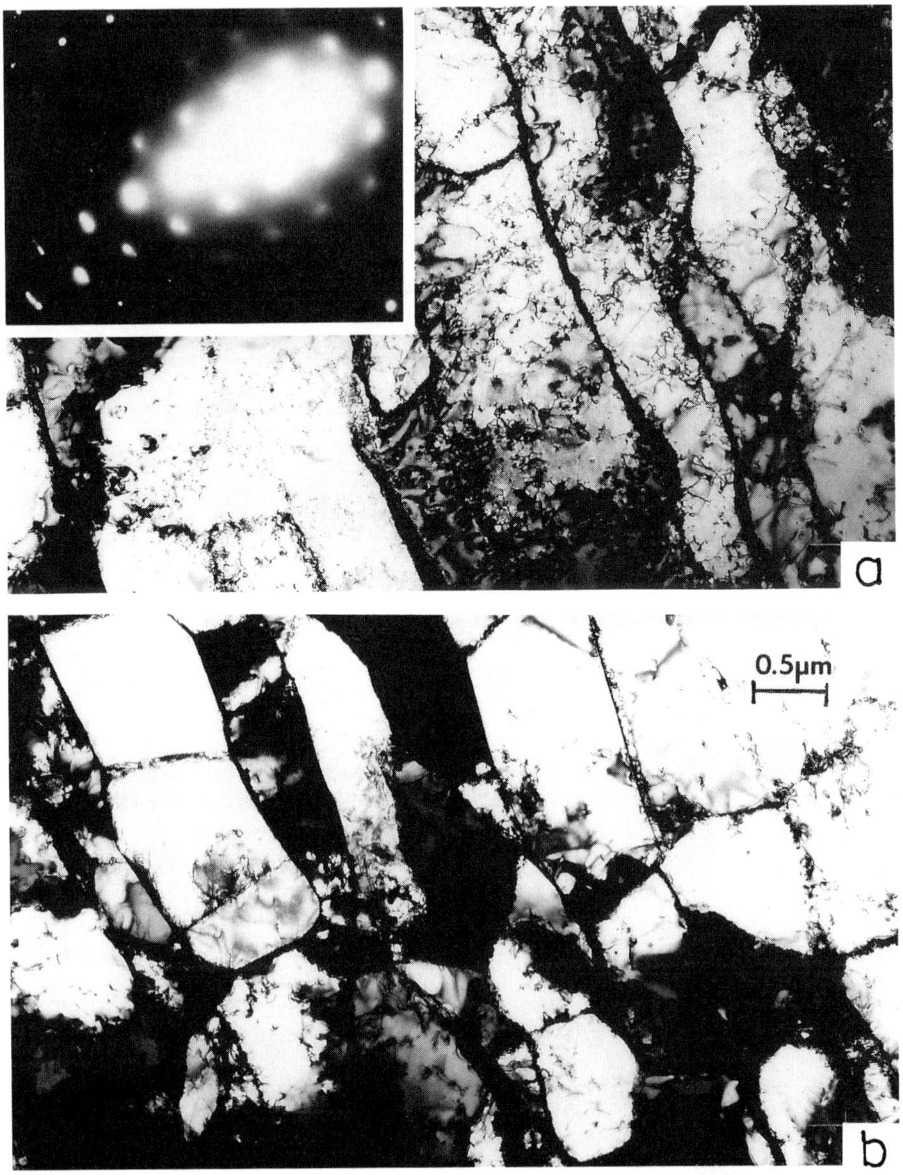

Figure 3. TEM bright-field images typical of heavily deformed regions including 2 and 3 in Fig. 1. (a) Elongated dislocation cells exhibiting misorientations of several degrees (SAD insert). (b) Equiaxed grains typical of zones within extreme deformation regions in 3 in Fig. 1.

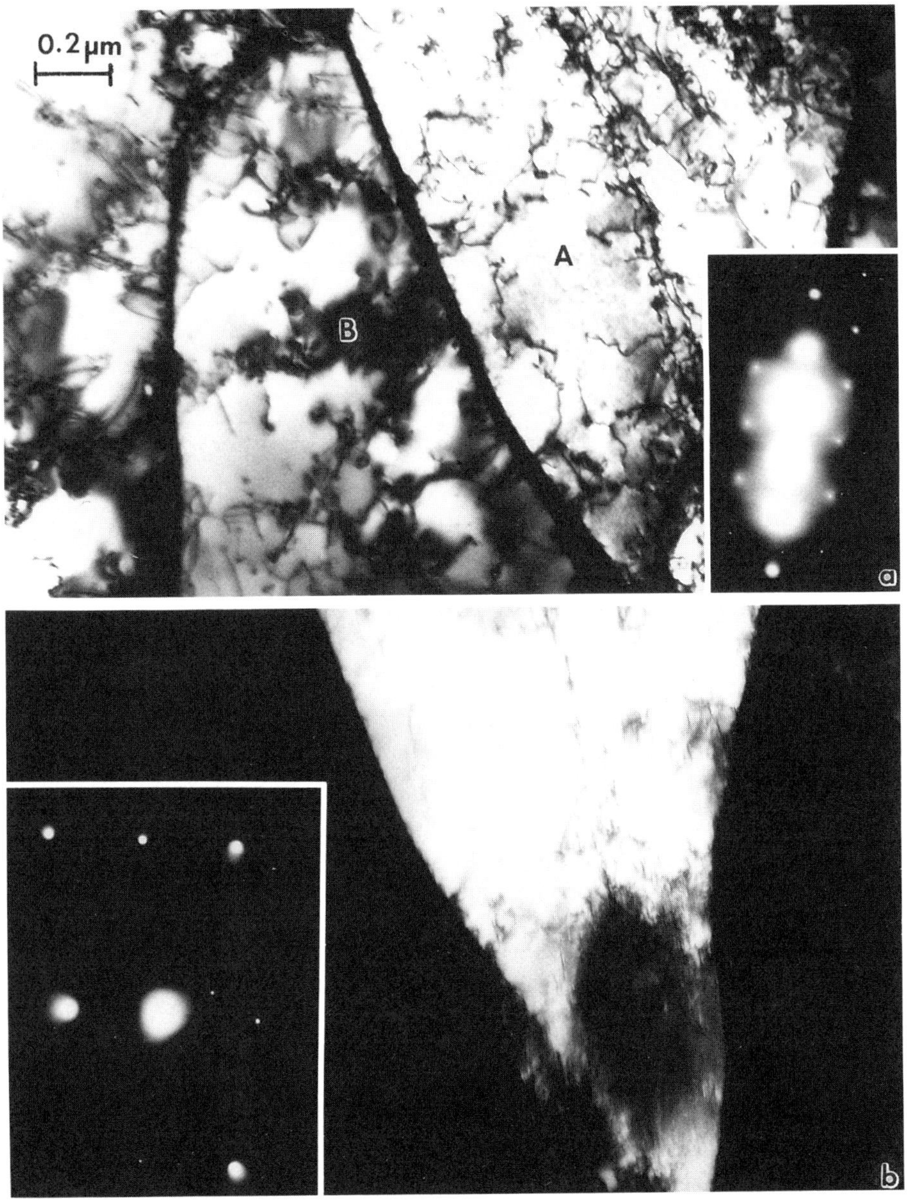

Figure 4. Bright-field (a) and dark-field (b) TEM image sequence for high deformation zone in tantalum EFP. The SAD pattern in (a) is specific to the grain under examination (A). The SAD pattern in (b) is specific to grain B.

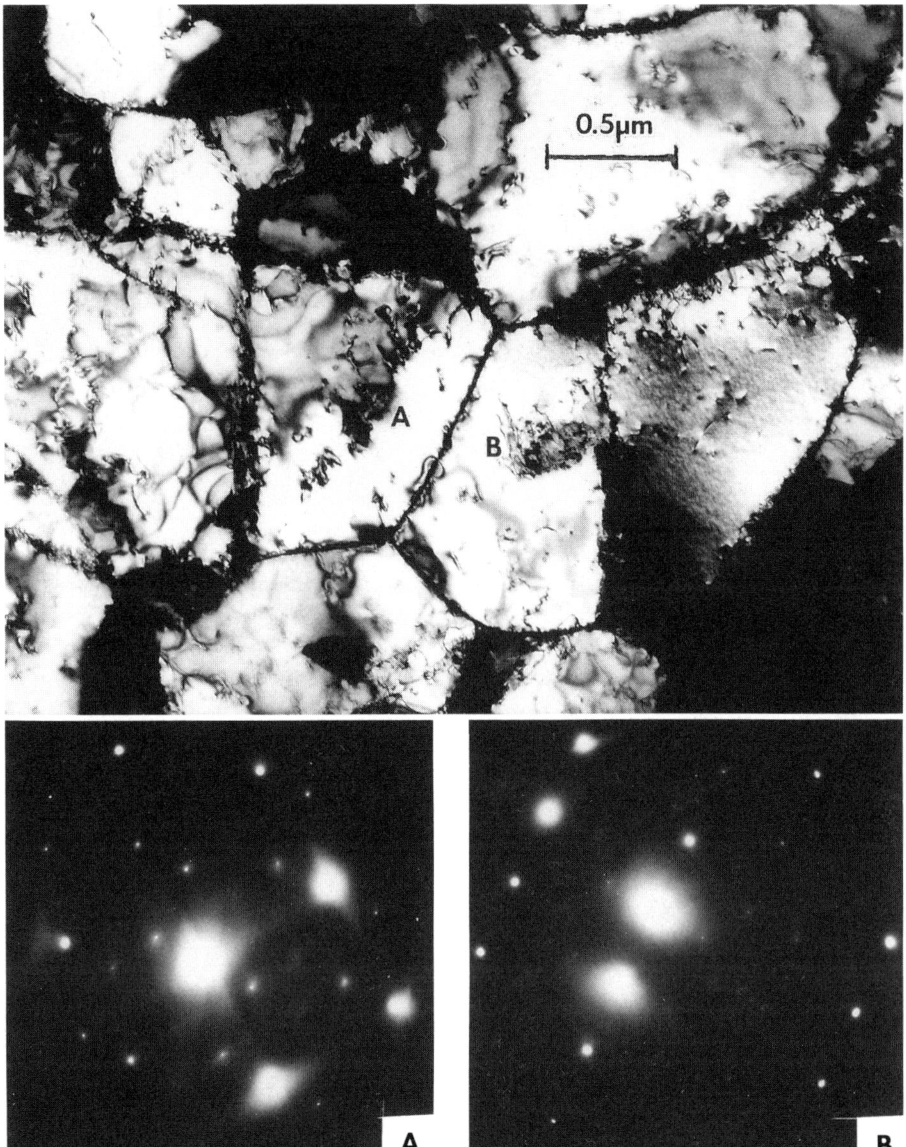

Figure 5. TEM bright-field analysis of two equiaxed grains in a heavily deformed zone in the tantalum EFP shown in Fig. 1(c). Corresponding SAD patterns are shown for A and B above. The grain surface orientations in A and B are (111) and (112) respectively.

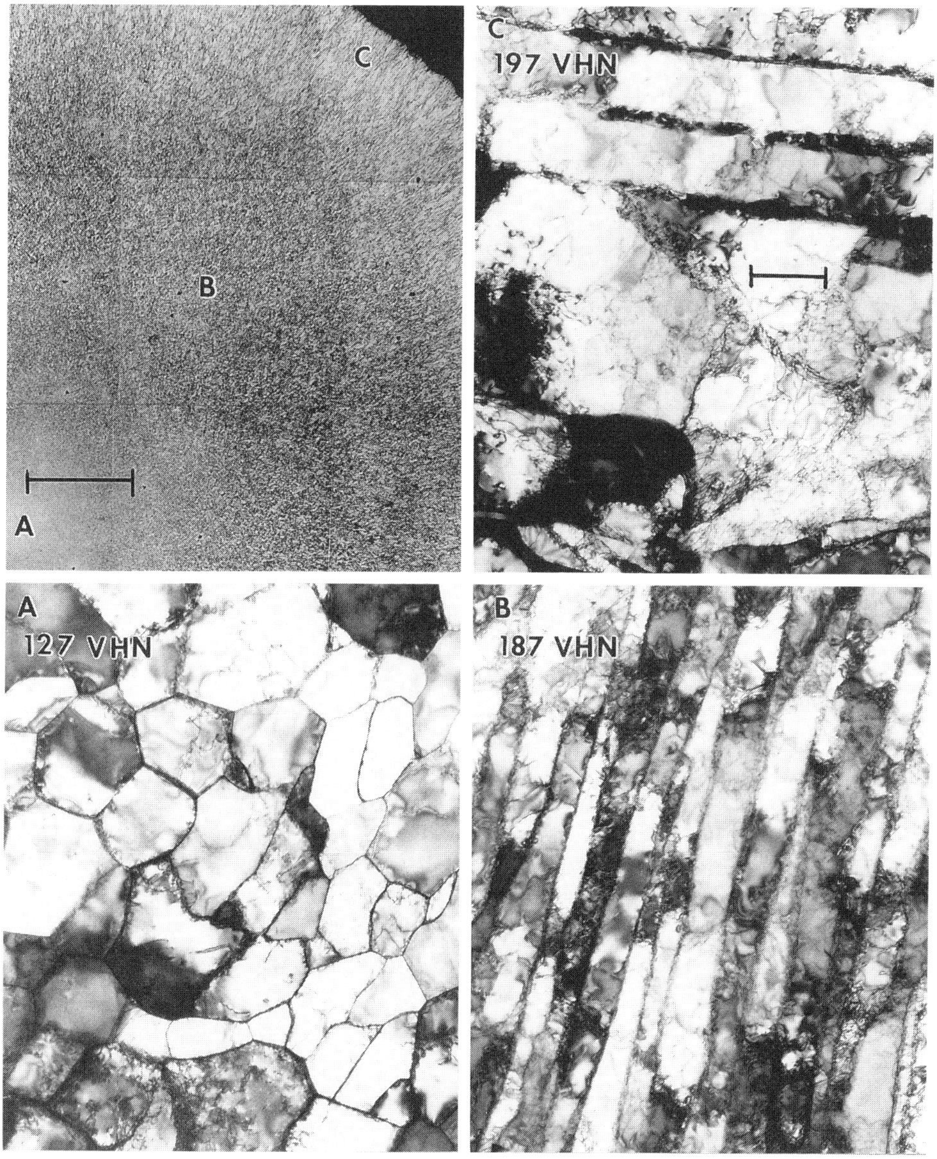

Figure 6. Comparison of microstructures in a recovered tantalum shaped charge slug cross-section. Slug cross-section showing prominent zones. TEM bright-field images showing details of regions A, B, C.

similar in some respects to the interplay or intermixing between microstructures shown in various EFP locations (Figs. 3-5). However, while there is some symmetry and regularity to the distribution of unique microstructural regions in the EFP, there are variations throughout the EFP geometry, characteristic of the variations of strain within the EFP.

The implications for irregular and strain-dependent microstructural regions in the EFP (consisting of dislocation cells, cell walls, dense walls or cell block microstructures, sub-boundaries, and recrystallized boundaries) are that the microstructure evolution, while tacitly strain dependent, will follow regular constitutive equations (9, 10), without some functional adjustments to account for the abrupt transition between dynamic recovery and recrystallization microstructures as recently described by Andrade, et al. (11). This is because the EFP microstructure evolution, unlike the shaped charge jet seems to follow a regular pattern for low-energy dislocation structure evolution with increasing plastic deformation (3, 4).

DISCUSSION AND CONCLUSIONS

The microstructures observed in recovered tantalum EFP's demonstrate a rather consistent variation of dislocation arrangements/geometries with variations in the degree of deformation (or strain). Dislocation tangles lead to cell structures which decrease in size, elongate, and then break up into very fine "blocks". As this microstructure refinement occurs, the relative misorientations of the associated boundaries increase, and for the regions of maximum deformation (maximum strain) which contain the very fine blocks, the orientations often differ significantly, rendering a virtual grain refinement. We have considered these microstructural variations and evolutions to be characterized generally by dynamic recovery (2), and in the most extreme cases, dynamic recrystallization, since similar features have been observed in the deformation of tantalum shaped charges (1). However, unlike the shaped charge, the EFP is characterized by much lower maximum strains, and these are very localized, although systematically distributed over the EFP volume. As a consequence, the EFP temperature is likely not to be significantly elevated, even in the regions of maximum strain. Surely temperatures which are requisite for dynamic recovery in a classical sense ($T>0.7T_M$ or about 2000°C in tantalum) are not achieved.

It is interesting to note that the dislocation structures and grain refinement features observed in Figs. 2-5 can be explained on the basis of the so-called low-energy dislocation structures (LEDS) hypothesis (3, 4, 12) which represents a consistent application of the second law of thermodynamics wherein dislocation structures strive to approach minimum energy subject to limitations of slip systems and dislocation mobility. In the LEDS principle, which incorporates the mesh length theory (3) dislocations assemble into structures in which the dislocation stresses are screened to the level of τ_0 at near neighbor distances. Without the intervention of climb (the absence of high temperature annealing effects), the density of the trapped dislocations is a simple function of the previously highest imposed tractions, and plastic deformation takes place virtually and instantaneously through glide. In terms of the mesh length theory (3), the concept of dynamic recovery is meaningless, and the energy per unit length of dislocation line rises with the density of interaction jogs. Consequently, individual grains break up into volume elements called cell blocks (CB's) in which the selection of simultaneously acting slip systems differs among neighboring CB's. As a result, straining causes different lattice rotations from CB to CB, and the resulting misorientations among neighboring CB's are accommodated by geometrically necessary boundaries (GNB's), or commonly called dense dislocation walls (DDW). These are longer, sharper, and can exhibit large angular misorientations. For example, in tribological processes in OFHC copper involving sliding wear and erosion, dislocation structures associated with the plastically deformed material have been identified as LEDS with CB boundaries having misorientations as high as 60° (13).

It is also important to recognize that the concept of misorientation is a simplistic view of lattice variance since a grain boundary is ideally characterized by 5 degrees of freedom (6). Consequently, misorientation will often have no meaning for boundaries separating blocks of different crystallographic orientation as illustrated, for example, in Figs. 4 and 5. It should also be recalled that in general, straight boundaries are low energy boundaries (6), and many of the boundaries observed in Figs. 4 and 5 are straight. These features are consistent with LEDS, and this concept makes more sense than dynamic recrystallization which in principal, in the case of the elongating shaped charge jet, represents a net volume energy increase if the recrystallized grains are high energy grain boundaries. While it is speculative to believe that the shaped charge slug may also represent

LEDS characteristics (Fig. 6), this may represent the significant difference observed between the slug and recovered jet fragment microstructures in tantalum (1), especially if the slug temperature is low (<0.5 T_M) in contrast to the elongating jet.

Finally, we should point out that virtually nothing is known about dynamic recrystallization at the high strain rates characteristic of both the shaped charge and the EFP. In addition, temperatures, which are crucial to supporting any arguments for significant dislocation climb required for any consideration of classical annealing phenomena, are really not known, even for the shaped charge. Consequently, the LEDS theory provides a much more tractable explanation for the associated plastic deformation, and the residual microstructures observed in recovered EFP's.

ACKNOWLEDGEMENTS

This research was supported in part by a Mr. and Mrs. MacIntosh Murchison Endowed Chair (L.E.M.) and the U.S. Army Research, Development and Engineering Center, Picatinny Arsenal, New Jersey. We are grateful for enlightening discussions with Dr. Doris Kuhlmann-Wilsdorf.

REFERENCES

1. L.E. Murr, C-S. Niou and S.H. Shih, *Materials Characterization*, 32 (1994).
2. L.E. Murr, C-S. Niou and C. Feng, *Scripta Metall. et Materialia*, 31 (3), 297 (1994).
3. D. Kuhlmann-Wilsdorf, *Mater. Sci. Engr.*, A113, 1 (1989).
4. D. Kuhlmann-Wilsdorf and N. Hansen, *Scripta Metal. et Materialia*, 25, 1557 (1991).
5. L.E. Murr, Chap. 37 in Shock Wave and High-Strain Rate Phenomena in Metals, (Ed.) M.A. Meyers and L.E. Murr, Plenum Press, New York, 1981, p. 537.
6. L.E. Murr, *Interfacial Phenomena in Metals*, Addison-Wesley, Reading, Mass, 1975 (reprinted by Tech Books; Herndon, VA, 1991).
7. R.W. Cahn, *Recovery and Recrystallization*, Physical Metallurgy, Part II, Ed. R.W. Cahn and P. Haasen, Elsevier, New York, 1983, p. 1595.

8. A.D. Rollett, M.J. Cutton, and D.J. Srolovitz, *Acta Metall*, 40(1), 43 (1990).
9. R.W. Armstrong and F.J. Zerilli, *J. Physique*, 49, 105 (1988).
10. G.R. Johnson and W.H. Cook, in Proc. 7th Int. Symposium on Ballistics, The Hague, Netherlands, 1983, p. 541.
11. U.R. Andrade, M.A. Meyers, and A.H. Chokshi, *Scripta Metall. et Materialia, 30(7)*, 933 (1994).
12. D. Kuhlmann-Wilsdorf, *Phys. Stat. Sol.* A, 104, 121 (1987).
13. A. Rigney, M.G.S. Naylor, and R. Divakar, *Mater. Sci. Engr.*, 81, 409 (1986).

ON FIELD METALLOGRAPHY APPLICATIONS

Iain Le May[1], Peggy Cheung[1] and Tito Luiz da Silveira[2]

ABSTRACT

The paper discusses the wide applicability of field metallography in the assessment of plant and equipment condition, and as a quality control or nondestructive evaluation (NDE) tool. Examples are given of various microstructural examinations. The need for proper and well defined preparation procedures is emphasized in order to prevent misleading results being produced. In this connection the need for training procedures for NDE personnel is discussed.

INTRODUCTION

Field metallography is a procedure that is becoming more widely used, although it is still not as accepted as a standard method of microstructural evaluation in many laboratories as it could be. The purpose of this paper is to review some of the applications for which it is particularly suited and to discuss some of the problems that can arise in its use.

Components that require metallographic examination without destruction can be examined through local grinding, polishing and etching of relevant areas. Polishing may be done either by means of mechanical or electrolytic procedures depending on the particular material to be examined, the availability of equipment and the accessibility of the area. The type of equipment used and the general procedures have been discussed previously [1,2]. The use of remote metallographic procedures for examination of otherwise inaccessible areas has been discussed by Bagnall [3].

Examination of the microstructure disclosed can be made in several ways. First, it may be made directly using a field microscope, which can also be fitted with a camera; second, plastic replicas may be made and examined subsequently in the laboratory; finally, use may

---

[1] Metallurgical Consulting Services Ltd., P.O. Box 5006, Saskatoon, Canada, S7K 4E3

[2] Tito Silveira Engenharia e Consultaria Ltda., Rua Couto Maghalães 744, 20930-090 Rio de Janeiro, RJ, Brazil

be made of a miniature video camera set up as a remotely operated microscope, and the image observed and recorded at a distance [3].

APPLICATIONS OF FIELD METALLOGRAPHY

Applications of field metallography are many. The first use made by one of the authors was in evaluating the changes in microstructure caused by local overheating in the rotor of a large generator in a power station [4]. This had been inadvertently connected to the power grid, causing the generator to operate as a motor and accelerate the steam turbine from its off-load condition to some two-thirds of its synchronous speed in around 30 seconds. The use of replicas was necessary rather than direct observation because the surface was curved and, with progressive removal of the material, eventually a shallow hollow was formed. Because of the extent of the local overheating, the rotor could not be used and was replaced after an extensive delay while awaiting the new component.

A similar use of in situ metallography and replicas was made in the examination of the bodies of valves from a natural gas line that had suffered rupture and a resulting fire. Evaluation showed that the heating had not been severe and the steel had not suffered any significant microstructural changes. This was confirmed from hardness measurements using a portable instrument. As a consequence of the evaluation the valve bodies could be reused in rebuilt valves.

Widespread use has been made of field metallography in the evaluation of superheater tubes and headers in steam generators. The work of Neubauer and Wedel [5] and Auerkari [6] is particularly noteworthy in this respect. Subsequently, representative microstructures have been published to illustrate the extent of creep damage and its classification [7]. The primary objective in the work of Neubauer, Auerkari and others working in this area has been to determine the degree of creep damage, the extent of the safe life remaining and the requirement for repairs.

Recently, da Silveira and Le May [8] and Samuels et al [9] have shown that damage, in the form of apparent voids, may be disclosed in creep damaged material if successive polish-etch steps are used but, if the polishing is done with great care, voids may not be disclosed. The explanation lies in the nature of the damage, which, in Cr-Mo steels at least, appears to originate as decohesion at carbide-matrix interfaces of carbides that precipitate on grain boundaries. The polish-etch cycles serve to open up and exaggerate these cavities so that they appear as voids. Hence, disclosure of creep damage requires well regulated preparation procedures to be used, and this applies whether or not the metallography is done in situ or using specimens prepared in the laboratory.

Figure 1 illustrates the situation of apparent voids and their absence as a function of the polish-etch procedure used, for Cr-Mo steel.

Field metallography can be a valuable nondestructive evaluation (NDE) procedure in a wide range of components. The examples already given are all NDE applications in one way or another. It is a regularly used technique in examining the state of high temperature nickel-base alloy sheet metal used in the combustion chambers of jet engines, small microstructural changes being readily distinguished. Other examples include materials evaluation in petrochemical vessels and reactors (Figures 2, 3 and 4); and in the evaluation of the integrity of a pipeline. In the latter case, the authors were faced with the evaluation of a pipeline whose

Figure 1: Effect of polishing and etching procedure on disclosure of creep damage. (a) Mechanically polished and etched in Nital — no voids or microcracks apparent. (b) Same area after repeated polish-etch cycles showing apparent voids and microcracks.

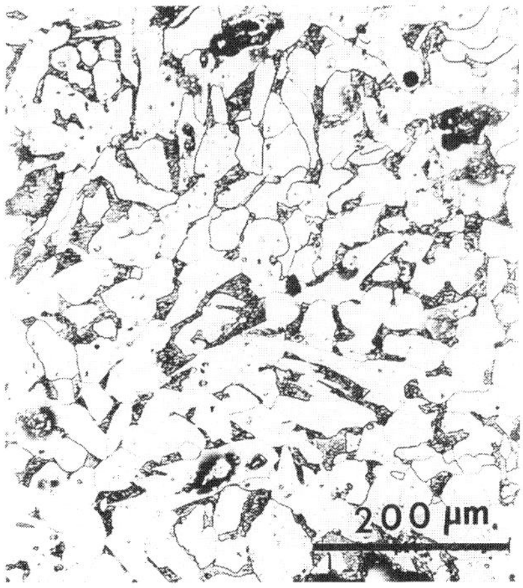

Figure 2: Replica from the internal surface of a catalytic reformer in a refinery after 320,000 h operation at a nominal temperature of 454°C. Material: carbon steel, A201 grade A. Ground to 600 grit, electrolytic polishing in 94% acetic, 6% perchloric acid, Nital etch. Metallized replica, disclosing the absence of appreciable damage from aging.

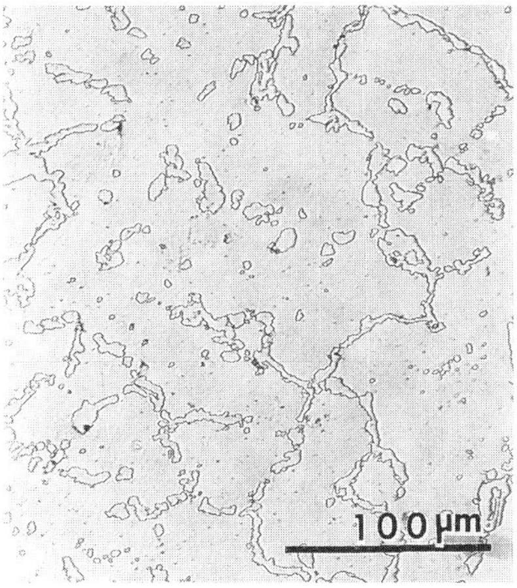

Figure 3: Replica from the external surface of a reformer furnace column used to produce ethanol, after 17,000 h of operation. Material: HP40 modified. Ground to 600 grit, polished with 1 μm diamond, etched with glyceregia. Metallized replica, showing extensive microstructural changes corresponding to a temperature of around 1100°C, indicative of improper operation.

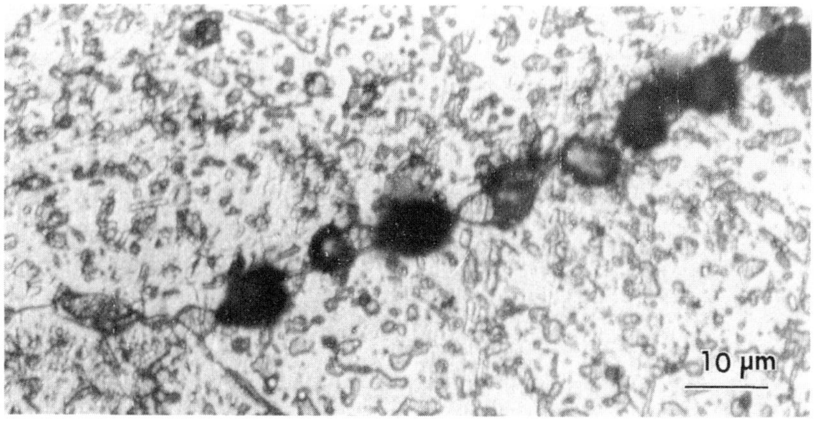

Figure 4: Replica from the welded joint of a collector from a superheater in a unit producing styrene. Operating time 108,000 h; operating temperature 812°C; material 304H stainless steel. The metallography was to determine the origin of cracking detected by liquid penetrant. Ground to 600 grit, polished with alumina 1 μm and 0.3 μm; glyceregia etch; metallized replica. It shows that the cracks are the result of the interlinking of damage centers produced by creep.

materials of construction were unknown, as were the welding procedures. In situ polishing using a Struers electropolishing unit and etching, allowed the microstructure to be disclosed in representative areas (Figure 5). This was one of the techniques that allowed for continued operation of the line [10].

Field metallography can be very useful in evaluating failure mechanisms. Creep damage detection has already been discussed, but the nature of other damage such as stress corrosion cracking (SCC) can be disclosed by replication around areas of cracking or possible cracking. Figure 6 shows such an example.

TRAINING OF PERSONNEL

The writers have observed a number of examples of field metallography where no useful information could be obtained because of the poor techniques used in preparation. Indeed, completely erroneous conclusions can be drawn from poorly prepared microstructures or replicas. Examples seen have been surface scratches, apparent voids that are nothing more than points where there was an absence of contact between the surface and the replicating material, and dirt that has been trapped during replication.

The very real problem involved in disclosing creep damage indicates that it is not simply a matter of doing the polishing and etching very carefully, but of using it in the most appropriate way. To this end it seems odd that we look for certification of NDE personnel in

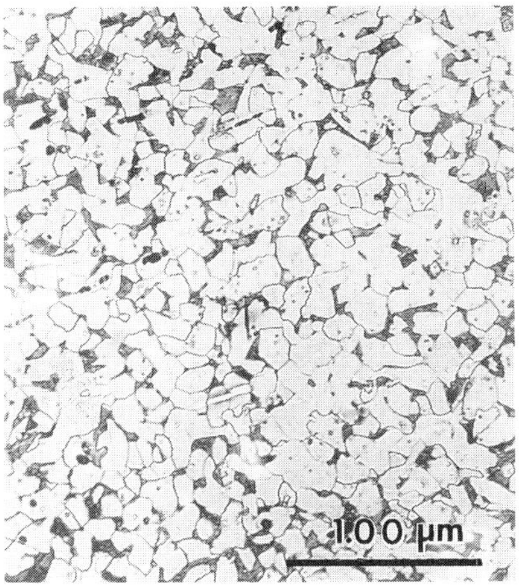

Figure 5: Microstructure of pipeline steel. Replica from electropolished Nital etched surface. Unmetallized.

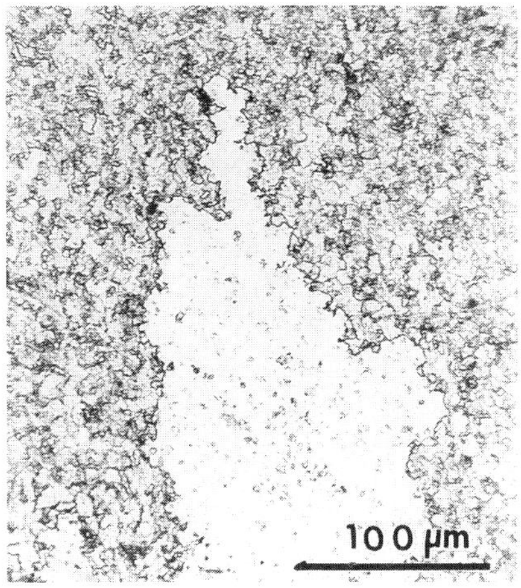

Figure 6: SCC in a deaerator vessel. Ground to 600 grit, electropolished and Nital etched. Metallized replica. This shows the end of a principal propagating crack.

most commonly used techniques, but there is no such certification (or training) provided for metallography. Considering the importance of the information that field metallography can provide and the critical nature of decisions that may be based on it, it seems that strong efforts should be made to remedy this.

Before contracting services for field metallography, particularly one involving replication, it is recommended that, in the absence of a certification program, the customer should require that appropriate samples be prepared and the capabilities of the metallographers involved be demonstrated to be satisfactory.

REFERENCES

1) E.V. Sullivan, "Field Metallography Equipment and Techniques," in Microstructural Science, Volume 15, M.E. Blum, P.M. French, R.M. Middleton and G.F. Vander Voort, Eds., p.3 IMS/ASM, Metals Park, Ohio, 1987.

2) J.F. Henry, "Field Metallography: The Applied Techniques of In-Place Analysis," in Microstructural Science, Volume 12, D.O. Northwood, W.E. White and G.F. Vander Voort, Eds., p. 537, IMS/ASM, Metals Park, Ohio, 1985.

3) C. Bagnall, "Development of Tooling and Techniques for Remote Field Metallography," in Proc. 5th Brazilian Symposium on Piping and Pressure Vessels, p. 533, ABCM Salvador, BA, 1988.

4) M.W. Lui, I. Le May and R.I. Coote, "Nondestructive Metallurgical Examination of a Damaged 100 MW Generator Rotor," Praktische Metallographie, Vol. XII, p. 319, 1975.

5) N. Neubauer and U. Wedel, "Restlife Estimation of Creeping Components by Means of Replicas," in ASME International Conference on Advances in Life Prediction Methods, D.A. Woodford and J.R. Whitehead, Eds., p. 307, ASME, New York, 1983.

6) P. Auerkari, "Remanent Creep Life Estimation of Old Power Plant Steam Piping Systems" in ASME International Conference on Advances in Life Prediction Methods, D.A. Woodford and J.R. Whitehead, Eds., p. 353, ASME, New York, 1983.

7) P. Auerkari, K. Borggren and J. Salonen, Reference Micrographs for Evaluation of Creep Damage in Replica Inspections, NT Techn. Report 170, Nordtest, Espoo, Finland, 1992.

8) T.L. da Silveira and I. Le May, "Effects of Metallographic Preparation Procedures on Creep Damage Assessment", Materials Characterization, Vol. 28, p. 75, 1992.

9) L.E. Samuels, R.W. Coade and S.D. Mann, "Precracking Structures in a Creep-Ruptured Low-Carbon Cr-Mo Steel: Their Nature and Detection by Light Microscopy and Scanning Electron Microscopy", Materials Characterization, Vol. 29, p. 343, 1992.

10) I. Le May and T.L. da Silveira, "Fitness-For-Purpose Evaluation of an Older Pipeline", in Proceedings of the First International Symposium on Process Industry Piping, p. 27-1, NACE, Houston, Texas, 1993.

DIGITAL SEM and EDS IMAGING IN CHARACTERIZING DEBONDED PORCELAIN FUSED TO METAL INTERFACES

H.J. Mueller[1]

ABSTRACT

Debonded porcelain fused to alloy samples were characterized by digital SEM and EDS imaging processing in order to detect and to quantify the area fraction of retained porcelain. A Pd-Cu-Ga-Au alloy showed selective Cu and Ga oxidation in producing surface oxide films for fusion and bonding with porcelain. The porcelain was detected by the presence of Si, K, Al, Na, Sn, and Ti. Debonded surfaces showed the presence of all elements from alloy and porcelain. X-ray maps revealed the associations of Si, K, Sn, and Al in one region, Pd, Cu, and Au into another region, and at higher magnifications, Na and Ga into a third region. Titanium, present in *low concentrations, was randomly located over the fractured* surface. Area fraction of retained porcelain was at least 50 %. This value was higher at lower magnifications, since at higher magnifications specific areas were selected that purposely contained lower amounts of porcelain.

INTRODUCTION

Fusion of a ceramic layer onto a metallic substrate occurs with a number of biomaterials. One application with a dental material involving

---

[1] Dental Materials Science, Research, Division of Science,
American Dental Association, 211 East Chicago Ave, Chicago, IL 60611.

porcelain fused to metal, is used for the restoration of teeth as crowns, veneers, and partial dentures. The advantages include aesthetics and corrosion resistance from the outer porcelain layer, and strength and rigidity from the underlying metal substructure. A major concern [1] is the potential for the restoration to fail during service by fracturing of the porcelain layer in a direction either perpendicular or parallel to the porcelain/metal interface. Laboratory evaluations made of the bond strength between porcelain fused to metal and of the fractography from debonded porcelain fused to metal interfaces are means to provide assurances that processing techniques and porcelain and alloy formulations meet requirements for long term service performance. Alloy surfaces that retain a large percentage of porcelain following porcelain debonding indicates fracture predominantly through bulk porcelain and strong alloy/porcelain bonding.

In order for porcelain to bond to metal, two requirements must be met; first, the porcelain must wet the metal and second, stresses resulting from coefficient of thermal expansion differences must not exceed the tensile stress of the porcelain. These requirements are satisfied when an adherent film of oxide covering the metal surface partially dissolves into the porcelain during fusion of the porcelain, and when the coefficients of thermal expansion of both metal and porcelain are the same over the entire temperature range from room temperature to the porcelain's glass transition temperature. [2]

Dental alloys utilized with porcelain bonding include the Ni-Cr, Co-Cr, Ag-Pd, high Pd-content, and Au-Pt-Pd-Ag systems. The base-metal alloys develop oxide films of NiO, Cr_2O_3, and others which promote bonding as described above. Minor additions of Al, Ti, Fe, Ga, and others also promote bonding by a dissolution of their metallic oxides with porcelain during fusion. Porcelain bonding with the noble metal alloys includes only a chemical bonding contribution from the formed oxides on the minor non noble metal additives, since Au, Pt, and other noble metals do not oxidize significantly. Increased concentrations of the minor alloying additions have been detected near the porcelain/metal interface. [3]

Cold worked and smeared surface layers and too large of a surface roughness prevent proper wetting of surface by porcelain during fusion, thereby generating porosity and decreased bond strength. For this reason, the ground and polished alloy is usually air blasted with alumina particles prior to air oxidation during a controlled heat treatment by heating at around 50 °C/min up to a temperature of around 1000 °C. Porcelain is applied by painting the oxidized alloy with a water slurry of the porcelain powder and fused to alloy with a similar heating schedule.

The purpose of this project was to characterize microstructurally alloy/porcelain interfaces following fracturing and debonding of its fused porcelain layer. The objectives were i) to be able to discriminate between regions containing and not containing retained porcelain, and ii) to obtain an estimate or to quantitatively determine the percent area of retained porcelain.

EXPERIMENTAL PROCEDURES

A high Pd-content ceramic alloy (Rx Naturelle, Jeneric/Pentron Inc., Wallingford, CT) along with porcelain (Synspar, Jeneric/Pentron) compatible with the alloy were used. The composition of the alloy (wt %) as given by the manufacturer is 79 Pd, 10 Cu, 9 Ga, and 2 Au.

Specimens of the cast alloy, 20 mm long x 5 mm wide, and x 0.4 mm thick, containing porcelain fused to one of its face surfaces were supplied by the manufacturer. The fused porcelain layer several millimeters thick consisted of an inner layer adjacent to the alloy of opaque porcelain and an outer layer of body porcelain. The opaque porcelain layer masks the metallic color of the alloy, while the body porcelain layer provides the life-like colorants.

The samples were bent lengthwise by hand around a 10 mm diameter stainless steel bar. After conforming to the 10 mm diameter arc, the samples were bent back to an approximate straight length position. Except for the porcelain at the extreme ends of the bar, the remaining porcelain within the middle span length (about 10-15 mm) fractured off of the metallic substrate. Visually with the unaided eye, doubt remained as to whether the debonded alloy surfaces contained any porcelain still attached to the alloy.

Light and scanning electron microscopies (SEM) were used to characterize the debonded surfaces. Because of the limited depth of field with light microscopy, SEM provided much more useful information to meet our objectives. A Leica-Cambridge MK II 250 SEM together with a Princeton Gamma Tech (PGT) energy dispersive spectrometer (EDS) were interfaced with a PGT microanalyzer capable for integrating both digital SEM imaging and digital EDS x-ray imaging (IMIX, version 7 with a Sun workstation). The SEM was operated at 20 kV. All EDS spectra were collected to correspond to a total number of counts of 200,000 $\pm$ 15,000. SEM and x-ray spectra were first obtained from the polished alloy surface, the oxidized alloy surface, the opaque porcelain, the body porcelain, and the porcelain/alloy interface after porcelain debonding as described above. Images and spectra were stored on hard disk, which permitted easy access and manipulations for computer comparisons. A useful capability was the ability to generate digital x-ray maps of up to 12 elements simultaneously during one characterization. The maps were generated with a pixel resolution of 128 and with total x-rays counts between 200-300 per pixel. This required an average time of about 25 minutes with the system and operating conditions used. The paint, x-ray math, and feature calculation programs provided additional means to compare, graph, and obtain quantitative information from the data.

RESULTS AND DISCUSSION

Figure 1 presents a secondary electron micrograph of a polished cross section from the porcelain fused to alloy interface, showing all three

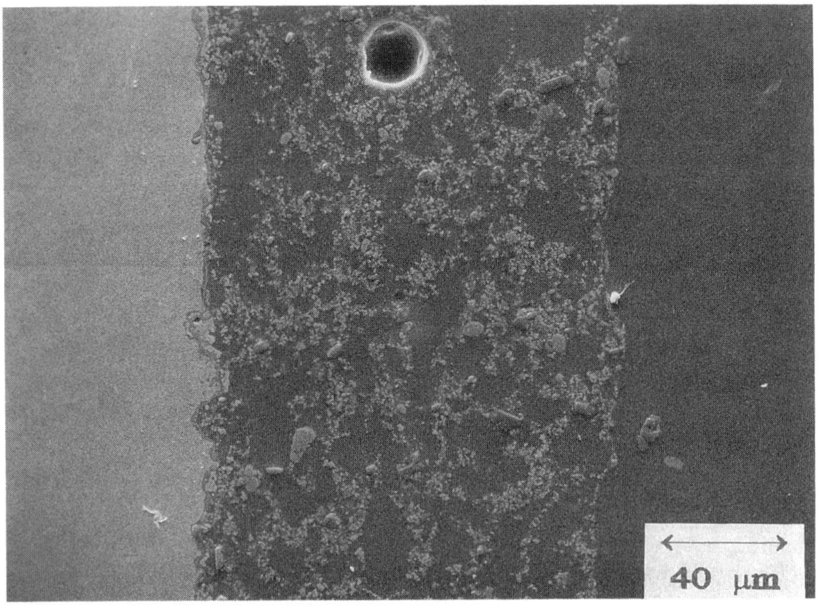

Figure 1. Micrograph of section through porcelain fused to alloy.

materials involved, the alloy (left), the opaque porcelain (middle layer), and the body porcelain (right). Utilizing the image display feature from IMIX, the mean length measurement for the thickness of the opaque porcelain layer over the entire length of the sample was obtained to be 0.12 mm. The back surface of the alloy sample was not coated with porcelain, therefore, this surface represents only the oxidized surface with no direc fusion to porcelain. Figure 2 is a micrograph of a polished cross section

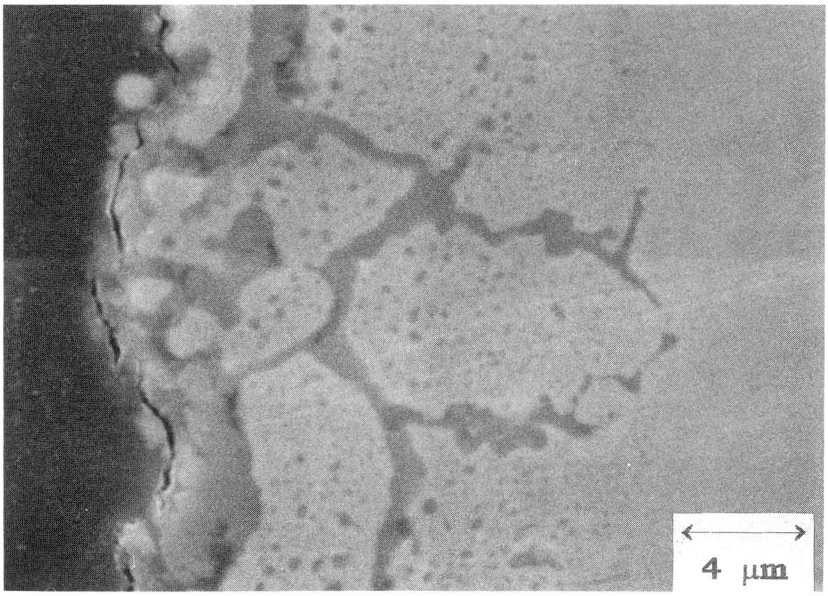

Figure 2. Micrograph of section through alloy and oxide film.

through oxide layer on bottom surface of sample. Oxidation on this alloy surface is shown to occur via crack formation mechanism with oxidation products being formed along the crack path length. Use of the image display feature for length measurement revealed that on average over the entire length of sample the mean thickness of alloy being affected by oxidation reaction was 15 μm. Figure 3 presents a higher magnification image of the alloy/opaque porcelain interface shown in Figure 1. Revealed here is the intimate contact between alloy (left), alloy oxide (middle layer), and opaque porcelain (right dark matrix material with lighter particles). The thickness of the alloy oxide affected layer is less here (≅ 8 μm) than on the back alloy surface. A polished cross sectional image of the alloy/porcelain interface following debonding of the porcelain is presented in Figure 4. It is to be noted that fracturing has occurred either at the oxide/porcelain interface or within the bulk porcelain. For the most part, even on cross sectional images, porcelain is visually detected covering the outer oxide surface. Needing consideration, is the possibility that the outer surface layer is actually an interaction zone between alloy oxide and porcelain.

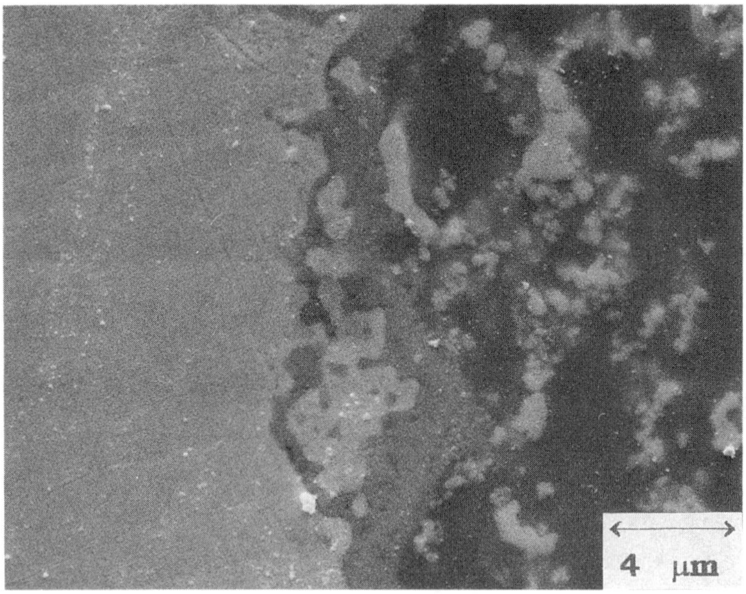

Figure 3. Micrograph of section through porcelain fused to alloy.

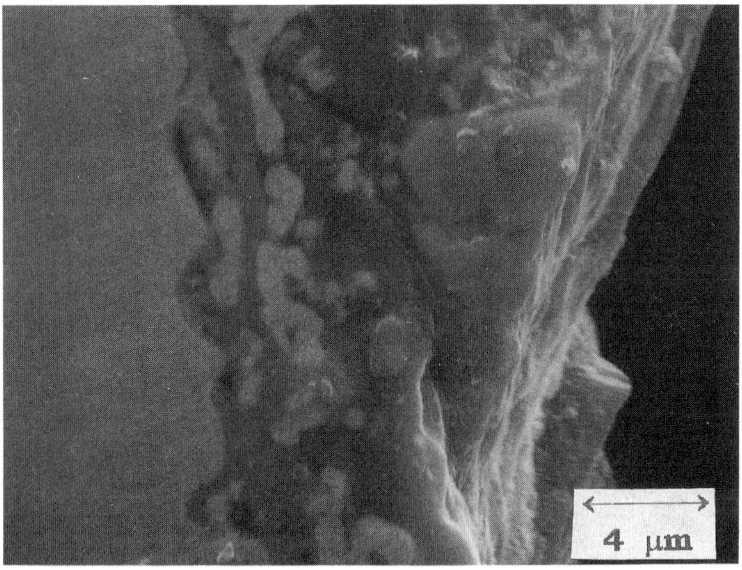

Figure 4. Micrograph of section through porcelain debonded alloy.

Figure 5 presents EDS spectra for various components from the alloy/porcelain system. All elements in the alloy as stated by manufacture

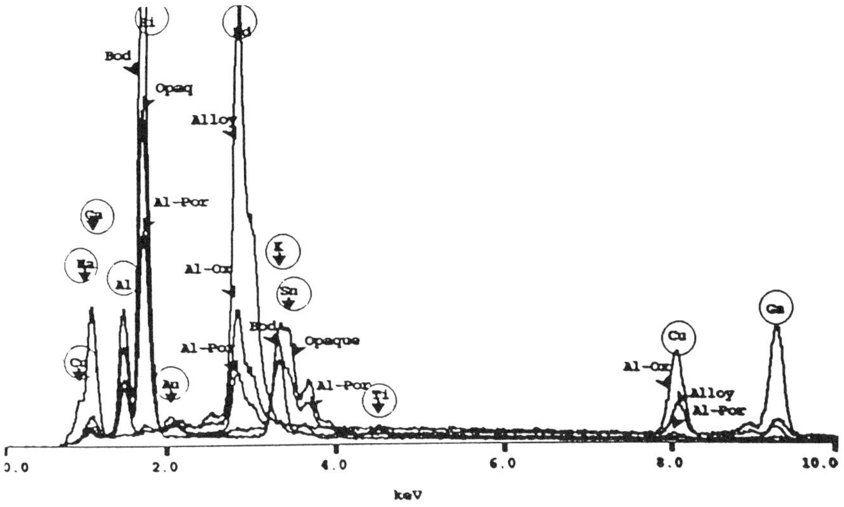

Figure 5. EDS spectra from alloy/porcelain system.

are detected. These are represented by a strong Pd-L line, by moderate Cu- and Ga-K lines, and by a weak Au-M line. The alloy in the oxidized condition (Al-Ox in figure 5) shows very much lower Pd intensity, but moderately higher Cu and Ga intensities. This infers that both Cu and Ga must preferentially be oxidized from the bulk alloy so that higher surface concentrations exist. Both the alloy and the alloy oxidized show negligible intensities for K, Sn, Ti, and Si, as to be expected.. The alloy oxidized, but not the alloy alone, shows a moderate Al-K line This may be the result from a diffusion process around the sides of the specimen, as put forth for the case of Na, but also from the effects of a air blasting process with alumina particles during initial surface preparation. The alloy oxidized reveals a significant Na-K level. Even though the bottom surface of the alloy specimen did not receive direct porcelain application, the real likelihood exists that Na diffusion occurred easily around the sides of the alloy bar to interact with the bottom surface. As will be shown with the results from the x-ray maps, Na preferentially occupies the same locations as does Ga, hence Na likely reacted with Ga. The body porcelain (Bod in figure 5) shows a very intense Si-K line, moderate K-K and Al-K lines, a weak Na Line, and as expected no Sn, Ti, Cu, Ga, and Au. The opaque porcelain (Opaq in figure 5) shows a reduced Si intensity, about the same Al, Na, and K levels, but with moderate Sn-K and weak Ti-K lines. Tin and Ti oxides are added to the opaque porcelain to hinder the passage of light. Figures 1, 2-4 show these additives as the small particle inclusions within the opaque porcelain layer. Finally, the surface from debonded porcelain fused to alloy (Al-Por in figure 5) reveals elements present in both alloy and porcelain, Hence, Pd, Cu, Ga, and Au from the alloy are detected, as well as Si, Al, K, Sn, and Na from the porcelain.

Digital x-ray maps were obtained from surfaces after the porcelain was debonded from the alloy substrate. Figures 6 (320x) and 7 (3200x) present

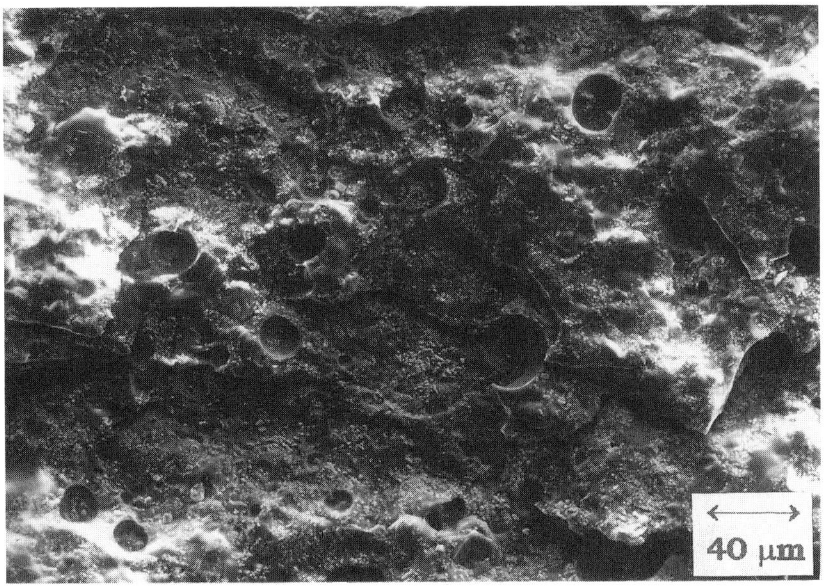

Figure 6. Micrograph of alloy surface following porcelain debonding.

two regions that were mapped for the various x-ray spectra. Results obtained from the surface region shown in figure 7 are presented in Figure 8. Twelve individual maps are shown. Inspection indicates that the Si and K distributions are much the same, while the Pd, Cu, and Au distributions are much the same and opposite to that for Si and K. Tin occupies regions common to Si and K. Titanium appears not to be localized, while Al is localized to specific small areas. Finally, both Ga and Na have similarities in location. This latter combination is interesting, since Ga is contained in the alloy and Na contained in the porcelain. Two

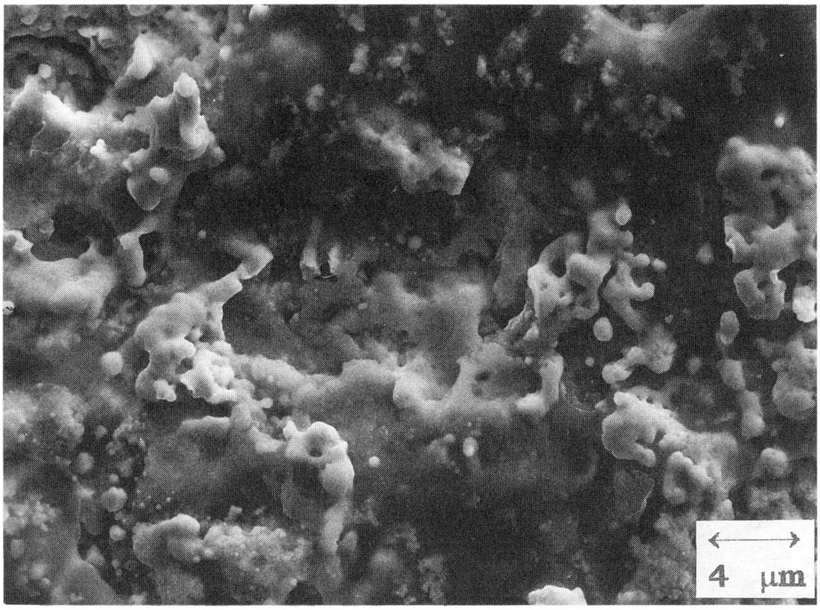

Figure 7. Micrograph of alloy surface following porcelain debonding.

additional maps contained in Figure 8, include total collected x-rays and the background, represented here as the Re-L line. Utilizing the image display feature with IMIX, intensity distributions were obtained for each of the elemental maps and related to an area fraction with the selection of a threshold intensity value. Utilizing this feature as well as the image math program, area fractions for elements contained within similar regions were region 1 (Si-K-Sn-Al) = 47.7%, region 2 (Pd-Cu-Au) = 38.2%, and region 3 (Na-Ga) = 15.2%.. Area fraction results obtained from the surface region shown in Figure 6 are region 1 (Si-K-Sn-Al) = 78.2%, region 2 (Pd-Cu-Au) = 11.1%, and region 3 (Na-Ga) = 8.0 %.. It was noted that as the magnification decreased, the area percent containing the porcelain elements (Si-K-Sn-Al) increased and the area percent containing (Pd-Cu-Au) decreased. This is further shown with the spectra obtained at three magnifications presented in Figure 9.. This is likely due to the fact that in selecting area regions at higher magnifications, regions were purposely selected that contained the least amount of porcelain. Also, as the magnification decreased, distinguishing the Na-Ga region as being a unique region was not possible.

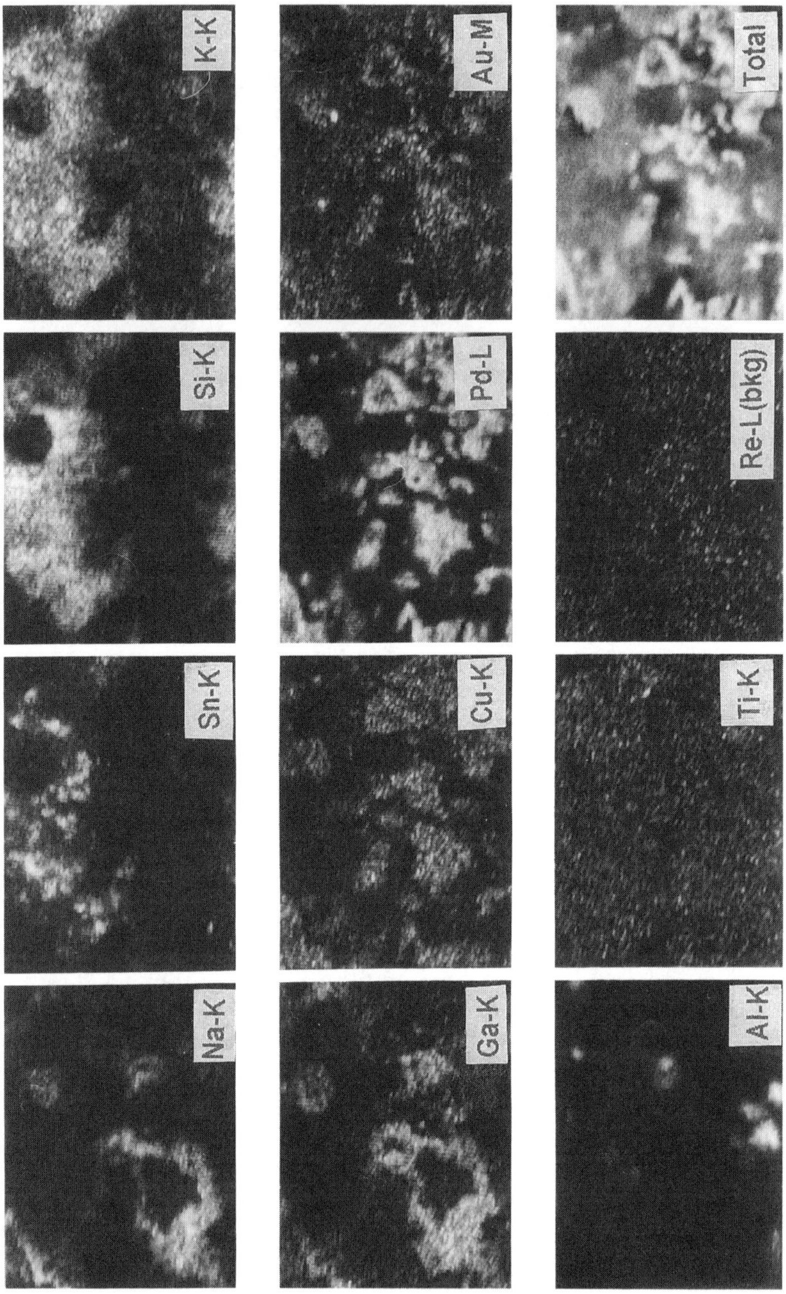

Figure 8. Elemental x-ray maps corresponding to figure 7.

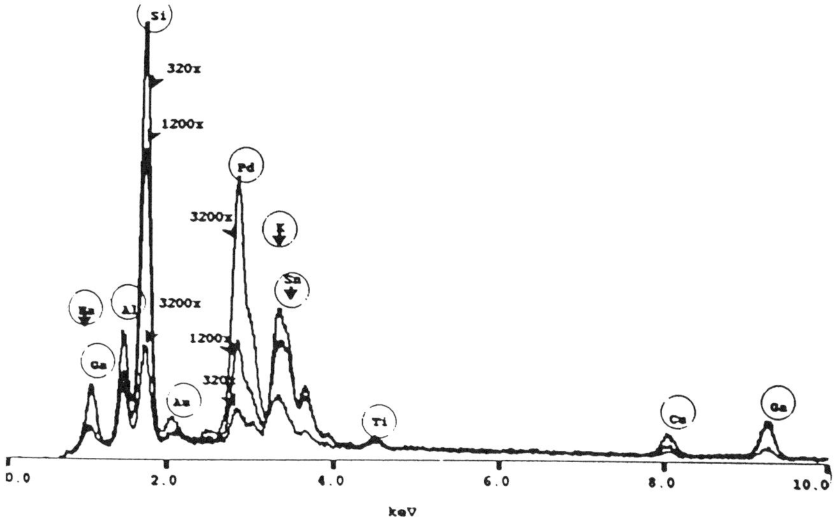

Figure 9. EDS spectra of alloy surface following porcelain debonding.

CONCLUSIONS

1. Oxidation of the Pd-Cu-Ga-Au alloy occurred by preferential Cu-Ga oxidation by producing increased concentrations of these elements within the surface oxide layer.

2. Fracture occurred either at the porcelain/oxide interface or within the bulk porcelain itself.

3. Gallium and sodium were located within the same surface regions. This implied that Ga and Na reacted during porcelain fusion to alloy.

4. The area fraction of retained porcelain was greater than 50 %.. Even at higher magnification on surface sites where apparent porcelain deficient regions were purposely selected, area fractions of actual retained porcelain were still high.

ACKNOWLEDGMENT

Support was provided by National Institutes of Health contract, Number DE 12584.

REFERENCES

1) I. Murakami and A. Schulman, Aspects of Metal-Ceramic Bonding, Dent. Clin. N. Am., 31(3), 333, 1987.
2) D.W. Jones, Coatings of Ceramics on Metals, Ann. N. Y. Acad. Sci., 523, 19, 1988
3) J.A. Pask and A.P. Tomsia, Oxidation and Ceramic Coatings on 80Ni20Cr Alloys, J. Dent. Res., 67, 1164, 1988.

Chapter 2

Metallographic Techniques

EXPLOITING THE OPTICAL MICROSCOPE

AN INSIGHT INTO DIFFERENTIAL INTERFERENCE CONTRAST (DIC)

Brian Bousfield

ABSTRACT

The series of papers, 'Exploitation of the Optical Microscope' have been undertaken to demonstrate the various optical techniques in an application mode that is specific to the materials field. This paper will be one such example specific to Differential Interference Contrast (DIC), sometimes called the Normaski technique named after the inventor. A completely new concept of combining techniques such as DIC and Dark Ground will also be introduced (the author is not aware of any previous publications of such material). On this occasion the optical theory of DIC will be used to develop the practical implications of what we can see and how best to exploit the technique.

For example, the need to:-

- Use objectives who's limit of resolution is less than the smallest specimen element.
- Identify by angles not steps.
- Define the maximum height of interphase relief most desirable for each microscope objective.
- Understand the vital importance in surface preparation by first achieving interphase planarity prior to achieving the optimum interphase relief.
- Avoid the use of etchants.
- Introduce the concept of 1st order grey DIC.
- Investigate optical staining.

OPTICAL PRINCIPLE

DIC can be described as phase contrast in colour. It converts grey levels caused by optical path differences into chromatic differences. In reflected light it converts **slopes, not** differences in height as is often thought, into

BUEHLER EUROPE LTD COVENTRY ENGLAND

colour differences. In transmitted light Normarski, it is the specimen refractive index difference that causes the colours. The term 'interference microscopy' usually relates to systems in which the incident ray is divided into two separate beams which interfere when recombined. This technique is most applicable to isotropic objects (metals) that are invisible by other optical techniques. Anistropic materials (minerals) where the specimen itself creates two beams do not require this technique of interference microscopy.

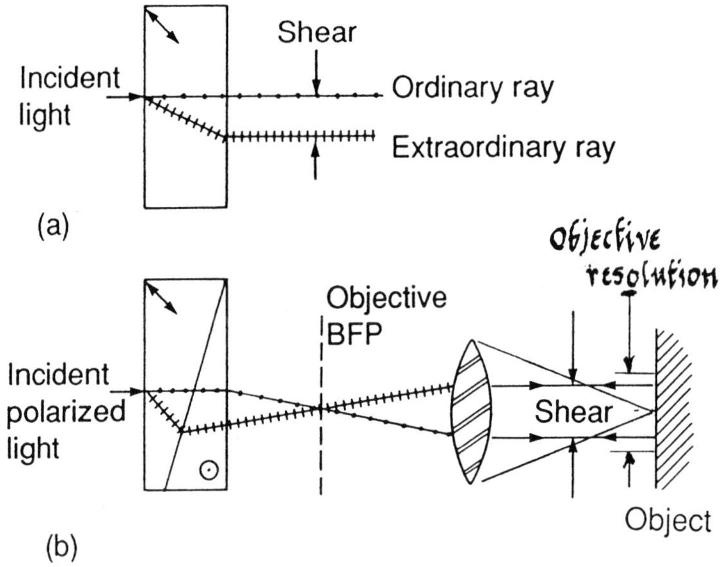

Figure 1 Creating the OPD. (a) Single beam-splitter and (b) modified beam-splitter (↘ and ⊙ indicate the direction of the optic axis)

DIC makes use of a birefringent beam-splitter (possessing two different refractive indices) where the incident beam is split into two; the ordinary ray and the extraordinary ray. From figure 1b notice the shear of separation between these two rays is less than the limit of resolution of the objective being used. Providing specimen detail remains greater than the objectives limit of resolution, double vision will be avoided. In order to restrict vibration directions to one axis and to introduce chromatic related optical path differences (OPD), the incident ray is polarized and the reflected ray passes through an analyser.

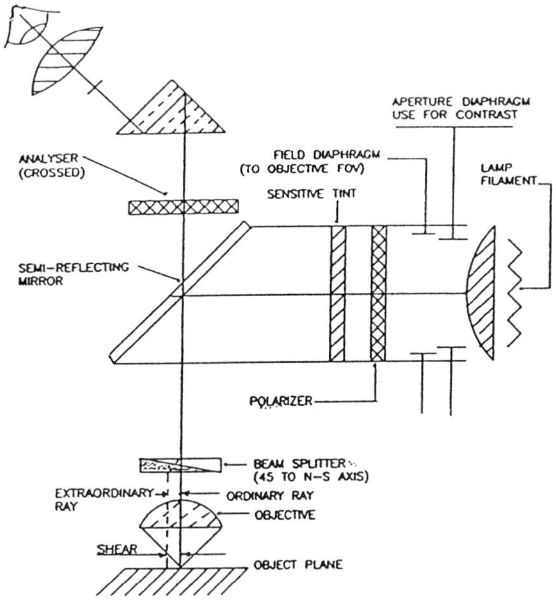

FIGURE 2 DIFFERENTIAL INTERFERENCE CONTRAST REFLECTED LIGHT

In principle the polarized vibration direction is East/West the analyser North/South i.e. polarizer and analyser are crossed, extinguishing the direct ray. (see figure 2). The beam splitter is positioned at 45º to the polarizer/analyser vibration direction. It is for this reason that non uniform specimens may need to be rotated in order to find the best position.

Reference figure 1a: light when passing through the modified beam splitter is split into two rays, both rays vibrating perpendicular with a polarizing direction at 90º. The modified beam splitter consists of two wedge - shaped uniaxial (one axis) prisms, whose optical axes are at right angles to each other. The incident polarized beam once split is caused to intersect at the objective back focal plane, prior to being imaged by the objective onto the specimen. This is the reason why each objective must have it's own carefully positioned prism in order to accommodate differing objective back focal plane positions. With a perfectly flat object as depicted in figure 1b the background interference colour will be a result of the OPD caused by the prism and not the specimen. In this optical configuration if we were to traverse the prism inwards or outwards we would change the OPD and therefore change the specimen colour. If in fact the specimen background colour should be multicoloured and not a single uniform colour this will signify an incorrect microscope objective/prism combination i.e., the prism will not be positioned in the objective back focal plane.

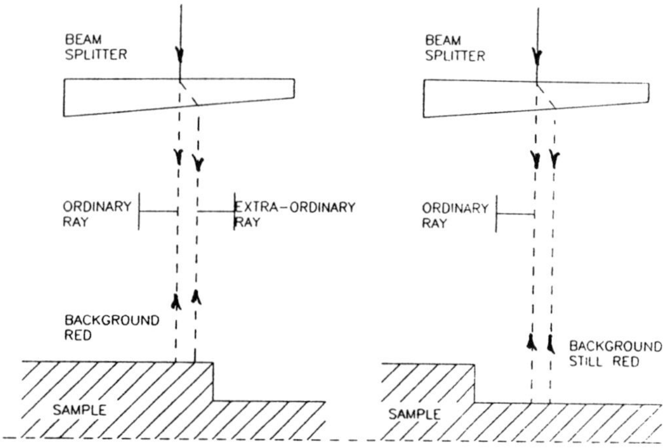

FIGURE 3 STEPS NOT DETECTED

Figure 3(A) shows the ordinary and extraordinary ray at a given OPD and shear, dictated by the prism manifesting say a RED background. At (B) consideration is given to the lower step which under identical optical conditions will also give a background colour of RED. This reinforces the introductory statement that DIC does not detect differences in height as in a step. What DIC does detect in our specimens is slopes which is quite fortunate since **all** grained specimens prepared by mechanical means will display a mass of different curves and angles. At this stage it is necessary for me to defend the situation from the metallographers' perception of a perfectly planar and flat specimen. In interference microscopy we look for angles associated with interphase height differences of $< 1\mu m$.

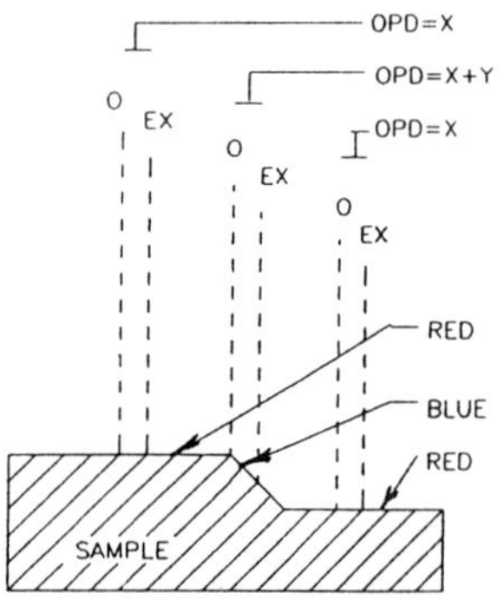

FIGURE 4 INDENTIFICATION BY SLOPE

Figure 4 illustrates how the introduction of a slope to our stepped specimen has created an increase in the OPD and therefore changed the colour for this area.

The length of this slope is important if the microscope is to be used for identification and not meaningless aesthetics. A short slope will add clarity and ease of identification, a long slope will confuse the interpretation. Consider a grained subject where the grain boundary can be highlighted by DIC, if the relief and hence slope be great, then the grain boundary will be confused within large coloured area. The degree of slope that can be tolerated is related to optical shear and resolution of each individual objective. This point is developed in figure 5.

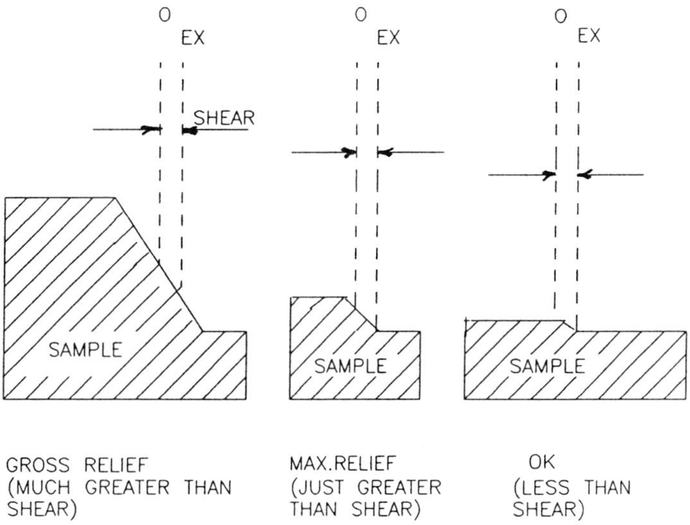

FIGURE 5 ACCEPTABLE SLOPE SIZE (RELIEF)

From figure 5 we can see that the most desirable maximum slope would be just greater than the shear of the birefringent Normarski prism. We also if possible need to define the minimum slope detectable in order to identify the sensitivity of such systems. The writer has on many occasions been placed on record as having said the system is sensitive in detection to within 0.1μm difference in interphase relief. Since the limit of resolution of the optical microscope is approximately 0.25μm this statement needs clarification.

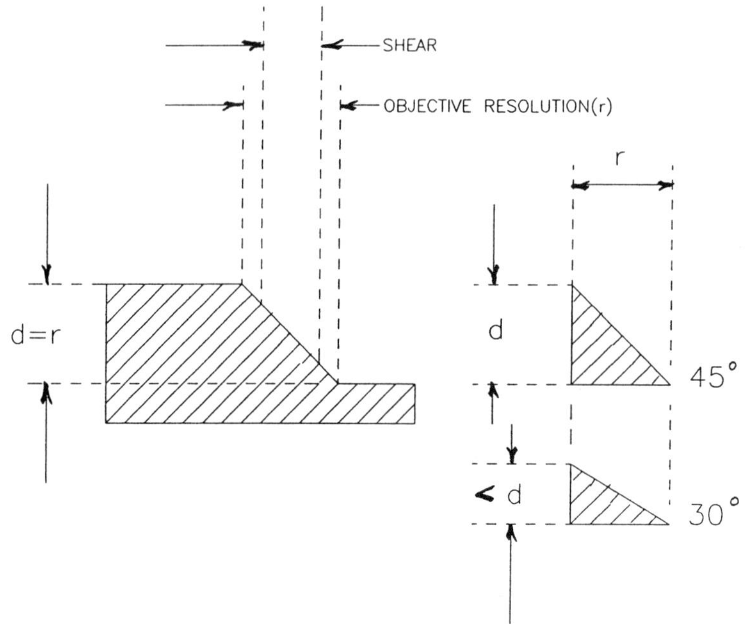

FIGURE 6 MAXIMUM INTERPHASE RELIEF

The maximum interphase relief assuming a 45° slope (according to figure 6) is when the depth 'd' = R the objective resolution. Since this dimension is based on a 45° slope it will obviously reduce as the slope lengthens, which is what usually happens when for example prolonged polishing takes place. We therefore have established two very interesting points (I) target for a relief dimension of no more than the objective resolution and (II) avoid prolonged polishing (the exception to the last statement is if the polishing is carried out on a vibratory polisher, in which case the slope is rarely lengthened and can with gross relief be reduced). Back to the question of the sensitivity or the minimum detectable slope by DIC. When we look at the MICHEL LEVY colour chart we see how the geologist or mineralogist can detect by colour OPD changes of as little as 20 *nm*. For the purpose of this discussion take as an example an optical difference (OPD) of 100 *nm* as an acceptable level of detection. To relate this into micrometres of interphase relief:-

$$(1\mu m = 1000nm) \qquad \frac{100nm}{1000\mu m} = 0.1\mu m$$

Therefore it is not unreasonable to say the optical detection level in terms of interphase relief would be 0.1µm and that this would be independent of objective resolution. This last point is quite interesting since we now have a detection level superior to the objective resolution and unlike the requirements of resolution we don't have to use the highest numerical aperture (NA) or magnification objectives to reap the benefits of critical detection, see chart No.1.

CHART NO 1 - OBJECTIVE CHARACERISTICS

Objective	NA	Resolution (µm)	DIC Sensitivity (µm)	Max. Interphase Relief (µm)
5	0.10	2.75	0.10	4
10	0.25	1.18	0.10	2
20	0.5	0.55	0.10	1
40	0.65	0.42	0.10	1
60	0.85	0.33	0.10	0.5
100 dry	0.90	0.31	0.10	0.5
100 oil	1.25	0.22	0.10	0.5

DIC IN USE

As an optical technique it is often thought to require some complicated setting up procedure requiring specialised equipment and at best only offers pretty pictures. In the past there has been an element of truth in that assumption. Objectives for use with DIC are special strain free versions and are usually supplied at a special (expensive) price, this however is not necessary since the level of strain existing in NORMAL objectives is much below the critical level for DIC.

Quite simply all that is required for reflected light is the birefringent prism be positioned in the objective BFP at 45° to the extinction position of the crossed polars. To remove either or both polarizer and analyser from the optic axis will revert the system back to bright field. Beware however from leaving the birefringent prism in the optic axis since this will ghost image all specimen detail.

COMBINED DIC / DARKGROUND

The ability to combine different techniques into a single image has challeneged the writer for many years. To have the facility to view a combined image of Brightfield / Darkground or DIC/Darkground can now be achieved as shown in figure 7 - viz:-

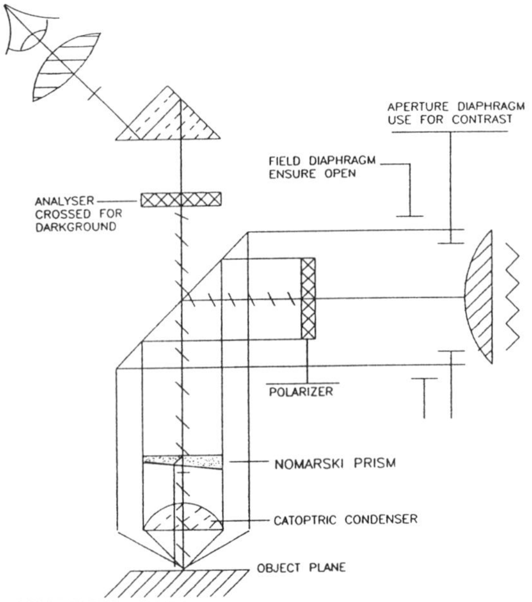

FIGURE 7 - COMBINED DARK GROUND - DIFFERENTIAL CONTRAST

- Combined DIC / Darkground - Polarizer and analyser crossed, Normaski prism introduced.
- Combined Brightfield / Darkground - remove Normarski prism, uncross the polars.
- Darkground - cross the polars.
- Brightfield - remove analyser, stop down field diaphragm to extremety of field of view.
- DIC - with polarizer stop/analyser and Normarski prism in position stop down field diaphragm to extremety of field of view.

This simple change will trasform the materials microscope, improving the versatility and combining vital information in a single field of view.

SURFACE PREPARATION FOR DIC

Materials Surface preparation for DIC as demonstrated above, requires an interphase relief within given limits, as has been shown in chart No.1. When a material is mechanically worked by abrasion, minute height differences are created between features or phases caused by different levels of abrasion. This condition is usually associated with different slopes at the interface of these features or phases.

This situation can be exploited by DIC when the width of these slopes is **just** greater or less than the resolution of the microscope objective.

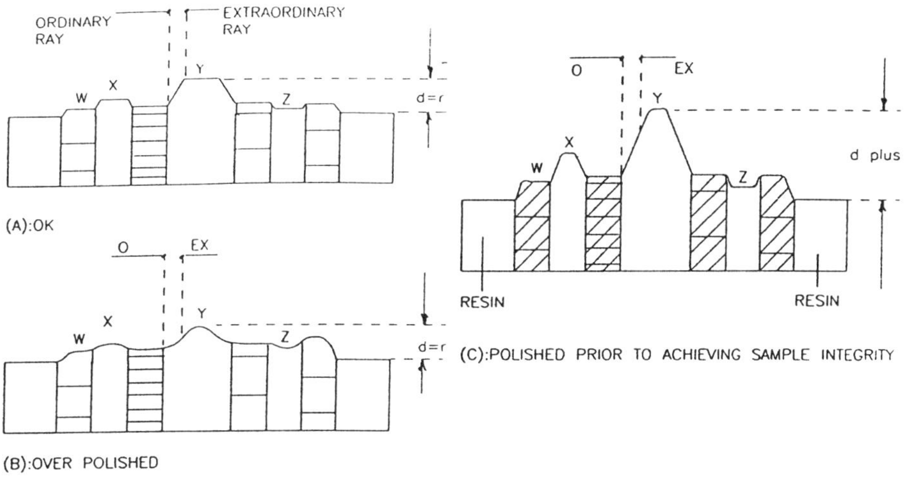

FIGUE 8 - DIFFERENTIAL ABRASION

To utilise the classical approach to metallography where the sample receives a course grind, fine grind, rough polish and final polish, will, inevitably lead to gross relief shown as "d plus" in figure No 8c. This is the result of polishing prior to achieving sample integrity. The target therefore in specimen surface preparation is to achieve sample integrity via stages of grinding or lapping prior to the one and only polishing step which must be short in duration. Reference to figure No 8b depicts the undesirable condition when having achieved sample integrity, the polishing stage is prolonged.

DIC IN ACTION

Micrograph No 1 & 2 x 500 - Copper, showing grain and slip planes(all magnifications are quoted as 'unity')

This polished sample is in the unetched condition and has, after polishing, been compression stressed. The phenomenon of material removal by slip plane dislocation is excellently illustrated in this Micrograph, it is also an excellent example of the 1st order grey birefringent prism setting. Micrograph No 2 illustrates the wonderful images created by colour i.e. careful withdrawal of Normarski prism.

Micrograph No's 3, 4 & 5 Micro indentation

These three micrographs are used to reinforce the technique of DIC for slip plane detection and also top surface 'ridging'. Micrograph No 3 is the bright field image of various Knoop hardness indents into a stainless steel polished sample. What makes Micrograph No 4 fascinating is the clear emergence of slip planes as a result of the shear stress caused when applying the indent load. Notice also the increased slip plane activity as the indent load increases.

Micrograph No 5 is a colourful slide depicting the area of ridging displacement. DIC has identified by colour the change in surface slope from the original planar purple surface.

Micrograph No's 6, 7 and 8 - Steel Mesh Filter x 50

The use of different optical techniques is to be encouraged since each technique can on occasions reveal hidden information. With this steel mesh filter we are able to reveal in Figure No 6 (bright field) the various mesh sizes within the filter.

Micrograph No 7 is showing the image when using DIC. Whenever the Normarski prism is within the optic axis, as is required for DIC, then a double ray (shear) is created in the object plane. A double image would not normally be visible, since the shear or separation is less than the resolving power of each microscope objective lens. Observe the larger mesh size and you will notice only one image of each subject, that subject being three dimensional. Now observe the smallest mesh size and you will have double vision of each subject. (It may be necessary to use an eye glass to confirm).

Some microscope designs require that the Normarski prism be left in the optic axis (for convenience) when not in use. It is true that to remove the polarizer or analyser from the optic axis will covert DIC to bright field.It will however still cause a double image which will still be visible with subjects smaller than the resolving power of the objective lens in use, as is shown in Micrograph No 8. What perhaps is more important is that a double image will **always** occur when Normarski prism is within the optics even if it is not apparent to the observer, This prism therefore should never be left in the optic axis for any optical technique other than DIC.

Micrograph No's 9 and 10 - Complex Aluminium Alloys x 500

Etching has long been employed in the identification of complex phases along with the advantages of anodising to reveal grains. DIC however will help in identification by clearly delineating differential abrasive characteristics (as well as grey levels) without the need for specific etching. Micrograph No.9 in bright field shows the extreme interphase planarity. DIC with this same field of view has highlighted the Al/Cu/Fe/Mn script phase. (Figure No 10)

Micrograph No's 11 and 12 - Bone (Hard Tissue)

Hard tissue such as bone is traditionally observed in transmitted light using a thin section. This classical approach lends itself to microtome thin section techniques. However, hard tissue in particular when accompanied by metal or ceramic implant material necessitates a totally different approach. When dealing with undecalsified tissue it will be necessary to embed with a methylmethacrylate or similar resin after dehydration. The specimen is then prepared using normal metallographic techniques, ensuring planarity at the sample integrity stage prior to a short interphase relief polish. Micrograph No 11 is a bright field image indicating what looks like excellent planarity. Micrograph No 12 reveals in reflected light DIC, the vascular tissue and offers information often sufficient, thus avoiding the need for thin section transmitted light application. Much information can be achieved with thick sections using reflected light optical techniques. There is however, no substitute for thin sections. To make a thin section of 'high integrity' it will be necessary to cement this finished surface onto a glass microscope slide, after which it will be sectioned, then ground to size (30μm) using the same technique as for the thick section.

CONCLUSIONS

The intention of this paper was to illustrate necessary actions in order to exploit this wonderful world of differential interference contrast. In the field of materialography (metallography plus) we are often judged by our micrographs, for the reader these had to be restricted to 12 only.

This paper is one of a series to be produced specifically for materials scientists wanting to EXPLOIT THE OPTICAL MICROSCOPE.

REFERENCES

This material has been sourced from the book 'Surface Preparation & Microscopy of Materials' by Brian Bousfield.

Fig. 1 - Copper (500X)

Fig. 2 - Copper (500X)

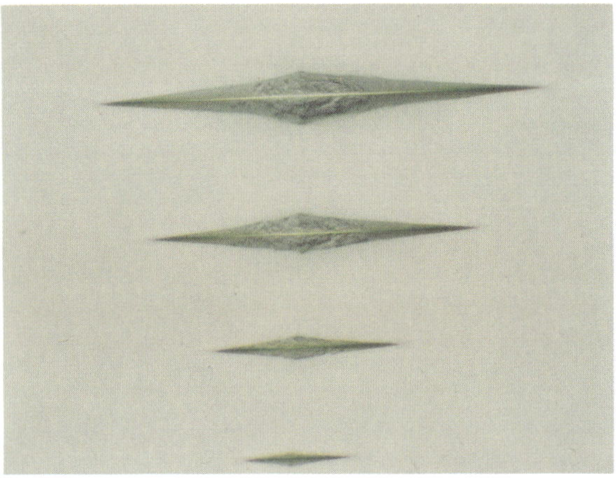

Fig. 3 - Micro Indentations (200X)

Fig. 4 - Micro Indentations (200X)

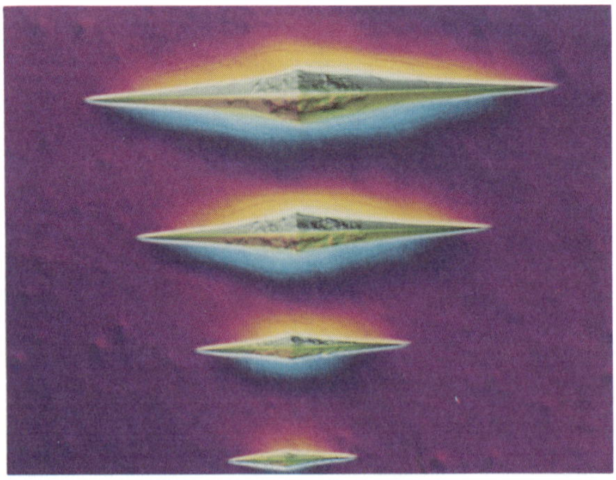

Fig. 5 - Micro Indentations (200X)

Fig. 6 - Steel Mesh Filter (50X)

'ig. 7 - Steel Mesh Filter (50X)

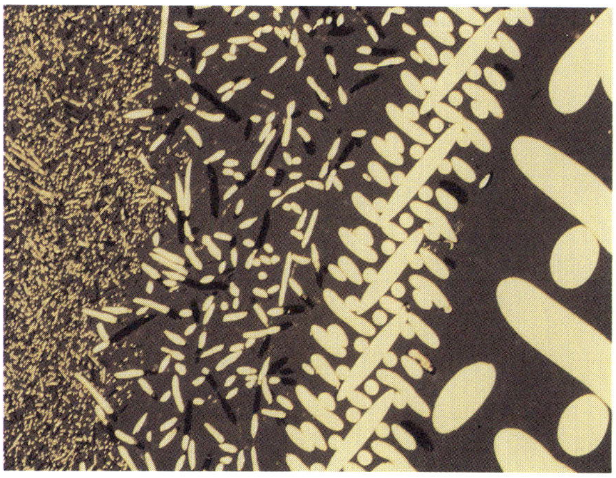

Fig. 8 - Steel Mesh Filter (50X)

ig. 9 - Complex Alum. Alloy (500X)

Fig. 10 - Complex Alum. Alloy (500X)

g. 11- Bone (100X)

Fig. 12 - Bone (100X)

APPLICATION OF DARK-FIELD AND D.I.C. ILLUMINATION TECHNIQUES IN THE EVALUATION OF ALUMINUM D.C. CAST INGOT AND DOWNSTREAM PRODUCT MICROSTRUCTURES

Andris B. Innus [1]

ABSTRACT

Examples are given where dark-field illumination can help to better visualize (in aluminum) such aspects as precipitate and/or dispersoid distribution, and for the use of differential interference contrast, D.I.C., such as for highlighting dendritic coring to better define cell boundaries, and other features that arise from polishing relief.

INTRODUCTION

In microscopy, the choice of type of illumination can be a factor in whether a given feature becomes evident, either during direct observation, or in photo-micrographs. From side by side comparison of the illumination options on our LECO 300 (bright-field, polarized

---

[1] Alcan Smelters and Chemicals Ltd., Casting Technology Service, P.O. Box 1500, Jonquière, Québec, Canada.

light, dark-field and Nomarski), certain types of microscopic features of the samples handled by our production metallography lab. were selected to be routinely evaluated and/or photographed specifically using either dark-field or D.I.C.

Dark-field Illumination

In dark-field illumination [1], the incident rays of light are routed 360° around the periphery of the objective, and arrive at the specimen's surface at an angle oblique to the axis of the objective lens, eg. Fig. 1. In this way, those rays of light reflected from planes perpendicular to the objective lens are not captured, but those reflected from oblique surfaces do pass back through the objective lens. The result is that surface irregularities appear bright, while the flat background is dark.

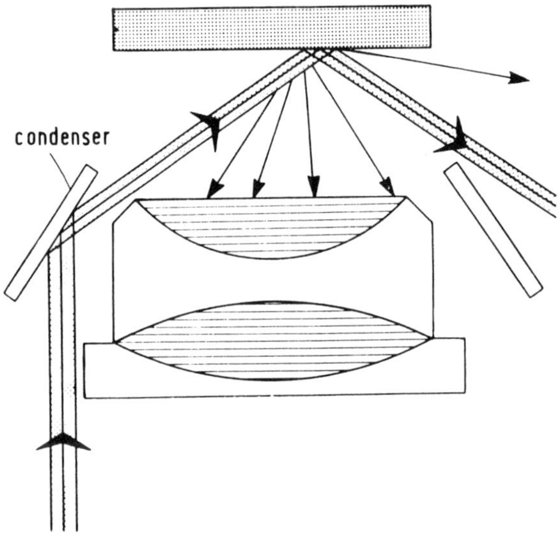

Figure 1. Schematic representation [2] of dark-field illumination.

From the literature, useful possible applications are described [2], [3], [4], such as:
- viewing very small features such as pores, cracks, inclusions (as they appear to be self-luminous, thereby improving resolution),
- improving contrast in colorless structures,
- observing colors of structures.

A few visual examples have been found: lamellar eutectic texture [1], grain orientation via etch pits [2], presence of annealing twins [3], distinguishing reimpressed from original SiC particles in an MMC [5].

D.I.C.

In differential interference contrast, D.I.C. [4], after Nomarski, eg. Fig. 2, upon passing through a Wollaston prism (biprism), a linearly polarized ray of light is split into two beams with vibration planes perpendicular to each other's. Upon passing through the objective lens, the path of the two beams becomes parallel. After reflection from the sample's surface, the two beams are superimposed by passing back through the Wollaston prism. By this method, small differences in height on the sample's surface produces a difference in path length between the two reflected beams, which affects how much interference will occur when the two beams are superimposed. The extent of interference appear as variations in brightness or color.

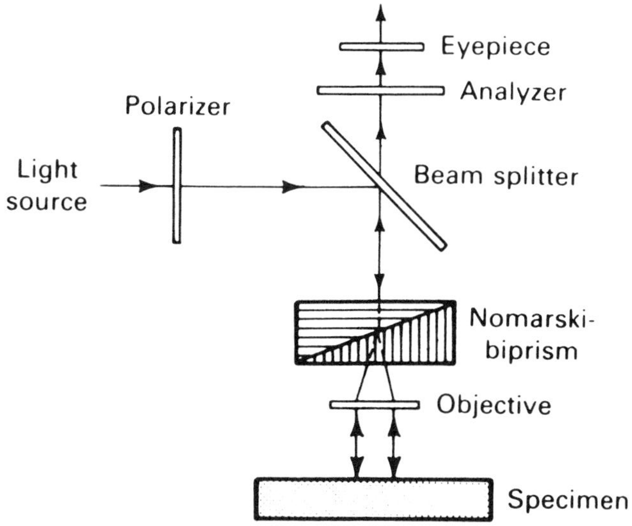

Figure 2. Schematic representation [4] of differential interference illumination, after Nomarski.

From the literature [3], [4], [5], the principle application described is the revealing of low contrast structures of different hardness (resulting from differences in polishing relief). Some particularly good visual examples are present in ref. [5].

PROCEDURE

The black & white dark-field photo-micrographs to follow are on Polaroid type 53 film (ISO 800), while bright-field and D.I.C. images are on type 55 film (ISO 50). A 150 watt xenon light source was used to avoid excessively long exposure times. The colour prints are on Polacolor 64 tungsten (ISO 64), using tungsten halogen illumination.

Note that for visual examination, tungsten halogen lighting is sufficient in most cases.

Under automatic exposure conditions, the light reflected from a greater number of point indications, eg. precipitate etch pits, would shorten the exposure time, and vice versa. In this way, with all other microstructural aspects being similar, ie. amounts of constituent, porosity, it is possible to obtain a semi-quantitative estimate of precipitate concentrations based upon exposure times, eg. Figs. 3A & 3B.

For photo-micrographs having similar etch pit concentrations, but with enough of other features as to influence exposure time, using the same exposure conditions is advisable to maintain a visual continuity. As our LECO 300 has the capability to memorize exposure times, the exposure conditions for the field of view with the least amount of superfluous light is memorized and used for the subsequent fields of view.

Also note that if there is porosity present in the field of view, when using automatic exposure there may be glare from the pores and even the precipitate etch pits, eg. Fig. 4A. If this is not acceptable it may be necessary to either manually cut back on the exposure time, Fig. 4B, or even put a green filter into the light path. However, the latter would substantially increase exposure time.

Of equal importance when comparing dark-field images of certain microsections is that the samples be prepared in exactly the same fashion. If not using a multiple sample holder of an automatic preparation system, it may be necessary to mount sections of the samples to be evaluated within the same mould. For features revealed by micro-etching, reproducing etching conditions and times is also necessary for similar alloys. Note that for 6063 ingot sections, a 90 second etch in 0.5% HF was used.

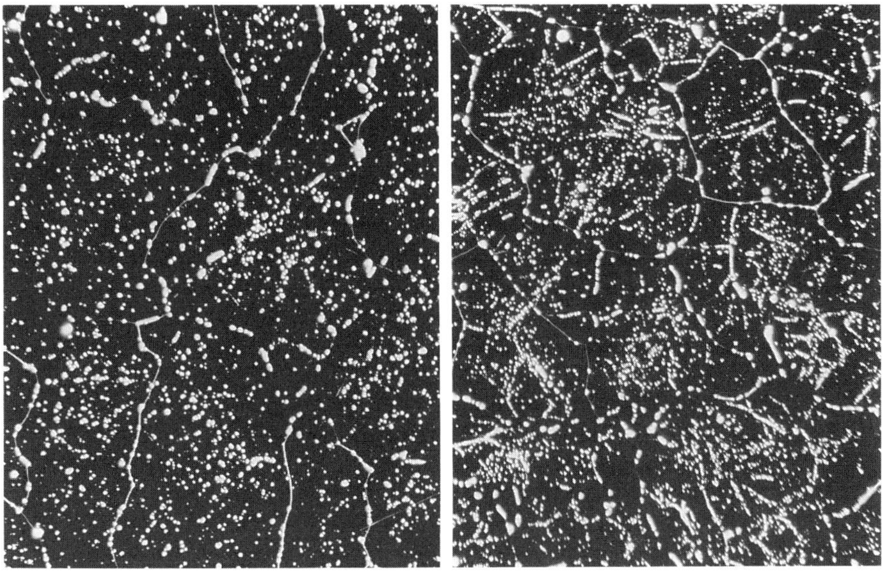

A = 9.37 sec. exposure 200 X B = 4.58 sec. exposure

Figure 3. Comparison of dark-field images of Mg_2Si precipitate. More precipitate yields a shorter exposure time.

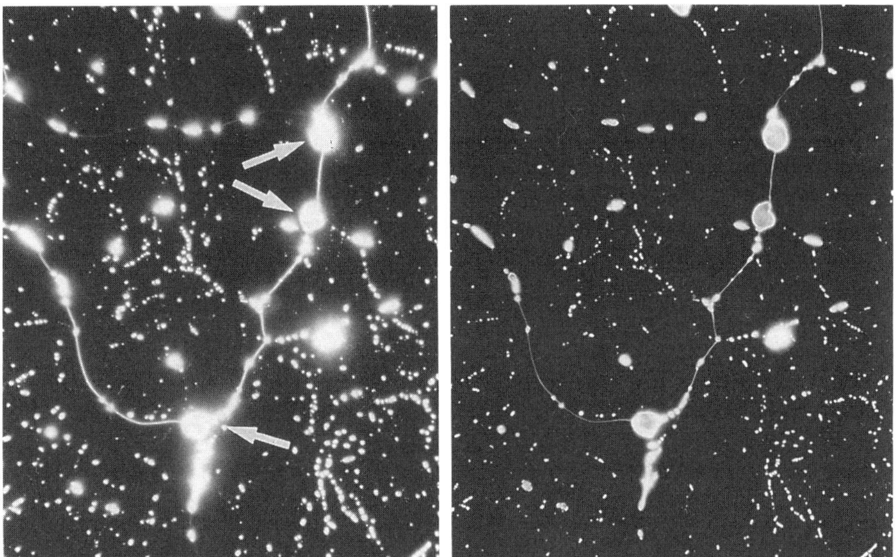

A = 2.70 sec. exposure 500 X B = 0.54 sec. exposure

Figure 4. Appearance of glare (A) from porosity (arrows), and the effects of reducing exposure time (B).

Depending upon the extent of relief produced during polishing (in part a function of the specimen's alloy), further micro-etching may or may not be required to help accentuate the relief obtained using D.I.C..

EXAMPLES - DARK-FIELD

Photo-micrography of Mg_2Si Precipitates and Mn Dispersoids

By far the most useful application of dark-field illumination in our lab. is to illustrate, via. photo-micrography, precipitate and / or dispersoid distributions and their relative concentrations in D.C. cast and homogenized ingot. Although some sense of distribution and concentration is apparent under bright-field, under dark-field the effect can be made to be more striking, eg. compare Figs. 5A & 5B.

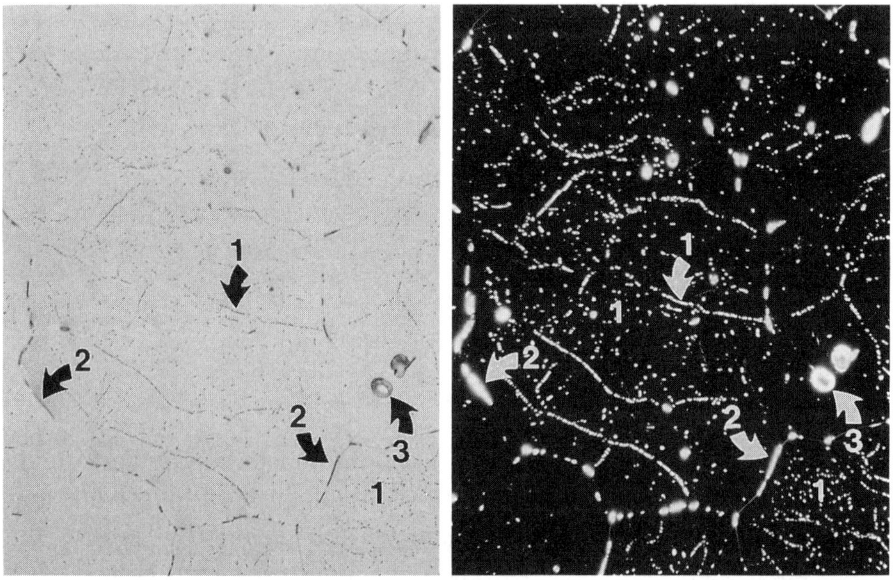

A = Bright-field 400 X B = Dark-field

Figure 5. Comparison of bright-field and dark-field images in a 6063 ingot: Mg_2Si precipitates (1), AlFeSi constituent phase particles (2), and porosity (3).

Note that for those accustomed to looking at bright-field photomicrographs, it may be advisable to give some orientation on the dark-field images by identifying common features, eg. constituent phase particles, porosity, etc., such as illustrated in Fig. 5B.

For those alloys containing both Mg_2Si precipitates and Mn dispersoids, it may be possible to distinguish between the two based upon the amount of light returned from their respective etch pits, as illustrated in Fig. 6, where the Mn dispersoid etch pits are smaller, and so reflect back less light. Naturally, the extent of such a difference (if in fact one can exist) is largely a function of the ingot's homogenisation and cooling conditions.

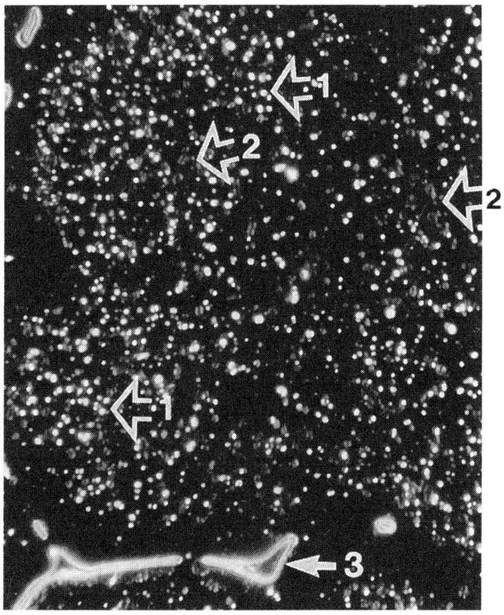

Fig. 6. From a 6005A ingot. There are two families of etch pits present, the brighter (1) are from Mg_2Si, whereas the fainter are from Mn dispersoids (2). Also apparent is an Al(Fe,Mn)Si constituent particle (3).

Because the interior of porosity can be illuminated under dark-field, it is possible to focus into pores, and see the texture of their walls and even their contents to a certain extent.

For 6XXX alloy extrusions, dark-field illumination has helped us to highlight the presence of transverse welds (between successive billets within the extrusion [6]), over extrusion (going too far into the extrusion cycle, causing ingot surface to appear within the extrudate [7]), and incomplete longitudinal welds of hollow sections [8]. All of these defects usually have either local differences in concentration of Mg_2Si precipitate or thin, linear indications such as oxide films or gaps. The technique has so far not been found to be useful for comparing the extent of Mg_2Si precipitation from the artificial aging of extrusions - the etch pits are by and large too small as to be optically discernible, except in the cases of extreme over-aging.

There is also a pitfall when using dark-field for examining extrusion microsections, in that there may be some assumption that etch pits present on recrystallized grain boundaries, eg. Fig. 7, are exclusively the result of precipitates that have nucleated and grown on those boundaries because of a slow press quench (cooling down of the extrudate exiting the press). It has been our experience that such etch pits may also be from the coincidental superimposition of the recrystallized grain boundary and undissolved Mg_2Si particles (extrudate exit temperature too low to dissolve those particles, or that they were too large to be returned into solution within the time provided). The presence of such undissolved Mg_2Si can be verified by looking at an as-polished specimen under the backscatter mode of an SEM (undissolved Mg_2Si appears as dark particles, confirmed by X-ray analysis, vs. the light appearing AlFeSi).

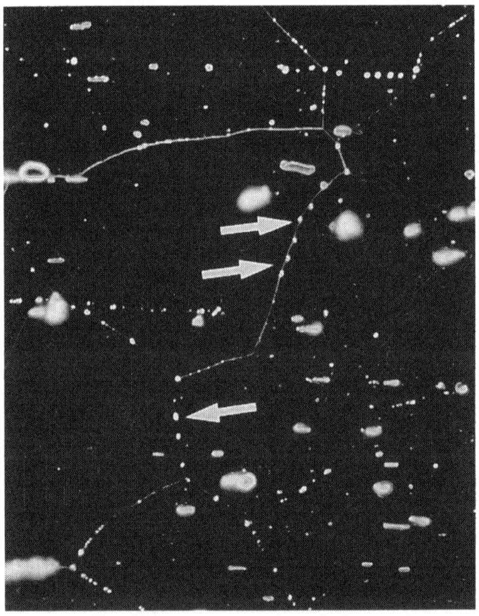

Fig. 7. Appearance of etch pits on an extrusion's recrystallized grain boundary (arrows), which may not be due to precipitation, but due to a coincidental superimposition of undissolved Mg_2Si on that boundary, 750 X.

Also in extrusions, under certain conditions, eg. extent of grain recrystallisation and of polishing practice, a grain orientation effect can be obtained, whereby the body of certain grains will appear lighter under dark-field illumination, eg. Fig. 8. This may give the impression of more precipitates within those grains than is actually the case.

Fig. 8. Appearance of grain bodies of different luminescences, ie. the body of certain recrystallised grains is brighter due to a grain orientation effect, 200 X.

D.C. Cast Ingot Inverse Segregation

For as-cast ingot, the boundary of certain types of diffuse inverse segregation can be difficult to clearly resolve under bright-field illumination. With the use of dark-field, relatively slight differences in constituent phase concentration (a function of the degree of alloying elements' macrosegregation, and their respective partition coefficients) become more apparent, eg. Fig. 9. It also becomes easier to tell when the pattern of the constituent networks become more ordered. Both of which help to locate an inverse segregation boundary. Note that this same principle applies for locating denuded regions within the ingot's shell zone.

For homogenised 6XXX type ingot, the constituent phase networks may have been sufficiently fragmented by the homogenisation process as to make an inverse segregation depth determination difficult based solely upon those particles. In such cases, the greater concentration of Mg_2Si precipitate within the inverse segregation layer, as highlighted under dark-field illumination, may help in establishing its depth.

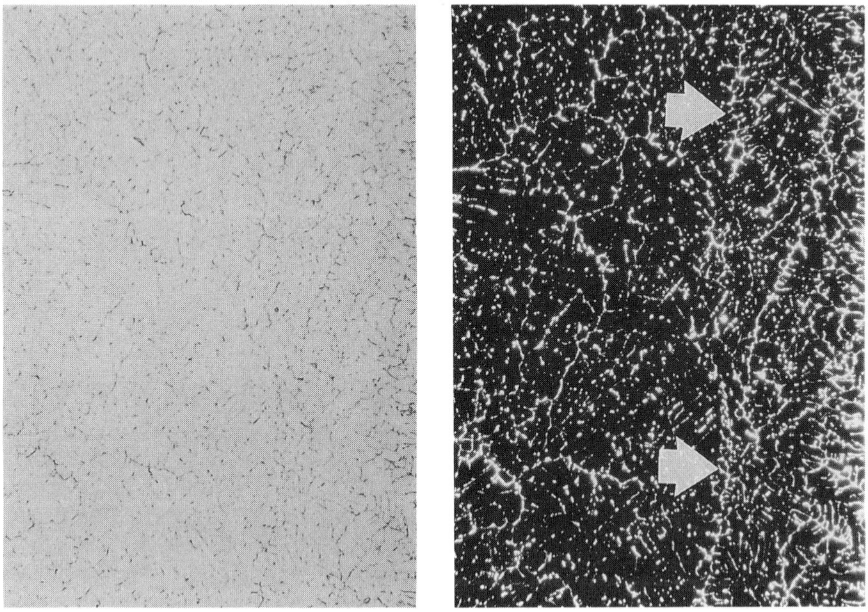

A = Bright-field 50 X B = Dark-field

Figure 9. Appearance of an inverse segregation layer with the boundary more easily resolved (arrows) using dark-field.

Unpolished Surfaces

The use of dark-field illumination has also helped us in the visualization and documentation of unpolished surfaces, eg. fracture surfaces, and as-extruded surfaces. For the former, Fig. 10, under bright-field illumination the oxidized fracture surface of a steel bolt becomes washed out, whereas by using dark-field, even the presence of copper particles on the surface becomes much more evident. For the latter, Fig. 11, more of a three dimensional effect is obtained using dark-field for die lines on an as-extruded surface, giving a better sense of die line depths without needing to use an SEM.

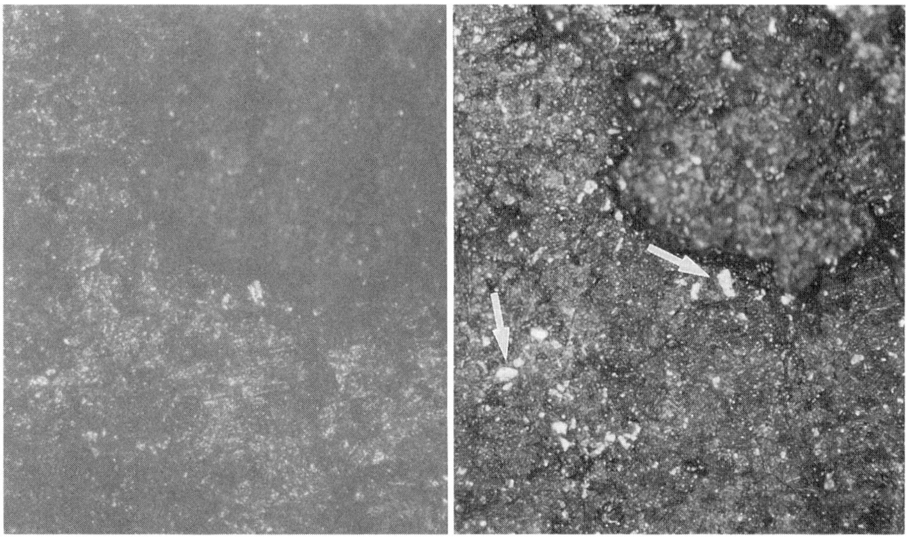

A = Bright-field 50 X Dark-field

Figure 10. Appearance of the oxidized fracture surface of a steel bolt. Note that more detail is apparent using dark-field, to the point that the color of the copper particles (arrows) is distinguishable.

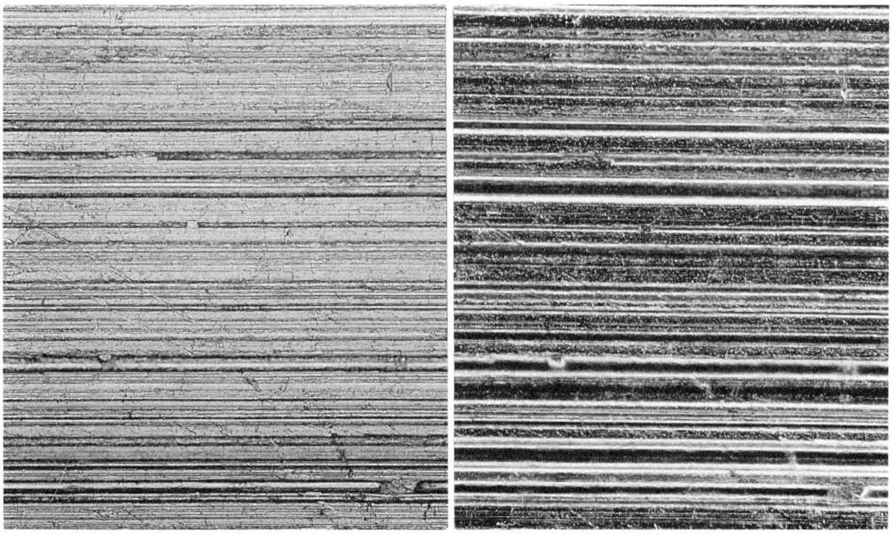

A = Bright-field 200 X B = Dark-field

Figure 11. Appearance of die lines on an as-extruded surface. By using dark-field illumination, more of a three dimensional effect is obtained.

EXAMPLES - D.I.C.

As-Cast Dendrite Cell Structure

The principle application in our lab. for D.I.C. is to help define cell boundaries in the as-cast microstructure of certain alloyed (usually Mg containing) ingot, for the purposes of dendrite cell size or arm spacing measurements. Since those constituent phase networks visible under bright-field, eg. Fig. 12A, may not always well define all of the cell boundaries, the use of Nomarski allows dendritic coring to become easily visible, Fig. 12B, so that more boundaries become apparent. In this way it is possible to extrapolate cell boundaries between constituent particles, and to see those not having constituent particles.

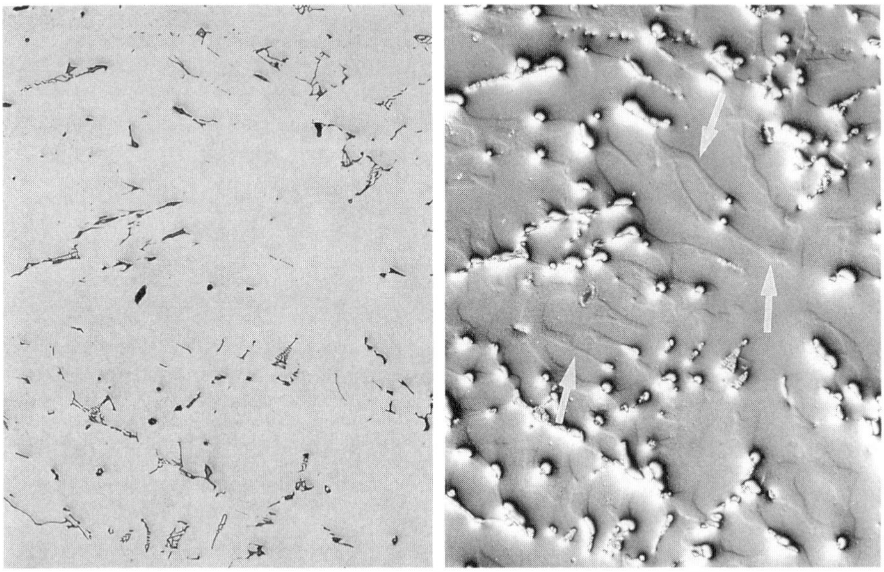

A = Bright-field 100 X B = Nomarski

Figure 12. Dendrite coring (arrows) is more apparent in this 3104 ingot section using Nomarski.

Microstructure Bands and Grain in Extrusions

Microstructural anomalies such as the bands shown in Fig. 13, were observed in longitudinal (extrusion direction) cross-sections of 6XXX alloy extrusions only after acquiring D.I.C. These bands are suspected to be related to solute microsegregation within the ingot that was not fully removed by heat treatment.

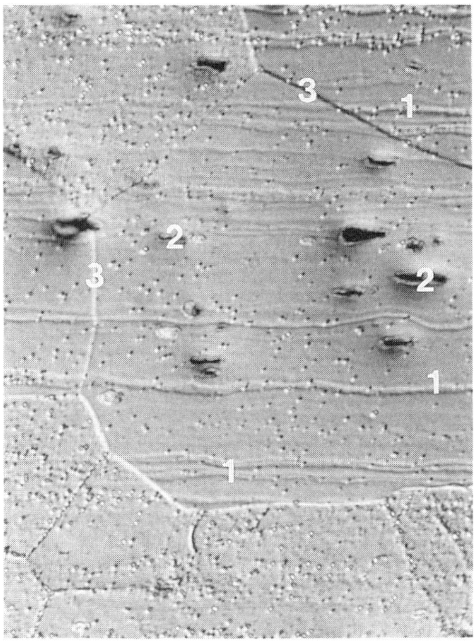

Fig. 13. Bands of material polished in relief (1) within the cross-section of a 6061 extrusion. Also visible are AlFeSi constituent phase particles (2), and recrystallized grain boundaries (3), 1000 X.

When presenting photo-micrographs taken with D.I.C., it should be kept in mind whether the feature of interest is in fact in relief or in recess, and so to take or orientate the photo-micrograph to give that optical impression, eg. by turning this page upside down, the bands shown in Fig. 13 can be made to appear as valleys rather than as plateaus. That the bands are in relief after polishing is determined by comparing their shadows with those of the hard constituent phase particles that must be in relief. The same applies for determining the relative elevations of the recrystallized grains shown in Fig. 13.

CONCLUSION

Examples have been given for the application of dark-field and D.I.C. illuminations, as well as for possible shortcomings to keep in mind when using those respective techniques. Although there is no intent implied here that those two techniques should fully replace the use of bright-field illumination, their availability and potential qualities (as well as those of other techniques) should be kept in mind to help make microstructural features as evident as possible.

REFERENCES

1) A. Tomer, "Structure of Metals Through Optical Microscopy", ASM International, 1990.
2) R.C. Gifkins, "Optical Microscopy of Metals", Pitman Press, 1970.
3) G.F. Vander Voort, "Metallography Principles and Practice", McGraw-Hill Book Company, 1984.
4) "Metallography and Microstructures", **Metals Handbook Ninth Edition**, Vol. 9, 1985.
5) B. Bousfield, 'Surface Preparation and Microscopy of Materials", John Wiley & Sons, 1992.
6) W.-D. Finkelnburg, G. Scharf, "Some Investigations on the Metal Flow During Extrusion of Al Alloys", **Proceedings: 5th International Extrusion Technology Seminar**, Vol. 2, p. 465, 1992.
7) M. Lefsta, O. Reiso, V. Johnson, "Flow of the Billet Surface in Aluminum Extrusion", **Proceedings: 5th International Extrusion Technology Seminar**, Vol. 2, p. 503, 1992.
8) R. Akeret, "Extrusion Welds - Quality Aspects Are Now Center Stage", **Proceedings: 5th International Extrusion Technology Seminar**, Vol. 1, p. 319, 1992.

PRINCIPALS AND PRACTICES FOR METALLOGRAPHIC PREPARATION FOR POWDER METALLURGY

G. A. Blann[1], D. J. Diaz, K. A. Leithner, R. Stone[2]

ABSTRACT

The field of metallography has evolved dramatically in recent years. Many changes in the manufacturing of powder metal parts have resulted in widespread use of complex alloys and new applications for high-strength, lightweight materials. As the processes and complexity of the materials have increased, metallography has proven to be a useful tool in the manufacturing process for evaluating product quality and characterizing microstructural defects. Microstructural analysis plays an important role in the development of new materials and product reliability in the powdered metal industry. Traditional sample preparation is commonly used today; however, new developments in abrasive and surface technology can lead to an improved metallographic sample preparation process. This paper will compare the traditional and improved metallographic techniques using light optical and SEM microscopy.

INTRODUCTION

New, innovative materials with unique properties are being developed and produced by powder metal manufacturers for a wide range of industrial applications. In order to properly evaluate these advanced powder metals, improved metallographic preparation techniques have been developed. Abrasive grinding and polishing techniques employed by early material scientists produce

---

[1]Buehler Ltd., 41 Waukegan Road, Lake Bluff, IL 60044
[2]Buehler Ltd.

deformation that prevented an accurate evaluation of the powder metal part being evaluated. These efforts to reveal the microstructural characteristics of the material were hindered by a misunderstanding of the existence and nature of surface deformation.

The purpose of metallographic sample preparation is to produce a polished surface that reveals the true microstructure when viewed on the light optical microscope. The basic preparation procedure are: *Sectioning, Mounting, Sample Integrity* (grinding & rough polishing), and *Final Polishing*. All preparation stages are important. Careless preparation will lead to erroneous interpretations and measurements during microscopic evaluation. The sectioning and grinding steps are characteristically coarser than the polishing steps; therefore, the risk of altering the microstructure is greater. Incorrectly chosen abrasive materials, surface (cloth), wheel speed, lubrication, or head pressure will cause mechanical deformation that will in turn alter the microstructure. As materials become increasingly more complex and unique, metallographic preparation techniques and equipment must rise to meet the challenge and provide a broader range of capabilities. Care must be taken at each stage of specimen preparation to assure a valid microstructural analysis.

SECTIONING

Two key concerns when sectioning powder metal parts are the reduction of cutting time and minimization of induced damage. When any given material is sectioned, a minimal amount of surface and sub-surface damage is induced. The amount of this damage varies with the material hardness and blade selection. Precautions can be taken to minimize this damage. The proper selection of cutting parameters such as wheel speed, feed rate, and coolant flow, plus proper blade selection will produce a clean cut with minimal surface and sub-surface damage. This in turn will substantially reduce the time spent in the subsequent stages of grinding, sample integrity, and polishing. The recommended abrasive blade for ferrous powder metal parts is aluminum oxide abrasive with resin-rubber bond, and for non-ferrous powder metal parts is silicon carbide abrasive with rubber bond. The use of a band saw or hack saw should be avoided due to the excessive damage created during the cutting procedure.

Adequate cooling of the sample is necessary to prevent thermal damage. Uniform cooling is important to prevent uneven wear of abrasive wheels resulting in improper cuts and possible breakage. Figure 1 shows the proper fluid flow with the sample/blade interface.

The powder metal samples evaluated in this paper were porous thus entrapping the cutting fluid in the pores. Prior to mounting the samples, they were immersed in acetone for five minutes allowing the acetone to penetrate the pores. The samples were then placed in a 120°F oven for fifteen minutes to make certain that the acetone was fully evaporated throughout the sample.

METALLOGRAPHIC SAMPLE MOUNTING

The purpose of encapsulating sectioned samples is to preserve edges and to protect delicate and/or friable samples from mechanical damage or erosion during metallographic preparation. Porous samples are encapsulated in a vacuum in order to impregnate the open, interconnected porosity with the castable epoxy. Although powder metal parts are able to withstand high temperatures and pressures during encapsulation, the thermosetting powder does not infiltrate the pores. Open pores are susceptible to becoming enlarged during the grinding and polishing steps and cause lidding that cannot be readily removed.

METALLOGRAPHIC SAMPLE PREPARATION

The objective of sample preparation is to obtain a representative microstructure that is free from surface and sub-surface damage thus revealing the true characteristics of the material. The use of multiple steps using SiC abrasive paper in decreasing grit sizes followed by diamond and Al_2O_3 abrasive has been the traditional method used for many years. Although this method is still considered to be acceptable, it is not the most cost effective. Recent developments in the field of metallographic sample preparation have produced new, cost-effective methods using surfaces and abrasives that provide flat, deformation-free microstructures. This method was used in the preparation of the powder metal gears and is shown in Table I.

The purpose of the planar grinding stage is to remove the gross deformation created during the sectioning operation, and this was accomplished by using a 180 grit SiC abrasive appear. In order to remove the deformation created by the planar grinding step, a napless, polyester weave cloth was charged with 6μm polycrystalline diamond which has a high material removal rate. The second sample integrity step substantially improved the surface appearance and further reduced the surface and sub-surface deformation. Once again, a napless cloth and finer polycrystalline diamond (3μm) was used which rapidly removed the deformation from the previous step in a short period of time. In order to remove the 3μm scratches and surface smear, a final polish was accomplished by using a napped cloth charged with a new Al_2O_3 formulation. This step clarified the microstructural detail. The preparation sequence is pictorially shown in Figures 2 and 3. A low and high density powder gear was used to show their response to the four step procedure.

MICROSTRUCTURAL OBSERVATIONS

The as-polished samples were etched with 2% Nital to reveal the microstructure and qualify the polishing procedure. The microstructural detail was clean and crisp with no evidence of scratches as shown in Figures 4 and 5. High magnification evaluation revealed lidding in the pore areas causing a misinterpretation of the amount of porosity when measured manually or automatically on an image analysis system (Figure 6). In order to more accurately evaluate the porosity percent, a combination of etch-repolish-ultrasonic clean cycles were conducted at 1, 2, 4, 8, and 16-minute ultrasonic intervals. The percent porosity was measured using an image analysis system after each cycle, and the results were graphed as shown in Figure 8. The combination of the etch and re-polish cycles along with the ultrasonic cleaning at extended time cycles removed the lidding and provided a more accurate picture of the amount of porosity in each sample. A four-minute ultrasonic cycle removes the lidding as shown in Figure 7. The change in amount of porosity from the as-polished condition to eight minutes of ultrasonic cleaning is shown pictorially in Figures 9 and 10. The high- and low-density samples were evaluated in the gear tooth and gear body area. The gear tooth had a higher porosity percent in both the high- and low-density gears due to the method of fabrication.

CONCLUSIONS

1. The use of new surfaces and abrasives when metallographically preparing samples produces true microstructural characteristics in a cost-effective manner.
2. Vacuum infiltration of castable epoxies will maintain pore integrity throughout the preparation process of powder metal parts.
3. Ultrasonic cleaning of powder metal parts for 4 minutes will remove pore lidding and reveal an accurate picture of the porosity.

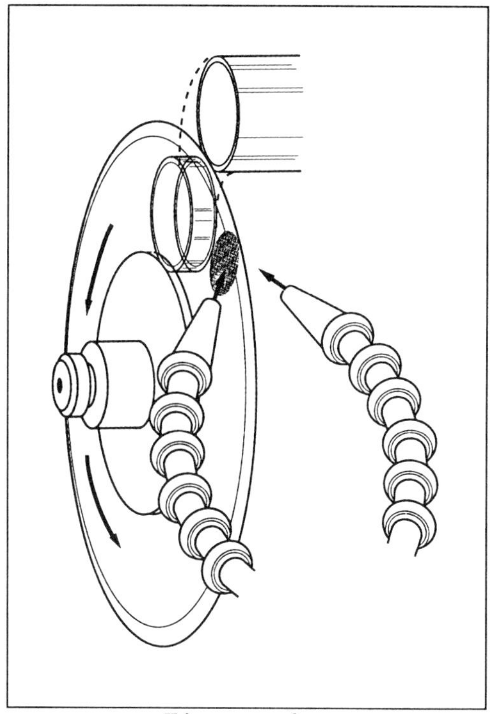

Figure 1

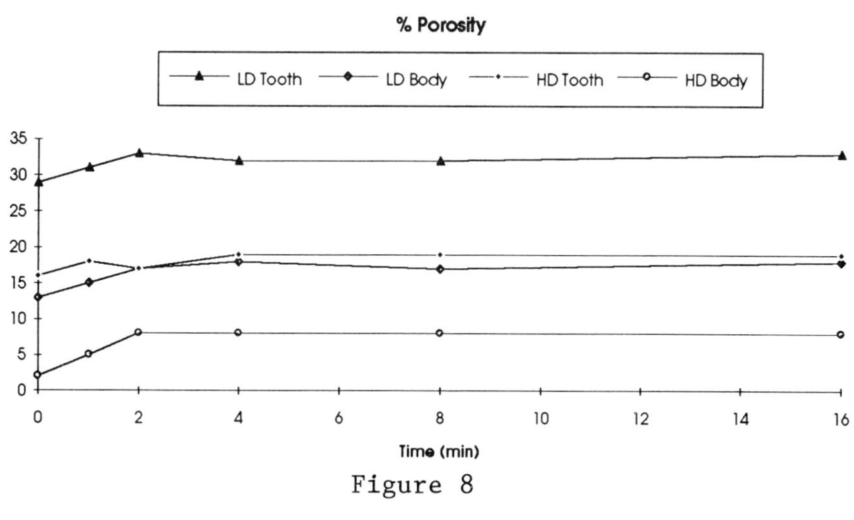

Figure 8

Table I

Surface	Lubricant	Abrasive Type/Size/ Carrier	Time/ Minutes	Force per Sample	Speed Setting	Relative Rotation
Planar Grinding						
Paper	Water	P220	25 seconds	25N	250 rpm	Comp.
Diamond Lapping						
Hard Woven Cloth	Water-base Suspension	6 µm polycrystalline diamond	4	25N	120 rpm	Comp.
Rough Polishing						
Non-woven Textile	Lapping Oil	3µm diamond	2.5	30N	120 rpm	Comp.
Final Polishing						
Synthetic Suede	Distilled Water	Colloidal Silica	2	35N	120 rpm	Contra

Low-Density Gear Tooth
Figure 2

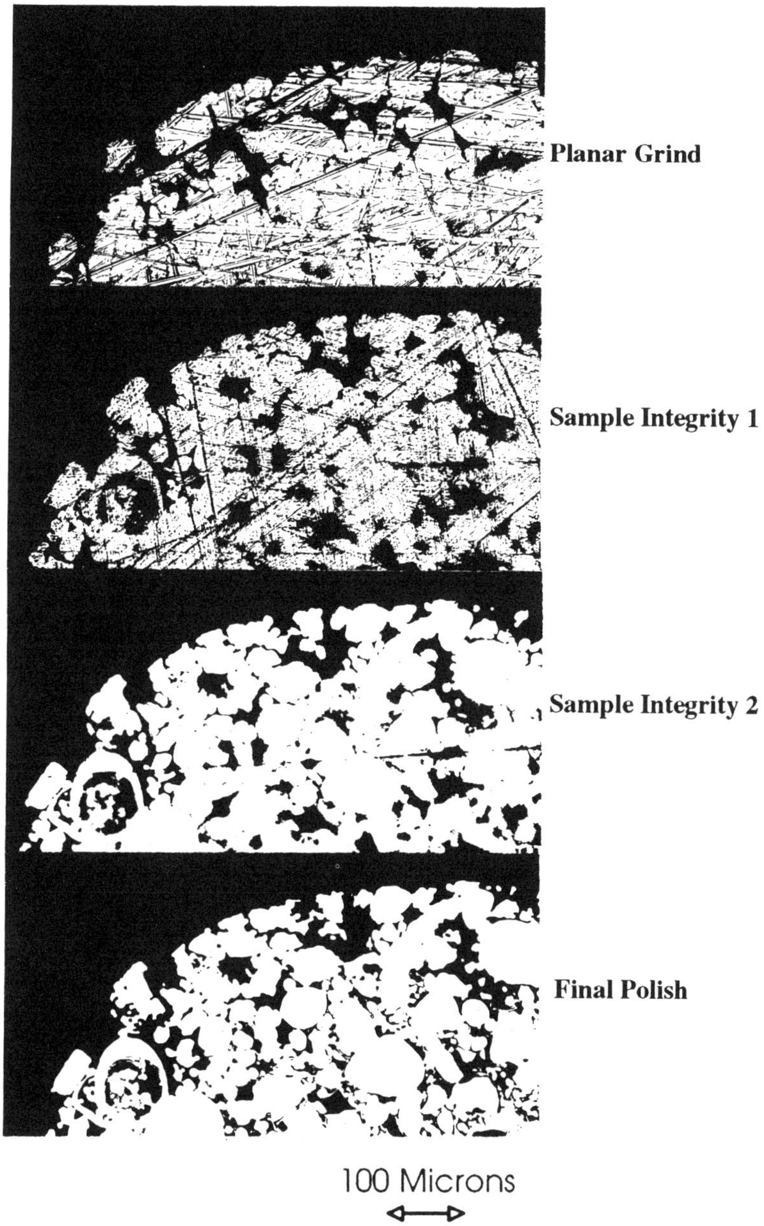

High-density Gear Tooth
Figure 3

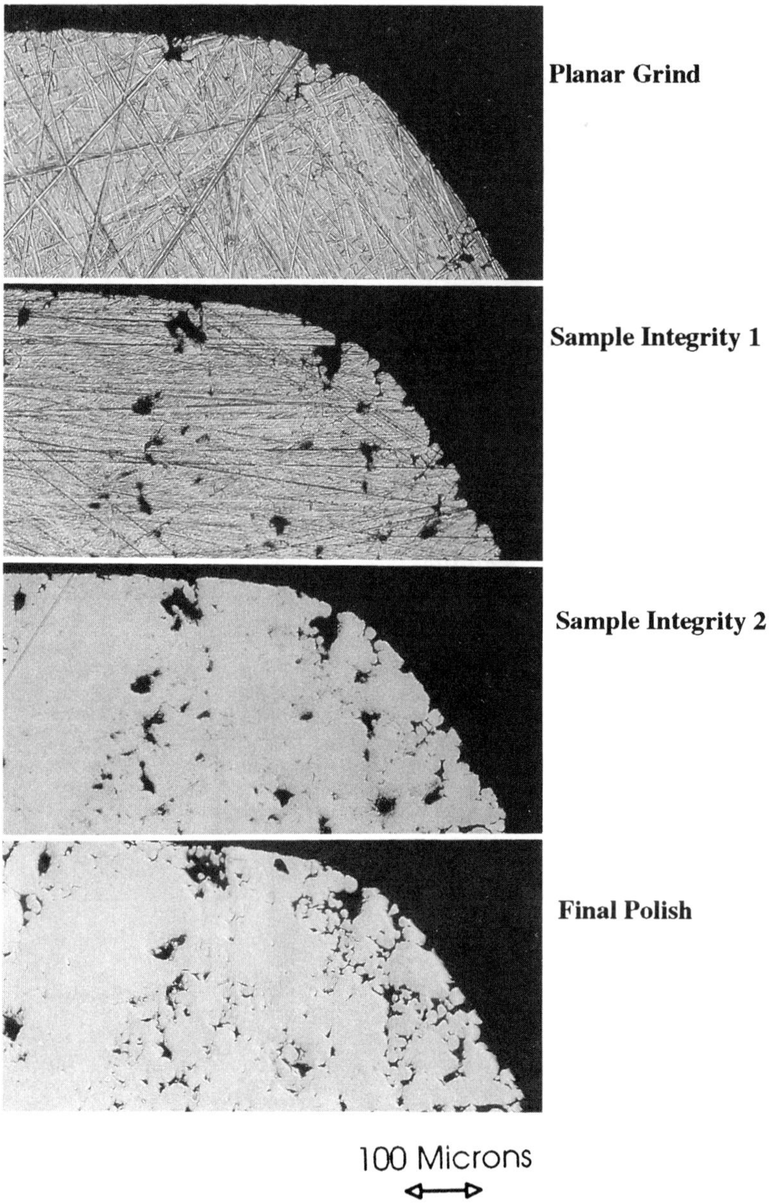

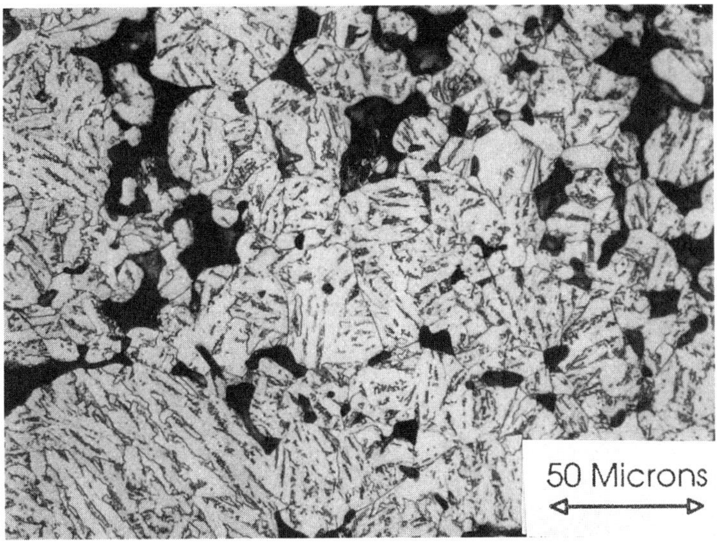

Body - 8 Minute Ultrasonic Low-density
2% Nital Etch - Figure 4

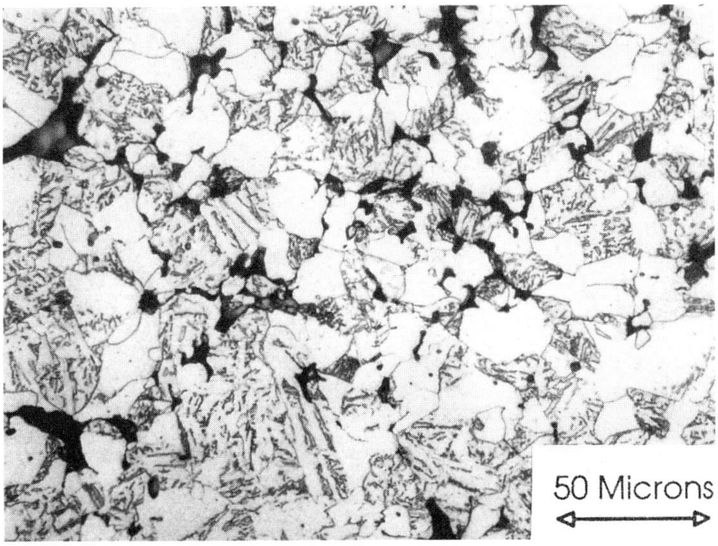

Body - 8 Minute Ultrasonic High-density
2% Nital Etch - Figure 5

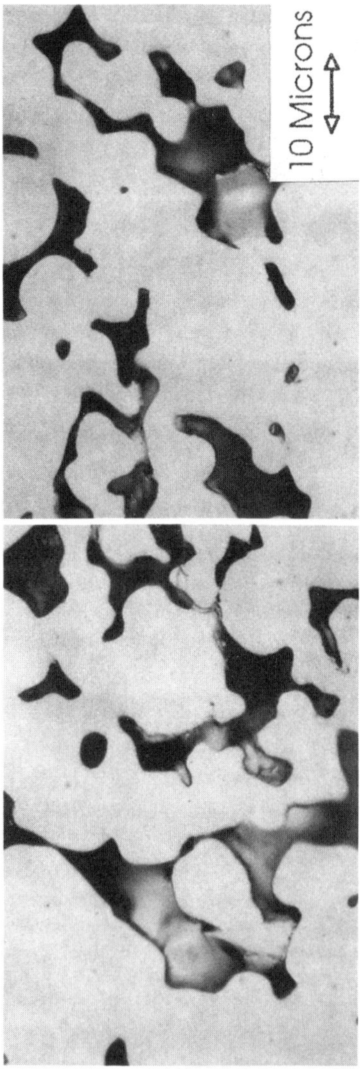

Figure 6. As polished

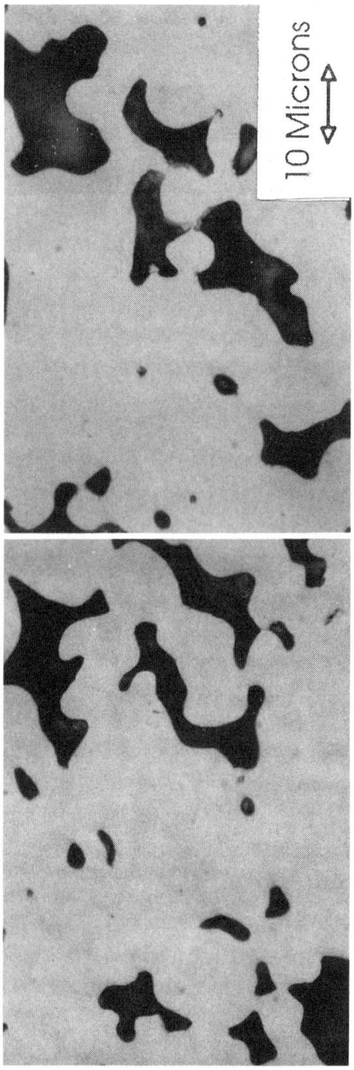

Figure 7. 4 minutes Ultrasonic

High-density P/M Gear

Tooth **Body**

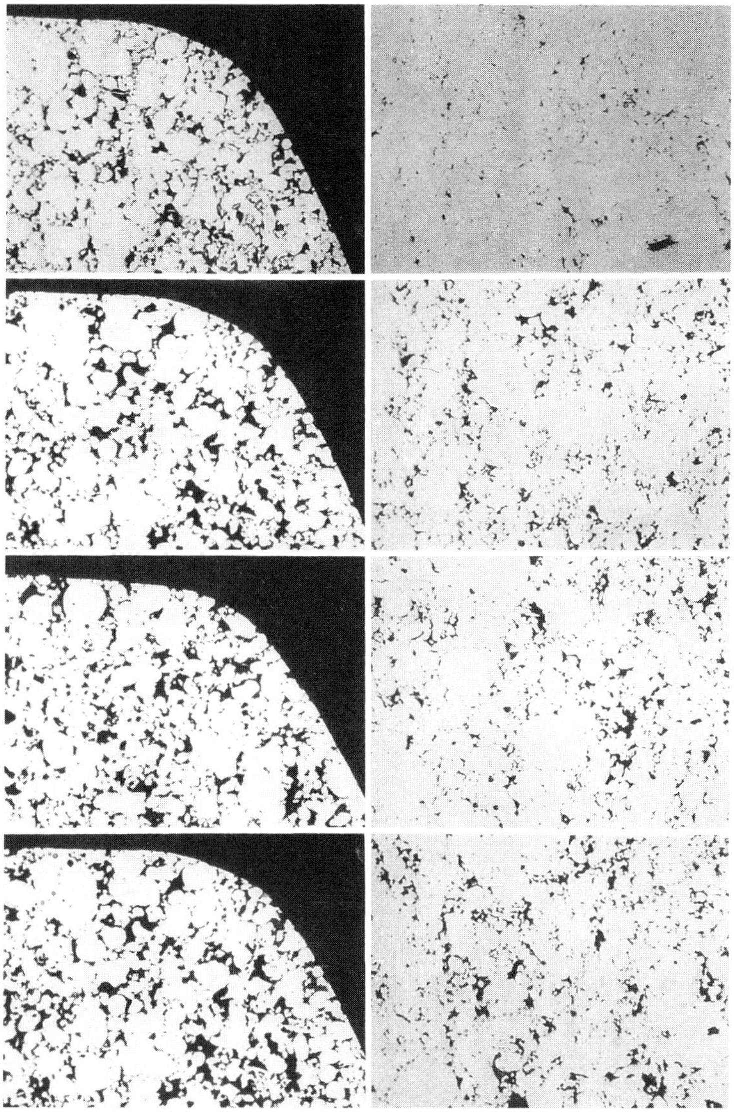

As polished

2 minutes

4 minutes

8 minutes

50 Microns

Ultrasonic Bath - Figure 9

Low-density P/M Gear

Tooth **Body**

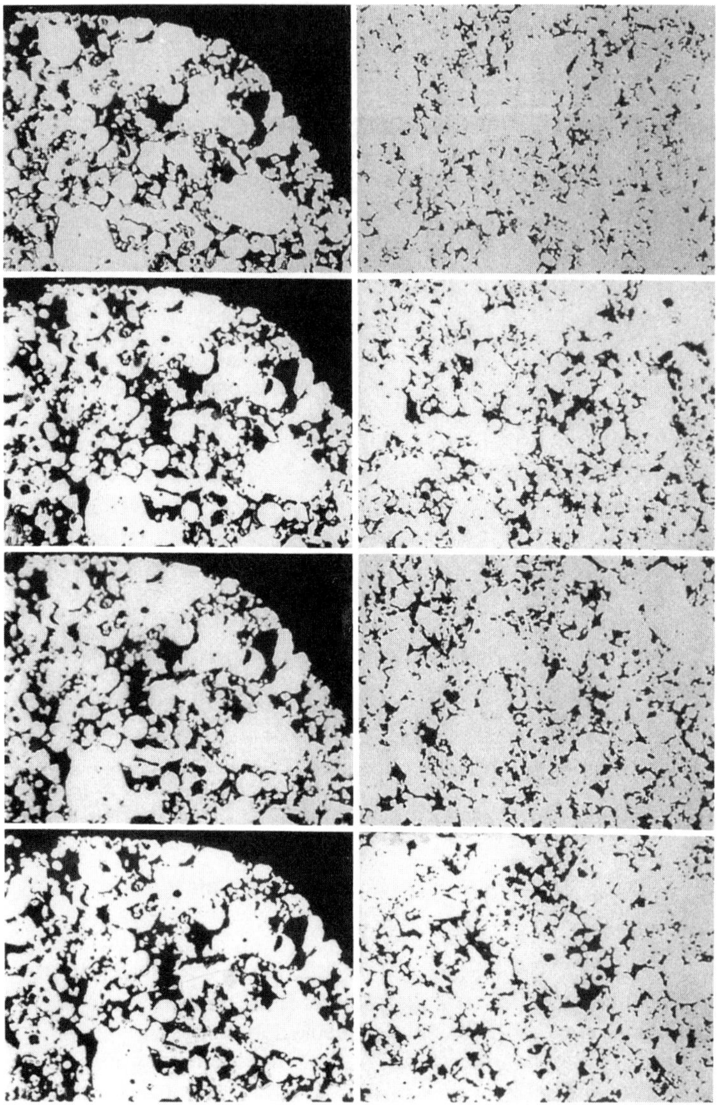

As polished

2 minutes

4 minutes

8 minutes

Ultrasonic Bath - Figure 10

50 Microns

A NEW METALLOGRAPHIC PROCEDURE FOR EDGE RETENTION OF ENCLOSED SURFACES

O. M. Katz*

ABSTRACT

Utilization of a low melting point, metallic alloy that expands on solidification has provided a reproducible means of preserving edges and accurately measuring deposits on enclosed surfaces mounted for metallographic study. In normal laboratory practice the electrically conducting mounting material generates no hazardous waste, needs no special equipment to prepare and is available commercially at a relatively reasonable price. Previous standard edge preservation techniques were found to be 90% inefficient from a time utilization view compared with the new procedure.

This new mounting procedure has greatly improved the quality and efficiency of microstructural studies of all materials, especially those on the inside diameters of heat exchanger components. These studies have included reaction products, shallow creep cracks and deposits on tubing, for which the procedure has proved indispensable.

BACKGROUND

In the 130 years since metallography was established as a means of studying microstructures, many techniques for specimen preparation have been devised[1]. Most recent progress has revolved around the use of high hardness, low shrinkage, thermosetting plastics cured rapidly under high pressure in standard mounting presses[2]. Recent attempts to preserve edges so that films, deposits and even substrate surface structure could be studied have resulted in numerous intricate mounting procedures[3]. One of these procedures involved the following steps: (1) slicing the alloy tube in half carefully to preserve the deposits on the inside diameter,

---

*Westinghouse Electric Corporation, West Mifflin, Pennsylvania 15122

(2) rotary gold coating the inside diameter in a vacuum evaporator, (3) copper (Cu) or nickel (Ni) electrolytic plating to build up a protective edge layer, (4) standard pressure mounting in a hard thermosetting plastic like Epomet (registered trade-mark of Buehler, Ltd.) at about 325°F for a few minutes and (5) polishing the mounted sample so as to portray intact areas where the edges were preserved. The plating material did not always retain adherence to the alloy and/or deposits on a particular cross-section at enough areas to warrant study, necessitating numerous cross-sectional preparations for a given tube sample. Figure 1 makes a general comparison with the new technique by showing a typical gap that resulted from the above procedure using a copper plating solution. Nickel plates produced similar results, and all of these processes produced liquid solutions that incurred disposal costs. Electroless deposits for us were even more unsatisfactory. Total sample preparation time was found to be a little over nine hours per mount.

Work to preserve the deposits and the surface features of tubing with a metallic material began with the use of Wood's alloy, which contains bismuth (Bi), lead (Pb), tin (Sn), and cadmium (Cd) and melts rather low at 158°F, considering potential heat-up during sample preparation. A number of others with 50 weight percent bismuth or less[4] were also available but either melted at equally low temperatures, contained hazardous elements like lead and cadmium or did not expand on solidification. All things considered, the procedure described herein utilized a eutectic alloy of 58 weight percent Bi and 42 weight percent Sn[5] which does expand on solidification just enough to retain all inside diameter deposits and features without apparent alteration.

PROCESS

The process consisted of the following steps:

(A) Chips from the as-received cast slab were heated in a pyrex beaker to about 50°F above their eutectic melting point of 281°F. Melting only what is needed limits the introduction of undesirable, excessive amounts of oxide dross in the casting. However, we have been able to skim the oxide dross to facilitate many reuses. The excess 50°F provides for good fluidity on casting.

(B) Whole tube sections from one-half to one inch long were held in a vertical position and the mounting alloy was cast into them. Usually the supporting base for the tubes was a poor thermal conductor to slow the freezing process down. Thin walled tubes from one-quarter to one inch diameter were mounted successfully, as were thicker walled sections with these same size holes.

(C) Samples whose outside surfaces needed careful preservation were also easily accommodated with this alloy mounting process by surrounding the sample with an appropriate diameter mold, three quarters of an inch high for standard automated met lab preparation.

(D) Mounting in some resin to facilitate metallographic preparation was not critical even though there has been great success with the Epomet process mentioned in the Background section. This process reaches a temperature above 281°F and remelts the alloy, but under controlled conditions in the high pressure resin mounting press. Slow cooling results.

(E) There appeared to be no special polishing precautions necessary even though cloths and techniques promoting minimal smearing and edge retention would be the best to use. Because the mounting alloy could be relied on to expand around practically 100% of the inside diameter circumference, it was usual to obtain all the data needed from a single, metallographic, cross-sectional preparation.

(F) Electrolytic etching was usually facilitated for thin walled samples because of the larger electrode contact area available with the metallic mounting alloy; insulating surface layers would, however, be a limitation in these cases.

(G) Chemical etching was easily performed also, but the greater attack of the bismuth/tin alloy mount in most cases meant that repreparation of the sample had to reinitiate at the grinding step to level the metallographic mount's surface.

(H) Long term studies of a given metallographic cross-sectional preparation should have no problems from the known dimensional changes of bismuth alloys on aging. The vendor (Indium Corporation of America, Utica, NY) provided data to show that this alloy was the most stable of all their alloys as far as growth and shrinkage after casting was concerned; however, some long term (>1 month) changes in the surface flatness of mounts has been observed.

RESULTS

Alloys of greater than 50 weight percent (wt. %) bismuth expand during solidification even filling the intricate crevices of corrosion deposits as seen in the lower photomicrograph in Figure 1. On the inside diameter of this tube, there were four distinct layers, whose compositions were analyzed by the energy dispersive x-ray analyzer (EDX) on the scanning electron microscope (SEM). Layer 4 was very interesting as it contained 46 wt % Bi and 21 wt % Sn as impregnated from the mounting alloy; analyses of the mount were fairly close to the expected 60 wt % Bi and 40 wt % Sn over larger areas. There was no indication of the mounting alloy in the other three corrosion deposit layers.

Figure 2 shows other tubes of various Inconel alloys which had grown layers on the inside diameters during testing. Light optical micrographs (LOM) at 500X and 1000X show the preservation of the edge of the layers although slight rounding of the mounting alloy produced some out-of-focus mount structure at the higher magnification. The SEM photomicrograph at 2000X further showed intimate contact.

Figure 3 shows micrographs typical of the inside diameter edge retention for zircaloy-4 (Zr-4) tubes. The eutectic structure in our alloy of choice, 58 Bi plus 42 Sn, has a relatively fine phase dispersion and was more resistant to pull-out and hence loss of edge retention than some of the other alloys. Possibly due to the improved wettability of indium, 67 Bi plus 32 IN was a good second choice of the alloys that were tried. This later compound usage was shown for a Zr-4 tube in the bottom of Figure 3.

Figure 4 compares two alloy 600 tubes from the same heat of material but drawn to different diameters which showed different grain size distributions due to different heat treatments. The mounting material retained both inside diameter (I.D.) and outside diameter (O.D.) edges and was not excessively attacked through this etching process.

Figure 5 shows results of additional application of the technique to high nickel-based alloy tubes using a 58 Bi plus 42 Sn metallographic mounting compound. An attempt was made to further improve the edge retention capabilities of the technique with hard ceramic balls and/or stainless steel springs against the tube surfaces. Although hardeners were not detrimental, their inclusion in the low melting point mount appeared superfluous. Similarly, the standard spring support and surface retainer was helpful from the handling aspect but superfluous if the Bi/Sn alloy was used. Fine surface cracks as shallow as a couple of microns were retained at the sample surfaces even though the fluidity of the Bi/Sn alloy was not quite sufficient to flow into them.

Data from 1992 showed that, in a period of high demand for tubing work especially, there was a manpower savings of over nineteen man-weeks.

CONCLUSIONS

(1) The old and new metallographic mounting procedures were compared as follows:

PROCESS TIME	OLD (hrs.)	NEW (hrs.)
CUT SPECIMEN	0.3	0
GOLD COAT	2.0	0
COPPER PLATE	6.5	0
CAST ALLOY	0	0.4
POLISH SPECIMEN	0.7	0.3
	9.5	0.7

(2) The new procedure utilizes a low melting point bismuth-based alloy, which expands on solidification, needs no special equipment to employ and creates no hazardous waste.

(3) The new process continues to offer significant manpower savings while being utilized for the metallographic study of even more intricately contoured samples.

ACKNOWLEDGEMENT

The following laboratory personnel aided in this development: A. Stinson, S. A. Vazquez, F. W. Page, W. Kinnen and D. L. Ward

REFERENCES

1) L. L. Wyman and R. J. Gray, Fifty Years of Progress in Metallographic Techniques, ASTM STP #430, 1966, p.1 to 62
2) J. A. Nelson and W. V. Ahmend, Microstructural Science, 1969, V. 17, ASM, IMS, Materials Park, Ohio, p. 217
3) G. F. VanderVoort, Metallography, McGraw Hill, New York, 1984, p. 71 to 93.
4) D. V. Miley and A. E. Colabra, "A Review of Specimen Mounting Methods for Metallograph", Metallographic Specimen Preparation: Optical and Electron Microscopy, Plenum Press, New York, 1974, p. 1-40
5) Metals Handbook; Properties and Selection of Metals, ASM, Materials Park, Ohio, Eighth Edition, V. 1, 1961, p. 863 to 864.
6) C. A. Hampel, The Encyclopedia of the Chemical Elements; Reinhold, New York, 1964, p. 58 to 65.

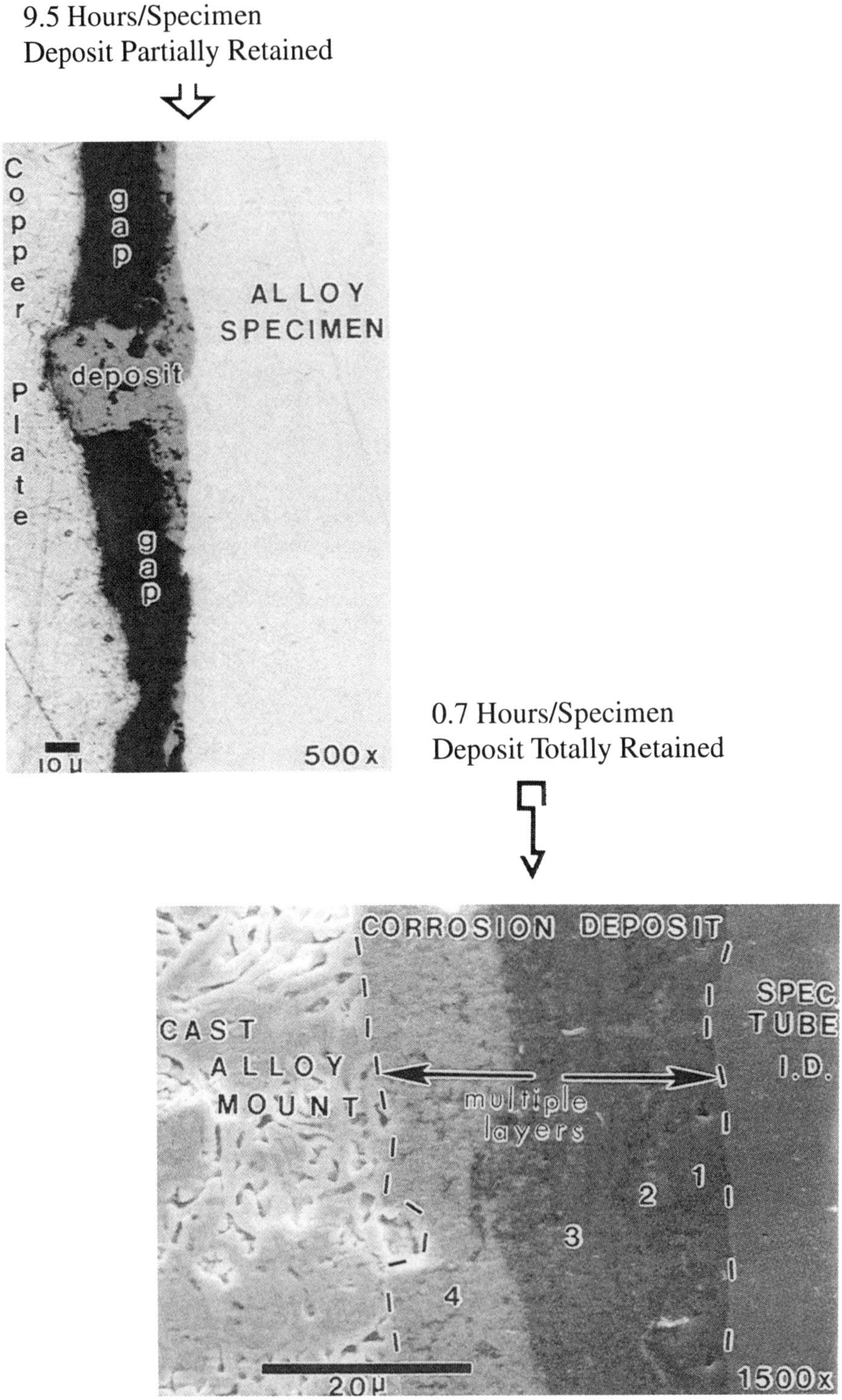

FIGURE 1 New Metallographic Preparation Technique Improves Quality and Efficiency

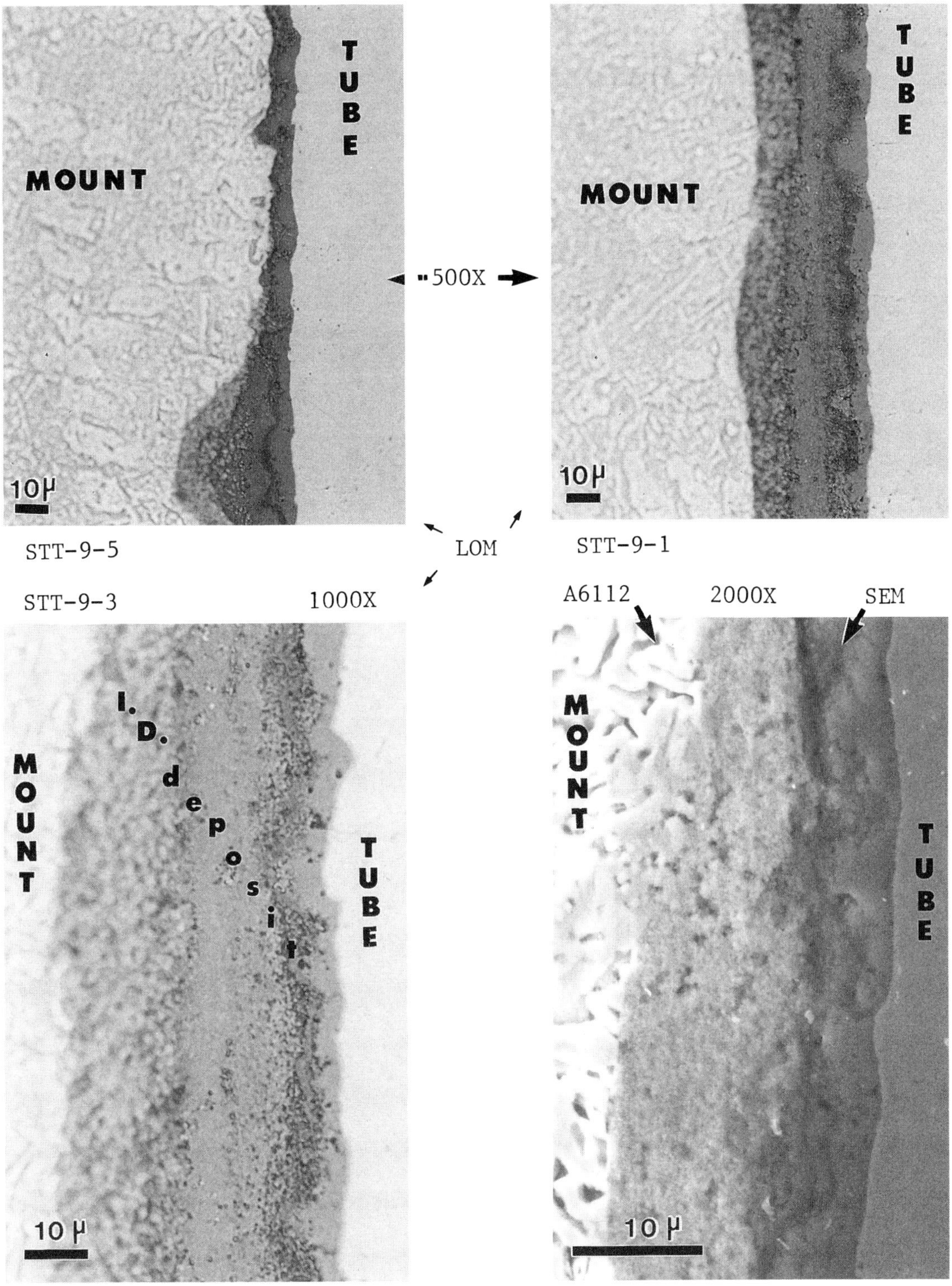

FIGURE 2 Preservation of deposit layers on the inside diameter of Inconel Alloy Tubes. Layer porosity decreases closer to tube surface.

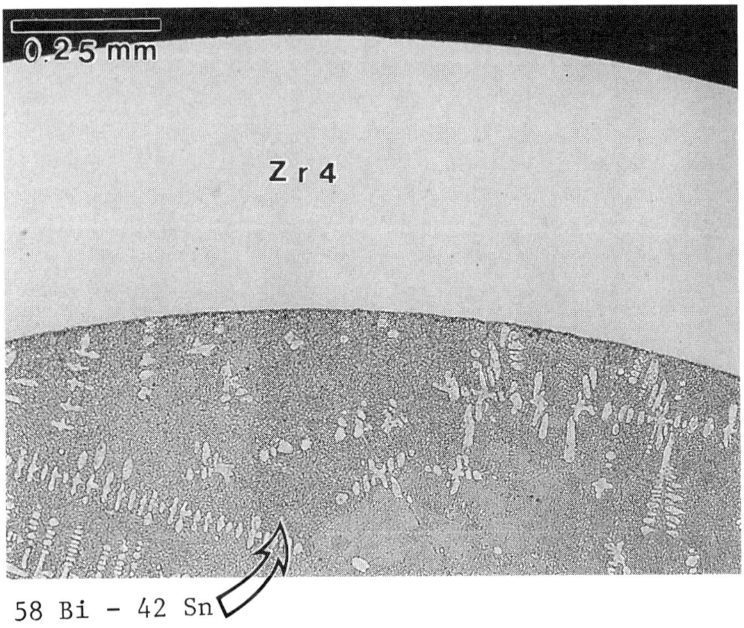

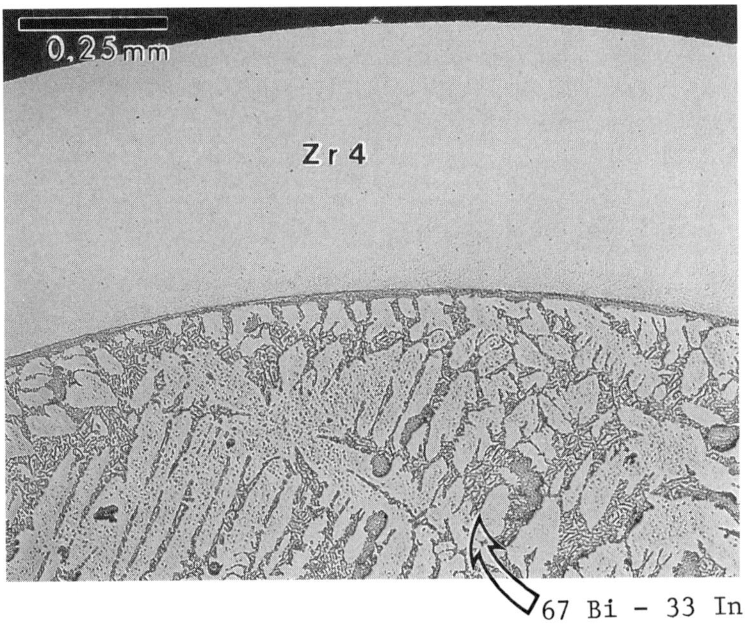

FIGURE 3 Very good edge adhesion and retention from high Bismuth, eutectic alloys used as mounting materials (80X). As polished tubes were 1/4" O.D. Zr-4 for which very dilute acid was used on final wheel.

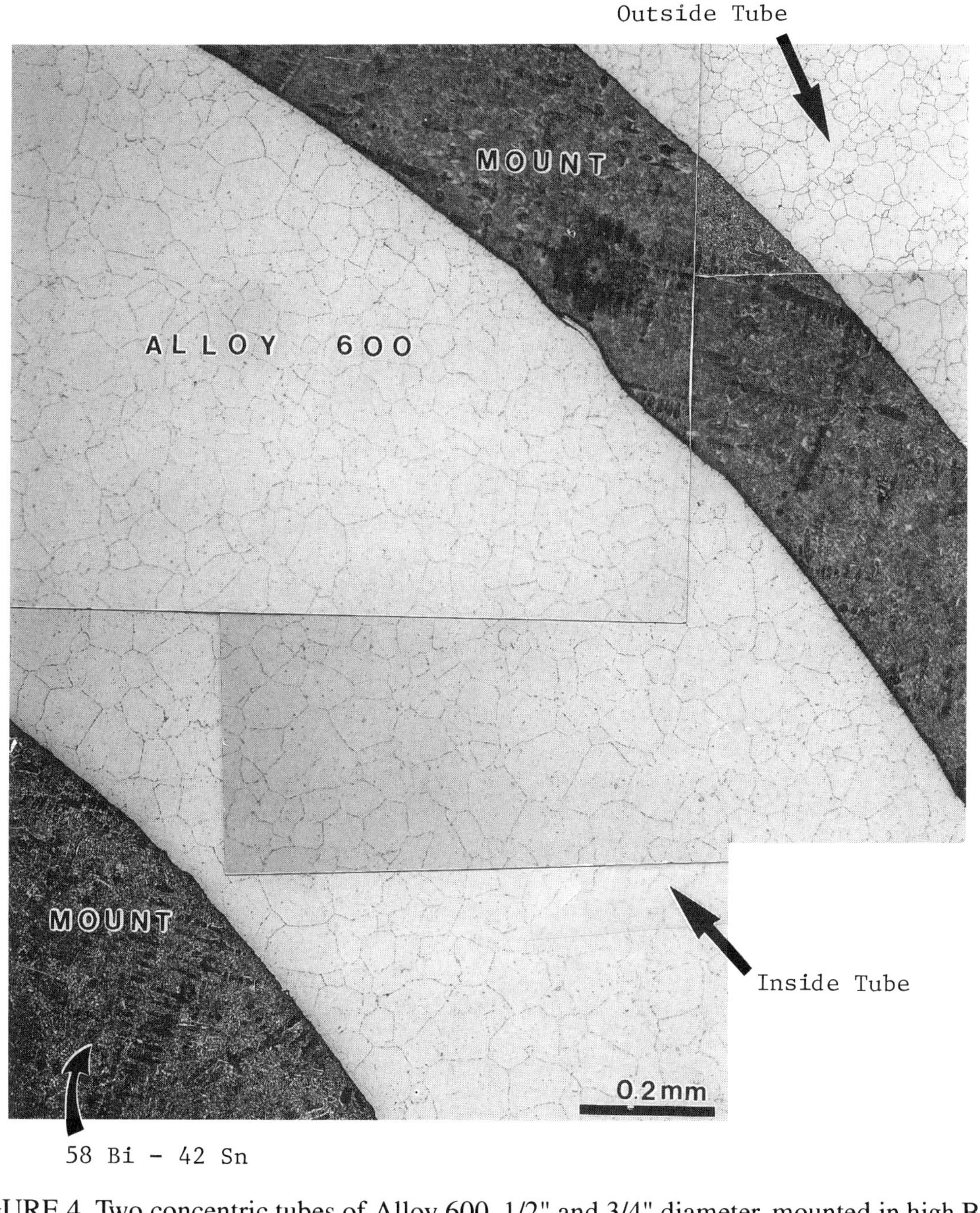

FIGURE 4 Two concentric tubes of Alloy 600, 1/2" and 3/4" diameter, mounted in high Bismuth alloy to preserve the I.D. edge around 360° of the circumference. Etched in 10% oxalic acid (100X).

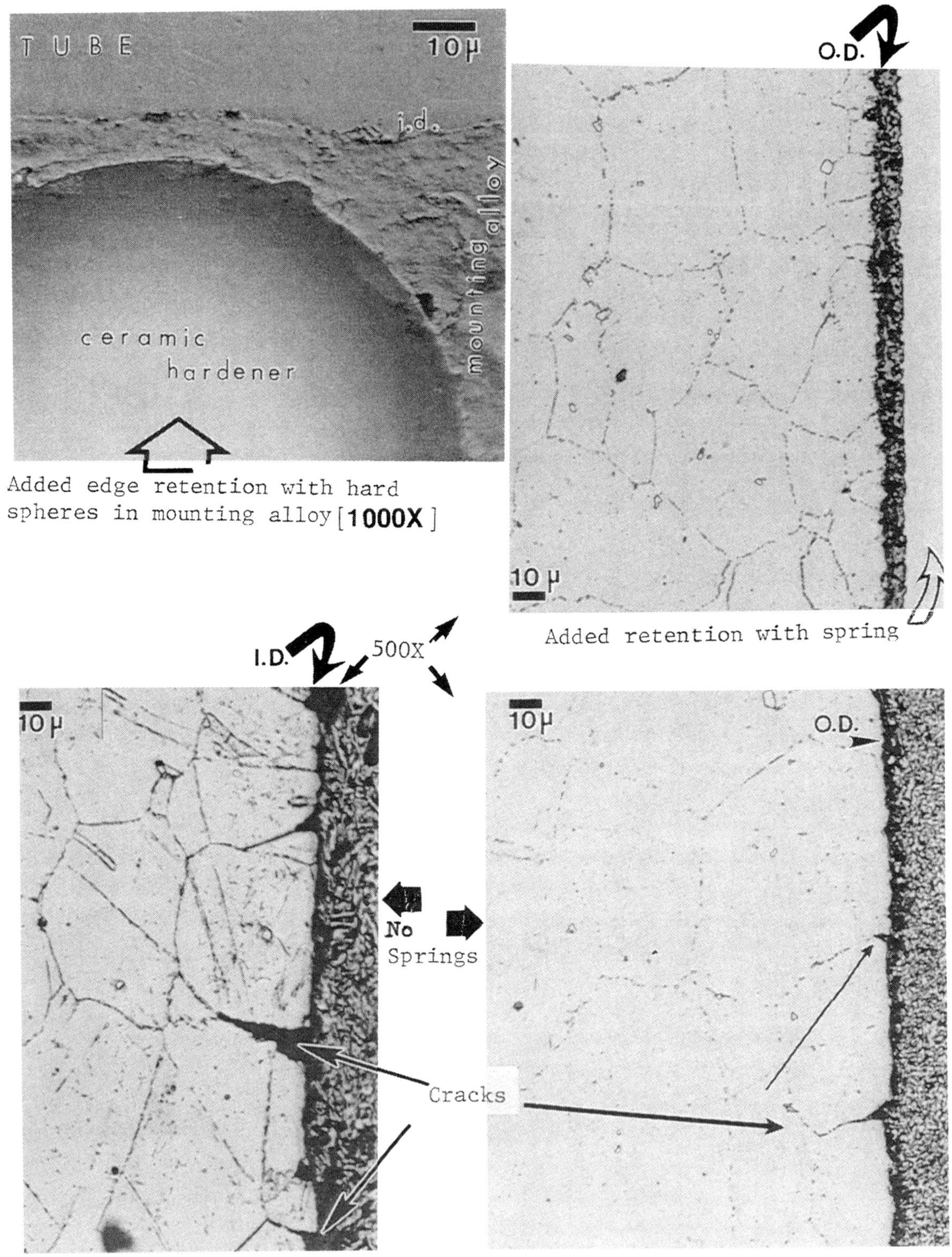

FIGURE 5 Tight adherence and protection of the surfaces of Inconel alloy tubes in high Bismuth alloy, 58 Bi - 42 Sn. Helped with hardeners and springs.

Analytical Threshold Settings For Image Analysis

Dennis W. Hetzner

ABSTRACT

One of the most common techniques for setting the gray level thresholds for phase discrimination using automated image analysis systems is the flicker method. This procedure involves switching back and forth between gray images and the detected image. The correct threshold setting is established by determining the gray level value where the detected phase and gray image appear to be identical. For best results several fields of view are observed to obtain an average specimen threshold setting. The major disadvantage of this technique is operator bias and poor reproducibility between different observers.

When considering the gray level histogram for a two phase system, there are two peaks in the distribution corresponding to each phase and a valley between the two peaks. Techniques are presented for mathematically representing the region between the peaks by a second degree polynomial equation. The proper threshold setting corresponds to the minimum of this curve. Similarly, the threshold setting can be found by determining the inflection point of a third order polynomial equation obtained from the cumulative gray level distribution function in the same region. Examples of applying these techniques to area fraction measurements of pearlite in plain carbon steel and delta ferrite specimens currently being evaluated by ASTM Committee E04 for metallography are discussed.

INTRODUCTION

The accuracy and repeatability of measurements made by using automated image analysis (I/A) systems is dependent upon numerous variables. The most common hardware variable that can influence these measurements is the optical microscope's illumination system. The response and resolution of the TV scanner used to convert the optical image into an electrical signal affect these measurements. These variables are fixed once a system

Timken Research Canton, Ohio 44706

is installed and operational. The quality and integrity of the specimens used for image analysis as well as the selection of the magnification to use when performing the analysis are of paramount importance. Once these concerns have been properly rationalized, the only remaining variable is what range of gray level should be selected to properly detect the phase or objects of interest in the appropriate specimen.

The most common method used to select the proper gray level range is the "flicker method" [1]. This procedure consists of switching back and forth between the live image and the detected image. The threshold limits are set so that the phase of interest appears to be correctly detected and the detected image does not appear to enlarge the phase boundaries. For objects such as sulfide or oxide inclusions, where their gray level or reflectivity is very dependent upon chemical analysis, it will be found that very little difference in the inclusion's gray levels will be observed among most specimens made by the same processing procedures. However, when chemical or electrolytic etching is required to reveal constituents of interest, somewhat large changes in gray level can occur between the same phases in two different specimens. In these cases preparation techniques may vary between specimens, and proper gray level thresholding procedures are required.

Another important factor that must be considered is the method by which the I/A scanner converts the optical image into a TV image. The scanner converts the image down into a matrix of small square pixels, Figure 1. The higher the density of pixels per unit area, the better is the scanner resolution. Consider the idealized dark gray square illustrated in Figure 1. Since the square completely fills each pixel containing it, a six bit gray level histogram of this object would have 25 pixels at the gray level corresponding to its reflectivity (23) and all the other pixels would be at a gray level of 63 (0=black, 63=white), Figure 2a. The cumulative gray level distribution for this object has a large plateau

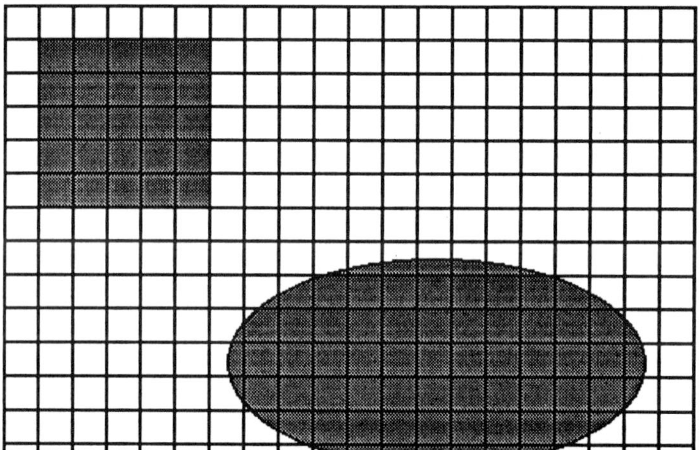

Figure 1. Uniform gray level square and ellipse.

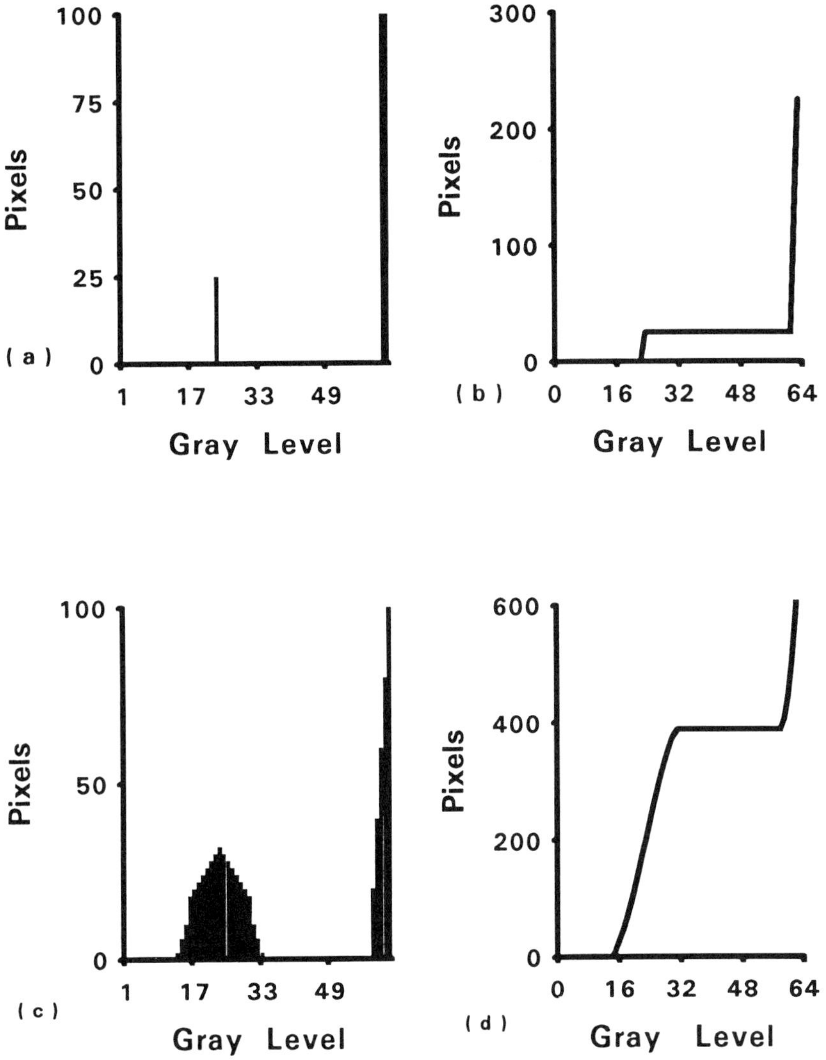

Figure 2. (a), (b) Gray level histogram and cumulative gray level distribution for the ideal square; (c), (d) ellipse histogram and cumulative gray level histogram.

extending from 23 through 63. Thus, any threshold setting including a portion of the plateau would properly measure the square. This set of circumstances would never exist in reality.

Now consider an ellipse having only one gray level, Figure 1. Clearly every pixel contained inside the ellipse has the same gray level. However, this is not the case for the pixels on the perimeter. Some of these pixels contain most of the ellipse while others contain virtually none of the ellipse. In this case the gray level distribution will tend to be spread out, Figure 2c. The corresponding cumulative gray level distribution, Figure 2d, will still contain a plateau, but its span will contain fewer gray level bins than the idealized square. This response has been reported for sulfide inclusions in steel [2].

For most metallographic specimens there is not a transition between the gray levels of the matrix and other phases. The gray level distribution is composed of two overlapping peaks, Figure 3a. Similarly, the cumulative gray level distribution may not contain a plateau but does appear to have an inflection point, Figure 3b. When these conditions are met, point X_0 corresponds to the proper threshold setting. This point is either the relative minimum of the gray level distribution curve or the inflection point of the cumulative gray level distribution curve. Analytical methods to use to determine this point will be discussed.

EXPERIMENTAL METHODS AND PROCEDURES

The specimens used to illustrate the procedures for analytic threshold settings were E-Beam micro-fabricated circles, pearlite in 1018 steel and four materials currently being used in an ASTM Committee E04 for metallography series of round robin tests, Figure 4. The E-Beam specimens were chosen because they can be considered to correspond to the ideal ellipse previously discussed. The hypoeutectoid steel is a specimen that is easy to prepare and this type of alloy is commercially available. The brass and delta ferrite specimens being evaluated by Committee E04 were selected because the results obtained by using these methods can be compared to the results obtained by a relatively large group of other labs using different I/A systems and standard threshold setting techniques.

The specimens were mounted in Bakelite, ground and polished using automated machines. The 1018 steel specimen was etched with 2% nital, and the CDA brass specimen was etched with NH_4OH plus H_2O_2. The stainless steel specimens containing delta ferrite were electrolytically etched in a solution containing 40 gm of NaOH and 100 ml. of H_20 for 30s at 3v.

Image analysis was performed with a Cambridge/Leica Q970 Quantimet equipped with a Chalnicon scanner and a Leitz Orthoplan microscope. An 80X objective lens and 400 fields of view were used for the MOT specimen. In all other cases 64 fields of view and a 32X objective lens were used for the field measurement. For each specimen, a gray level histogram was constructed from ten separate fields of view. Six bit gray level histograms were obtained by standard operating procedures. Customized Pascal software was developed and integrated into the user functions menu to obtain eight bit gray level histograms.

For these specimens, the gray level histograms contained two peaks and a relative minimum between the peaks. Letting X_i and X_j represent gray levels below and above the relative minima, X_0, this portion of the histogram can be represented by a least

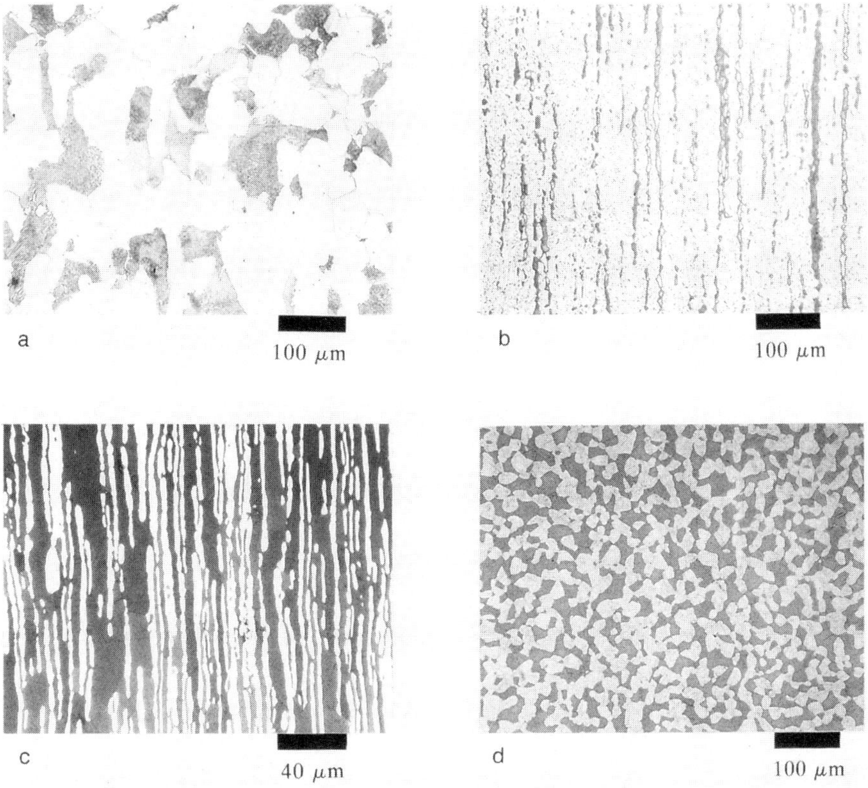

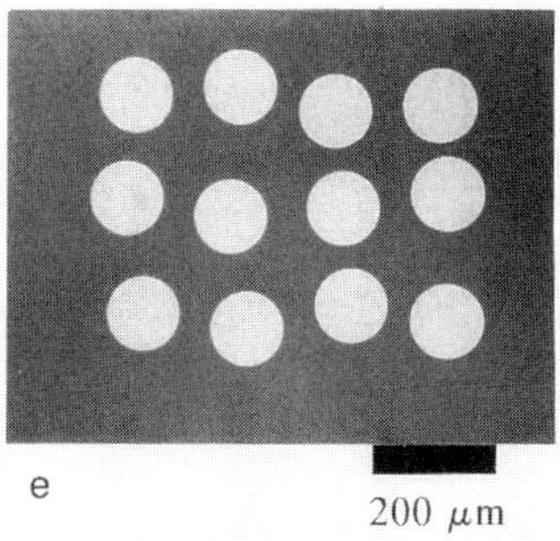

Figure 4. (a) 1018 Steel, (b) AM 350, (c) MOT, (d) CDA Brass, (e) E Beam Circles

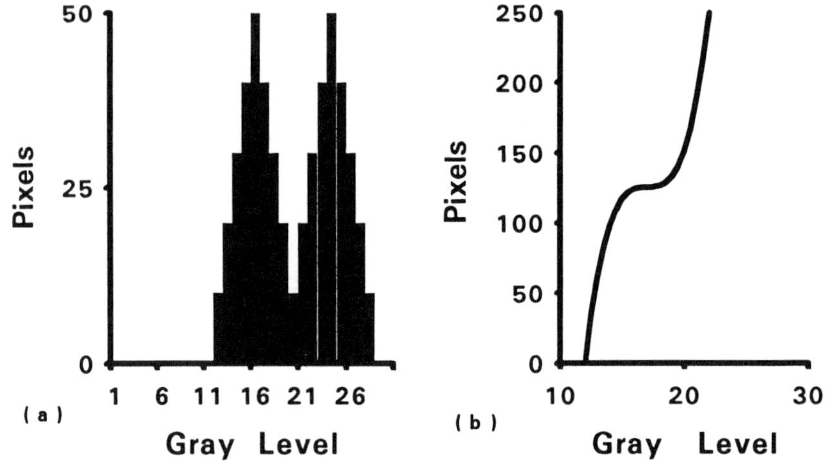

Figure 3. (a) Idealized gray level histogram; (b), Cumulative gray level distribution function.

squares fit polynomial equation of the form

$$y = a_0 + a_1 x + a_2 x^2 \quad (1)$$

The minima can be found by setting the first derivative equal to zero; i.e.,

$$X_0 = \frac{-a_1}{2a_2} \quad (2)$$

Similarly, the cumulative gray distribution can be represented by

$$y = a_0 + a_1 x + a_2 x^2 + a_3 x^3 \quad (3)$$

In this case, the proper threshold setting corresponds to the inflection point of the curve, i.e., the value of X_0 for which the second derivative is zero.

$$X_0 = \frac{-a_2}{3a_3} \quad (4)$$

This is the procedure used to establish the threshold settings for the specimens evaluated in this investigation. The data analysis was performed with a DEC VAX 60420 Vector computer system. A generalized Pascal least squares curve fitting program having linear coefficients similar to that described by A. R. Miller [3] was used to calculate the equations representing the gray level histograms. For best accuracy, the program was compiled using extended precision.

RESULTS

When observed visually, subtle gray level differences in metallographic specimens are generally not considered to be of much importance; however, when attempting to accurately determine proper threshold settings, these variations in gray level can have a significant effect on which things are measured and which things are not measured. The video response of the scanner can lead to changes in gray levels as the beam rasters across a specimen. For the Chalnicon scanner employed in this work, as a line scans across a dark background onto a bright E-beam circle, overshoot occurs. The brightness value near the edge of the circle is considerably higher than the interior of the circle, Figure 5a and b. (The horizontal scan line corresponds to 126 on the y-axis.) While the gray level of the circle appears visually uniform, some fluctuation above and below the mean gray level is detected by the scanner. Finally, as the scan line leaves the circle and moves back into the dark background, undershoot occurs prior to achieving a mean dark level. The same type of response is observed for a horizontal line that is scanned across the alternating dark and light regions in an AM350 specimen, Figure 5c and d.

The eight bit gray level histogram for the E-beam circles appears to contain two large, widely separated peaks, Figure 6a. Closer examination of the gray levels from 100 through 200 reveals a relative minima between the peaks at 143, Figure 6b. Similarly, when at least squares curve is fit for gray levels from 100 through 200, the cumulative distribution curve has an inflection point at 142, Figure 6c. This value would correspond to the proper threshold setting.

A six bit gray level histogram for the hypoeutectoid steel has a small peak at a gray level of 18. This corresponds to the dark portion of the pearlite colonies. The large peak at a gray level of 50 corresponds to the white ferrite phase, Figure 7a. There is a relative minimum in the gray level range of 20 through 45, Table 1 and Figure 7b. Using these data points, the curve can be expressed as

$$y = 20267 - 1116.6\,X + 18.318\,X^2 \qquad (5)$$

and the relative minimum is $X_0 = 30.5$. The cumulative gray level distribution can be represented by

$$y = 222200 + 19829X - 544.01\,X^2 + 5.9677\,X^3 \qquad (6)$$

which has an inflection point at $X_0 = 30.4$. In each case the upper detection threshold would be less than 31. Visual observation indicates a great deal of scatter in the data points in the histogram when compared to the cumulative distribution curve. The correlation coefficient, r^2, for the second degree polynomial is only 0.4714; $r^2 = 0.9958$ for the cumulative gray level distribution cubic polynomial.

Threshold analysis for the delta ferrite phase in the AM 350 specimen was performed with 6 bit and 8 bit gray level histograms, Figure 8 and 9. For the 8 bit histogram, the upper threshold was 136.5, Figure 8b. The upper threshold was 136.2 for the cumulative distribution, Figure 8c. For the six bit gray level histogram and cumulative distribution, the upper thresholds were 34.2 and 34.1, respectively. All these values are in excellent agreement with each other, especially considering that the six bit equivalent of 136 is 34.

The benefit of this methodology for determining the threshold setting becomes apparent when considering the preliminary results of an ongoing ASTM round robin test for this specimen. The mean delta ferrite value was found to be 12.97% when the outliers were removed, (■) Figure 10. Using the 6 bit gray level threshold, 13.22% delta ferrite was measured, (◆) Figure 10.

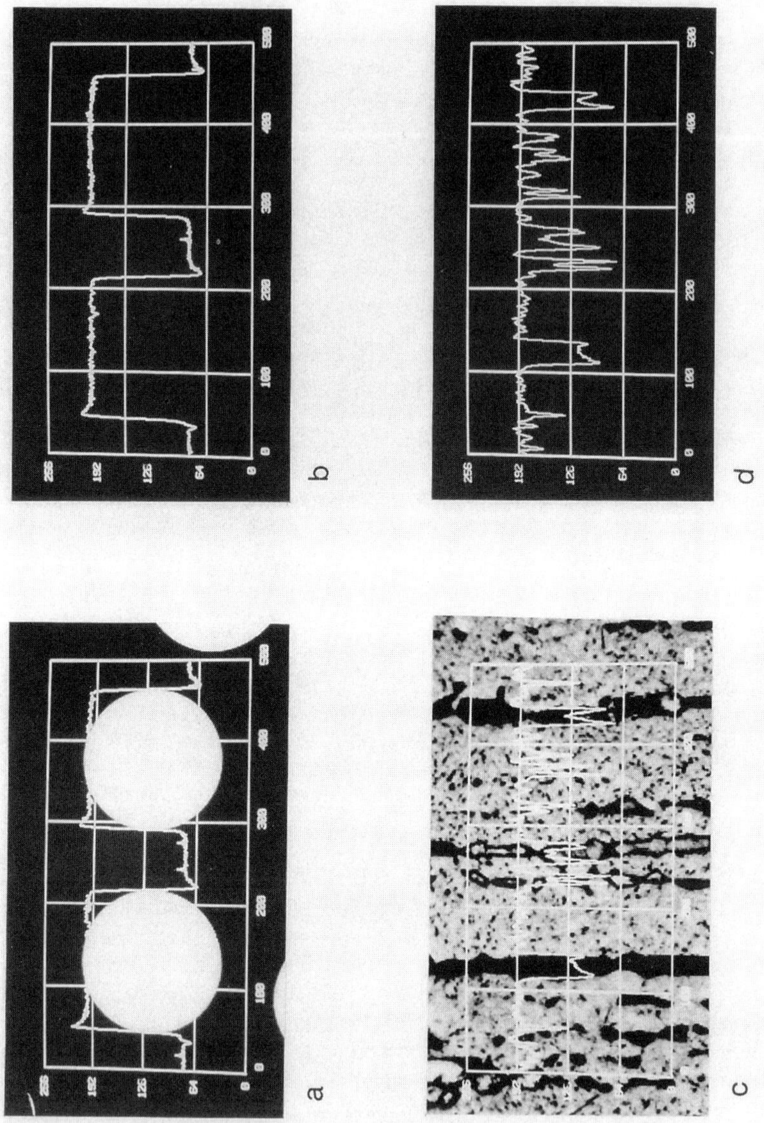

Figure 5. Detector Response. (a,b) Circles, (c,d) AM 350.

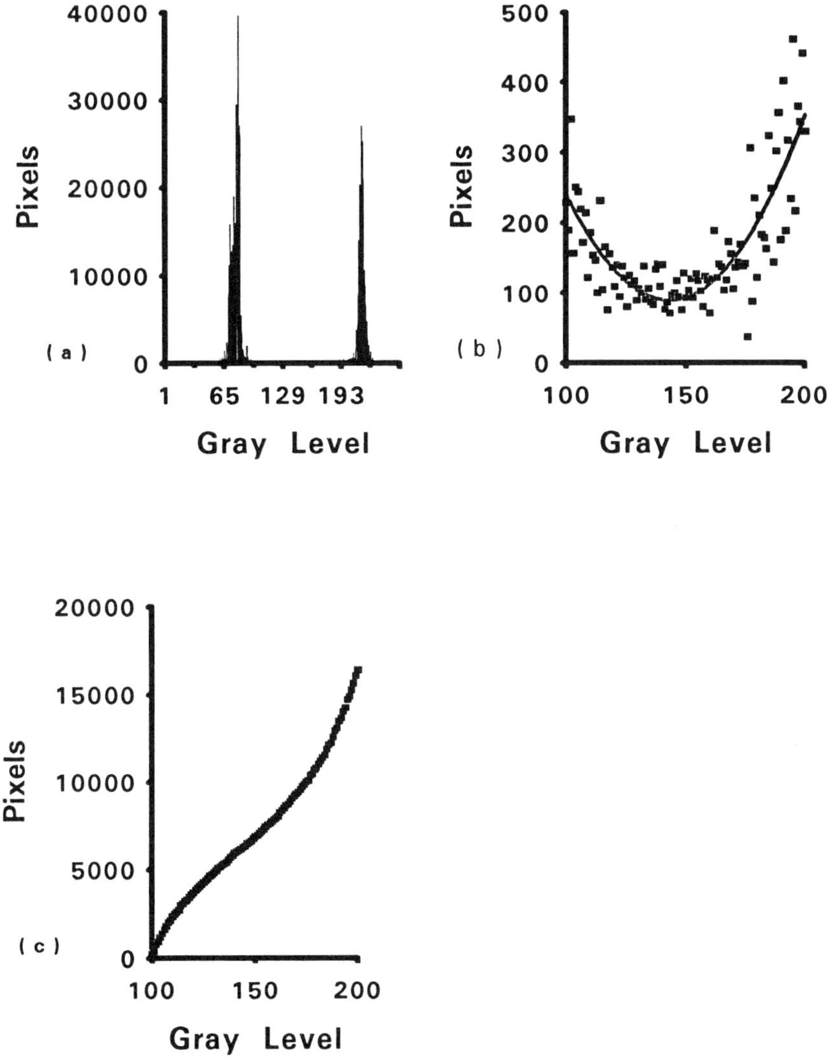

Figure 6. E beam circles; (a), Eight bit gray level histogram; (b), Second degree polynomial; (c), Cumulative distribution least squares

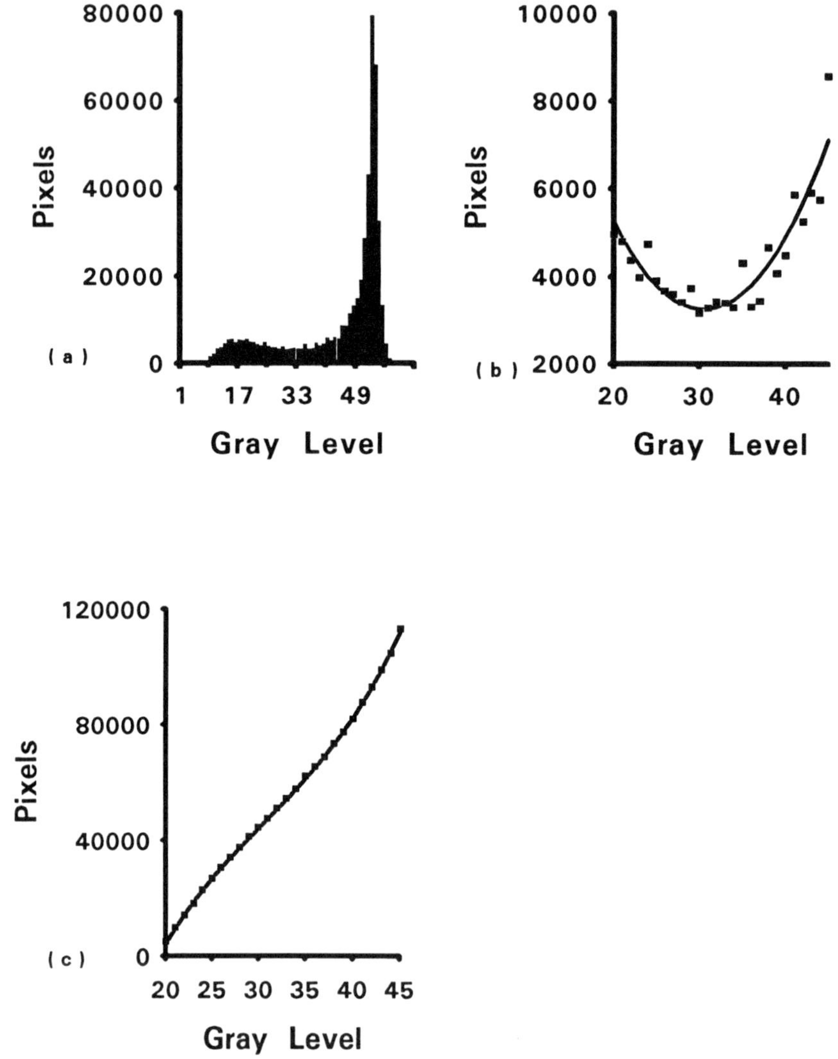

Figure 7. Hypoeutectic steel; (a), Six bit gray level histogram; (b), Second degree polynomial; (c), Cumulative distribution.

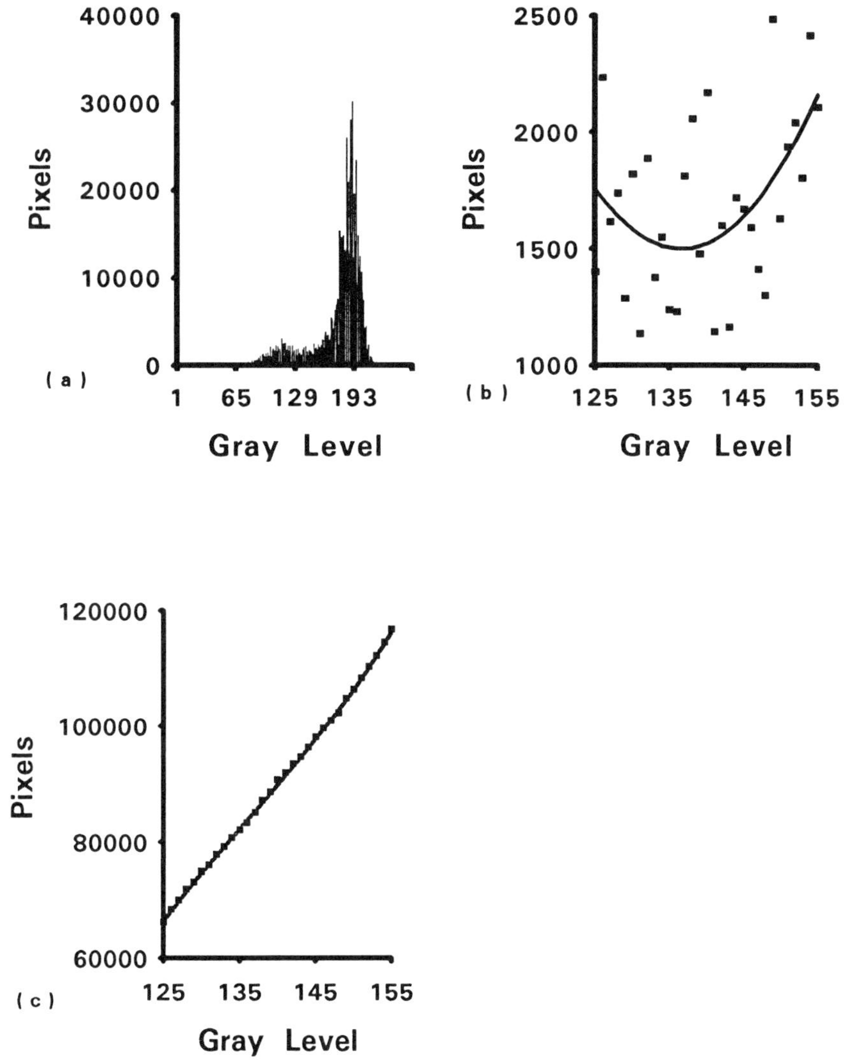

Figure 8. Specimen AM350; (a), Eight bit ray level histogram; (b), Polynomial curve fit; (c), Cumulative distribution.

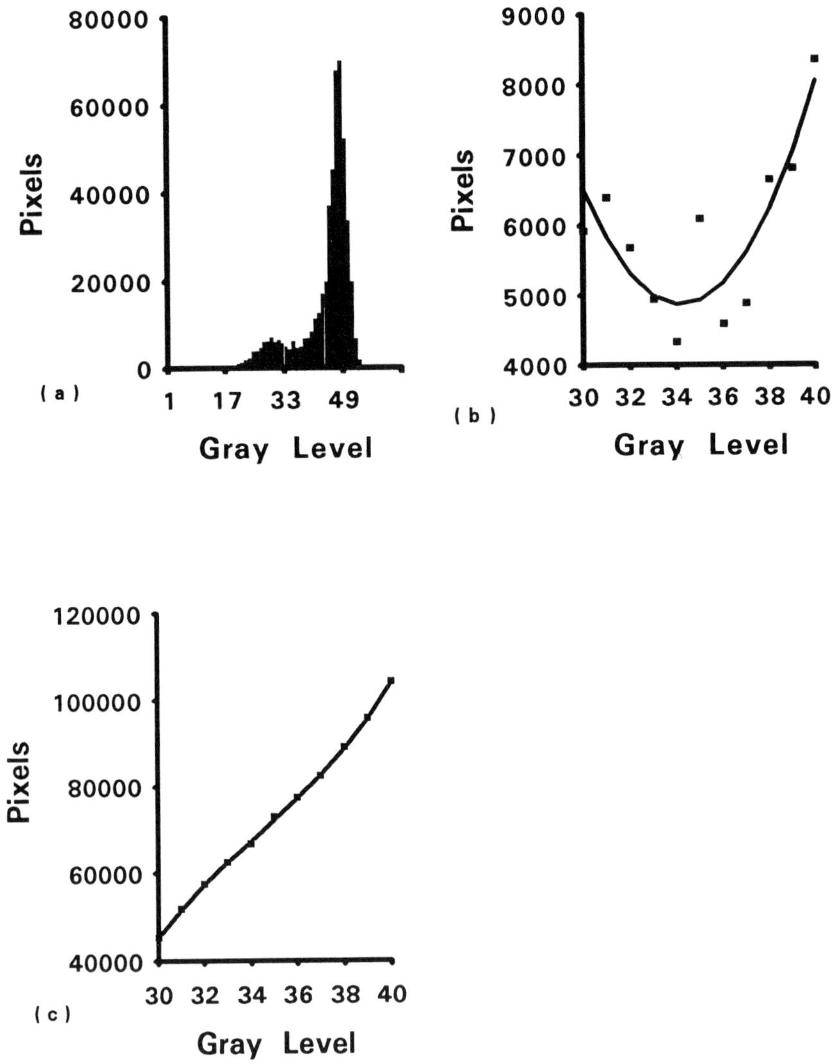

Figure 9. Specimen AM350; (a), Six bit gray level histogram; (b), Second degree least squares curve fit; (c), Cumulative distribution.

For the 8 bit gray level threshold, 13.41% delta ferrite was measured (▲). The standard deviations for these measurements were 2.60% and 2.30% respectively.

Table 1. 1018 Steel Gray Level Histogram

Gray Level	Pixels	Gray Level	Pixels
20	4950	33	3392
21	4789	34	3283
22	4365	35	4310
23	3978	36	3304
24	4734	37	3440
25	3906	38	4659
26	3669	39	4083
27	3590	40	4494
28	3400	41	5858
29	3731	42	5254
30	3154	43	5901
31	3267	44	5730
32	3430	45	8558

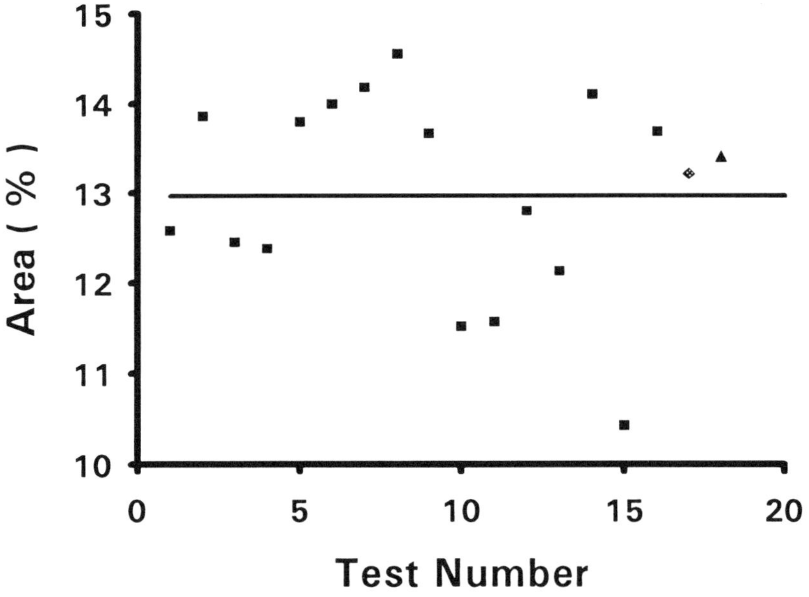

Figure 10. Comparison of δ ferrite measurements; (■) ASTM round robin test, (◆) 6 bit gray, (▲) 8 bit gray.

Dark etching delta ferrite was the major phase observed in specimen 755, Figures 4c and 11a. Using the six bit gray level histogram, the upper threshold was found to be 39.3, Figure 11b. The measured delta ferrite content was 72.5 %.

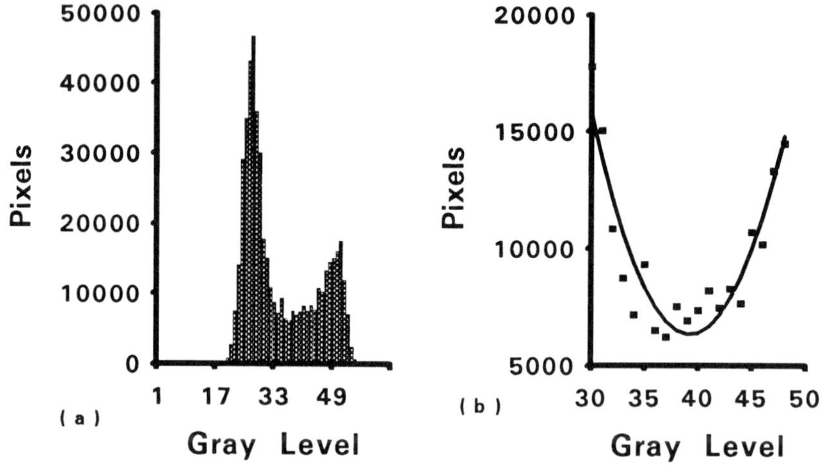

Figure 11. Specimen SS 755; (a), Six bit gray level histogram; (b), Second degree polynomial curve fit.

Using a standard green filter with the CDA brass specimen, a large degree of overlap between the gray level peaks of the α and β phases was observed, Figure 12a. However, by using a red filter, the overlap was greatly reduced, Figure 12b. For this specimen the dark β phase was measured. The upper threshold for a six bit gray image was 45, and the mean area and standard deviation were 50.0% and 1.72%, respectively. The upper threshold for the 8 bit gray image was 180, Figure 13. The measured area was 49.4% and the standard deviation was 1.70%.

The final specimen that was analyzed was the MOT stainless steel. A large valley was observed between the two gray level peaks, Figure 14a. In this case the upper threshold setting using the 6 bit gray image was determined to be 35.5 from the second order polynomial and 34.4 from the third degree polynomial for the cumulative distribution. The mean measured area was 60.19% and the standard deviation was 7.22%. A summary of these results is contained in Table 2.

DISCUSSION

The problems of scanner overshoot and undershoot that occur as the beam moves from a light phase to a dark phase clearly reveal one of the problems that can be encountered in accurately determining threshold settings, Figure 5. The normal slight variations in gray level observed in this example show another variable that can influence threshold settings. These factors help to explain the critical importance of specimen preparation for Image

Analysis Applications. Artifacts introduced from improper polishing or etching can lead to problems when Image Analysis is performed. Even for well prepared specimens, changes in the color or type of illumination can greatly affect the gray level histogram. This was clearly shown when CDA brass was illuminated with red and green filtered light, Figure 12.

Table 2. Field Measurements

Specimen	Bits	Mean Area (%)	Sdev (%)
1018	6	17.3	1.43
AM350-1	6	13.22	2.60
	8	13.41	2.30
SS 755	6	72.5	4.54
CDA Brass	6	50.0	1.74
	8	49.4	1.70
MOT	6	60.19	7.22

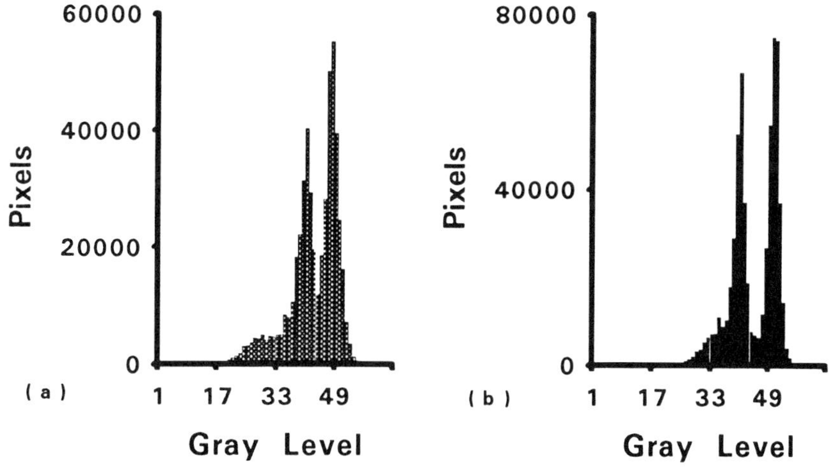

Figure 12. CDA brass; (a), Green filter; (b), Red filter.

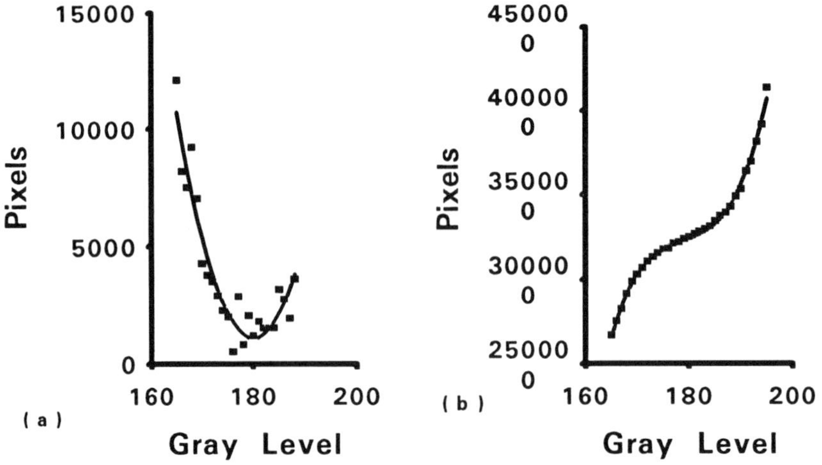

Figure 13. CDA brass, red filter; (a), Second degree polynomial curve; (b), Cumulative distribution.

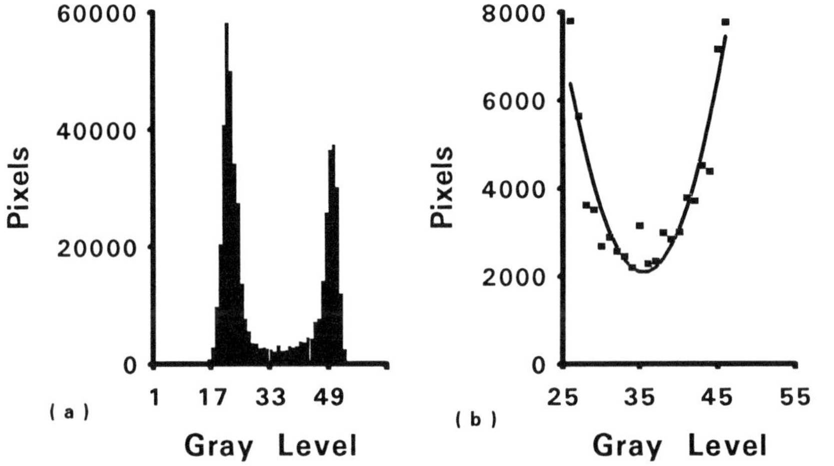

Figure 14. MOT stainless steel specimen; (a), Six bit gray level histogram; (b), polynomial curve fit.

The power of this technique is the removal of operator bias in determining the appropriate threshold settings. The excellent agreement for the results obtained from the AM 350-1 specimen using this methodology when compared to the results obtained from the round robin testing illustrate this point. On the basis of the examples cited, a better curve fit is obtained from the cumulative gray level distribution than from a curve fit to the data points between the peaks of the standard gray level histogram.

The reason for selecting the relative minimum in the gray level distribution or the inflection point in the cumulative gray level distribution as the threshold point was the uniqueness of this point. Mathematically it is the point where the first or second derivative is zero. By viewing the gray level histogram, a fair guess of the location can be made. For an eight bit gray image, a region of interest of 25 gray levels above and below this guess can be used for the analytic determination of the threshold setting. If the threshold setting is too far from the first trial, a second region of interest can be set up around the new point. Several iterations can then be used to precisely set the threshold. The least squares fit correlation coefficient and the standard error of the estimate can be used as measurements of the quality of fit of the polynomial equation.

The specimens evaluated in this investigation contained a relatively large area fraction of the phase being detected. This was purposely done so as to easily illustrate how to perform the techniques that were presented. In situations where only a small area fraction of a constituent is present, some modifications to these procedures can be implemented. First, the measuring frame size can be reduced so that the constituent of interest is dominant. Then a large number of the small frames can be used to create an appropriate gray level histogram. Second, a large number of line scans can be used to determine the proper gray levels to use for phase discrimination, Figure 5.

In this study the data analysis was not made with the I/A system computer. However, with the more advanced systems presently available, the entire analysis procedure could be interactively performed by using a mouse to select the regions of interest and the I/A system to fit the curves.

CONCLUSIONS

1. A new and powerful analytical technique to accurately determine threshold settings for Image Analysis was presented.
2. Using these procedures, good agreement with a group of specimens evaluated by other laboratories was obtained.

ACKNOWLEDGMENTS

The author would like to thank The Timken Company for permission to publish this paper, and to Saralyn A. Stevens for preparing the manuscript.

REFERENCES

1) ASTM Specification E 1245-89, Standard Practice For Determining Inclusion Content of Steel and Other Metals by Automatic Image Analysis.

2) D.W. Hetzner, "Quantitative Image Analysis Methodology for High-Sulfur Calcium-Treated Steels," *MiCon 90: Advances in Video Technology for Microstructural Control, ASTM STP 1094*, G.F.Vander Voort, Ed., American Society for Testing and Materials, Philadelphia, p.151, 1990.
3) A.R. Miller, *Pascal Programs For Scientist and Engineers*, SYBEX Inc., Berkeley, CA, 1981.

QUANTIFICATION OF PERIODIC PROPERTIES

IN THE MICROSTRUCTURAL ANALYSIS OF MATERIALS

S. John Louisnathan[1]

ABSTRACT

The statistical autocorrelation function is a powerful tool in classifying and quantifying microstructures with periodic properties. A simple and fast image analysis algorithm, herein called as the image autocorrelation, is revisited. The results of the image autocorrelation function (*IACF*) are compared with those of the conventional computational autocorrelation function (*CACF*). Using the data of the *IACF* of series of computer generated images, where the orientation and the spacing of microstructural elements vary from highly ordered to disordered tessellations, a single parameter, Δacf, is derived which appears to satisfactorily quantify the degree of preferred orientation. The *IACF* and *CACF* methods were used early in the development of fiber reinforced glass composites to understand the relationships between processes, properties and microstructure. Examples from this study are discussed to demonstrate the power and usefulness of autocorrelation function in the general area of image analysis for microstructure quantification.

INTRODUCTION

Periodic repetition of certain structural elements within a microstructure is a common phenomenon. Periodic properties in a microstructure are sometimes desirable and therefore one attempts to maximize that part of the process which introduces the periodicity. At other times, one may want the product to have isotropic properties, in which case a part of the process that inhibits periodicity is optimized. Image analysis is a powerful tool for

[1] Corning Incorporated, R,D&E Division, Corning, NY 14831

quantifying periodic properties in the microstructures of materials. Such a quantification then provides the directions for improving the process in order to obtain materials of desired physical properties

This paper briefly reviews the computational autocorrelation function (*CACF*) that is often used in time series analysis of stock market forecasting as well as in the general area of spectral analysis [1]. It then introduces the algorithm for image autocorrelation function (*IACF*). The algorithms for applying both *CACF* and *IACF* in the image analysis of microstructures is outlined. Using a series of computer generated images, where the orientation and the spacing of the microstructural elements vary from ordered to disordered tessellations, a preferred orientation parameter, called the Δacf, is defined which quantifies and classifies preferred orientation in a series of related microstructural bodies. Finally, examples of image analysis application of *IACF* and *CACF* for refining the processes of manufacturing silicon carbide fiber reinforced glass composites (*FRC*) are given [2, 3]

COMPUTATIONAL AUTOCORRELATION FUNCTION (*CACF*)

The correlation between successive measurements of image features observed at an interval or spacing $\boldsymbol{k}$ apart is quantitatively quantized by the *lag k sample autocorrelation coefficient*, given by [4],

$$r_k = \frac{\sum_n (y_i - \bar{y})(y_{i+k} - \bar{y})}{\sum_n (y_i - \bar{y})^2} \qquad \text{(EQ 1)}$$

where y_i is an array of some measured image analysis parameter, and $\bar{y}$ is the average value of the array, and the summations are taken over all $\boldsymbol{n}$ available products and squares. The ensemble of all $\boldsymbol{r_k}$, $f(r)$, is then the autocorrelation function.

IMAGE AUTOCORRELATION FUNCTION (*IACF*)

The correlation between an entire image $\{\mathbf{A}\}$ and successively shifted image of itself $\{\mathbf{A}\}_k$, for all $\boldsymbol{k}$-interval of shifts, is given by,

$$m_k = (m_i)\, of \{A\} \bullet AND \bullet \{A\}_k ; \alpha \qquad \text{(EQ 2)}$$

where $\boldsymbol{m}$ is a measured value, such as the area of the image, as the image $\{\mathbf{A}\}$ is sliding across itself in the angular direction α, logically *AND*-ing, and computing the overlap area as a function of relative offset $\boldsymbol{k}$ [5, 6]. The *IACF* or $f(m)$ is then the ensemble of all $\boldsymbol{n}$ number of $\boldsymbol{m_k}$ values.

CACF* vs. *IACF

The relationship between *CACF* and *IACF* is best illustrated by a simple example. Figure 1 is an image of a set, (0.1, 0.3, 0.5, 0.7, 0.9), of horizontal gray level bands each of 16 pixels height and each band is alternated by a null (gray level 0) band of the same height. The set repeats itself three times. Thus the y_i array in (EQ 1) consists of 0.1 repeating 16 times, then 0.0 repeats 16 times, followed by 0.3 repeating 16 times, an so on.

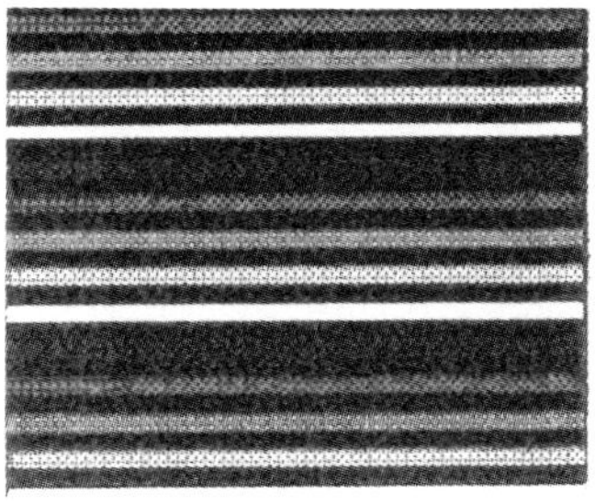

Figure 1. Two-dimensional image of one-dimensionally repeating gray bands.

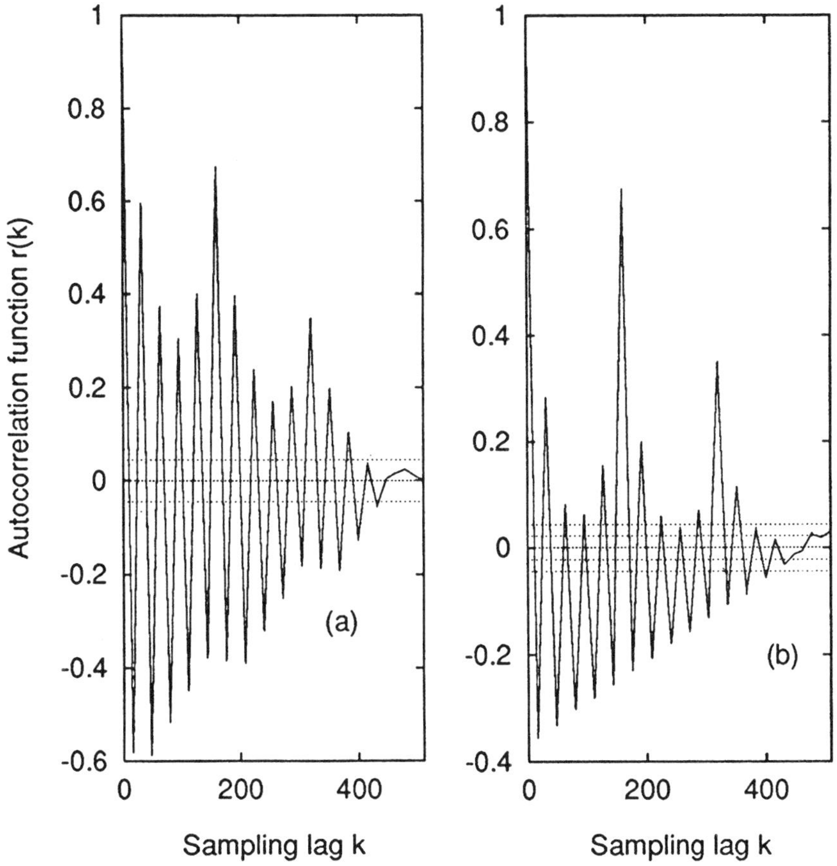

Figure 2. (a) The autocorrelation function (*CACF*) computed from the numerical array (*n*=512) of Figure 1. (b) The image autocorrelation function (*IACF*) of Figure 1 obtained by sliding the image across itself and plotting the resulting overlap area, which is normalized to unity.

The plot of *CACF* or $f(r)$ as computed by (EQ 1) is shown in Figure 2a and the plot of *IACF* or $f(m)$ as computed by (EQ 2) is shown in Figure 2b. The $f(r)$ and the $f(m)$ are remarkably similar although not identical. In *CACF* the self correlation value at r_0 is always unity. In *IACF* the negative linear slope resulting from null parts of the shifting image superimposing on the non-null parts original image is subtracted and the m_k are then normalized to unity. Both the *ACF*s quantify all three critical informations contained in the image in Figure 1, namely, (1) the correlation length, adjacent peak to peak or valley to valley distance, is exactly 32 pixels (2) both show the thrice repeating set of five gray level bands, major peak to major peak correlation distance is 160 pixels, and (3) because of the finiteness of the image both show degrading coefficients at large k lags (the so-called series termination effect). The dotted horizontal lines in Figure 2a represent the standard error of autocorrelation [4], which is approximately equal to $1/\sqrt{n}$. While all r_k are significantly larger (both positive and negative) than the error band in *CACF*, some of the m_k values of *IACF* barely make it to the $1/\sqrt{n}$, outermost pair of error bands. Thus *IACF* spectra (Figure 2b) do not fully bring out all the details of the image, unless one sets a different empirical error level. For this reason it is proposed that an approximate error band for *IACF* be taken at one half the value of CACF, namely at $1/(2\sqrt{n})$,as shown by the inner pair of dotted lines in Figure 2b. In microstructure quantification, the image analysis measurements required for *CACF* come from the analysis of a large number of fields (microscope fields of view) and the analytical time may be as high as several hours. On the other hand, it takes only a few minutes to compute the *IACF*.

MODELING THE DEGREE OF PREFERRED ORIENTATION

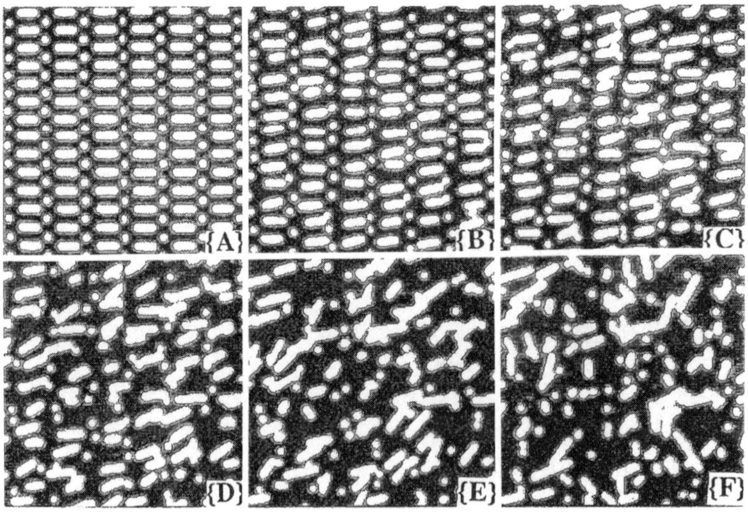

Figure 3. Computer generated images of progressively ordered to disorder tessellations. Image **{A}**, top left, shows a high degree of order in preferred orientation, images **{B}**, **{C}**, **{D}**, and **{E}** show increasing disorder while image **{F}**, bottom right, is highly disordered.

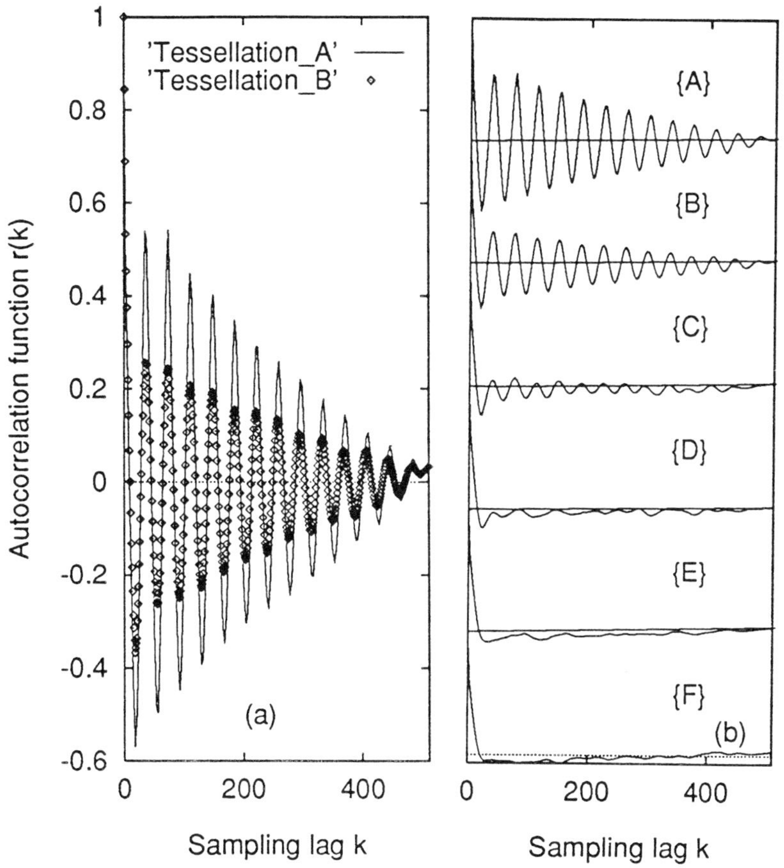

Figure 4. The *IACF* of the order-disorder series of synthetic images in Fig. 3.

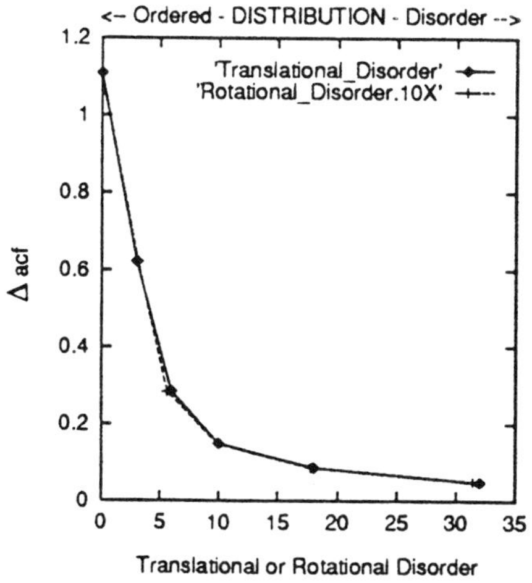

Figure 5. The preferred orientation or order-disorder indicies, Δacf, plotted as a function L_l, L_s, and L_θ.

The preferred orientation in the microstructures of materials is a striking example of periodically repeating elements. In order to find a well defined and progressively varying and *known* (a priori) degrees of preferred orientation, a series of microstructure-like images were synthesized as shown in Figure 3. In the highly ordered image, {A}, one set of lath-shaped features is parallel to the plane of the image and the other set (shown as white dots) is also lath-shaped, but is oriented normal to the plane of the image. In {A} all laths are of equal length, $L_l = 11 \times 4$ pixels, equally spaced, $L_s = 13$ pixels, and the orientation L_θ (with respect to the normal to the plane of the image) is either 0 or $\pi/2$ radians. In order to create progressively disordered tessellations, L_l, L_s and L_θ were *randomly* varied, but the maximum random values were limited as shown in Table 1. The IACF of the individual images in Figure 3 were

Table 1: Randomization Coefficients and Δacf for Images in Figure 3.

Image	L_l (pixels)	L_s (pixels)	L_θ (radians)	Δacf $\alpha=0^o$	Δacf $\alpha=90^o$
{A}	0	0	0.00	0.992	1.110
{B}	3	3	0.31	0.475	0.623
{C}	6	6	0.56	0.169	0.285
{D}	10	10	0.99	0.154	0.150
{E}	17	18	1.77	0.110	0.086
{F}	31	32	3.14	0.066	0.049

calculated by the algorithm shown in (EQ 2). The IACF or *f(m)* curves of these images are shown in Figure 4. In Figure 4a the *f(m)* of {A} and {B} are plotted superimposed on top of each other to show how the amplitudes of r_k in the slightly disordered tessellation {B} has rapidly deteriorated. In Figure 4b it is shown that both the correlation lengths or the periodicity (peak to peak or valley to valley spacings) and the amplitude of r_k deteriorate as disorder increases. Because the amplitude of r_k becomes systematically and significantly smaller as disorder sets in, an order-disorder index or a preferred orientation index, Δacf, is defined as the maximum ***absolute*** difference, $|-r_{valley1}-(+r_{peak2})|$, between the first valley to the second peak (omitting peak at r_0) in the *f(m)* spectrum. The Δacf values of the ordered-disordered series of images are given in Table 1for both $\alpha=0^o$ (horizontal sliding of the image) and $\alpha=90^o$ (vertical sliding) and are also plotted as a curve, for $\alpha=90^o$, in Figure 5.

APPLICATION OF *IACF* AND *CACF* FOR THE MICROSTRUCTURE QUANTIFICATION OF FIBER REINFORCED COMPOSITES

The fiber reinforced composites (*FRC*) described in this paper consist of long silicon carbide fibers embedded in a glass matrix. All the *FRC* discussed in this paper have the fibers in single orientation, normal to the image plane as shown in Figures 6 and 7. For the purpose of microstructure quantification, four phases are recognized by the image analysis algorithm: (1) the fibers, in Figures 6 and 7, the highly reflective circular features of approximately 10 to 15 μm in diameter, (2) the glass matrix, the light gray background, (3) voids, the very dark irregular pore space, and (4) holes in the matrix, a morphologically derived feature where at least 3 more fibers should be present such that the average fiber to fiber distance is 30 μm. Six sets of experimental conditions, with 3 to 4 composites in each set, were studied. Each set represents a different processing temperature condition, and within each set other processing conditions (not pertinent to current discussions) were varied. Microstructures of four of these six sets are shown in Figure 7. One of the undesired properties in some of the composites was the slate-like fracture pattern arising from layering of the fiber distribution in the matrix. Such layering is clearly visible in FRC-1 in Figure 7, and the layering decreases in the order FRC-2, FRC-4, and FRC-6. The layering runs normal to the vertical axis, $\alpha=90^{0}$, which is the direction in which the composite is built up during the process. Hence, the *IACF*s were performed in the $\alpha=90^{0}$ direction. The *f(m)* spectra of the *IACF* transforms are shown in Figure 8.

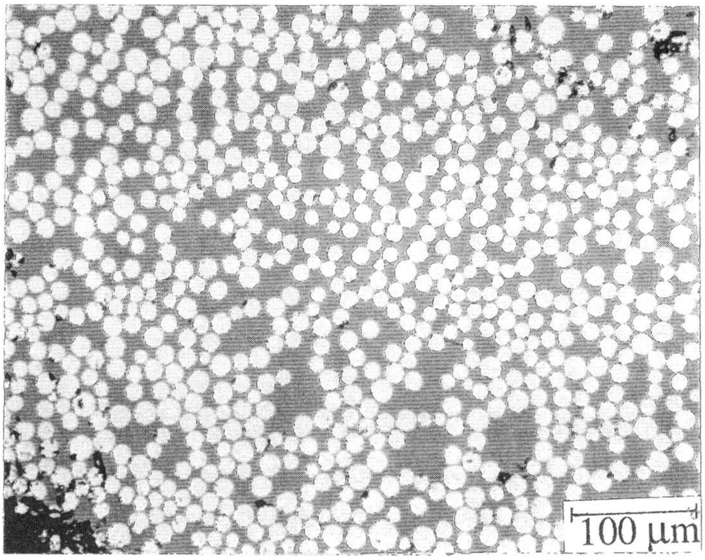

Figure 6. Optical microscope view of the polished cross section of a *FRC* sample. Such high magnification images were used for microstructure quantification of the four different phases (see text) and the numeric data of such analysis were used to compute the *CAFC*.

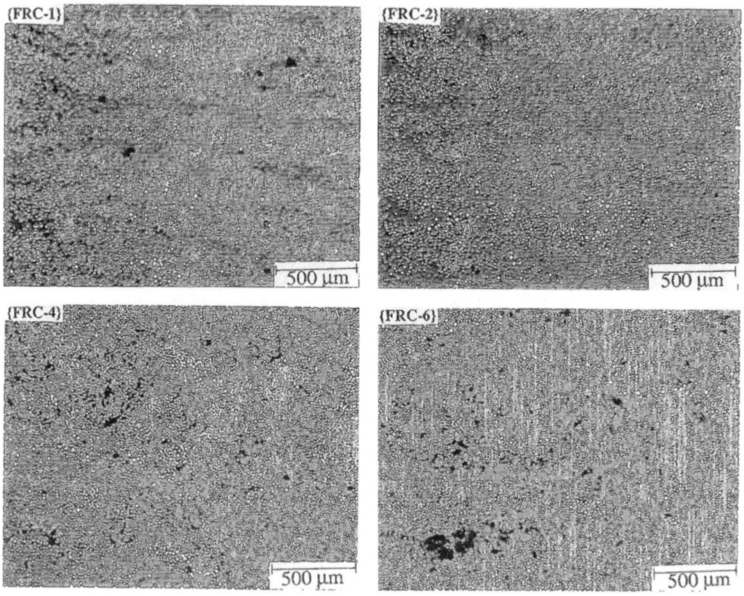

Figure 7. Optical microscope views of four FRC samples that exhibit varying degrees of periodic microstructure, the layering in the vertical direction. The *IACF* were performed in the α=90° direction (vertical sliding of the images).

In Figure 8a the *IACF* of FRC-1 and FRC-4 are superimposed to demonstrate the remarkable reduction of periodicity and r_k amplitudes as fiber distribution becomes more uniform. In Figure 8b a series of six *f(m)* spectra with a systematically reducing periodicity and r_k amplitude is shown for six composite sets included in the experiments. The preferred orientation indices, Δacf, (average of 3 to 4 samples from each set) are given in Table 2.

Table 2: Preferred Orientation Indices, Δacf for the *FRC* Samples

Sample	Temperature (^{o}C)	Δacf *IACF*	Δacf *CACF*
FRC-1	865	0.986	0.987
FRC-2	945	0.633	0.623
FRC-3	1015	0.480	0.419
FRC-4	1105	0.278	0.281
FRC-5	1295	0.180	0.182
FRC-6	1325	0.030	0.038

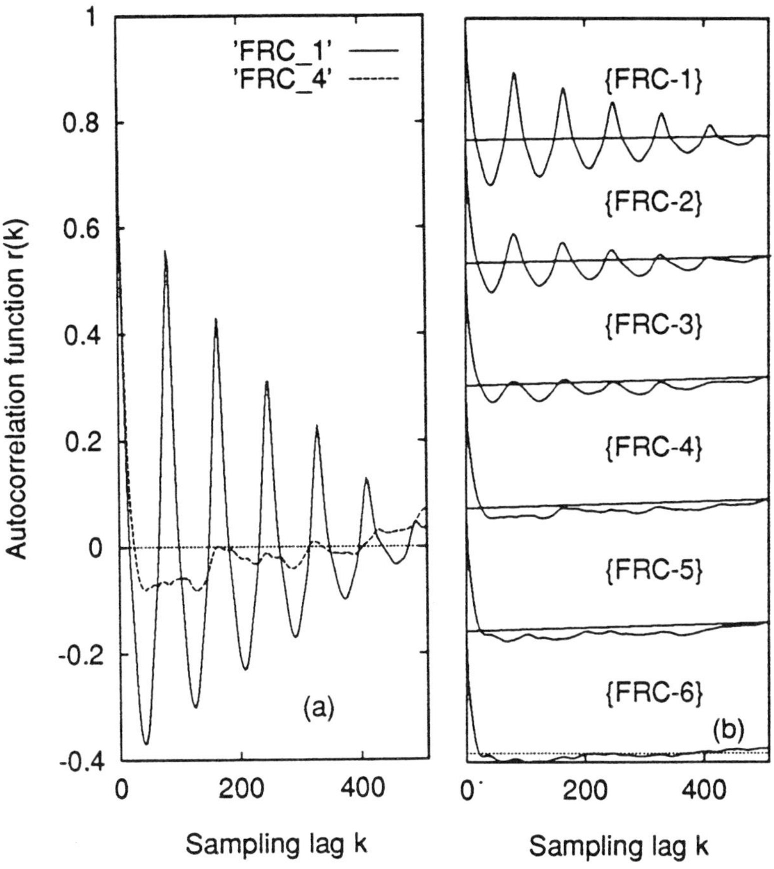

Figure 8. The *IACF* of six *FRC* samples. The periodicity and the amplitude of the autocorrelation decreases as the fiber distribution becomes more uniform.

We now turn our attention to more detailed microstructural analysis and the *CACF* of the *FRC* samples. One of the microstructural parameters that is commonly measured is the volume fraction of phases in a multicomponent system. In addition to measuring the volume fraction of the four phases, an additional parameter herein called the ***tessellation index, T,*** was also measured. In attempting to measure the inter-stellar distances, Chandrasekhar [7] defined that the probability of a particle existing in spherical shell of radial measure between $\boldsymbol{R}$ and $\boldsymbol{R}+\boldsymbol{dR}$ as,

$$P(R)\,dR = \left[1 - \int_0^R P(R)\,dR\right] 4\pi R^2 N_V dR \qquad \text{(EQ 3)}$$

where N_V is the average number of particles per unit volume, and from this equation he derived the *nearest neighbor* average distance, $\overline{d_{3D}}$, as

$$\overline{d_{3D}} = 0.544 \cdot \sqrt[3]{N_V} \qquad \text{(EQ 4)}$$

Following the procedure of deriving (EQ 4) from (EQ 3), Gurland [8] and Underwood [9] derived the *nearest neighbor* average distance, $\overline{d_{2D}}$, for the two-dimensional case as,

$$\overline{d_{2D}} = 0.5 \cdot \sqrt{N_A} \qquad \text{(EQ 5)}$$

where N_A is the average number of points in two dimensions (such as an image). The integration constant in (EQ 4) or (EQ 5) can be a variable, (**1/*T***), if one allows for varying probabilities in (EQ 3). This is equivalent to saying that the average particle to particle, or in the present case the *nearest neighbor* average fiber to fiber or hole to hole distance, in a given field of view varies. Thus (EQ 5) can be empirically rewritten as,

$$\overline{d_i} = \frac{1}{T_i}\sqrt{N_i} \qquad \text{(EQ 6)}$$

where the subscript i refers to individual fields of view sampled by the optical microscope. In order to define the tessellation index (EQ 6) is inverted and multiplied by the *nearest neighbor* coordination number, C_i, of the particle,

$$T_i = C_i\frac{\sqrt{N_i}}{\overline{d_i}} \qquad \text{(EQ 7)}$$

All three quantities of the right hand side of (EQ 7) are measurable by computerized image analysis [2, 3]. In this paper T_i is evaluated as field-average measured quantity. The tessellation index describes the particle (fiber) distribution much better than the volume fraction. For example, it contains the nearest neighbor information and distinguishes fields of same volume fraction, but with differing fiber distribution, as shown in Figure 9.

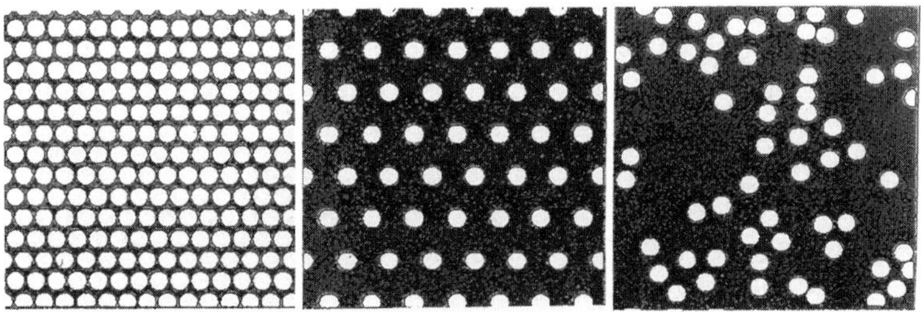

Figure 9. Three theoretical fiber tessellations. In the left is the periodic hexagonally coordinated tessellation with ideal fiber to fiber distances with T_i value of 2.46. In the middle is also a periodic hexagonal tessellation with large fiber to fiber distances and hence with a lower T_i of 0.65. On the right is a random tessellation with a T_i of 0.31. The middle and the right tessellations have identical fiber count (60) and fiber volume fraction (16%), but yield different tessellation indices.

Using a computerized image analysis system, the six sets of *FRC* samples were analyzed for the T_i of fibers and holes. On each sample data was gathered from a matrix of 16 x 8 adjacent fields. The successive steps in the automated analysis were as follows.

1. Move scanning stage to the next adjacent field.
2. Adjust light level using integrated reflected light intensity.
3. Perform coarse and fine autofocus.
4. Compute gray level histogram and threshold levels.
5. Threshold the image.
6. Measure volume fraction of all four phases.
7. Isolate the touching fibers by conditional thinning and region growth.
8. Count fibers and holes.
9. Transform the feature image to coordination polygon image by isotropic and conditional (growing regions of adjacent features do not coalesce) region growth.
10. Measure the average diameter of the polygons to obtain $\overline{d}_i$.
11. Isolate and map each polygon to a different image memory.
12 Perform the perimeter image transform on the coordination polygons.
13. Remove corner points of the polygon to isolate the edges.
14. Count edges to obtain the nearest neighbor coordination number, C_i.
15. Compute field-average T_i
16. Compute *CACF* for T_i of fibers and holes.

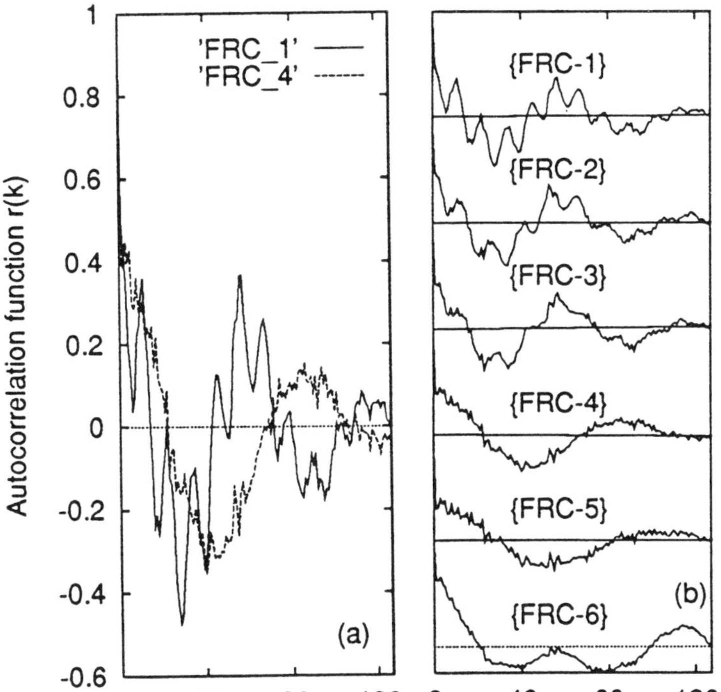

Figure 10. The *CACF* of six *FRC* samples. There are dual periodicities in the first three composites not revealed by the *IACF* (see Figure 8). The periodicity and amplitude of the *f(r)* decrease as the fiber distributions become uniform.

The *CACF* of the T_i of fibers for the six *FRC* samples are shown in Figure 10. In Figure 10a the *CACF* of FRC-1 and FRC-4 are shown superimposed, while Figure 10a compares the *f(r)* spectra of all six *FRC* samples. These plots are to be compared with the *IACF* plots given in Figures 8a and 8b. There are some striking similarities and some significant and informative differences as well. First of all, the lag k distances are different, mainly due the higher magnification of the image (compare Figure 6 with Figure 7) and the x-axis scaling. Both *IACF* and *CACF* bring out the periodicity due to layering with approximately the same correlation lengths. In the first three composites (FRC-1, FRC-2 and FRC-3) of Figure 10, there are two periodicities. The low frequency envelope curve represents the layering periodicity. The high frequency periodicity was a surprise revelation. Upon closer analysis of the data and the micrographs, it turned out that this high frequency component arises from the approximately periodic fiber clusters within the fiber rich layers. We had originally missed the presence of this, both in recognizing them in the micrographs (see Figure 7) and in the *IACF* spectra. The presence of this high frequency component was further verified by Fourier transform analysis. It turns out that these fiber clusters are due to undispersed fibers from the fiber tows as received from the fiber manufacturer. The systematic increase in the fiber distribution is very clearly visualized in the *CACF*s where the correlation lengths increase and the high frequency periodicity gradually disappears from FRC-1 to FRC-6. There is a strong linear correlation between the r_k of T_i of fiber *CACF* and the r_k of T_i of hole *CACF* as shown in Figure 11. From the microstructural point of view this not a surprising or significant finding, because when there is fiber layering there is concomitant hole layering as well. However, the significance of such a correlation lies in the fact that there a self-consistency in the autocorrelation method of microstructure analysis.

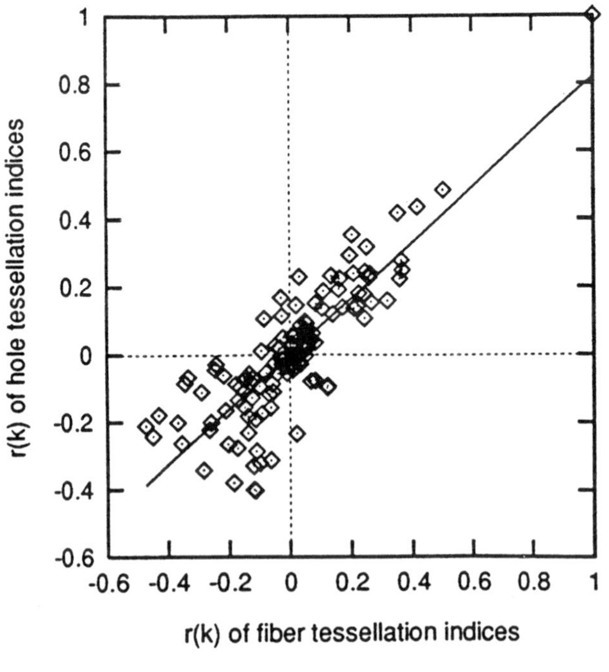

Figure 11. Inter-correlation or cross-correlation of the r_k of fiber and hole tessellation CAFC analyses.

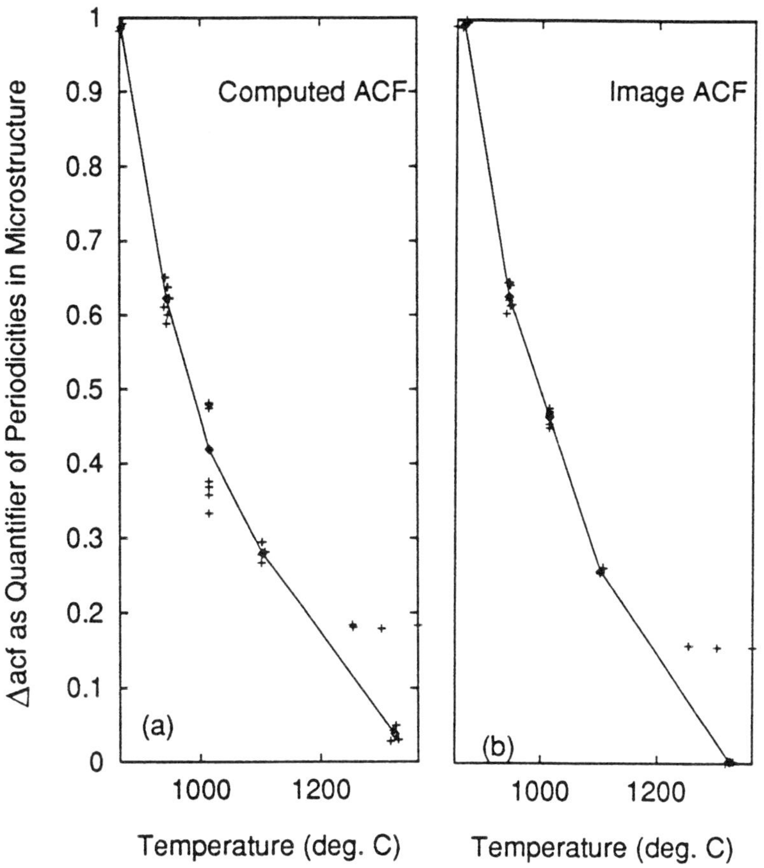

Figure 12. Plots of Δacf as a function of process temperature. (a) Δacf from the *f(r)* spectrum from *CACF* and (b)Δacf from the *f(m)* spectrum from *IACF*.

Finally, there is a good correlation between the microstructural parameter, Δacf, and a process parameter, namely the processing temperature, as shown in Figure 12. It is important to note that there is a larger scatter in the *CACF* than in the *IACF*, because *CACF* is more sensitive to minor variations in the microstructure. For example, the samples with a constant temperature show larger Δacf variation in the *CACF* than in the *IACF*. There appears to be some systematic meaning to the vertical scatter observed in Figure 12a, if process parameters other than temperature are taken into account. It should also be mentioned that *FRC* processed around 1000 to 1100°C show very good fiber distribution ($\Delta acf \sim 0.02$), if the matrix composition is different. Thus the Δacf vs. process curve for a different series of materials, with varying periodic texture, will have the same general shape and trend, but will fall on a curve of different slope and intercept.

CONCLUSION

It is shown that the autocorrelation method is a powerful tool for quantifying periodic properties and preferred orientation in the microstructural analysis of materials. The technique has been used in this paper in a manner that provides a paradigm for correlating process, property and microstructure. Further, the technique has been successfully applied by the author to discern variations in periodicity in extruded ceramics, metal or polymer coatings on optical waveguide fibers and in the study of petrographic textures.

In practice the image autocorrelation (*IACF*) is faster and often requires no image processing (image enhancement or thresholding etc.) prior to measurement. When very fine details in the periodic properties of microstructure need to be quantified, one may have to resort to computational (*CACF*) autocorrelation analysis. It should be mentioned that autocorrelation is not the only available tool to the microstructure analyst for quantifying periodic properties. There are instances where the author had to resort to techniques like least squares curve-fitting, fractal theory or Fourier transforms.

ACKNOWLEDGEMENTS

The author is thankful to Drs. M. P. Taylor and K. C. Chyung of Corning Incorporated, Prof. N. Ahuja of University of Illinois for critical and helpful suggestions throughout this work. Colleague Tanya A. Jacobus was responsible for writing several software modules and some of the microstructure analysis reported in this paper. The author acknowledges William H. Schuler for critically reading the manuscript and Susan K. Halstead for various chores in getting the manuscript to its final form.

REFERENCES

[1] C. Chatfield, *The Analysis of Time Series*, Chapman and Hall, New York, third edition, p. 60, 1984

[2] S. J. Louisnathan, T. A. Jacobus, M. P. Taylor and K. C. Taylor, "Characterization of Fiber Reinforced Glass Composites", NASA/DOD Conference on Metal Matrix, Carbon, and Ceramic Matrix Composites, Cocoa Beach, Florida, *Abstract only*, 1984.

[3] S. J. Louisnathan and T. A. Jacobus, "Microstructure Quantification by Mathematical Morphology", Collected papers XIV International Congress on Glass, New Delhi, India, p.302, 1986.

[4] G. E. P. Box, W. H. Hunter, and J. S. Hunter, *Statistics for Experimenters*, John Wiley & Sons, New York, p.584, 1978.

[5] J. Serra, *Image Analysis and Mathematical Morphology*, Academic Press, New York, p.271, 1982.

[6] J. C. Russ, *The Image Processing Handbook*, CRC Press, Boca Raton, Fl, p. 165, 1992.

[7] S. Chandrasekhar, "Stochastic Problems in Physics and Astronomy", Rev. Mod. Phys. **15**, p.86, 1943

[8] J. Gurland, "Spatial Distribution of Discrete Particles", *Quantitative Microscopy*, Ed. R. T. DeHoff and F. N. Rhines, McGraw-Hill, p.278, 1968.

[9] E. E. Underwood, *Quantitative Stereology*, Addison-Wesley Publishing Company, Reading, MA, p.80, 1970.

PRINCIPLES AND PRACTICES OF VIBRATORY USE

IN METALLOGRAPHIC SAMPLE PREPARATION

N. T. Saenz[1] and D. J. Diaz[2]

ABSTRACT

The field of metallography has evolved drastically in recent years. Many of the changes have resulted in widespread use of complex alloys and new applications for high-strength, lightweight materials. As the complexity of materials has increased, the use of vibratory polishing becomes an integral mechanism in the preparation of these materials by taking the best approach in reducing the deformation/damage produced during metallographic sample preparation. For many applications in material analysis, vibratory polishing outperforms mechanical or rotary polishing. This paper will address the principles and practices for vibratory use in metallographic sample preparation.

Introduction

Vibratory polishing has been around for a long time. It is possible to produce excellent flatness with a superior finish that can't be matched with rotary or mechanical polishing. The vibratory polishing bowl is stationary which is one of the major reasons for the lack of deformation produced in comparison to rotary or mechanical polishing. Let us take a closer look at how the movement is created. Following is a schematic of an old vibratory polisher as shown in Figure 1.

---

[1]Pacific Northwest Laboratory, Richland, Washington
[2]Buehler Ltd., Lake Bluff, Illinois

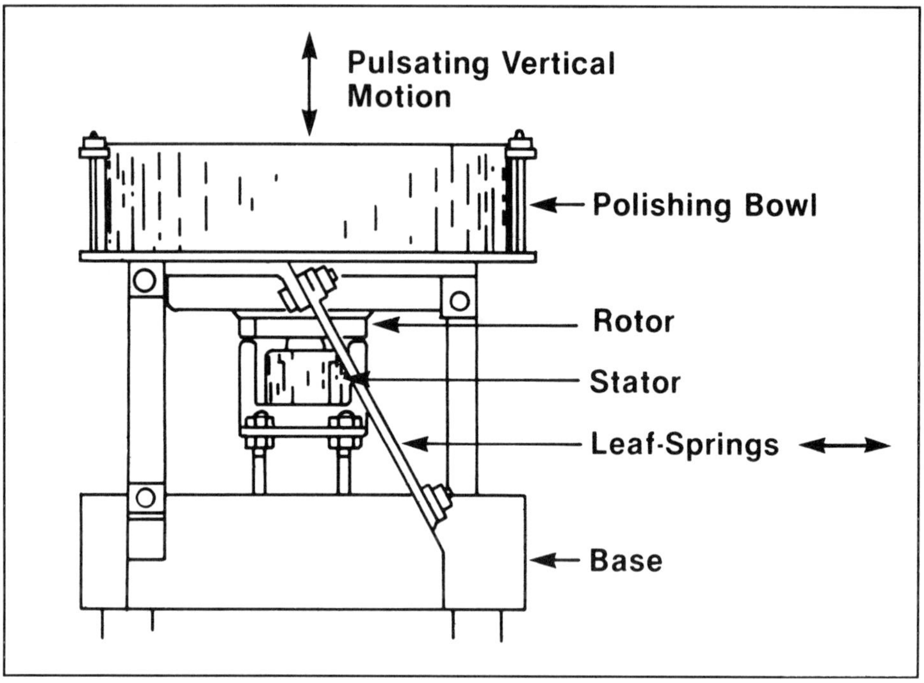

Figure 1. Basic Components of the original vibratory polisher.

This polisher consists of a laminated E-frame (stator) wound with a magnet wire which is pulsed with a half-wave rectified current. This results in a pulsating magnetic field. The rotor is separated from the stator by a small air gap. This pulsating unit was placed below the center of a 12 inch platen. With each pulse, the rotor/stator is attached and produces a pulsating vertical motion. By attaching the base at four locations to leaf springs which are in a 75 degree angle to the base, the polishing surface moves up and counterclockwise when the pulse dies. While the sample or mount is moving upward, the bowl drops away from the sample or mount allowing it to fall by gravity to a place on the bowl a small distance from where it had started, Figure 2 (points 1a - 2a). While the distance involved is only a few thousandths of an inch, it is repeated 3600 vibrations per minute. This makes the sample or mount appear to be in a continuous motion. Weighted sample holders are used to increase this motion. In this logic, the older units placed a greater efficiency on weight whereas the greater the weight, the more vertical stress that is necessary to drive the sample or mounts forward. Two variables affect the polishing speed: 1. The weight of the sample, and 2. The amplitude of vibration. The rheostat setting is very critical on these older units. If the units were set too low, slow polishing rates result; if the setting was to high, the samples would bounce too much causing other polishing defects. With experience, however, the operator could combine the setting with the proper polishing cloth/pads and abrasives to achieve an excellent polish.

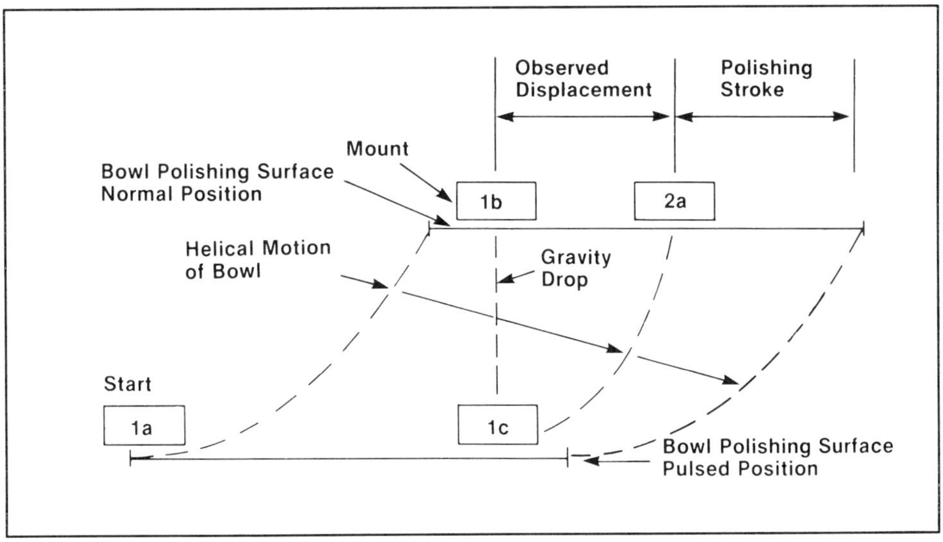

Figure 2. Schematic showing the polishing action for the original vibratory polisher.

A schematic of the newer designed vibratory polishers is shown in Figure 3. It consists of a two mass vibrating system: a bowl and a bowl crossarm coupled to an electromagnetic drive on the side or vertically. The base is isolated from the mounting area by the rubber isolator springs (mounting feet). The drive assembly consists of a core and armature. The core is connected directly to the base. The armature assembly is located opposite the core assembly and is directly connected to the bowl mounting crossarm. This spring assembly consists of stack springs similar to the old polisher except that it is larger in size and connected directly to the crossarm at the top and bottom of the base. The bowl and crossarm are joined and now become one complete unit that floats with the support of the leaf spring assembly. The power to the vibrator base is supplied through a separate control. This control in its simplest form consists of a fuse, control rheostat with gear, and a triad. The triad supplied with these units permits current to flow in both directions, which in turns provides control over the duration of flow during each cycle.

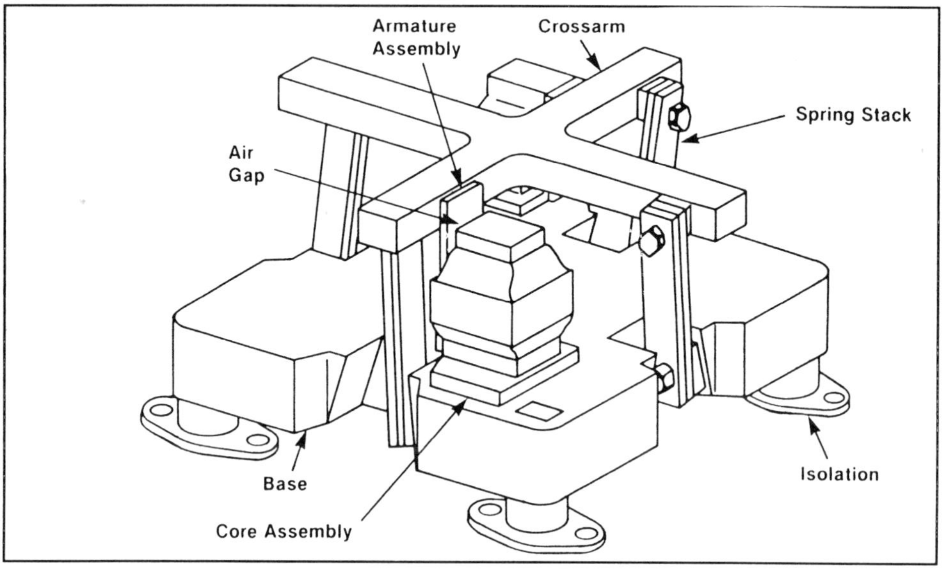

Figure 3. Basic components of the new vibratory polisher.

Figure 4 illustrates the action of a sample mount moving along the bowl surface. During a vibrating stroke, the bowl surface travels between its lowest point (A) to its highest limit (C). Although still traveling upward and forward, the bowl will decelerate between B and C. On the upward stroke, the sample mount will come into contact with the bowl from A to B. At point B, the velocity of the object becomes grater than the bowl, and the object leaves the bowl surface on a free flight trajectory from B to D. The sample lands back on the bowl surface at a position further forward (D) from where it started. While this distance is only a few thousandths of an inch, it is repeated 7200 vibrations per minute.

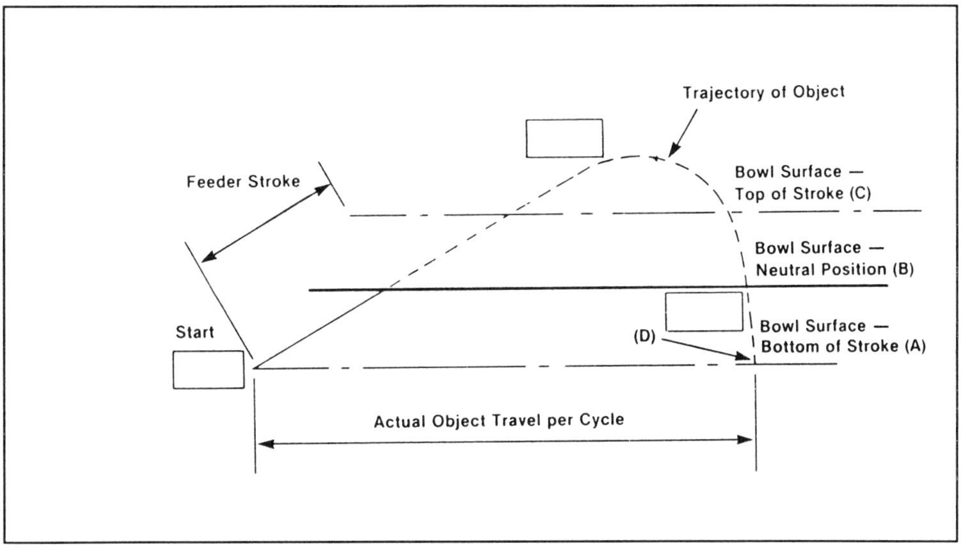

Figure 4. Schematic showing sample movement on the bowl surface of the new polisher.

Applications and Practices of Vibratory Polishing

Automation of the polishing process is attractive from many points of view. The likelihood of sample integrity producing a representative microstructural surface is greatly enhanced by the removal of the deformation layer produced by vibratory polishing. Also standardization of microstructural preparation of many types of materials using vibratory polishing techniques can produce uniform microstructures in multi-sample preparation as well as producing a high output rate of prepared metallographic samples. Vibratory polishing techniques, using a combination of abrasives, cloths, and pads help to maintain surface flatness and artifact-free samples. The use of diamond abrasives is indispensable for the preparation of brittle and fragile materials. The polishing rates using diamond abrasives are equal to, or better than, other polishing abrasives such as Al_2O_3, MgO_2, SiC, and CrO_2. The types of polishing cloths or pads used have an extremely important bearing on the finished specimen. Napless cloths and pads should be used when relief between the micro constituents of varying hardnesses and sample mount interfaces are to be held to a minimum. Some examples of

MgO_2, SiC, and CrO_2. The types of polishing cloths or pads used have an extremely important bearing on the finished specimen. Napless cloths and pads should be used when relief between the micro constituents of varying hardnesses and sample mount interfaces are to be held to a minimum. Some examples of napless polishing cloths are nylon, stainless steel mesh, and silk. Some of the pads used are TEXMET®, POLIMET®, CHEMOMET®, and TEXMET® 2000[3].

VIBRATORY PRACTICES

Planar Grinding

Planar grinding of the mounted sample eliminates or helps minimize the depth of any surface damage that might have occurred in the cutting of the material. Planar grinding can be accomplished with either fixed or rolling abrasives. With fixed abrasives, the lap material becomes charged with abrasive particles or the particles have been permanently fixed as in the case of ceramic, resin, or metal bonded belts or wheels. The cutting action with loose abrasives is due to the rolling of the abrasive particle making contact with the surface of the material being polished. Diamond abrasives, due to their high hardness, inertness, and low coefficient of friction, are used for ceramic material. Diamond particles also retain their shape and size during the abrasion process to produce a uniform, high rate of material removal with minimal induced surface damage, while Silicon carbide or Aluminum oxide is commonly used for ferrous and non-ferrous materials Since this is the most aggressive preparation step, care must be taken to ensure that new damage is not introduced to the material being polished.

All metallographic resin mounted specimens must be planar ground using either abrasive grinding paper/discs from 120 grit to 600 grit or diamond planar grinding discs down to 30 µm diamond abrasive size before the vibratory polishing can occur. It is necessary to maintain medium-to-heavy uniform pressure until all grinding scratches from the previous step have been removed. After each grinding step the mounted samples should be thoroughly cleaned to prevent contamination from the previous step. The mounted samples are then placed in a polishing weight assembly, shown in Figure 5.

---

[3]® Registered Trademarks of Buehler.

Figure 5. Polishing weight assemblies.

Vibratory Polishing is an effective polishing technique for a variety of materials that are difficult to polish with rotary polishing. Samples are placed in weight assemblies face down on the cloth/pad. Inertia from the sample weight produces movement which induces polishing. Figure 6 shows a polished weight assembly for ultra-thin ceramic preparation. Since the removal rate of material is low, several minutes to hours of polishing may be required. These materials in combination with an abrasive cloth/pad gently polish the sample surface while slowly removing material and creating very little thermal or mechanical shear loading to the material being studied.

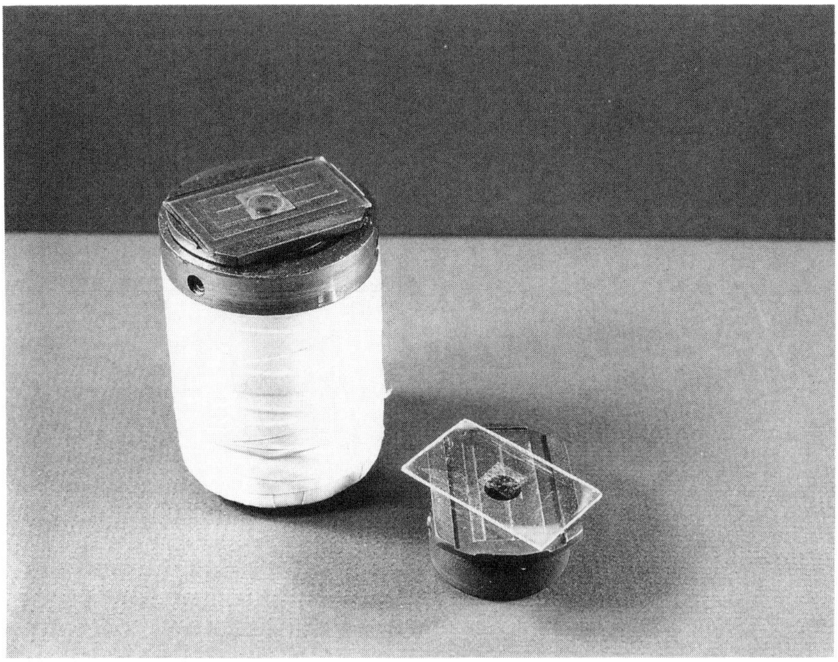

Figure 6. Polishing weight assembly for ultra-thin ceramic samples.

Ferrous and non-ferrous materials

Most ferrous and non-ferrous materials are commonly polished using an Aluminum oxide abrasive slurry consisting of Al_2O_3 powder (Linde A), distilled water, and a napless polishing pad such as TEXMET®. Polishing times depend on sample preparation prior to vibratory polishing and material hardness, 1 to 8h is typical, followed by a final polish using a colloidal silica or magnesium oxide slurry on a napped cloth for a few minutes to remove any fine scratches.

Vibratory polishing outshines other polishing methods in the preparation of materials that have inclusions such as corrosion products, multi alloyed, or materials with layered coatings of different hardnesses (Figure 7). Figure 8 shows the microstructure of grade 80-55-06 pearlitic ductile iron as-cast. The dark graphite nodules are retained in the matrix material and not pulled out using vibratory methods.

Figure 7. The layered coating is retained with flat interfaces between the matrix/coating layer.

With the increasing use of ceramic materials in many industrial areas, applied metallographic sample preparation techniques are used to produce true microstructural detail in ceramic materials. Due to the extreme hardness, porosity, and brittleness of ceramic materials; improper ceramic preparation techniques tend to cause intergranular relief, grain plucking, grain fracturing, or other undesirable effects. Because of these inherent characteristics, care must be taken in the preparation of ceramic material. For example, defects such as cracking of the specimen can be initiated during the sectioning step and can also occur if improper mounting procedures are used. Pullout of grains which results from improper sectioning or a grinding techniques that are too severe can also cause problems in later stages of preparation. To minimize any defects or a false microstructure, the metallographer must not only perform each step of sample preparation accurately but must also have a clear understanding of the intended purpose for each specific sequence of preparation steps.

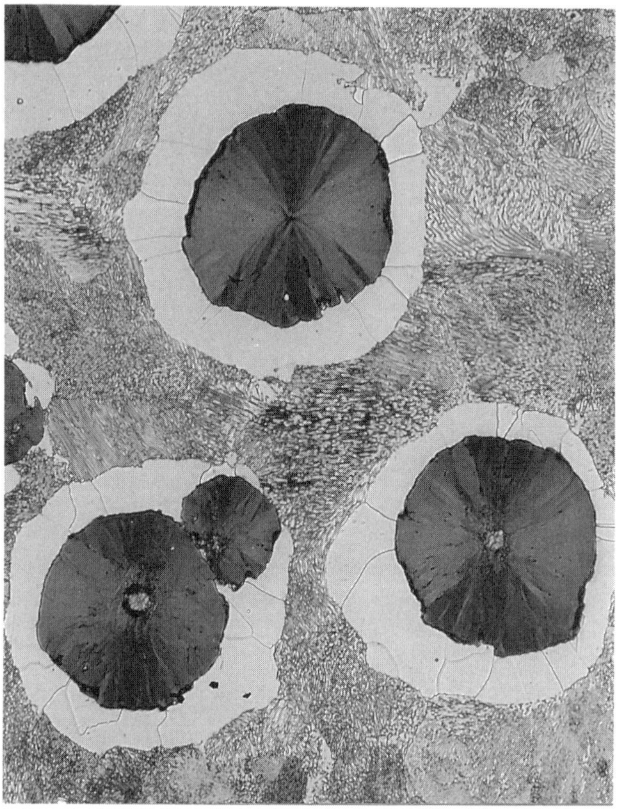

Figure 8. Grade 80-55-06 pearlite ductile iron.

For ceramics, the first vibratory polishing step involves the use of 3 μm or 15 μm diamond abrasives in conjunction with an oil-based or water-based suspension on a stainless steel mesh polishing cloth. This combination of polishing medium and cloth produces samples with excellent interfacial flatness between the micro constituents as well as excellent edge retention. The amount of time involved in this rough polishing or flattening step depends on the hardness variations within the specimen being polished; 1 to 4 h is typical.

The second vibratory polishing step consists of 1 μm or 3μm diamond abrasives in conjunction with an oil-based or water-based suspension on a synthetic polishing pad. Again the time depends on the removal rate of the material being studied; approximately 1 - 12 h of polishing time is required to remove any artifacts (Figure 9).

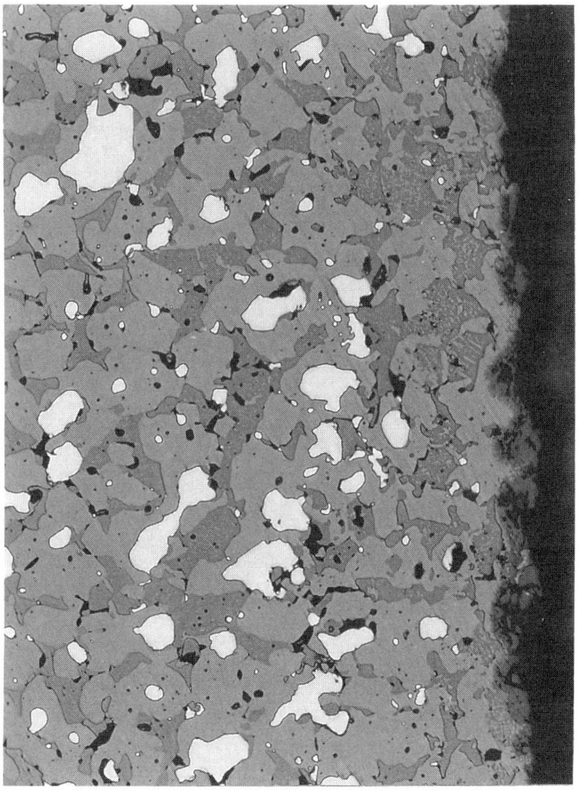

Figure 9. Sample (Cermet NiO_2|Cu) illustrating the edge retention established from the second stage of vibratory polishing in conjunction with 1μm diamond suspension on a napless pad.

Care must be taken to ensure that all surface deformation is removed (Figure 10). If not, deformation in the form of scratches or excess porosity may still be apparent in the as-polished state, and more polishing time may be needed. Figure 11 demonstrates how micro constituents with different hardness remain optically flat using vibratory polishing techniques.

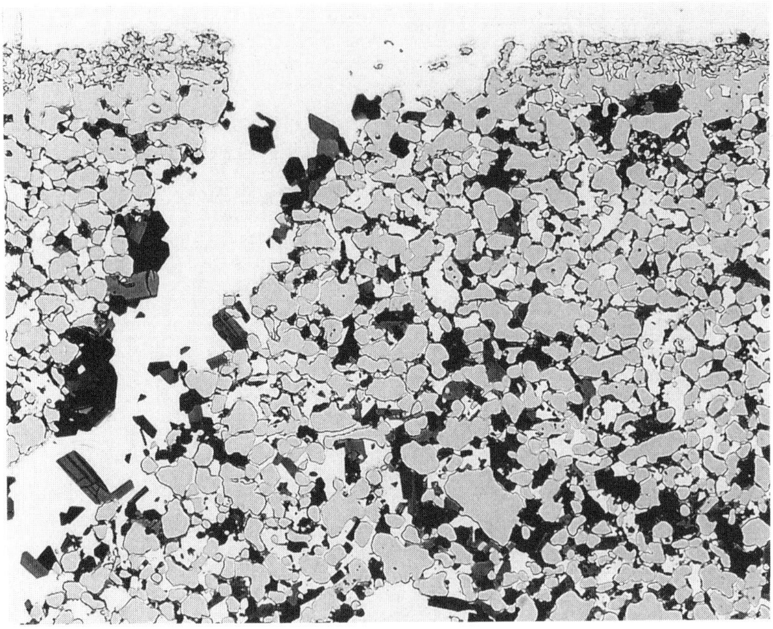

Figure 10. Ceramic, Glass, and Composites

Ceramic material can also be examined in transmitted polarized light such as that used in mineralography (1). For this purpose, the preparation of ultra-thin sections (2) are required. Ultra-thin sections are polished materials mounted on a glass slide and mechanically lapped. Ultra-thin sections are on the order of

1 to 5 μm in thickness producing a whole new order of optical access to microstructural detail. Information regarding crystal symmetry, phase identification, and crystal structure can be determined by transmission light optical microscopy. Figures 12a and 12b illustrate ultra-thin sections of SiC fiber - SiC ceramic composite material using reflective and transmitted polarized light techniques.

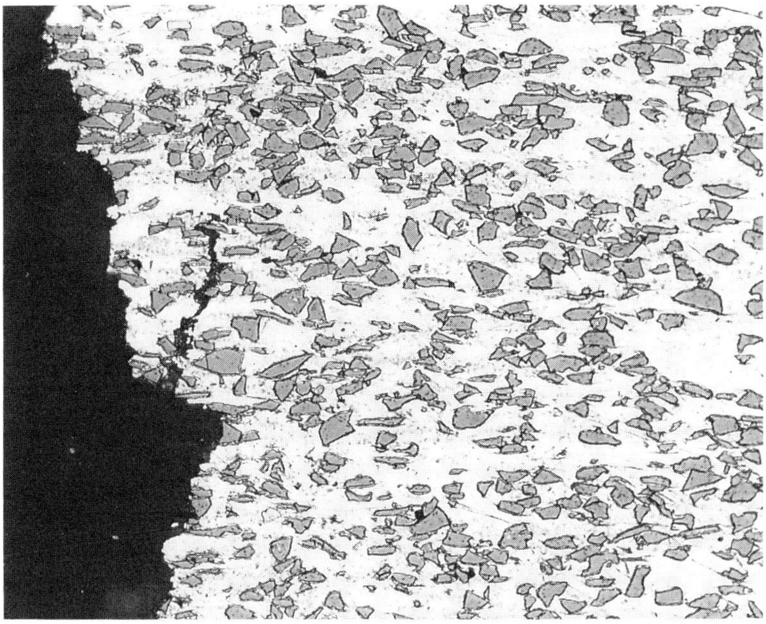

Figure 11. Mg (White) B4C (Gray) Metal Matrix Composite

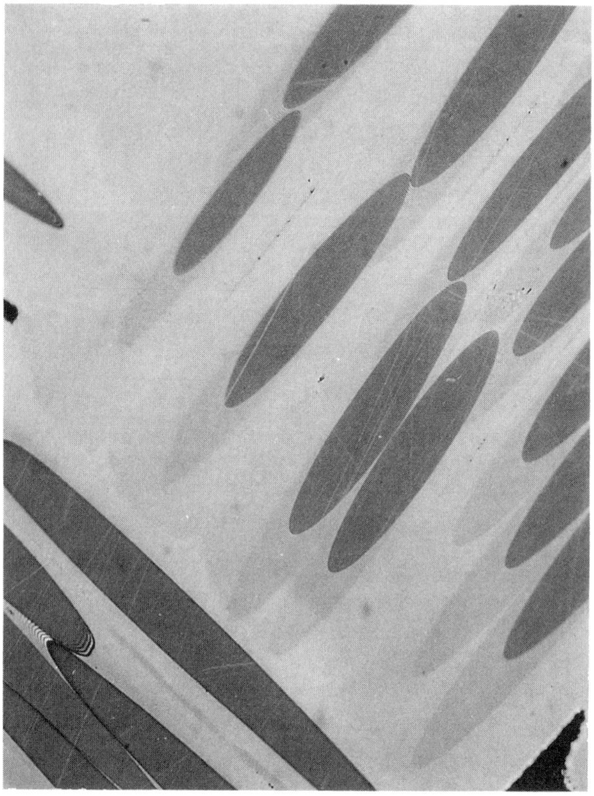

Figure 12a. SiC/SiC(f) Composite using reflective light techniques.

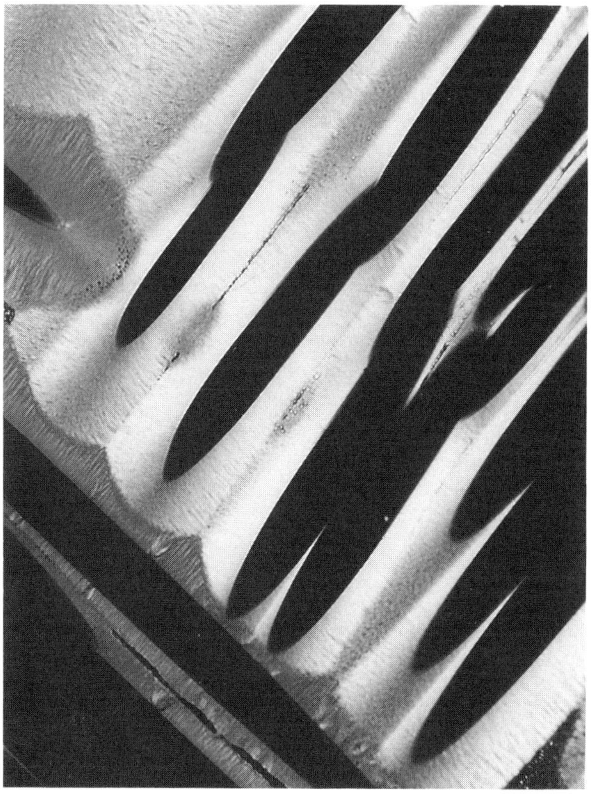

Figure 12b. SiC/SiC(f) Composite using transmitted polarized light techniques.

Conclusion

Vibratory polishing offers a viable means of obtaining true microstructures on various traditional and advanced materials. While introducing little or no deformation while obtaining excellent flatness, and a superior optical finish, vibratory polishing plays a very integral part in the materials science field. By understanding the mechanism of how vibratory action is produced in conjunction with the modern abrasives, cloths, and pads that are used in our industry today, vibratory polishing is a key in producing a true microstructure that is necessary for characterization in the materials science field.

References

1. Kerr, P. F., 1977. Optical Mineralogy, McGraw-Hill Co., New York, NY.
2. Saenz, N. T., 1989. "Ultra thinning Section Techniques For the Characterization of Brittle Materials".

Chapter 3

Composites and Polymers

Metallographic Studies on Metal-Ceramic Joints and Composite Implants

G. Elssner and D. Korn

Max-Planck-Institut für Metallforschung, D-70174 Stuttgart

Abstract. Metallographic methods were used in manufacturing and testing of sapphire/niobium-joints and of titanium/alumina composites for dental implants. Information on grinding and polishing operations and on the obtained surface quality of the starting metal and ceramic prior to diffusion bonding is given. The interface fracture of the diffusion bonded niobium-sapphire bicrystals was studied. Glide steps at the metal side of the interface fracture were detected and the operating glide systems were determined using light optical differential interference contrast and interference microscopy methods. Thereby the influence of crystallographic orientation and impurity content on the interface fracture energy could be explained. For dental implants prepared by diffusion bonding the transition zone between titanium-tantalum alloys and an alumina ceramic was investigated using light optical and SEM micrographs and WDX and EDX analysis. In first experiments the interface between the dental implant and the tissue was studied to control mechanical and chemical interactions.

Introduction

Some details of our metallographic studies on diffusion bonding and on diffusion bonded dental implants are given in this paper. In the first part we will report on the surface preparation of ceramic and metal single crystals and on fractographic observations of interface fracture using metallographic techniques such as differential interference contrast, laser scan microscopy , interferometry and stylus profilometry. For the bonding studies we have taken niobium and sapphire. They possess nearly the same expansion behaviour and do not form new intermediate phases during diffusion bonding. In the second part the manufacture of dental implants and the control of their microstructure will be described. Indications of interactions between dental implants and bone tissue will be presented.

Diffusion Bonded Niobium/Sapphire Bicrystals

Slabs of an area of 15 to 10 mm and a thickness between 1 and 2 mm are sputter cleaned by argon ion bombardment in a UHV diffusion bonding apparatus (Fig. 1) [1,2]. In a second step the slabs can be doped with impurity atoms. The formation of a Nb/sapphire bicrystal of a defined crystallographic orientation by diffusion bonding occurs in a vacuum of $1 \cdot 10^{-10}$ mbar at 1400°C for 2h under a mechanical pressure of 10 MPa [3,4]. At the top of the bicrystal a slab of polycrystalline niobium and at both sides pieces of polycrystalline alumina are added in this step of diffusion bonding. From the bonded pieces bend test specimens are cut. Particular attention must be paid to the surface preparation of the single crystals, in particular to the soft yet tough niobium crystals. For diffusion bonding flat surfaces of minimum surface roughness and curvature are needed. Surface deformation of the metal must be as far as possible avoided, since this can lead to the undesirable formation of multi-crystals. The metallographic preparation was carried out on an Abrapol machine with 16 specimens adhesively bonded to the carrier disc. The applied force was 10 N and the disc speed 150 rpm. Details of the preparation are given in Fig. 2. In preparation 1 final polishing of the niobium slabs was performed with a colloidal suspension of silica whereas in method 2 vibratory polishing with alumina was used [5,6].

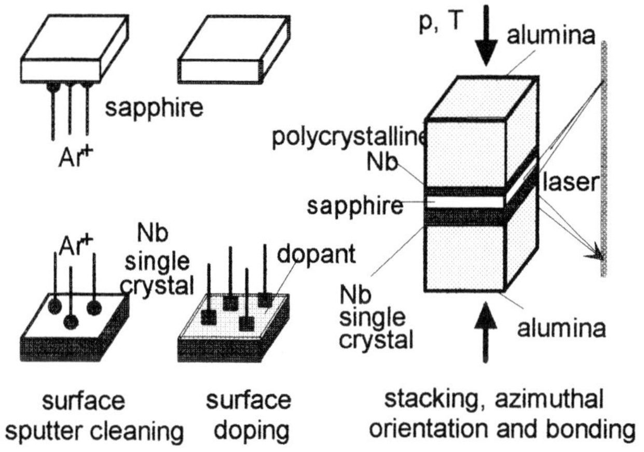

Fig. 1. Cleaning, doping and UHV diffusion bonding of sapphire to niobium single crystals.

Aim: To produce a polished, flat surface of average surface roughness < 10 μm and a curvature of < 4μm over a specimen length of 15 mm

	Niobium single crystal		Sapphire
Cutting	Spark erosion		Diamond saw
Grinding	1000 SiC, Abrapol, 5 min		20 μm diamond (D) disc
Polishing	Preparation 1 → RAM 6 μm D → RAM 3 μm D Alternate 4 to 8 h RAM 1 μm D, 15 min OPS on OP-Chem, 1 min	Preparation 2 DP-Dur 6 μm D, 4 to 6 h DP-Dur 3 μm D, 30 min 2 h vibratory polishing on Microcloth, using 0,25 μm alumina	DP-Plan 6 μm D, 8 to 12 h DP-Plan 3 μm D, 2 to 3 h DP-Plan 1 μm D, 15 min

Fig. 2. Metallographic preparation of niobium and sapphire single crystals.

Fig. 3 informs on the quality of surface preparation. Optimum roughness data were measured for the niobium crystals after the use of method 2. The curvature is defined as the maximum elevation in microns over a distance of 15 mm. Minimum values of curvature of 2 microns over 15 mm could be obtained. In the sample preparation Nb(110) surfaces of VHN 53 present more difficulties than Nb(100) surfaces of VHN 57 as demonstrated by Figs. 4 and 5. The (110) surface is considerably rougher and more scratched than the (100) surface after polishing on a RAM cloth with 6 μm diamond paste. An optimally polished niobium (110) surface after preparation 2 is shown in Fig. 6 [5,6].

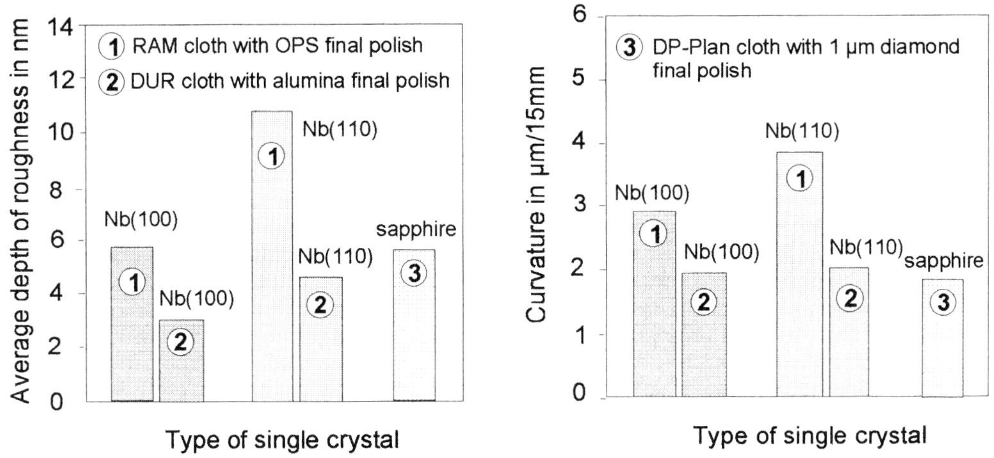

Fig. 3. Surface roughness and curvature of polished niobium single crystals and sapphire.

Fig. 4. Surface of an (110) niobium single crystal after polishing with 6 μm diamond paste on RAM cloth, DIC.

To control the quality of the bonding slightly scratched niobium crystals with a (100) surface were welded to sapphire slabs (Fig. 7). Optimum bonding is obtained when the interface appears after welding as in Fig. 8 which was photographed through the sapphire in differential interference contrast. After bonding the interface fracture energy of doped and undoped bicrystals was determined using interfacially notched bend test specimens. Ductile interface fracture (Fig. 9) was observed for undoped and slightly doped bicrystals.. We used metallographic techniques to study plastic deformation effects on the metal side of the interface fracture.

Fig. 5. Surface of an (100) niobium single crystal after polishing with 6 μm diamond paste on RAM cloth, DIC.

Fig. 6. Surface of an (100) niobium crystal optimally polished according to method 2, DIC.

Fig. 10 shows a DIC micrograph of a niobium side of an interface fracture. The notch is situated at the bottom. As can be seen white appearing glide steps were formed on the niobium during interface fracture. Fine black lines are also apparent. The height of glide steps can be evaluated using an interference microscope. The interference pattern in Fig. 11 represents part of the upper right section of the fracture surface given in Fig. 10. The height of the steps at the white lines is about 0.2 μm or approximately equal to the width between two dark lines of the interference pattern. The results obtained from interference patterns were confirmed by stylus roughness measurements (Fig. 12) [4,6,7].

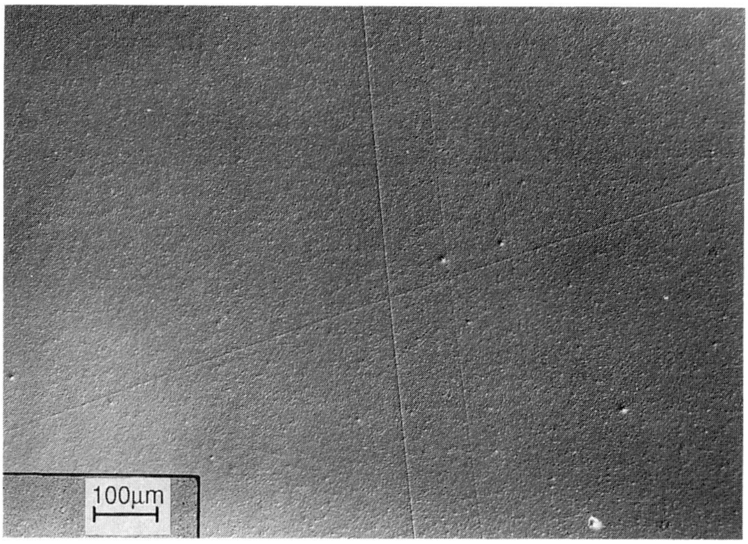

Fig. 7. Surface of an (100) niobium crystal before diffusion bonding at 1400°C, DIC.

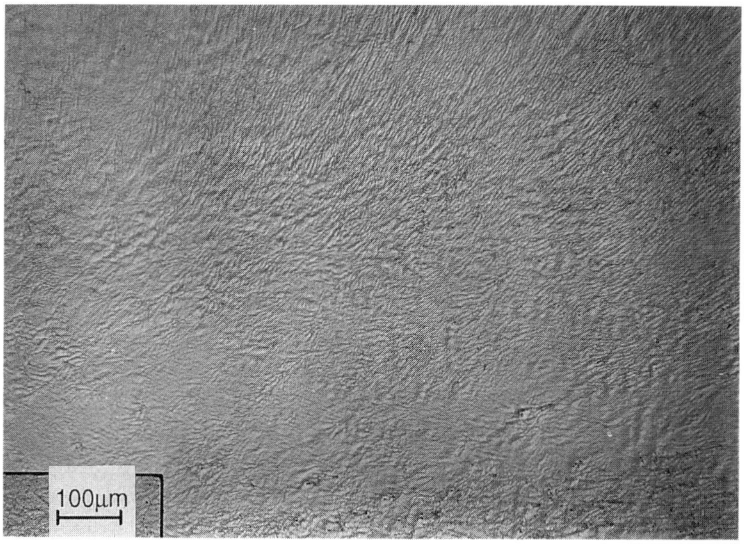

Fig. 8. Surface of an (100) niobium crystal after diffusion bonding at 1400°C photographed through the sapphire, DIC.

The laser scan microscope (LSM) [8] can also be used to reveal details of the topography of the niobium fracture interface (Fig. 13). Again glide steps are visible. The hardness impressions are additionally introduced into the surface. By using the confocal

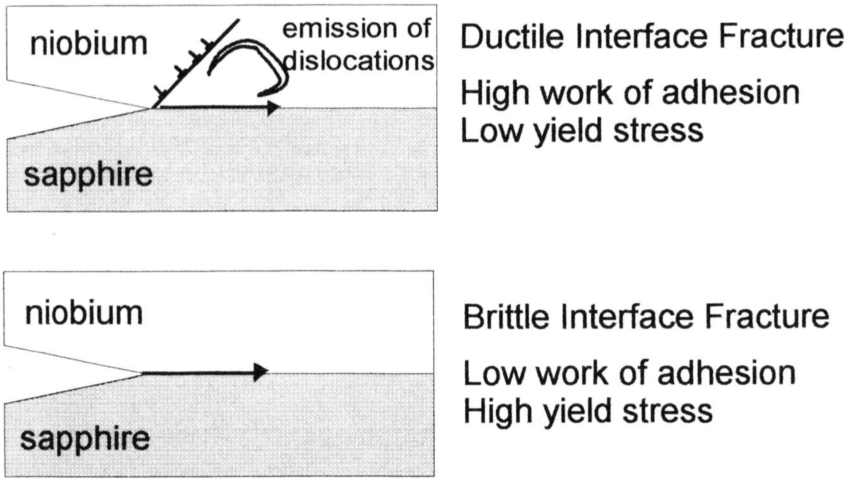

Fig. 9. Ductile and brittle interface fracture.

Fig. 10. Niobium side of an interface fracture, DIC.

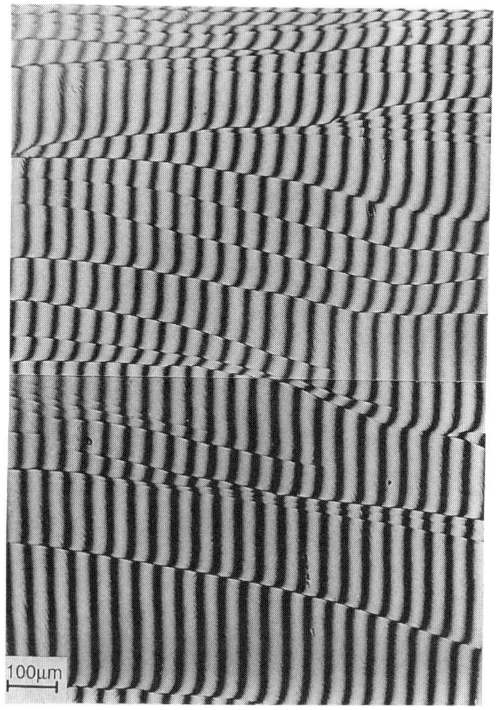

Fig. 11. Interference pattern of the niobium side of an interface fracture.

principle the surface can be optically cut plane by plane at a minimum distances of 20 nm. In the LSM the confocal optics with its pinhole diaphragm is responsible for the confocal effect (Fig. 14). The pinhole is exactly arranged where the image of the light spot from the specimen is formed. If the free diameter of the aperture is reduced to the size of the spot image it is only the photons originating from the focus plane which reach the detector unhindered. Photons from planes out of focus are blocked. Fig. 15 represents a three-dimensional reconstruction of a glide step on the niobium fracture interface. The image was obtained from sections digitally stored in the LSM computer. On the bottom of Fig. 15 a perpendicular section through the glide step is additionally given.

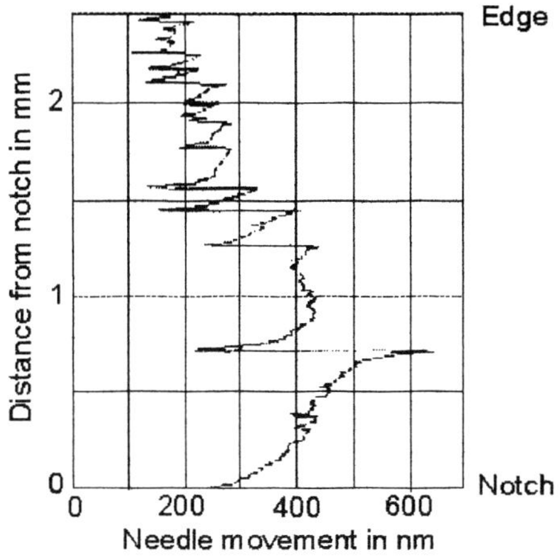

Fig. 12. Roughness of the niobium side of an interface fracture.

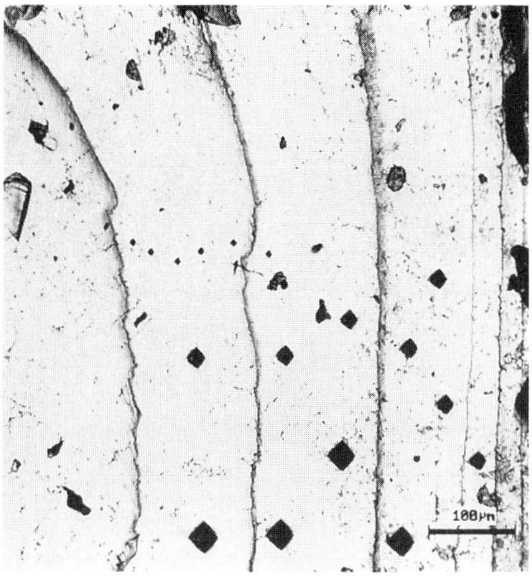

Fig. 13. Niobium side of an interface fracture, laser scan micrograph.

Fig. 16 shows glide lines between two glide steps in differential interference contrast. They inform on operating glide systems in interface fracture (Fig. 17). The microscopic analysis was used to explain and calculate contributions of the plastic deformation energy in the metal to the measured interface fracture of the bicrystals. Values of the interface fracture energy are given in Fig. 18 as a function of the work of adhesion. The amount of visible plastic deformation depends on the crystallographic orientation of the bicrystal and will be reduced if the fractional interface coverage of certain impurity atoms, e.g. sulphur or silver, is increased and concurrently the work of adhesion reduced [3,4,7].

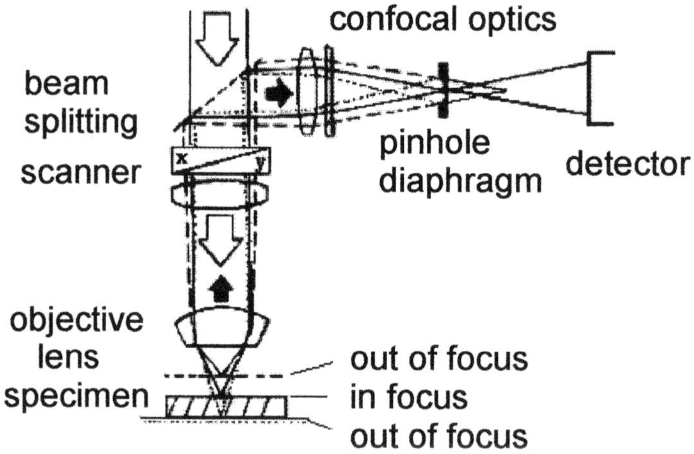

Fig. 14. Principle of confocal scanning.

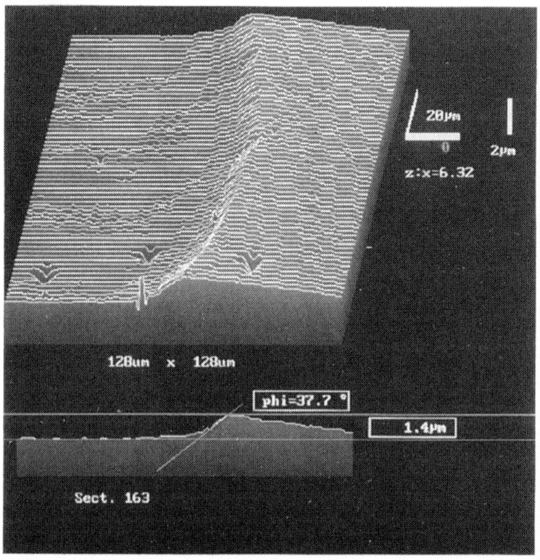

Fig. 15. Three-dimensional reconstruction of part of a niobium fracture surface by means of the laser scan microscope.

Diffusion Bonded Titanium/Alumina Implants

Dental implants manufactured from a diffusion bonded composite of titanium and alumina are developed in our institute in co-operation with dental surgeons of the University of Tübingen. The type of composite implant shown in Fig. 19 is thought to be optimal because of both the biocompatibility of titanium and the resistance against plaque formation of alumina. Metallographic studies are needed to control the quality of the bond between the metal and the ceramic. Metallographic and histological observations on sections through explants serve to reveal interactions between the tissue and the implant material.

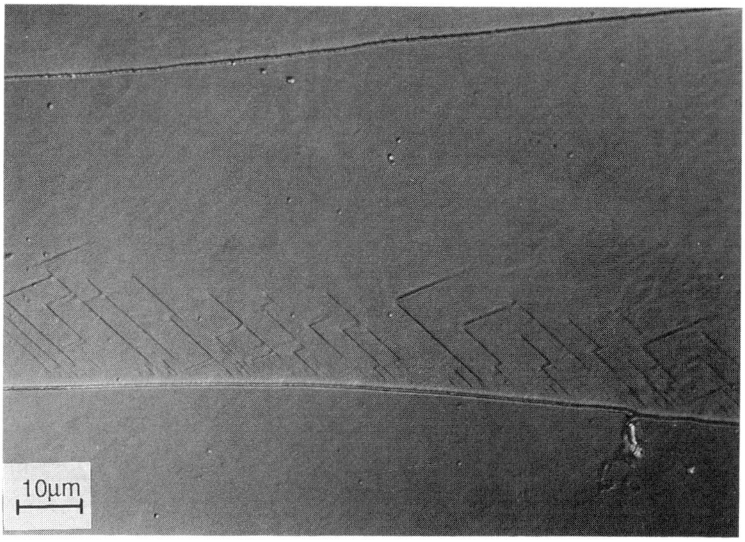

Fig. 16. Glide lines on the niobium side of an interface fracture, DIC.

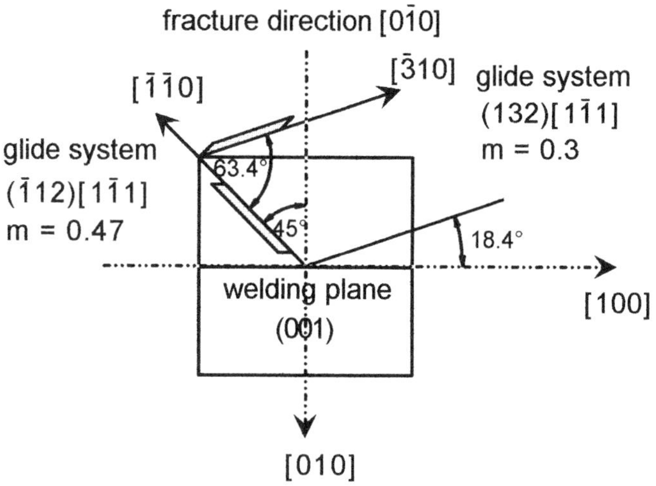

Fig. 17. Analysis of the glide lines shown in Fig. 16.

Fig. 20 shows the manufacture of an implant which consists of an alumina ring welded onto a titanium cylinder by means of a Nb interlayer. First of all the alumina ring is diffusion bonded to a niobium cylinder. The obtained composite is machined and then diffusion bonded, now within a boron nitride mould to a Ti30Ta cylinder. From the bonded piece the conical implant is manufactured. Finally, the supraconstruction is added which bears later on the artificial tooth. After insertion of the implant the conical titanium part is situated in the spongiosa whereas the alumina ring penetrates the gingiva [9].

Fig. 21 is a SEM micrograph of a cross section through alumina bonded to titanium by an 10 μm intermediate niobium foil. Adjacent to the alumina at the left the Nb-rich interlayer is situated. At the right a Nb-Ti solid solution has been formed during diffusion bonding of niobium to titanium. At a larger distance of the metal ceramic interface the α–β microstruc-

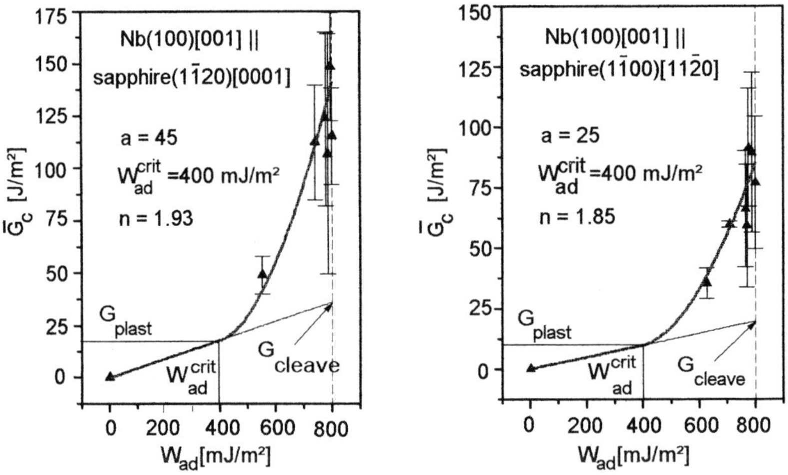

Fig. 18. Interface fracture energy $\overline{G}_c$ of niobium/sapphire joints as a function of the work of adhesion W_{ad}.

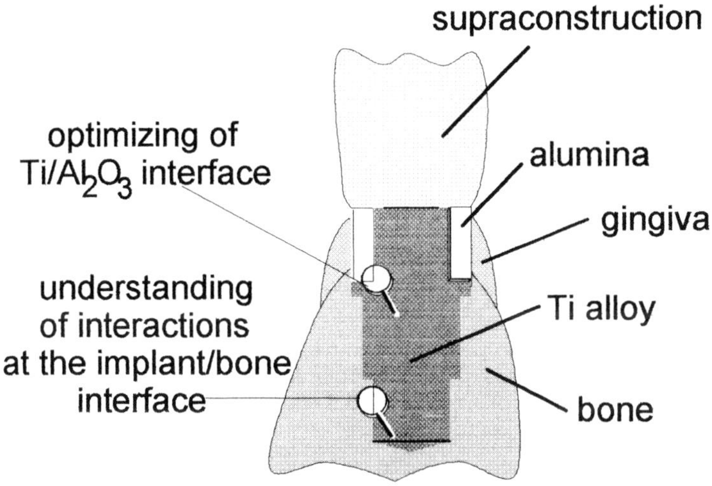

Fig. 19. Schematic of a titanium/alumina dental implant.

ture of titanium is visible. The results of a WDX analysis of such an interface area is given in Fig. 22. A decrease in Nb content occurs by Nb diffusion into the Ti matrix. Nb rich β-Ti needles cause the peaks in the Nb plot and the concurrent depressions in the Ti plot. By subdividing the diffusion bonding into the two steps of alumina-niobium bonding at 1500°C and bonding of niobium to titanium or titanium alloy the oxygen content could be further reduced to avoid immoderate solid solution hardening of the metal by oxygen. Fig. 23 is a photograph of two dental implants manufactured in our laboratory. It shows a titanium implant and a composite implant with an alumina ring onto a cone-like titanium body [10].

After diffusion bonding not only the end faces of the alumina ring but also the internal surfaces of the ring must be well bonded. This condition is controlled by cross sections through the niobium-bonded alumina ring as demonstrated in Fig. 24. An interface without any voids is visible [9].

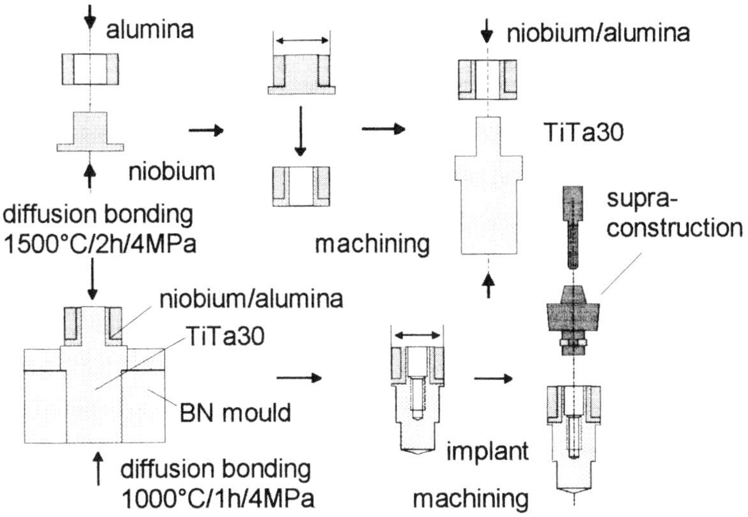

Fig. 20. Manufacture of dental implants.

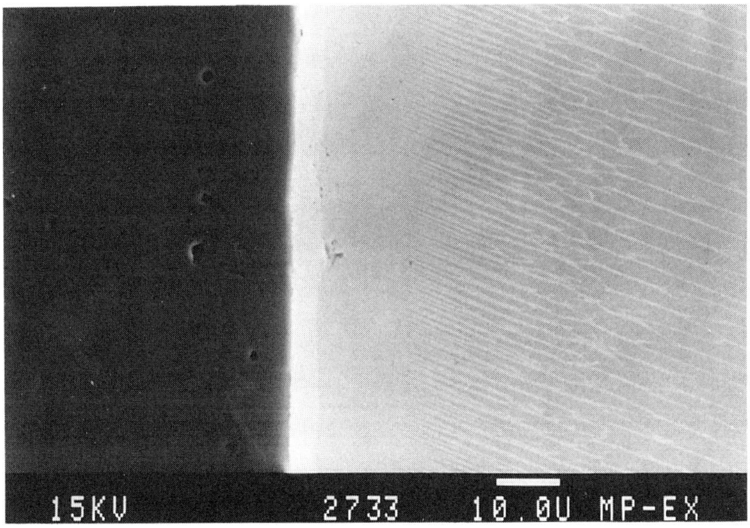

Fig. 21. Microstructure of titanium bonded to alumina by a 10 μm niobium foil, SEM.

Interdisciplinary research is required to control interactions between the implant and the bone tissue. Metallographic techniques uncommon in histology have to be applied to obtain sections through the hard metal and ceramic explants and the softer tissue. Fig. 25 shows the bone tissue at the top, then a soft tissue interlayer and at the bottom the titanium implant. A microanalysis of this cross section (Fig. 26) reveals that the biocompatible titanium interacts with the tissue and is present in the tissue interlayer [10].

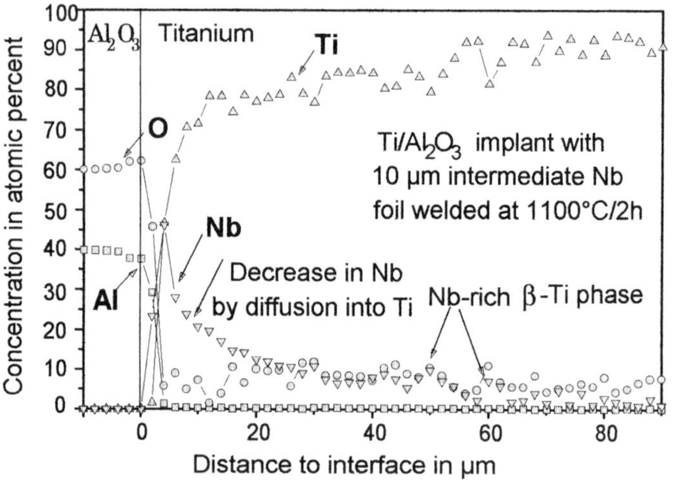

Fig. 22. WDX interface analysis of a $Ti/Nb/Al_2O_3$ bond.

Fig. 23. Dental implants made from a Ti/Al_2O_3 composite (left) and from titanium (right).

Lacunas are sometimes visible at the interface between alumina and bone tissue and a tissue interlayer is formed (Fig. 27). Fig. 28 is a SEM fractograph through the alumina implant and the bone tissue. As found by microanalysis measurements (Fig. 29) again a constituent of the implant, aluminum, penetrates from the implant into the tissue interlayer. through the niobium-bonded alumina ring as demonstrated in Fig. 24. An interface without any voids is visible [9].

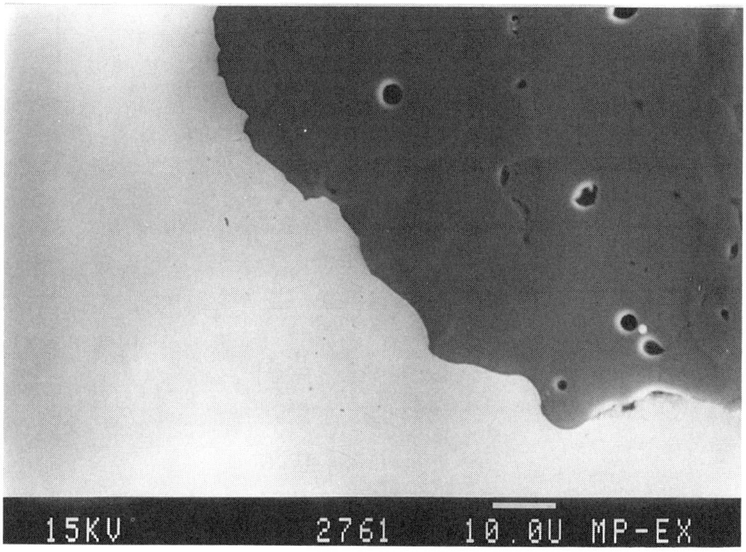

Fig. 24 Nb/Al_2O_3 interface region of a dental implant, SEM.

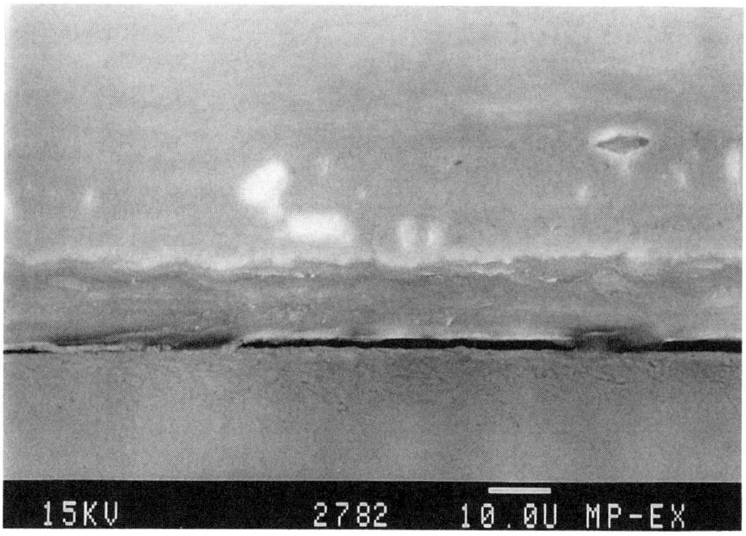

Fig. 25 Interface region between titanium and bone tissue, SEM.

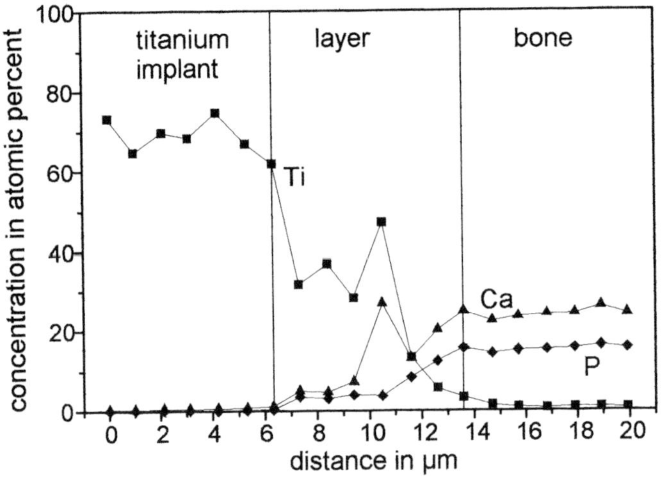

Fig. 26. EDX analysis of an interface region between titanium and bone tissue.

Fig. 27. Interface region between alumina and bone tissue, SEM.

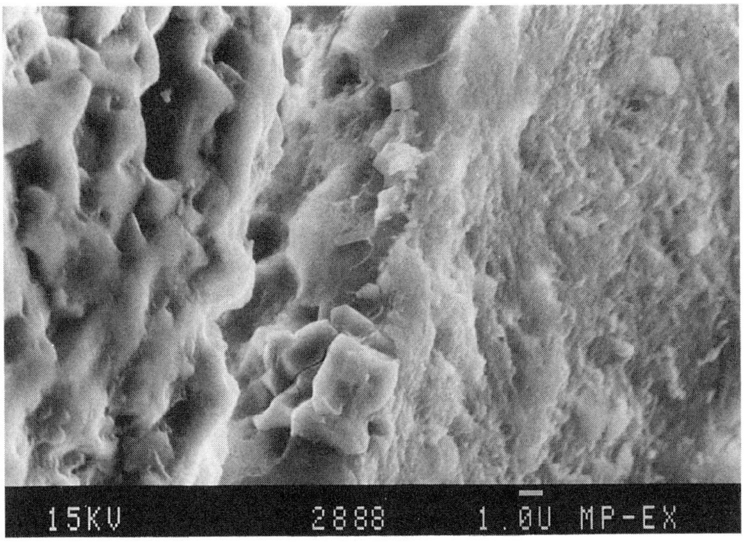

Fig. 28. Fractograph through an interface region between alumina and bone tissue, SEM.

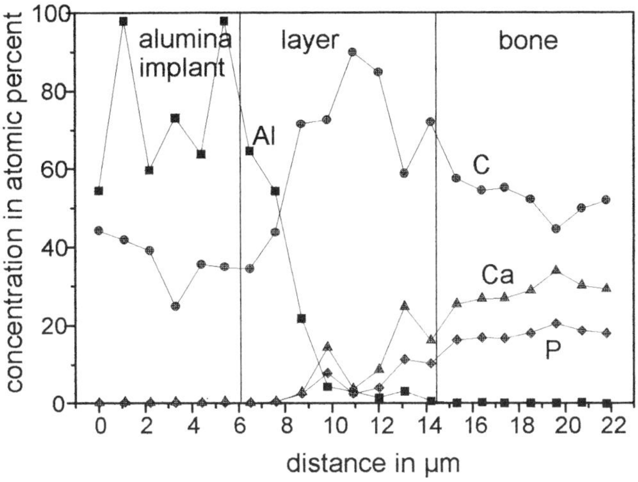

Fig. 29. EDX analysis of an interface region between alumina and bone tissue.

The research on this subject will be continued to obtain more results on interactions between biomaterial and tissue as a function of the duration time of the implant within the bone. First transmission electron microscopy studies were also performed to get a deeper metallographic and histological insight into the microstructure of this interface region

References

1. H.F. Fischmeister, G. Elssner, B. Gibbesch, K.-H. Kadow, D. Korn, F. Kawa, W. Mader, M. Turwitt, "Solid-State Bonding of Accurately Oriented Metal/Ceramic Bicrystals in Ultrahigh Vacuum, Rev. Sci. Instrum. 64 (1), 234-242 (1993)

2. H.F. Fischmeister, W. Mader, B. Gibbesch, G. Elssner, "Preparation, Properties and Structure of Metal-Oxide Interfaces", Mat. Res. Soc. Proc. 122, 529-540 (1988)

3. D. Korn, G. Elssner, H.F. Fischmeister, M. Rühle, " Influence of Interface Impurities on the Fracture Energy of UHV-bonded Niobium-Sapphire Bicrystals", Acta metall. mater. 40, S355-S360 (1992)

4. D. Korn, PhD thesis, VDI-Fortschrittsberichte Reihe 5 Nr. 320 (1993)

5. G. Kiessler, D. Korn, G. Elssner, "Metallographic and Fracture Surface Investigations of Metal-Ceramic Bicrystals", Pract. Met. 29, 597-610 (1992)

6. D. Korn, G. Kiessler, G. Elssner, "Fracture Surface Morphology of Niobium-Sapphire Bicrystals" (in German), Prakt. Met. Sonderband 24, 133-137 (1993)

7. G. Elssner, D. Korn, M. Rühle, submitted to Scripta metall. mater.

8. G. Elssner, G. Kiessler, H.-G. Kapitza, "Metallographic Surface and Interface Investigations using the Laser Scan Microscope", Pract. Met. 28, 443-454 (1991)

9. D. Korn, G. Elssner, G. Petzow, W. Schulte, "Optimizing of Titanium/Alumina Composites for Dental Implants" (in German), Z. Zahnärztliche Implantologie 8, 136-144 (1992)

10. V. Biehl, J. Breme, D. Korn, G. Elssner, G. Petzow, W. Schulte, B. d'Hoedt, W. Donath, submitted to Proc. SEM Conference, Saarbrücken (1994)

SURFACE CHARACTERIZATION OF TITANIUM DIBORIDE-ALUMINIUM COMPOSITE

Jacques Boutin and Amir Mirtchi

Alcan International Limited, Arvida Research and Development Centre,
P.O.Box 1250, 1955 Mellon Boul., Jonquière, Québec, Canada, G7S 4K8

ABSTRACT

Titanium Diboride ceramic (TiB_2) and Titanium Diboride-Aluminium (TiB_2-Al) ceramic-metal composites are excellent chemically resistant materials for use in liquid aluminium and molten cryolite (Na_3AlF_6) environments. The fabrication process of a TiB_2-Al composite and its chemical reactions with liquid aluminium can be understood, provided the microstructure is properly characterized. The relatively high hardness and brittleness of TiB_2 adds to the challenge of preparing damage free metallographic sections, permitting proper characterization. This paper presents successful techniques developed for metallographic preparation of TiB_2 and TiB_2-Al composite materials.

INTRODUCTION

The use of Titanium Diboride (TiB_2) ceramic and more particularly the TiB_2-Al ceramic-metal composite, presents a special interest for the aluminium industry because of its good electrical and thermal conductivity, its fairly high resistance to corrosion by molten cryolite (Na_3AlF_6) and its good wettability by liquid aluminium.

On the other hand, the elaboration of large rods of TiB_2-Al composites is particularly challenging. Indeed, the physical and mechanical properties of these composites will be drastically modified by a slight contamination of the virgin powder used to produce the composite, by a small variation in the volume fraction of porosity, by subtle fluctuation of the powder aggregate size distribution, or also by possible chemical reaction of liquid aluminium alloys with TiB_2 during the infiltration process.

Therefore, it is of prime importance that the fabrication process be very well controlled, and that proper metallographic specimen preparation techniques be used on the composite material without modifying the microstructure.

This work describes the metallographic techniques developed by the Alcan International Limited metallographic laboratory at the Arvida Research and Development Centre, to prepare damage free surfaces of TiB_2 ceramic and TiB_2-Al composite material. This microscopic evaluation gave information on opened and closed porosity, aggregate distribution size, material homogeneity, aluminium infiltration efficiency and the chemical reaction of TiB_2 with aluminium or with other impurities originating from the fabrication process or the molten infiltrated aluminium.

SOURCES OF TiB_2 POWDER

For this work, two different sources of TiB_2 powder were studied [1]; the first powder was obtained from titanium tetrachloride ($TiCl_4$) and boron trichloride (BCl_3) by a plasma arc technique and the second powder was obtained by a carbothermic method.

The arc-plasma grade is a high purity TiB_2 powder (Table 1). The virgin powder was coated with a special wax which prevents the TiB_2 particles from oxidizing upon contact with air moisture. Because of the wax coating process, the arc-plasma TiB_2 powder appears at low magnification, as more or less spherical agglomerations of approximately 500 microns in size (Figure 1a). But when observed at higher magnification, these agglomerations are shown to be made of very small particulates and aggregates of less than 1.5 microns in size (Figure 1b). The particulates obtained by the carbothermic technique have a tabular platelet shape, with dimensions averaging from 10 to 12 microns (Figure 2).

TiB_2 CERAMIC ELABORATION

The the TiB_2 powder was compacted out in an isostatic press and samples of 15 mm in diameter and 6 mm in length were produced. The range of pressures varied from 20 to 380 MPa [1].

Because of the pyrophoric nature of the TiB_2 powder, all handling operations of arc-plasma powder before sintering were carried out in an inert argon atmosphere. The TiB_2 bars were sintered for two hours in a vacuum furnace at about 2000ºC. By using a special pore forming agent and by controlling the various sintering parameters, TiB_2 rods were produced with a volume fraction of 2 to 35% open porosity (Figure 3) and a densification degree ranging from 60% to 99%.

After sintering, the TiB_2 rods were infiltrated by high purity (>99.999%) liquid aluminium in a vacuum furnace at about 850-1100ºC.

RESULTS AND DISCUSSION

The metallographic specimen preparation steps are: sectioning, lapping, polishing and etching. Practicing good specimen preparation will permit the the surface of the TiB_2 ceramic and the TiB_2-Al composite materials to be revealed without cracks, pull-out, polishing contaminants, etc. Furthermore, a good comprehension of each preparation step will help minimize the surface distortion of the specimen which would otherwise lead to misinterpretation of the ceramic behaviour during the fabrication process and during the operation. The TiB_2 is a very hard and extremely brittle ceramic. Only natural diamond is harder than TiB_2 material and diamond abrasive was used for all the preparation steps from cutting up to the final polish.

SECTIONING

In non-equiaxed polycrystalline TiB_2 microstructure and more particularly, for high density TiB_2 ceramics (>95%), the thermal expansion coefficients may cause residual stresses in the material when cooling at room temperature after sintering. Therefore, the sectioning step will be extremely critical. Vibration of the cut-off wheel machine, stress caused by the clamping device that holds the specimen in place during the cutting and the wheel in-feed pressure could produce spontaneous micro-cracks in the material (Figure 4). Once formed, these micro-cracks will propagate deeper into the material during the subsequent lapping and polishing steps, making it almost impossible to remove these artificial defects on the final surface that will be observed under the microscope.

To minimize vibrations during sectioning, stacking wax commonly used in machine shops was used to stick the TiB_2 specimen to a special "V" shaped specimen holder (Figure 5). The stacking wax melts at approximately 100ºC and solidifies at room temperature, firmly maintaining the specimen in place during cutting. The "V" shaped specimen holder is clamped to the cut-off machine vice without transferring the stress to the ceramic specimen. The use of the Exicut in-feed program available on the Exotom cut-off machine from Struers prevents excessive heating of the specimen and premature wearing of the blade. Following some of the advice given by König and Wemhöner [2], the sectioning conditions were optimized after many trials.

External contaminants observed on the metallographic specimen surface could be mistaken for impurities in the virgin TiB_2 powder, or introduced during the forming process of the TiB_2 ceramic. To avoid such external contaminants during sectioning, particular care was taken and a strict cutting procedure was established and followed in our laboratory. A copper bounded diamond blade was preferred over steel bounded blades because no copper was expected to be found in the virgin TiB_2 material, nor introduced during the fabrication of the ceramic. Any copper particle found could only come from the cutting blade. Also, the copper bounded diamond blade was used only for cutting TiB_2 specimens, thus minimizing the risk of contamination residue coming from other materials. Furthermore, the lubricant is known to transport residue generated from material previously cut on the cut-off machine. To avoid contamination of the TiB_2 ceramic by the lubricant, the cut-off machine was carefully cleaned before cutting the TiB_2 specimens and only freshly prepared lubricant was used.

MOUNTING

Hot-mounting in bakelite was not used because the high pressure set in the hot press could introduce cracks in the ceramic. A cold castable dental acrylic mounting media (Jet Pink from Lang Dental MFG.Co. Inc, Chicago, Ill, 60647, USA) was preferred over the common resin, because of its greater hardness. However, the acrylic mounting media contains carbon and oxygen that could smear over the specimen surface, leading to an erroneous detection of carbon and oxygen with the scanning electron microscope (SEM) by light energy dispersive X-ray (EDX) technique.

LAPPING AND POLISHING

The metallographic specimen surfaces of TiB_2 ceramics and TiB_2-Al composites were prepared according to the Buehler Dialog System described by Bousfield and Bousfield [3,4] and a polishing procedure established on the Buehler Ecomet-4 automatic polisher is presented in Table 2.

The two first lapping steps use natural diamond grade abrasive in an oil suspension which provides efficient cutting action of the hard TiB_2 ceramic. The specimen holder and the lapping wheel were set to rotate in a complementary direction to prevent a too severe stock removal rate, which would result in depth surface damage. The two lapping steps have the following objectives: 1. to remove the surface damage created during the sectioning step and 2. to flatten the specimen surface. The ceramic Metlap wheel insures excellent flatness and permits good edge sharpness when porosity is present.

The third step is called intermediate polishing, and is performed on a perforated Texmet cloth which is a synthetic, napless chemotextile mixture of rayon and nylon. The fourth step or the final polishing uses a DP-Mol type cloth, which is a coarse, woven, wool fabric. These napless polishing cloths keep the specimen surface flat until the end of the polishing. For both polishing steps, a mixture of diamond suspensions of 3 and 1 microns was used with 0.05 micron Mastermet colloïdal silica (SiO_2). The diamond abrasive eases the polishing of the TiB_2 ceramic and the Mastermet gives the shiny finish we expect to see for the aluminium phase. The final polishing is done for about four minutes, then the cloth is thoroughly washed under running water, and the polishing is continued with only colloïdal silica and the pressure is gradually reduced to zero. Prolonging the polishing time at steps 3 and 4 will result in an over-polishing of the aluminium phase, creating relief between the hard TiB_2 ceramic and the soft aluminium metal.

Because diamond and SiO_2 abrasives are used, the specimen could eventually be contaminated with carbon, silicone and oxygen, which could be detected by SEM light-EDX analysis. The diamond abrasive could be recognized as individual carbon particulates and the SiO_2 abrasive would be found as a dark residue present in small pores or cracks. If the polishing is done according to the state-of-the-art, very few or no polishing residue will remain on the specimen surface. Some of the cloth used may contaminate the specimen with chlorine. This was the case of the perforated Texmet whose fabrication process probably involved chlorine as a whitening agent. Care must be taken in interpreting the chlorine detection by SEM-EDX.

POROSITY MEASUREMENT

The evaluation of the open porosity was performed on the virgin ceramic by an immersion technique in kerosote according to the British Standard No. BS1902::Part 1A:1966 [5]. This bulk measurement was used to define the sintering parameters of the TiB_2 ceramic. On the other hand, this immersion technique cannot measure the porosity in the infiltrated TiB_2-Al composite.

The image analysis technique was used on the metallographic specimen surface to evaluate both the total porosity of the non-infiltrated TiB_2 ceramic and the efficiency of the aluminium infiltration in the open porosity of the infiltrated TiB_2-Al composite. This measurement was particularly useful to estimate the perfection of the polishing of the TiB_2 ceramic (Figure 6). Indeed, when the volume fraction of porosity measured by the image analysis technique was larger than that evaluated by the immersion technique, this was an indication of a significant pull-out of TiB_2 particulates or crack creation during polishing. Moreover, the measurement of the remaining porosity in infiltrated TiB_2-Al composite was a criteria of the infiltration efficiency (Figure 7).

GRAIN STRUCTURE ENHANCEMENT BY CHEMICAL ETCHING

The evaluation of the average TiB_2 grain size was a good indication of the grain growth as a function of the sintering temperature, pressure and time. Many etching techniques are proposed in the literature [6,7,8] to reveal the grain structure of TiB_2 ceramics, but not all of them give satisfactory results. After many trials of solution etching, electrolytic etching and plasma etching, the solution of 10 mL of nitric acid and 100 mL of hydrochloric acid appears to give the best results on our specimen. The specimen are etched for periods of 15 seconds at a time and observed under the optical microscope in Differential Interference Contrast (D.I.C.) illumination mode.

The grain size was evaluated according to the line intercept method described in the ASTM E-112 method [9]. A semi-automated method was used on an IBAS-Kontron image analysis system. In the case of TiB_2 ceramic obtained from the Arc-plasma grade powder, rather large aggregates are observed and measured (Figure 8a), while the ceramic produced by the carbothermic technique has a sharp grain structure (Figure 8b).

SEM LIGHT-EDX ANALYSIS

A Kevex QUANTUM light element EDX (energy dispersive X-ray) thin window detector, coupled to a JEOL-840A scanning electron microscope was used. To take good SEM pictures of the TiB_2 ceramic microstructure, a thin layer of gold-palladium was deposited (20 seconds) on the specimen's polished surface. When performing light-EDX analysis, an acceleration voltage varying from 5 to 15 kV was used. At 5 kV, light elements (more particularly carbon) could be detected and at 15 kV, heavy elements such as iron, calcium or other were detected. It is virtually impossible to detect the boron by SEM-EDX analysis in these TiB_2 ceramics.

Aluminium carbide was one of the most often encountered inclusions in this infiltrated TiB_2-Al composite (Figure 9a). In high purity TiB_2 powder, the carbon may come from the carbon crucible, in which the TiB_2 ceramic was infiltrated by the aluminium, or to be residual carbon from the coating wax which has not degassed during sintering. The aluminium oxide was the second inclusion most often found in the infiltrated TiB_2-Al composite. In Figure 9b, it appears that the formation and growth of the Al_2O_3 inclusions may lead to a crack formation and propagation inside the TiB_2-Al composite.

The presence of metallic impurities (Cr, Fe, W, Cu, Si) and their reaction to aluminium were only found in the carbothermic grade TiB_2 powder (Figure 10a and

10b). This is the main reason why the ceramic made from the carbothermic powder does not present as good physical properties as that made from the arc-plasma grade powder.

ACKNOWLEDGEMENT

The authors would like to thank M. Jules Bergeron for carrying out the laboratory test for the elaboration of the TiB_2 rod; Mrs. H. Dufour, Messrs. M. Belley, M. Lebleu and Y. Raymond for all metallographic preparation of TiB_2 specimens and SEM-EDX analysis; Mrs. V. Karl from the Max-Plank-Institute for Metals Research, Germany, for valuable discussion on metallographic preparation techniques for ceramic, and finally, the Buehler Albrecht Educational Centre Illinois, USA, for judicious advice on how to use the DIALOG System.

REFERENCES

1. A. Mirtchi, private communication.
2. W. König, and J. Wemhöner, Optimizing Grinding of SiSiC, Ceramic Bulletin, Vol 68, No. 3, pp 545-548 (1989)
3. B. Bousfield, A Systematic Approach to Sample Preparation, Metals and Materials, 758-761 (1988)
4. B. Bousfield, and T. Bousfield, Progress Towards a Metallography Standard, Metals and Materials, 146-148 (1990)
5. British Standard No. BS1902::Part 1A: Methods of Testing Refractory Materials (1966)
6. G. Petzow, Metallographic Etching, American Society for Metals, (1978)
7. G. Vander Voort, Metallography: Principles and Practice, McGraw-Hill Book Company, Materials Science and Engineering Series (1984)
8. Metals Handbook, Desk Edition, Edited by H.E. Boyer and T.L. Gall, American Society for Metals (1985)
9. ASTM E-112, Standard Test Methods for Determining Average Grain Size (1988)

Table 1: Physico-Chemical Characteristic of Both Source of TiB_2 Powder

(wt%)	Ti	B	Fe	W	Si	Ni	O_2	C	Cl
Arc-plasma	64.2	27.3	<0.005	---	<0.005	---	<0.20	0.80	<0.20
Carbothermic	67.8	27.3	0.020	0.003	0.004	<0.003	0.50	0.25	---

Table 2: Polishing procedure for TiB_2 ceramic and TiB_2-Al ceramic-metal composite

STEP	Cloth or Ceramic Wheel	Abrasive	Time (min)	Load (psi)	Speed (RPM)	Rotation
1	Metlap-10	Diamond 45μm Suspension-20 sec	until flat	maximum 12	120	Comp.
2	Metlap-6	Diamond 9μm Suspension-20 sec	2	maximum 12	120	Comp.
3	Perforated Texmet	Diamond 3μm + Mastermet	5	2	80	Cont.
4	DP-Mol	Diamond 1μm + Mastermet	7	2	80	Cont.

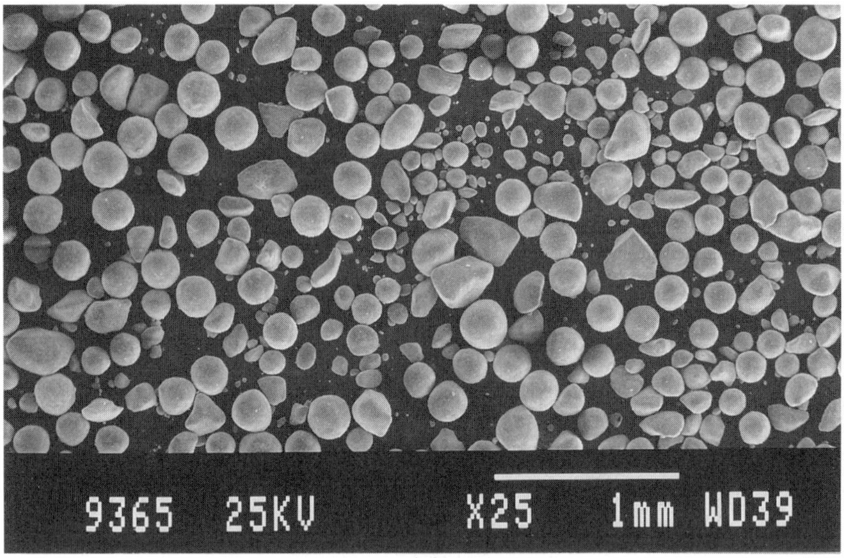

(a)

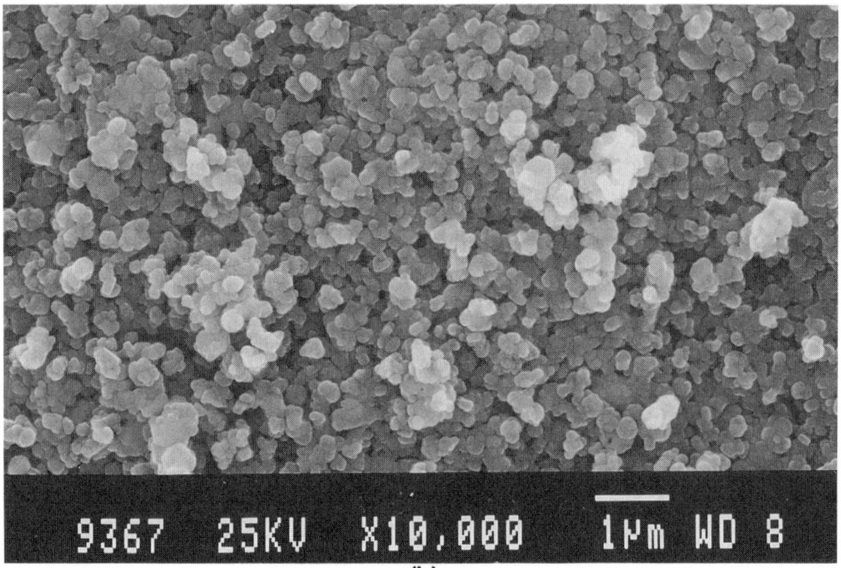

(b)

Figure 1: TiB_2 powder obtained by the arc-plasma technique shows (a) large agglomerations of up to 500 μm in size made of (b) small particulates of only a few microns (<1.53 μm) in diameter.

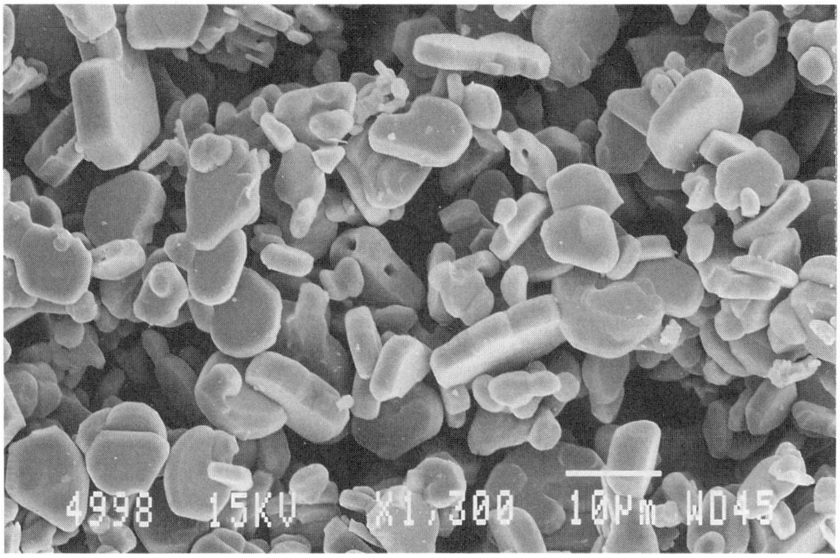

Figure 2: TiB_2 powder obtained by carbothermic technique presents large tabular platelets of 10 to 12 μm in size

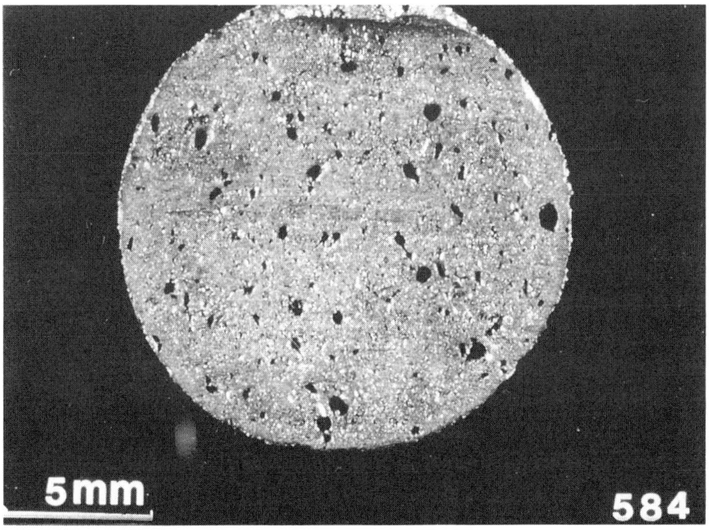

Figure 3: Cross-section of a TiB_2 bar with approximately 35% open porosity

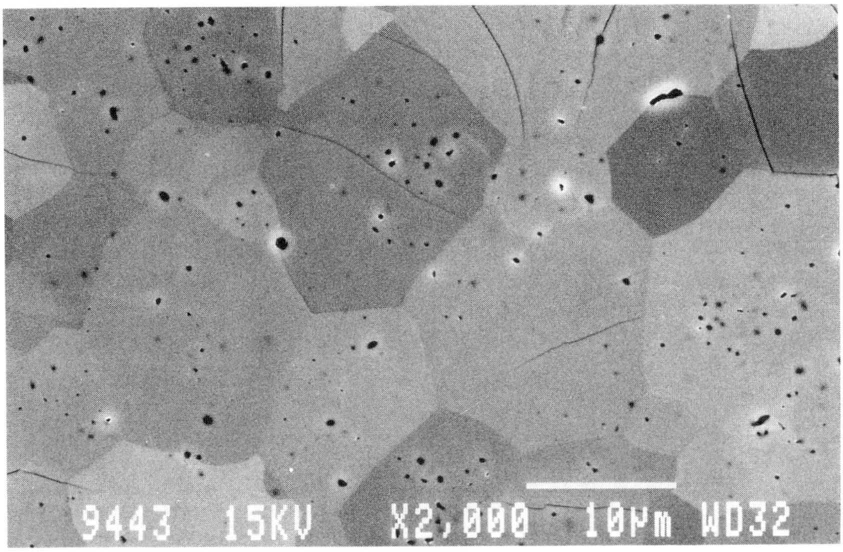

Figure 4: SEM micrograph showing micro-cracks created in the TiB_2 ceramic during cutting

Figure 5: "V" Shape holder used to prevent the clapping device of the cut-off machine to induce stresses in the TiB_2 ceramic specimen during sectioning

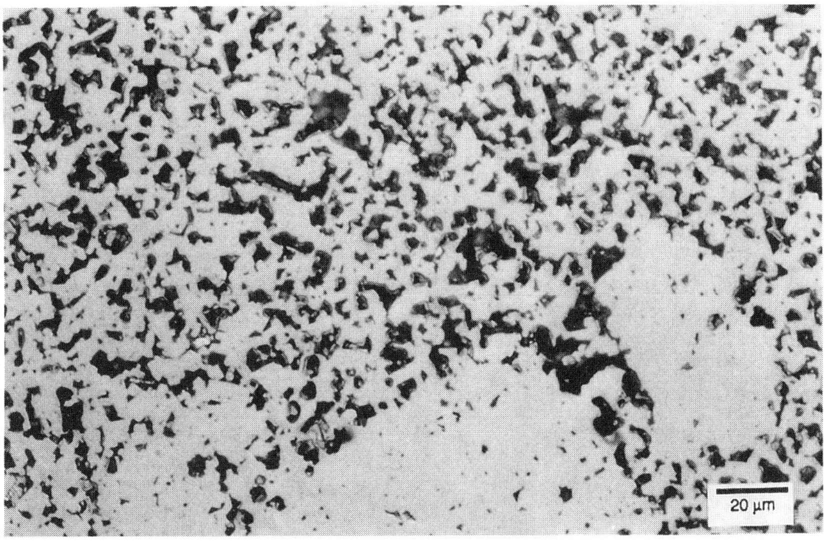

Figure 6: Image analysis technique estimated this TiB_2 ceramic at 34.8% porosity, while the immersion technique measured 35% porosity.

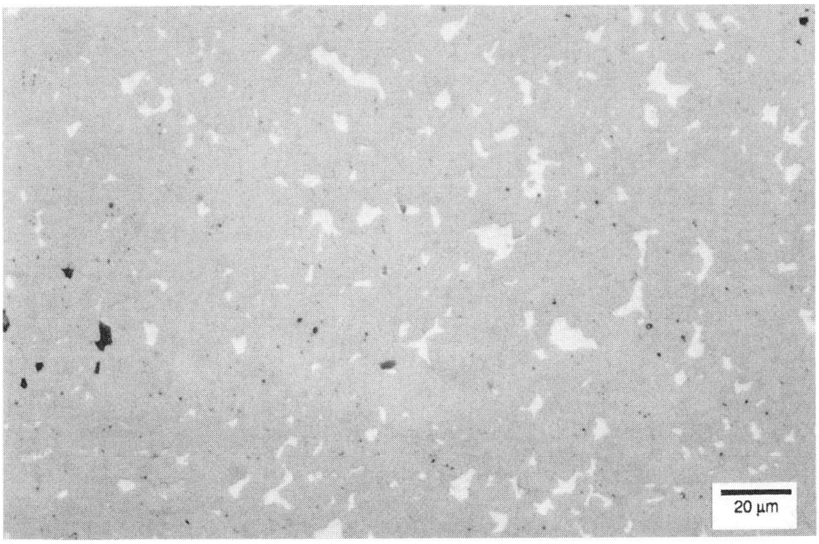

Figure 7: Aluminium inflitrated TiB_2 composite. Image analysis technique estimated 1.8% remaining porosity establishing an infiltration effeciency of 98.2%

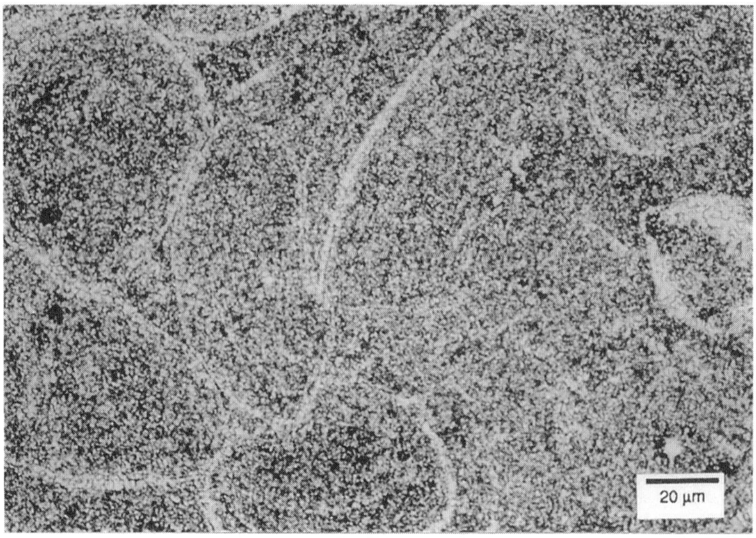

(a)

(b)

Figure 8: (a) TiB_2 ceramic produced with the Arc-plasma grade powder after chemical etching shows large aggregate and (b) produced with the carbothermic grade powder presents rather small grain size.

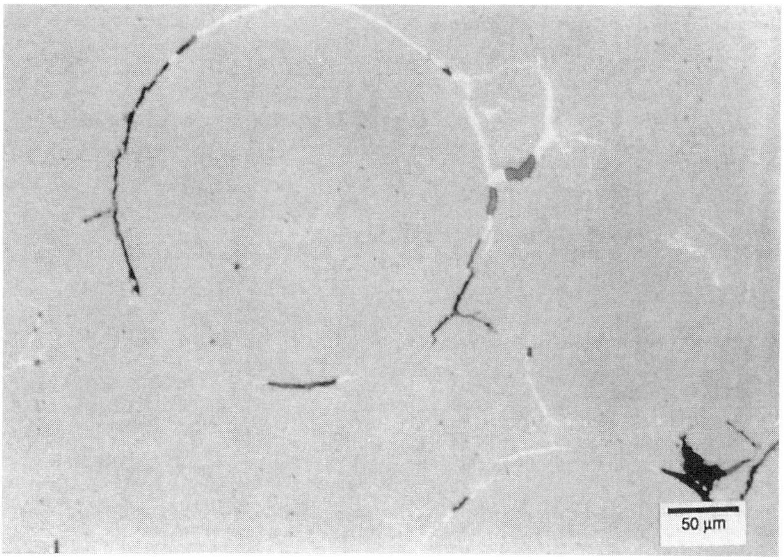

(a)

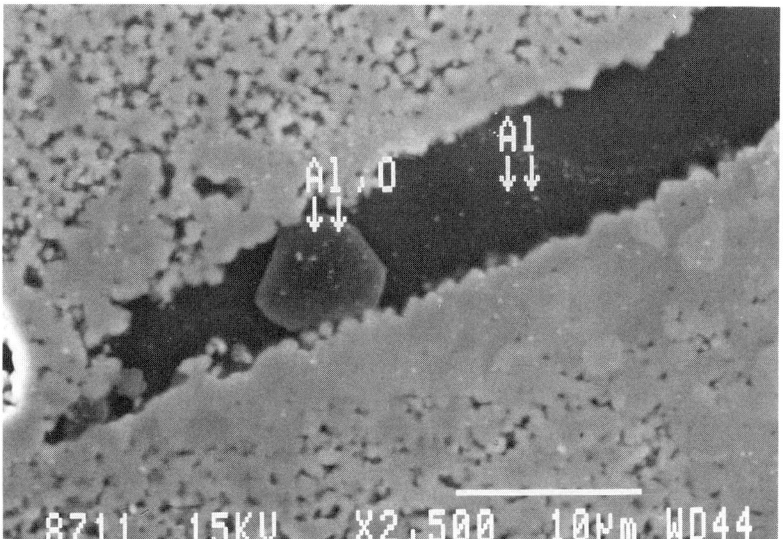

(b)

Figure 9: (a) Al_4C_3 and (b) Al_2O_3 inclusions were the only undesirable particulates encountered in TiB_2-Al composite produced from the Arc-plasma grade powder

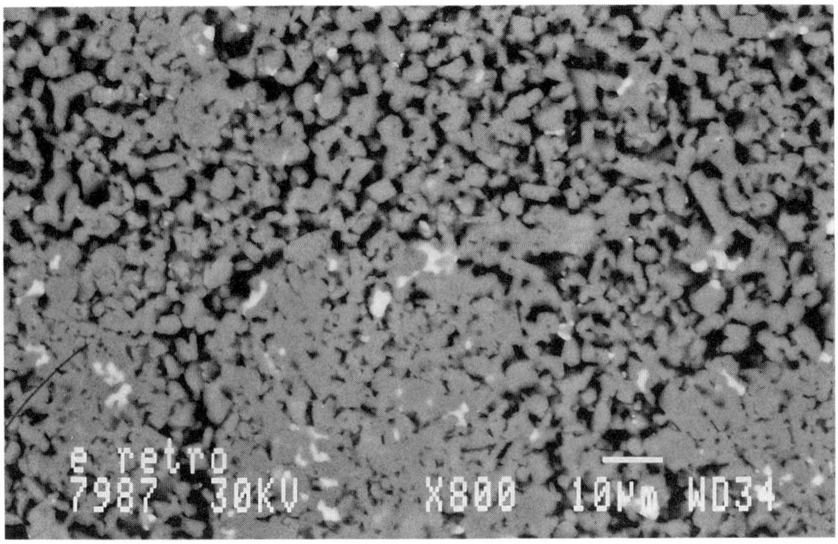

(a)

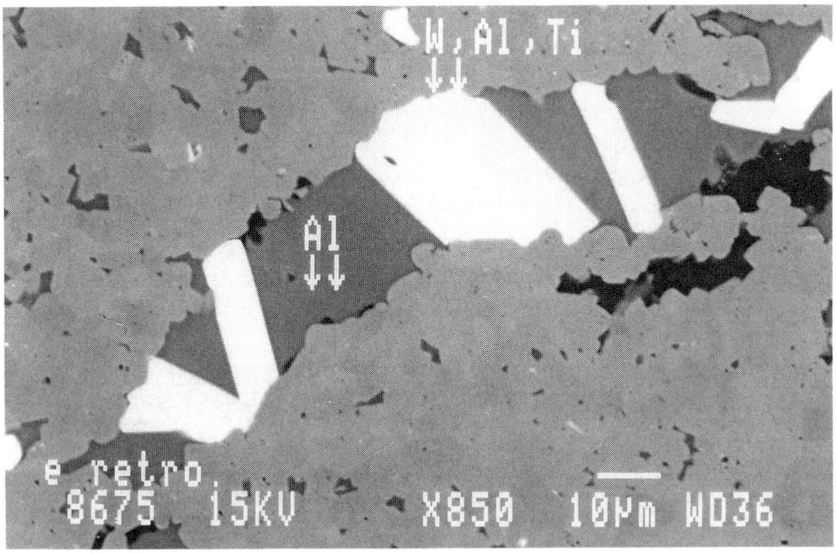

(b)

Figure 10: (a) Chromium and (b) Tungsten metallic impurities observed by electron backscattered mode on SEM in TiB_2-Al composite produced from the carbothermic grade powder

MICROSTRUCTURAL AND INTERFACIAL CHARACTERISTICS OF SIC REINFORCED AZ91D MAGNESIUM BASED COMPOSITE

J. Lo[1], M. Shehata[1], G. Carpenter[1], A. Luo[2] and M. Pekguleryuz[2]

ABSTRACT

A magnesium based composite comprising silicon carbide (SiC) particulates embedded in an magnesium AZ91D alloy was produced by the pressure infiltration technique. The microstructures of the composite in an as-cast condition was examined with Scanning and Transmission Electron Microscopy. Special attention was placed on the reaction products in the bulk alloy and at the interfaces between the SiC particle and the AZ91D matrix alloy when alumina binder was used. The possible effects on the mechanical properties due to these reaction products were discussed.

INTRODUCTION

Microstructural characteristics in a composite play an important role in determining the mechanical properties of a composite. In theory, reinforcing magnesium with SiC particulates should provide an improvement in strength and modulus of the composite over the unreinforced counterpart. However, many metallurgical or

---

1 MTL/CANMET, Dept. of Natural Resources Canada
568 Booth St., Ottawa, Ontario K1A 0G1, Canada.

2 Institute of Magnesium Technology, Inc.
357, rue Franquet, Sainte-Foy, Quebec G1P 4N7, Canada.

microstructural factors resulting from the fabrication process can cause degradation rather than improvements to the mechanical properties of a SiC/AZ91D composite. In evaluating the mechanical properties of a composite, it is very useful to examine the microstructure prior to conducting extensive mechanical tests. This should avoid the labor intensive steps of sample preparation for mechanical property evaluation, which is particularly difficult due to the high hardness of the composite materials. In addition, factors which influence the mechanical properties of a composite are mostly evident in the microstructures of a composite.

In examining the microstructure of a composite, information to look for in the reinforcing particles includes the volume fraction of particles, particle size, shape, aspect ratio and distribution. Wettability between particles and matrix alloy; and the presence of reaction products or layers are pertinent features to look for at the particle/matrix interface. Finally, grain size, precipitates and dislocation structures of the matrix alloy are important parameters to note. And there is substantial published information available for relating the above mentioned factors to the mechanical properties of a composite [1, 2, 3, 4]. For example, clustering of SiC particulates in the matrix alloy of a composite material is a well known phenomenon contributing to a fracture-initiation process which leads to premature failure of a composite material [1-2]. Excessive interfacial reaction between particles and the matrix alloy would definitely degrade the mechanical properties of both the reinforcement and the composite material. Interfacial and SiC particle cracking [4] are also undesirable phenomena which had been observed in SiC particulate reinforced magnesium composites[4]. The presence of Mg_2Si phase in SiC/AZ91D has an adverse effect on the mechanical properties of SiC/AZ91D, since Mg_2Si particles tend to fracture in an extremely brittle manner with frequent cleavage along the edge of the phase [3].

Considering the mechanical properties of a composite are closely linked to the fabrication process and fabrication conditions used, it is only through the understanding of the metallurgical factors which govern the properties of a composite material that one can fabricate a composite material with optimized properties, through the proper selection of materials and processing parameters.

In this paper, magnesium AZ91D alloy reinforced with silicon carbide (SiC) particulates was produced by the pressure infiltration technique. Alumina (Al_2O_3) was used as a binder material for fabricating SiC particulate preform. The microstructures of the composite in as-cast condition was first examined with optical microscopy, followed with scanning and transmission electron microscopy. Special attention was paid on the reaction products formed both in the bulk and interfaces between the SiC particle and the AZ91D matrix alloy.

EXPERIMENTAL PROCEDURES

AZ91D alloy was selected for this investigation. Table 1 shows the chemical composition of this alloy. The reinforcement particles used in this work are high purity SiC supplied by Superior Graphite Co. The average particle size of SiC was reported to be 7 μm.

Table 1. Chemical composition (wt.%) of AZ91D alloy.

Mg	Al	Zn	Mn	Si	Cu	Ni	Fe
bal.	8.5-9.5	0.5-0.9	>0.15	<0.02	<0.015	<0.001	<0.005

The pressure infiltration technique was used to fabricate the AZ91D composite. This technique consisted of two stages, the preform fabrication and the infiltration of molten AZ91D into the preform under pressure. In the preform fabrication stage, the SiC particulates were held together by an alumina binder to form a rigid three-dimensional network of inter-connecting particulates. The volume fraction of SiC particulate in the preform was 40 vol.%. The infiltration stage began by placing a preform in a die preheated to (~600°C) and the superheated AZ91D alloy (~100°C above its melting temperature) was introduced into the die. A punch was then lowered into the die at a speed of 1 mm/s, forcing the molten metal to infiltrate the preform. The pressure exerted by the punch on the molten metal was steadily increased to a pre-determined level of approximately 40 MPa. The applied pressure was then maintained until the molten metal was completely solidified, which normally took ~5 minutes. On completion of the pressure infiltration operation, the punch was retracted from the die and the composite product was removed.

In this study, both the SiC particulate preforms and SiC/AZ91D composites were fabricated at MTL/CANMET. To evaluate the microstructures of the composite materials, sectioned and polished tensile specimens were examined with a scanning electron microscope (SEM) using back-scattered electron (BSE) imaging to provide atomic number contrast. Specimens for TEM were cut in the form of sections ~200μm thick with a diamond wheel. Discs 3mm in diameter were then ground mechanically to a thickness of ~80μm and dimpled on one side using a Gatan Dimple Grinder, until the central region was ~10μm thick.

Final thinning to perforation was carried out in a Gatan Precision Ion Polishing System (PIPS) by argon-ion beam sputtering at 4keV at an incident angle of 4°. The TEM specimens were examined at 200kV in a Philips CM20FEG TEM/STEM fitted with an Oxford Instruments thin-window energy dispersive X-ray (EDX) detector with an eXL analyzing system, as well as a Gatan 666 parallel electron energy loss spectrometer (PEELS).

RESULTS

The microstructures of the SiC/AZ91D composites were composed of a distribution of SiC particulates surrounded by the AZ91D matrix as shown in figure 1. The particles were uniformly distributed in the specimen despite the low volume fraction of binder (2% Al_2O_3) that was used to hold the particles together during the pressure infiltration operation. Since the sample contains 40 vol% of SiC particulates, the

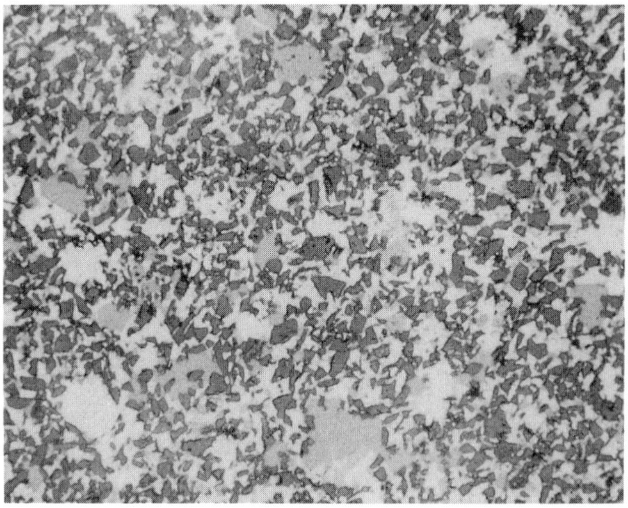

Figure 1. Optical micrograph showing the microstructure of pressure infiltrated SiC/AZ91D composite. (Magnification 500X)

sample is clustered with SiC particulates. The light grey chunky crystalline phase in the sample is Mg_2Si phase. The composition of this phase was confirmed using scanning and transmission electron microscopy. Other metallurgical aspects of the microstructure of SiC/AZ91D are illustrated by a scanning electron micrograph in figure 2. In this figure, the dark particulates are SiC reinforcement; the white needle-like phase was identified by EDXA to be an Mn/Al intermetallic; and the grey sharp-edge particles are crystals of Mg_2Si, which were the reaction products produced during pressure infiltration. In addition, some fine particles were observed at the SiC/AZ91D interfaces. Such particles were examined with transmission electron microscopy and the result is discussed in the following section. According to previous investigation [4], a discontinuous layer of microcrystals at the SiC/AZ91D interfaces was also reported. The SEM analysis showed two peaks of aluminum and carbon and a small one of oxygen. It was then postulated that such crystals could consisted of a complex ternary Al-C-O phase.

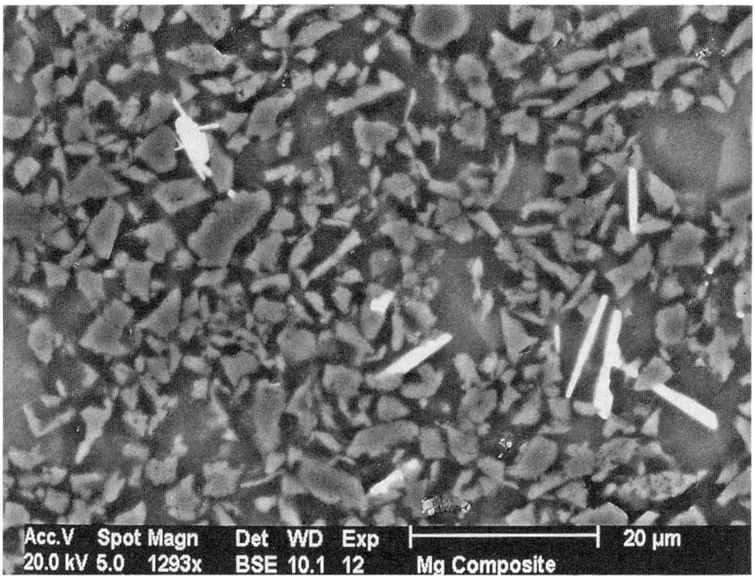

Figure 2. Scanning electron micrograph of SiC/AZ91D in back scattered electron mode showing white Mn/Al needle-like phase.

The SiC particles were readily distinguished from the matrix Mg alloy by the presence of faults or microtwins and few dislocations (figures 3 and 4), when evaluated with transmission electron microscopy. In contrast, the Mg matrix contained a high density of dislocations, probably caused by differential contraction of the two phases during cooling after melt infiltration. This has resulted in the building up of residual stresses in the matrix surrounding the particulates. The SiC particles were surrounded by a film that was mainly composed of fine (10~50 nm in diameter) oxide particles (e.g. figure 3), although some carbon may also have been present. This oxide film had a thickness of ~200 nm. Selected area electron diffraction and PEELS analysis showed this phase to be composed mainly of magnesium oxide, MgO. No evidence for the presence of the Al_2O_3 binder was observed in the TEM sample.

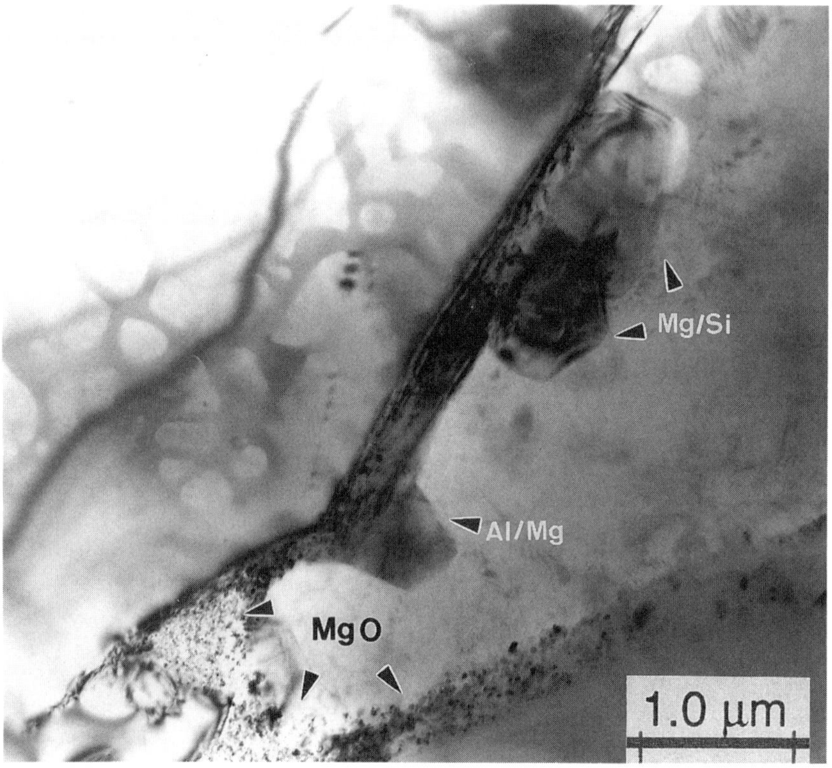

Figure 3. Sample showing a Mg/Al intermetallic and Mg_2Si precipitate, as well as fine-grained MgO at the particulate/matrix interfaces.

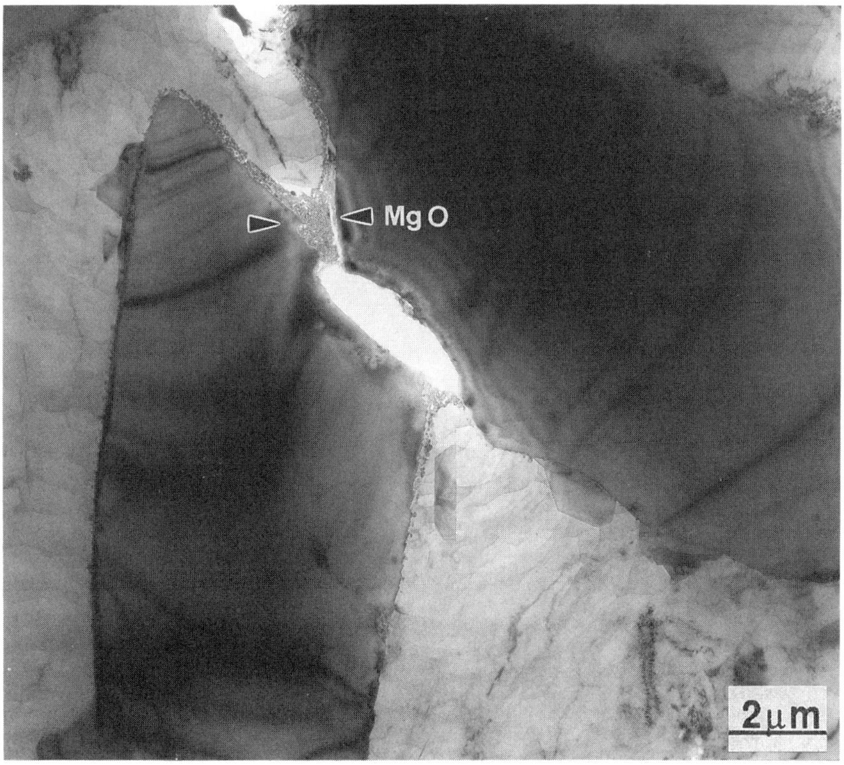

Figure 4. Fine-grained MgO phase at both SiC/matrix and SiC/SiC interfaces. Intermetallic phases are also present.

TEM also confirmed the presence of second-phase precipitates at the particle/matrix interfaces (figure 3). The majority of precipitates were composed of Mg and Al. Preliminary analysis using EDX and convergent beam diffraction (CBED) revealed the presence of Mg_3Al_2, $Mg_{17}Al_{12}$ and $Mg_{32}Al_{49}$. Also present were occasional "rectangular" Al/Mn precipitates and particles that were identified as Mg_2Si.

To summarize the above findings with the scanning and transmission electron microscopes, the infiltration of AZ91D into SiC particulate preform which Al_2O_3 was used as a binder had resulted with the formation of $Mg_{17}Al_{12}$, $Mg_{32}Al_{49}$, Mg_3Al_2, Mn/Al, Mg_2Si, MgO and possibly Al_4C_3. These reaction products found in the SiC/AZ91D composite can be summarized in figure 5. The presence of Mg/Al and Mn/Al phases is expected when AZ91D was solidified from a liquid state. However, the Mg_2Si, MgO and (possibly) Al_4C_3 are reaction products formed during the processing operation.

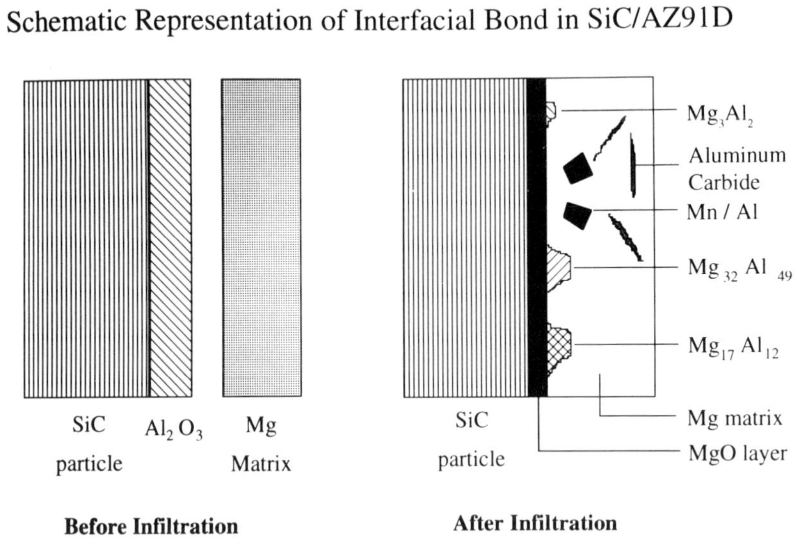

Figure 5. Schematic representation of interfacial bond in SiC/AZ91D composite before and after pressure infiltration.

In order to explain the formation of the above reaction products during the infiltration of AZ91D into SiC particulate preform, chemical reactions must have taken place between the AZ91D alloy with the Al_2O_3 binder and SiC particulates. It is postulated that when AZ91D was infiltrated into SiC preform, the magnesium reacted with the Al_2O_3 binder to form MgO as follows:

$$3\,Mg + Al_2O_3 = 3\,MgO + 2Al \quad (1)$$

Thus a skin of MgO is commonly found on the surfaces of most SiC particulates. Subsequently, the excess aluminum rejected into the solution reacted with magnesium to form Mg_3Al_2, $Mg_{17}Al_{12}$ or $Mg_{32}Al_{49}$. These Mg/Al phases are quite commonly found in AZ91D alloy. However, it is probable that the following reactions were also promoted:

$$4\,Al + 3\,SiC = Al_4C_3 + 3\,Si \quad (2)$$

$$Si + 2\,Mg = Mg_2Si \quad (3)$$

thus causing the formation of brittle Mg_2Si and Al_4C_3 phases.

Figure 6. Scanning electron micrograph showing a view of interparticle mode of failure in the SiC/AZ91D composite.

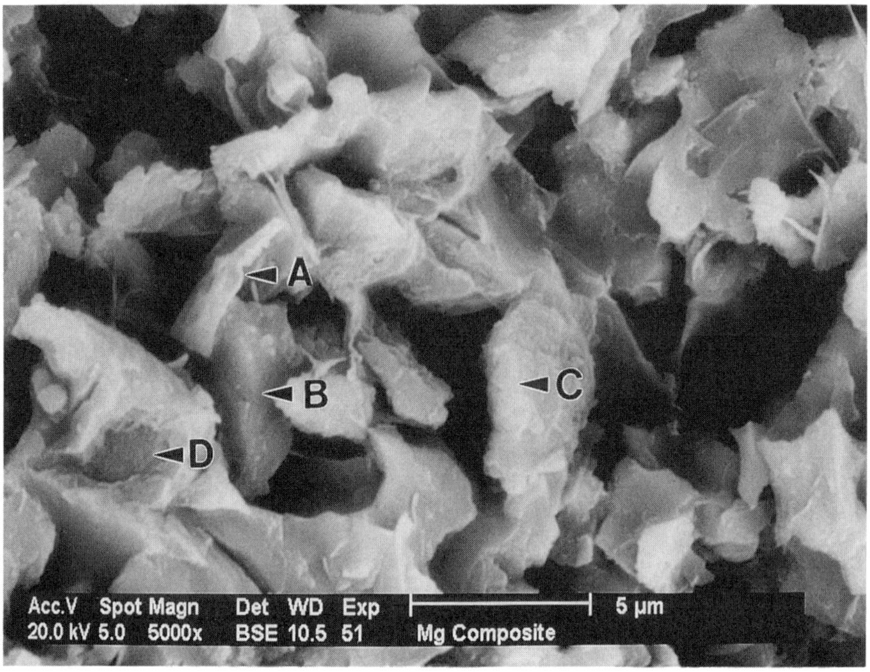

Figure 7. Scanning electron micrograph showing features such as (a) MgO film, (b) SiC particle, (c) SiC particle covered with MgO film and (d) fractured MgO film exposing SiC.

To acquire some understanding on the effects of the bulk and interfacial reaction products on the mechanical properties of SiC/AZ91D, the fracture surface of the composite was evaluated. Generally, the failure mode of the SiC/AZ91D was interparticulate fracture as shown in figure 6. This mode of failure is not commonly found in commercial aluminum or magnesium based SiC particulate reinforced composites. Although one should take into consideration that the employment of particulate preform is also not very common.

Making a more careful examination of the fracture surface, figure 7 provides a clearer picture of the fracture behaviour of SiC/AZ91D composite. In view of the fact that most of the SiC particulate were covered with oxide layer, such weak interfacial bond between the oxide layer and the AZ91D matrix was very prone to fracture, thus resulting in an interparticle mode of failure. In figure 7, a silicon carbide particles covered with oxide film; a silicon carbide particle with partially fractured oxide film; as well as a fractured oxide film are all apparent. Thus, the fracture surface of the composite revealed the brittle nature of the composite. At this point, one can safely concluded that the use of Al_2O_3 is not a suitable binder for making SiC particulate preform for fabricating magnesium based composites if pressure infiltration technique is employed. The use of other binder systems or different fabrication techniques (such as squeeze casting which provide a shorter interaction time between the reinforcement and the molten metal) should thus be considered.

CONCLUSIONS

Based on the observations made in both the optical, scanning and electron microscopy, the following conclusions are made:

1) Transmission electron microscopy showed that the particulates generally appeared to be well bonded to both the reaction layer and the matrix.
2) The large difference between the coefficient of thermal expansion of the matrix and the particulates generates sufficient stress to form a high density of dislocations in the matrix.
3) The presence of alumina binder at the particle/matrix interfaces results in the formation of a MgO layer around the SiC particles. The by-product aluminum from this reaction reacts with magnesium to form Mg_3Al_2, $Mg_{32}Al_{49}$ and $Mg_{17}Al_{12}$. Occasionally, Mg_2Si was also found.
4) Mn/Al phases were found present in the matrix of the AZ91D composite.
5) Extensive twinning was found in all SiC particles.
6) When the SiC/AZ91D sample was fractured, it failed by inter-particle mode of fracture, caused by the presence of the MgO interfacial layer on the SiC particles.
7) From the above observations, Al_2O_3 would not be a suitable binder for making SiC particulate preform, if the preform is used for fabricating magnesium based composites by pressure infiltration.

ACKNOWLEDGMENT

The authors would like to thank the following colleagues at MTL/CANMET: R. Dos Santos for the preparation of specimens, B. Casault and M. Charest for assisting with the scanning and transmission electron microscopy.

REFERENCES

[1] W.H. Hunt, O. Rihmond and R. D. Young, Proceedings of the Sixth International Conference on Composite Materials, Second European Conference on Composite Materials, London, July 1987, edited by F.L. Matthews, N.C.R. Buskell, J.M. Hodgkinson and J. Morton, (Elsevier, Amsterdam, 1987) vol. 2, 209.

[2] D.J. Lloyd, "Aspects of Fracture in Particulate Reinforced Metal Matrix Composites" Acta Metall. 39 (1) (1991) 59.

[3] K. Hicks, G. Love, V.D. Scott and R.L. Trumper, "Degradation of Nicalon Fibers in Magnesium-Based Composite", Proceedings of the 9th International Conference on Composite Materials, edited by Antonio Miravete (University of Zaragoza, Woodhead Publishing Ltd., 1993) vol. 1, 325.

[4] B. Inem and G. Pollard, "Interface Structure and Fractography of a Magnesium-Alloy, Metal Matrix Composite Reinforced with SiC Particles" Journal of Materials Science, 28 (1993) 4427.

EFFECTS OF PARTICULATE SPACING DISTRIBUTION ON INTERNAL STRESS AND DAMAGE IN MMCs

S.Tao*, N. Townley* and J.D. Boyd*

ABSTRACT

Methods of characterizing the spatial distribution of particulate in Metal Matrix Composites (MMCs) are reviewed. The degree of particle clustering is shown to vary with stirring time in the melt, mean volume fraction and extrusion ratio, and to correlate well with fracture toughness and the magnitude of the residual internal stress.

INTRODUCTION

Distributions of Particulate in MMCs

The processing of particulate reinforced MMCs produces an inherent non-uniform particle distribution [1]. In molten metal mixing techniques, particle clustering originates in the melt, and can be intensified by the solidification process. The processing is closely controlled to minimize these effects, nevertheless most cast MMCs have visibly clustered particle distributions (e.g. Fig. 1). Subsequent hot working of cast MMC billets homogenizes the as-cast structure somewhat, but anisotropic deformation processes such as extrusion or rolling produce another variation in particle spacing with orientation. The reinforcement is typically 10-20 volume percent of irregular shaped particles having a mean diameter of 10-15 μm and a mean aspect ratio of about 2. Hence, the spacing between adjacent particles in a cluster can be as small as a few microns, and it is postulated that the local stress in these regions can be as high

*Department of Materials and Metallurgical Engineering, Queen's University, Kingston, Ontario, Canada K7L 3N6

as 3 times the average stress (either external applied stress or internal residual stress resulting from processing) [1, 2]. Consequently, properties of these materials, notably fracture toughness, depend strongly on the spatial distribution of the particles, in addition to the mean values of their volume fraction, size and shape [3].

Methods of Characterizing the Spatial Distribution of a Dispersed Phase

The effects of spatial distribution of impurity phases on the fracture properties of alloy systems, such as inclusions in steels, have been studied by several authors [4 - 6]. In these systems the volume percent of the dispersed phase is low (~ 0.1%), and the spatial distribution can be accurately characterized by considering the particles as an array of points. Thus, the centroid of each particle is located by measurement and the spatial distribution is evaluated by calculating the number density, the nearest neighbour spacings or parameters derived from a spatial tessellation. The details of these methods are given in the review by Shehata [7]. (The basic problem of characterizing the spatial distribution of points or other geometric shapes exists throughout the natural sciences, and has been considered independently by crystallographers, geographers, astronomers and many other disciplines [8]).

The methods based on particle centroids are less applicable for materials such as MMCs because they do not accurately measure the real (i.e., edge-edge) inter-particle spacings when the volume percent is high and the particle shape is irregular. Modifications to particle centroid methods have been reported for fibre reinforced composites of constant fibre diameter [9], and for particulate reinforced composites with ellipsoidal shaped particles [10]. However, the most direct method to measure edge-edge inter-particle spacings is to employ the progressive dilation routine which exists on most image analysis systems, and record the number of intersections each step. In addition to obtaining a direct measure of the distribution of edge-edge spacings, a spatial tessellation can be constructed by creating a cell boundary at each point of impingement, and other parameters such as local volume fraction distribution and cell area distribution can be determined [7]. Since the automatic dilation routine follows a particular grid, such as square, hexagonal or octagonal, the resulting tessellation pattern will be different from the Dirichlet tessellation (Voronoi diagram) which is based on the normal bisector line of the line between each pair of centroid points. The difference between the patterns produced by the two types of tessellations becomes vanishingly small as the number of sides of the dilation polygon increases.

Quantitative Measure of the Degree of Particle Clustering

The distribution of particle spacings results from the physical processes which produced the distribution, e.g., molten metal mixing and solidification for

MMCs. An extremely clustered particle dispersion should exhibit a bimodal frequency distribution of particle spacings, with one peak corresponding to the small spacings within clusters and the second peak corresponding to the larger spacings of particles between clusters. This is rarely observed in real systems. Even when metallography shows obvious clustering of particles, the measured particle spacing distribution is usually unimodal, but the spread of the distribution increases with the degree of clustering. Hence the standard deviation of the measured distribution (edge-edge spacing, cell area or local volume fraction) is a good parameter to use to determine the effects of different processing variables on particle clustering.

It is also important to determine how much the measured spatial distribution differs from that expected for a random distribution of particles. Random distributions of particles can be generated by computer algorithms and compared with the measured distribution [7, 9], but it is generally sufficient to assume that a random distribution can be represented by a Poisson distribution. Thus, the measure of degree of clustering is the parameter $R = \sigma_M/\sigma_P$, where σ_M is the standard deviation of the measured distribution and σ_P is the standard deviation of the equivalent Poisson distribution.

EXPERIMENTAL PROCEDURES

In the present work particle distributions in particulate reinforced MMCs were analyzed by optical metallography with a KONTRON VIDAS image analysis system. An octagonal dilation routine was used to produce frequency distributions of nearest neighbour edge-edge spacings (λ_{e-e}) and to produce a tessellation of the particle distribution in each field. An example of the cell structure resulting from a tessellation is given in Fig. 2. Frequency distributions of tessellation cell area (A_{tc}) and local volume fraction of particulate (V_{lf}) were computed from the resulting cell structures. The standard deviation of the frequency distribution of one of the parameters λ_{e-e}, A_{tc}, V_{lf} was used as the measure of degree of clustering. The equivalent Poisson distribution was assumed to represent a random distribution for each parameter.

FRACTURE TOUGHNESS OF CAST MMCs

The material studied was aluminum A356 alloy reinforced with SiC particulate. Cast billet of A356 + 15 vol. pct. SiC obtained from Alcan was remelted, mechanically stirred and squeeze cast as 80 mm diameter × 35 mm height cylindrical blocks [2]. To produce a range of particle spacing distributions, a series of castings were prepared with the stirring time in the melt varied from 0.5 - 3.5 min. A second set of castings was prepared with the particle volume fraction ranging from 0 - 15 pct., by diluting the original melt with unreinforced A356 alloy and stirring for a constant time of 2 min.. A typical microstructure of the squeeze cast material is shown in Fig. 1.

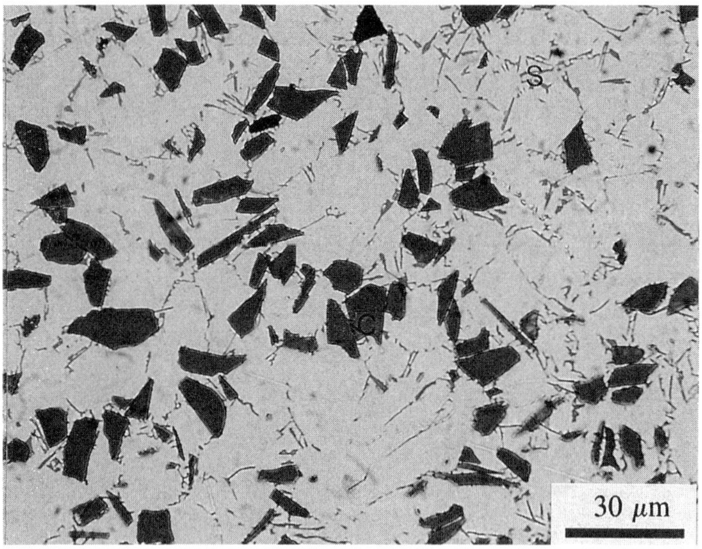

Figure 1. As-cast A356 + 15 vol. % SiC.

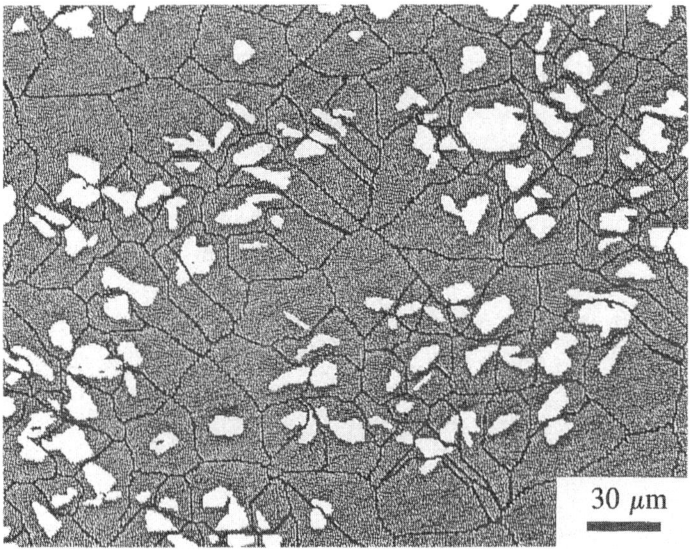

Figure 2. Tessellation network resulting from octagonal dilation procedure for A356 + 15 vol. % SiC.

The mean values ($\overline{\lambda}_{e\text{-}e}$) and the standard deviations (σ_λ) of the measured distributions of particle spacings for each condition are given in Table 1. The values of the R parameter indicate that all the particle distributions are clustered ($R_\lambda > 1$). Furthermore, there is an apparent decrease in the degree of clustering with increased stirring time at constant nominal volume fraction and with increased volume fraction at constant stirring time. This is illustrated in Fig. 3, which compares the frequency distributions of $\lambda_{e\text{-}e}$ for the 5.1 vol. pct and 14.2 vol. pct. materials and the equivalent Poisson distributions.

Table 1. Measurements of particle spacing and fracture toughness for squeeze cast A356 + SiC.

Stir time(min.)	V_f(%)	$\bar{\lambda}_{e\text{-}e}$($\mu$m)	σ_λ(μm)	σ_P(μm)	R_λ	K_Q (MPa-m$^{1/2}$)
0.5	12.2±4.1	19.9	19.8	10.0	1.98	11.8
1.0	12.8±1.8	16.9	18.1	9.2	1.97	14.3
1.5	13.9±3.0	18.9	12.3	9.7	1.27	19.1
2.0	14.2±1.9	17.2	14.2	9.3	1.53	14.1
2.5	13.5±1.8	17.6	15.6	9.4	1.66	15.5
3.5	14.1±1.9	15.6	11.7	8.8	1.32	17.8
2.0	0	----	----	----	----	30.0
2.0	5.1±1.1	26.8	22.7	11.6	1.96	14.0
2.0	5.8±2.3	19.5	18.3	9.9	1.85	17.0
2.0	9.9±0.6	23.2	18.4	10.8	1.71	15.8
2.0	12.3±2.2	23.3	19.3	10.8	1.79	17.9
2.0	14.2±1.9	17.2	14.2	9.3	1.53	14.1

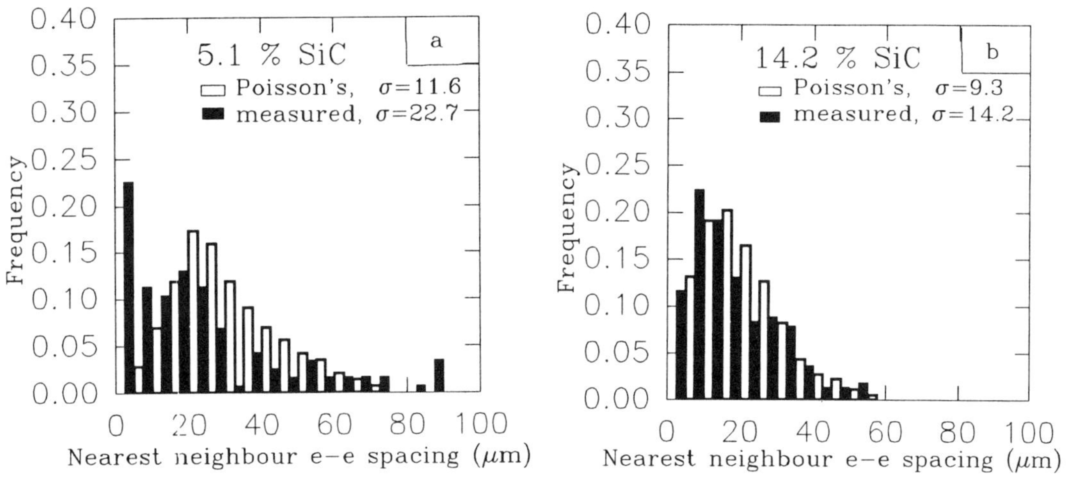

Figure 3. Frequency distributions of nearest neighbour edge-edge spacing and equivalent Poisson distribution (a) 5.1 vol. %SiC, (b) 14.2 vol. %SiC.

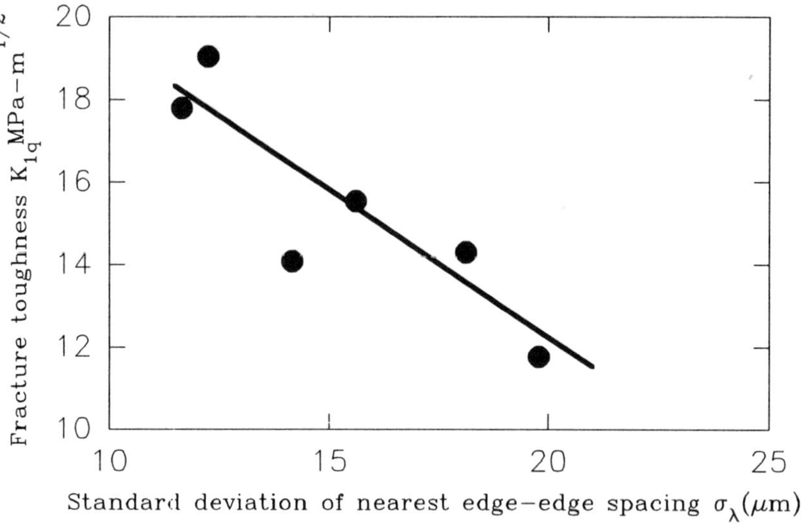

Figure 4. Correlation between fracture toughness and degree of particle clustering (σ_λ) for constant nominal V_f (15%) SiC.

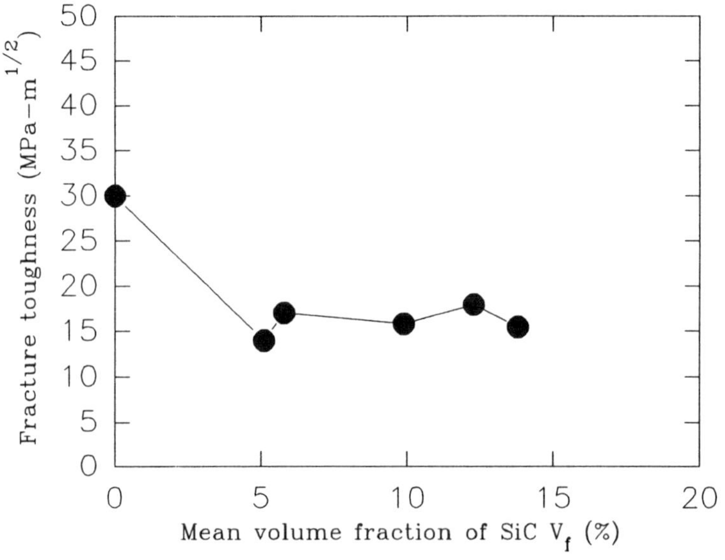

Figure 5. Variation in fracture toughness with mean V_f of SiC.

Fracture toughness tests were carried out on double-notch 4-point bend samples machined from each casting[2]. The K_Q values obtained from a minimum of 2 tests are given in Table 1. In Fig. 4, the K_Q values are plotted against σ_λ for the set of castings containing constant nominal volume fraction of SiC. The linear regression line fitted to this data (correlation coefficient = 0.88) shows a significant decrease in K_Q with increasing σ_λ. That is, at constant volume fraction, fracture toughness decreases with increased degree of clustering. For the set of castings with variable volume fraction, a plot of K_Q against mean volume fraction gives the result shown in Fig. 5. This shows that fracture toughness is insensitive to variation in mean volume fraction when the particle distribution is clustered. The fracture toughness correlates with the local parameter λ_{e-e} which is relatively constant while the mean volume fraction varies from 5-15%. These results are explained by observations that damage (particle cracking and decohesion) initiates preferentially in particle clusters and the fracture path links the damage clusters[2].

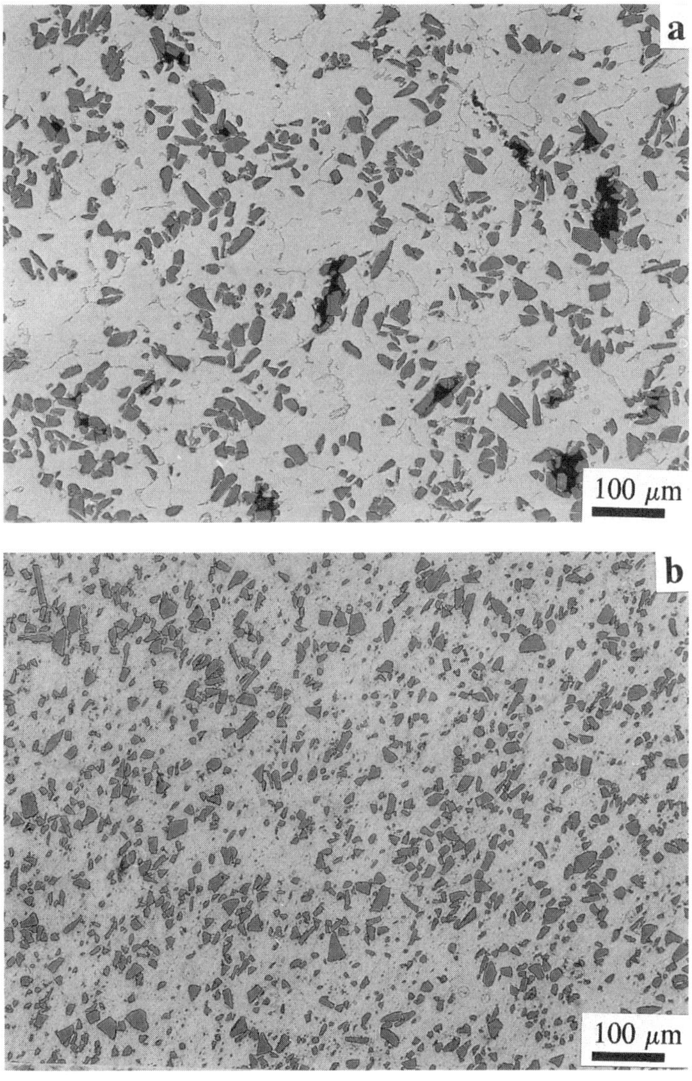

Figure 6. 2014 + 20 vol. % SiC (a) as-cast, (b) extruded 20:1 (transverse).

INTERNAL STRESSES IN CAST AND EXTRUDED MMCs

The material studied was cast and hot-extruded samples of aluminum 2014 + 20 vol. pct. SiC particulate obtained from Alcan. From the microstructures shown in Fig. 6, it is seen that the particle distribution in the as cast material is highly clustered, whereas in the extrusion, it appears less clustered (transverse section). The particle distributions were characterized by the distributions of cell areas which resulted from the octagonal dilation procedure. Typical cell patterns for the as-cast and extruded samples are shown in Fig. 7 and the frequency distributions of cell areas are given in Fig. 8.

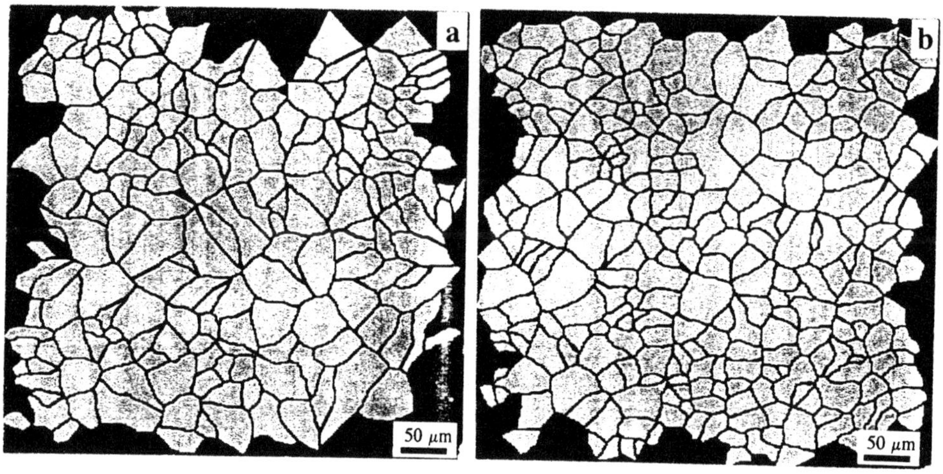

Figure 7. Tessellation networks resulting from octagonal dilation procedure for 2014 + 20 vol. % SiC (a) as-cast, (b) extruded 20:1 (transverse).

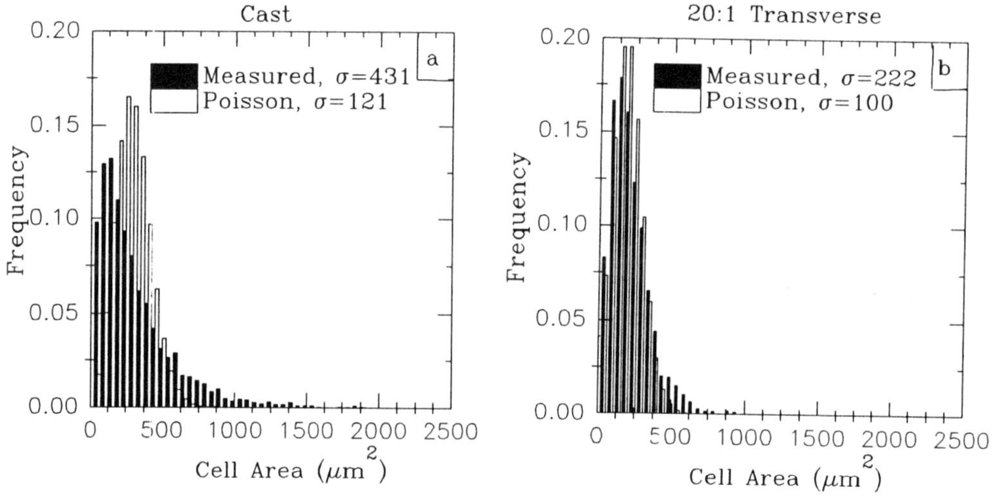

Figure 8. Frequency distributions of cell area and equivalent Poisson distribution (a) as-cast, (b) extruded 20:1 (transverse).

The measurements (Table 2) confirm that the particle distributions are clustered in all samples, and that the degree of clustering is sharply reduced by the extrusion process. This is also illustrated by comparing the frequency distributions of A_{tc} for the as-cast and extruded samples with the equivalent Poisson distributions shown in Fig. 8.

Table 2. Measurements of cell area for cast and extruded 2014+SiC.

sample	V_f(%)	$\overline{d}$(μm)	$\overline{A}_{tc}$(μm^2)	σ_A(μm^2)	σ_P(μm^2)	R_A
as-cast	19.0	12.1	291	431	121	3.57
extruded 5:1	20.0	10.4	230	286	107	2.67
extruded 20:1	18.4	10.2	200	222	100	2.22

Internal residual stresses in the 2014 matrix and the SiC particulate were determined by neutron diffraction measurements at AECL Research. Internal strains are determined by comparing measured lattice parameters in the MMC samples with strain-free 2014 and SiC, and stresses are calculated assuming a hydrostatic stress state [11]. The forces in the 2 phases must balance which permits the particulate volume fraction to be calculated if the stresses in each phase are known. The results are given in Table 3.

Table 3. Measurements of internal stresses in cast and extruded 2014+SiC.

sample	σ_{2014}(MPa)	σ_{SiC}(MPa)	Calc. V_f(%)	Meas. V_f(%)
as-cast	-6 ± 37	-612 ± 42	----	19.0
extruded 5:1	144 ± 20	-574 ± 40	20.0	20.0
extruded 20:1	177 ± 28	-526 ± 17	25.2	18.4

The primary source of internal residual stress in MMCs is the difference in thermal expansion coefficients and elastic moduli between the matrix and particulate phases. Assuming zero residual stress at the solidus temperature and no stress relaxation during cooling, the maximum stresses at room temperature are calculated to be 83 MPa for the matrix and -331 MPa for the SiC. The stress measurements are reasonable for the SiC in the as-cast and extruded samples, and for the matrix in the extruded samples. However, the as-cast sample has a negligible measured matrix stress. The particulate volume fraction calculated from the stress measurements for the extruded samples are close to the actual values. The anomalously low value for the matrix stress in the as-cast sample is attributed to the highly clustered particle distribution. There is a bi-modal strain distribution comprising the large particle-free volumes which have zero strain and the matrix in the particle clusters which must be highly strained to balance the stress in the particles. However, the large particle-free volumes determine the observed diffraction peak conditions, and zero strain is measured for the matrix. The extrusion process reduces the size of the particle-free volume (Figs. 6 and 8, Table 2), thus increasing the volume of constrained matrix and the measured matrix tensile strain. For this reason, measurements of internal residual strains in the matrix must be analyzed carefully, but the measurements of particulate strains are more straightforward. Comparing the SiC stresses for the as-cast and extruded samples indicates that the extrusion process causes a small net decrease in residual stress.

CONCLUSIONS

1. The spatial distribution of particulate in MMCs can be characterized by the standard deviation of frequency distributions of edge-edge spacing, local volume fraction or tessellated cell area. A quantitative measure of the degree of particle clustering is the ratio (R) of the standard deviation of the measured distribution to that of the equivalent Poisson distribution.
2. For the molten metal mixing and casting technique, the degree of particle clustering decreases with increasing stirring time in the melt and with increasing volume fraction of particulate.
3. Fracture toughness decreases with increasing degree of clustering. Damage initiates preferentially in particle clusters and the fracture path links regions of highly localized damage. Hence, fracture toughness is strongly influenced by spatial parameters such as edge-edge spacing and local volume fraction of particles, and less by mean volume fraction.
4. Hot extrusion of cast MMCs decreases the degree of clustering, but increases the anisotropy of the spatial distribution of particles. The amount of stress relaxation in the matrix during cooling from the casting temperature increases with increasing degree of clustering. Hot extrusion causes a small net decrease in the residual stress.

ACKNOWLEDGEMENTS

The authors are indebted to Dr. D.J. Lloyd (Alcan International, KRDC) and Dr. T.M. Holden (AECL Research) for their contributions to this research. Financial support from an industrial consortium (Alcan, Inco, Ontario Hydro, Sherritt Technology, Comdev), the Natural Sciences and Engineering Research Council, and the Ontario Centre for Materials Research is gratefully acknowledged.

REFERENCES

1. D.J. Lloyd, "Particle Reinforced Aluminum and Magnesium Matrix Composites", International Materials Reviews, Vol. 39, p. 1, 1994.
2. S. Tao and J.D. Boyd, "Mechanisms of Damage Accumulation and Fracture

in Particulate Reinforced Metal Matrix Composites", Mechanisms and Mechanics of Composite Fracture, R.B. Bhagat, S.G. Fishman and R.J. Arsenault, Eds., ASM, p.29, 1993.
3. D.J. Lloyd, "Aspects of Fracture in Particulate Reinforced Metal Matrix Composites", Acta Metall. Mater., Vol. 39, p. 59, 1991.
4. J.D. Embury and G. Burger, "The Influence of Microstructure on Toughness", 7th Int. Conf. on the Strength of Metals and Alloys, Pergamon, p. 1893, 1985.
5. W.A. Spitzig, J.F. Kelly and O. Richmond, "Quantitative Characterization of Second Phase Populations", Metallography, Vol. 18, p.235, 1985.
6. M.T. Shehata and J.D. Boyd, "Measurement of Spatial Distribution of Inclusions", Inclusions and Their Influence on Material Behaviour, R. Rungta, Ed., ASM, p. 123, 1988.
7. M.T. Shehata, "Application of Image Analysis in Characterizing Dispersion of Particles", Short Course on Image Analysis Applied to Mineral and Earth Science, Mineralogical Association of Canada, Chapter 14, p.119, 1989.
8. A. Okabe, B. Boots and K. Sugihara, "Spatial Tessellations - Concepts and Applications of Voronoi Diagrams", John Wiley and Sons, 1992.
9. R. Pyrz, "Microstructure Morphology for Composite Materials", Second Canadian International Composites Conference and Exhibition", W. Wallace, R. Gauvin and S.V. Hoa, Eds., Can. Assoc. for Composite Structures and Materials, p. 391, 1993.
10. J.J. Lewandowski, C. Liu and W.H. Hunt, "Effects of Matrix Microstructure and Particle Distribution on Fracture of an Aluminum Metal Matrix Composite", Mater. Sci. Eng., Vol. A107, p.241, 1989.
11. H.M. Ledbetter and M.W. Austin, "Internal Strain (Stress) in an SiC-Al Particle-Reinforced Composite : an X-ray Diffraction Study", Mater. Sci. Eng., Vol. 89, p. 53, 1987.

MICROSTRUCTURAL ANALYSIS OF Al-$TiAl_3$ ALLOYS PRODUCED THROUGH REACTIVE PLASMA FORMING

M. Entezarian*, J-P. Immarigeon**, P. Tsantrizos*** and R.A.L. Drew*

ABSTRACT

New materials can be produced by a novel process known as reactive plasma forming. This technique was used to make Al-$TiAl_3$ alloys by injecting both Al powder and $TiCl_4$ vapor into a D.C. plasma torch. Al-$TiAl_3$ alloys were also produced by casting, a powder metallurgy technique (PM) and vacuum plasma spraying (VPS) to assess differences in microstructure between the reactively plasma formed material and materials produced by the other more established processing routes. Microstructure was strongly dependent on processing method. The cast microstructure exhibited an acicular morphology of $TiAl_3$ in an Al matrix. Alloys produced by PM and VPS contained coarse equiaxed particles of the intermetallic phase in aluminum. The microstructure of the reactive plasma sprayed material was very uniform and exhibited a very fine dispersion of $TiAl_3$ in an Al matrix.

INTRODUCTION

Conventional high strength Al alloys derive their strength from precipitation hardening. There are very few elements that have the

----

* McGill University, Dept. of Mining and Metallurgical Eng., 3450 University St. Montréal, Québec, Canada, H3A 2A5,
** NRC Intitute for Aerospace Research, Montreal Road, Ottawa, Ontario, Canada, K1A 0R6,
***Pegasus Refractory Materials Inc. 1744 William, Montréal, Québec, Canada, H3J 1R4

required solubility in Al to form strengthening precipitates. Typically, only Cu, Mg, Zn, Si and Li have been used for commercial purposes. Special techniques like rapid solidification (RS) or mechanical alloying (MA) have been developed to broaden the range of alloying elements by extending limits of solid solubility in Al to other elements. With these techniques, new non-equilibrium strengthening phases can be obtained. In addition, the materials produced by RS and MA contain highly uniform (non-segregated) microstructures with very fine grain sizes which provide added benefits.[1,2]

In this regard, titanium has very low solid solubility in Al (maximum 0.8 at.%), Figure 1.[3] Cast alloys containing up to 36 wt% of Ti form large particles of primary $TiAl_3$ which grow during solidification because of the wide solidification range (600 to 700 °C) of the $TiAl_3$ + Liquid portion of the phase diagram. Mechanical alloying has been used to dispersed Ti in Al such that upon consolidation of MA powders by hot extrusion, a fine dispersion of strengthening $TiAl_3$ particles is formed in the Al matrix. Such materials have been shown to have attractive high temperature mechanical properties when compared to high temperature Al alloys (e.g. Al-Fe-X) or titanium alloys (e.g. Ti-6242).[4]

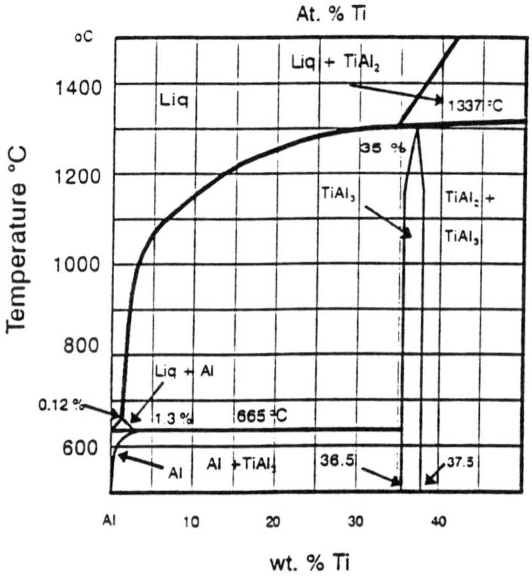

Figure 1. Al-Ti Phase diagram

REactive Plasma FORMing (REFORM) is another process that could be used to produce Ti-Al alloys. In the REFORM process, chemical precursors are introduced into a high temperature DC plasma where they react to form compounds. The reaction involves gas-liquid or gas-gas reactions which are not as well understood as the liquid-solid or solid-solid reactions associated with the more established processes, including RS and MA.

Reactive plasma synthesis has already been used to produce a variety of intermetallic phases such as VSi_2, $NbSi_2$, $NbGe_2$, Cr_3Si and Mo_3Si.[5] Tsunekawa et al have used elemental powders of Al and Fe to synthesize iron aluminide by reactive low pressure plasma spraying. They have also produced a TiAl matrix containing Ti_2AlN particles by injecting elemental Al and Ti powders into a dc plasma gas containing N_2.[6] The feasibility of producing titanium aluminide intermetallic has already been explored by Tsantrizos et al [7] using Al powder and $TiCl_4$ as precursor materials and a DC plasma.

Chemical thermodynamics predict that when Al is exposed to $TiCl_4$ the latter is reduced according to the following reaction:

$$TiCl_4(g) + 2Al(l) = 2AlCl_2(g) + Ti(l) \qquad (1)$$

Thermodynamically, this reaction proceeds to the right so long as the temperature is above 550 °C, when the change in standard free energy associated with reaction (1):

$$\Delta G0_R = 95212.38 - 171.625\,T + 7.87 \log T \qquad (2)$$

is negative, as shown in Figure 2. Equations (2) is derived from the standard free energy of formation of all compounds involved in reaction (1). Interesting industrial possibilities arise if Al is supplied in excess and is allowed to react with Ti obtained by the reduction of $TiCl_4$. The final product could be an intermetallic of the Al-Ti system, possibly $TiAl_3$, immersed in an Al matrix. The objective of the present study was to explore this possibility.

In this paper, the microstructure of material produced by the REFORM process, using Al powder and $TiCl_4$ as precursor, is compared to the microstructure of Al-$TiAl_3$ alloys produced by other techniques including casting, a powder metallurgy (PM) technique and conventional vacuum plasma spraying (VPS).

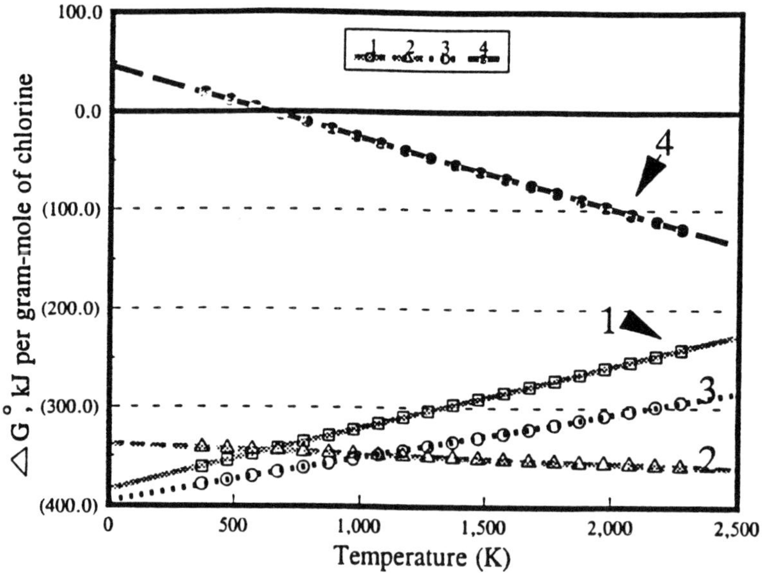

Figure 2. Change in standard free energy associated with reactions involving reduction of TiCl4 by Al. Curve 4 corresponds to reaction (1) in text; other curves correspond to the standard free energy of formation of $TiCl_4$ (curve 1), $AlCl_2$ (curve 2) and $AlCl_3$ (curve 3).

EXPERIMENTAL PROCEDURE

The procedures to produce the alloys evaluated in this study are described below.

Cast alloys. Pure Al was melted, super-heated, alloyed with Ti (5, 10 and 15 wt%) and cast in a water cooled copper mold.

PM alloys. Elemental powders of Al and Ti were well mixed, die-pressed and subjected to pressureless liquid phase sintering in a vacuum furnace at 700, 800 and 1100 °C.

VPS alloys. Elemental powders of Al and Ti were mixed and injected into a DC argon plasma and deposited on the surface of both stainless steel and Ti substrates at different substrate temperatures ranging from 400 to 900 °C.

REFORM alloys. A DC plasma torch, a cylindrical reactor, and a graphite substrate were used in this study. The reactants, Al powder and $TiCl_4$ vapor, were fed through two concentric tubes into the plasma region at the rate of 10 g/min and 20 ml/min, respectively. A simplified schematic of the REFORM process is shown in Figure 3. The plasma gas was pure argon and the products were collected as loose powder or in the form of solid layers on the surface of a graphite collector. Both the powder and the spray deposited layer materials were characterized by optical microscopy, Scanning Electron Microscopy (SEM), thermogravimetry analysis (TGA) and Energy Dispersive Spectroscopy (EDS).

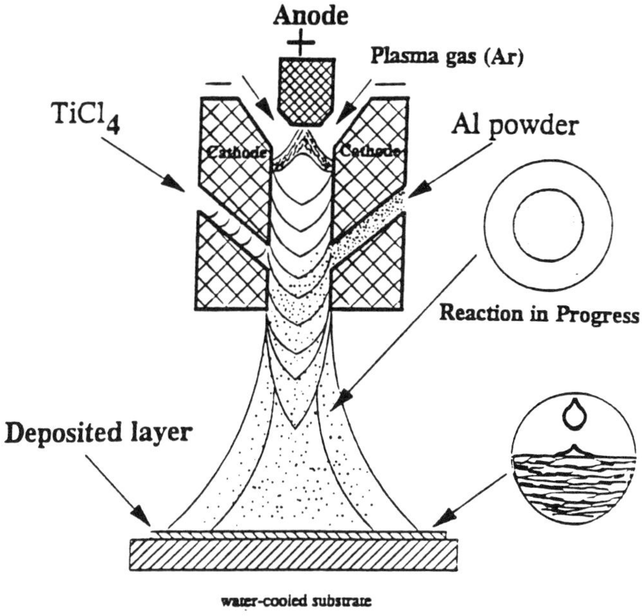

Figure 3. Schematic of the REFORM process.

Specimens from each of the above products were cut, mounted, ground and polished for metallography following standard procedures. Some difficulties were experienced with the reactively produced materials which tended to spontaneously etch during polishing owing to the presence of residual chlorine in the material. This problem was eleminated by using oil instead of water as a polishing coolant.

Results and Discussion

Cast alloys. A typical cast microstructure for the Al-5%Ti alloy is shown in Figure 4. The material contains primary $TiAl_3$ with a needle-like morphology in an Al matrix. The needles have an average

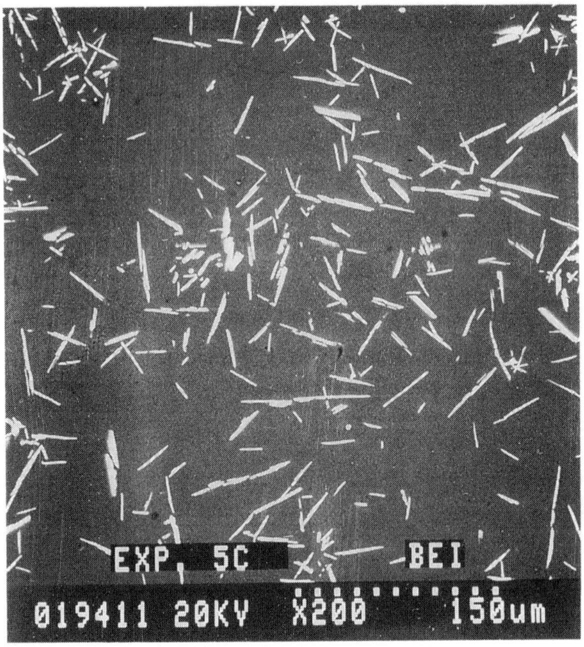

Figure 4. Typical microstructure of a chill-cast Al-5 wt.%Ti alloy. The white needles are $TiAl_3$

The size of the $TiAl_3$ particles was considerably finer than the size of the Ti powder used in the mixture. The average particle size for the alloys, containing 10 wt.% Ti and sintered at 700 and 800 °C, were 6 and 9 μm, respectively as compared to 60 μm for the starting Ti powder. This apparent refinement can be explained by the following sequential steps;

(a) reaction at the interface between solid Ti powder and molten Al to form a $TiAl_3$ envelope around the powder particles,

(b) break up of the brilttle $TiAl_3$ from the surface of the Ti, owing to the brittle nature of the intermetallic phase and its different CTE from that of the supporting solid Ti,

(c) redistribution of the broken-up envelope of $TiAl_3$ into liquid Al,

(d) continued reaction of as yet unreacted Ti with the molten Al until all Ti is consumed.

length of 30 to 50 μm, with an aspect ratio of around 10 to 20, and they are evenly distributed in the Al matrix. It is not clear how this material would behave mechanically. For instance, it might offer high wear resistance. On the other hand, the needles could have an embrittling effect. Further work is needed to determine the mechanical properties of this material. Depending on the Ti content of the melt, the $TiAl_3$ precipitates form during solidification at temperatures ranging anywhere from 1300°C to 660°C. Over this range, the liquidus rapidly increases with increasing Ti content, as shown in Figure 1. For instance, 2 wt.% Ti addition increases the liquidus temperature to above 1000 °C. This rapid increase in liquidus temperature is detrimental to castability as noted experimentally for the 15% Ti alloy composition.

Powder Metallurgy alloys. A typical microstructure for a powder processed Al-10%Ti alloy is shown in Figure 5(a). The sintering temperature was 800°C and the size of the $TiAl_3$ particles was about 6-9 μm. There was no significant change in the size of the $TiAl_3$ particles in the alloys sintered at 700°C and 800°C. In contrast, the $TiAl_3$ particles of the alloy sintered at 1100°C were coarser (average size 30 μm). The effect of sintering temperature on the average particle size of the intermetallic phase is shown in Figure 5(b). Above 700°C, liquid phase sintering is the dominant process responsible for coarsening, since formation of liquid Al would facilitate mass transfer.

Cracking of the intermetallic phases was observed in all samples. The extent of cracking was found to increase with increasing sintering temperature. This may be attributed to the difference in coefficient of thermal expansion (CTE) between the Al-matrix and the intermetallic phase and the poor ductility of this phase. Internal microcracks have been previously reported for $TiAl_3$, although no explanation for crack formation has been proposed.[8]

The $TiAl_3$ phase appears in the Al-matrix as coarse particles, a morphology that differs from that observed for cast alloys. The difference in morphology can be explained by differences in the mechanism of formation between the two processing routes. In casting, the intermetallic phase precipitates at high temperature from the liquid phase and this promotes needle growth. In contrast, in powder metallurgy, the intermetallic phase is formed as a result of diffusion controlled reaction between liquid Al and solid Ti powders which is a much slower mass transport mechanism than the mechanism which occurs during casting.

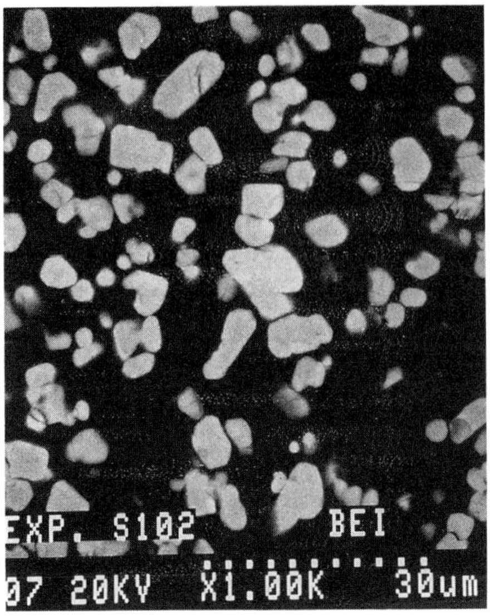

(a)

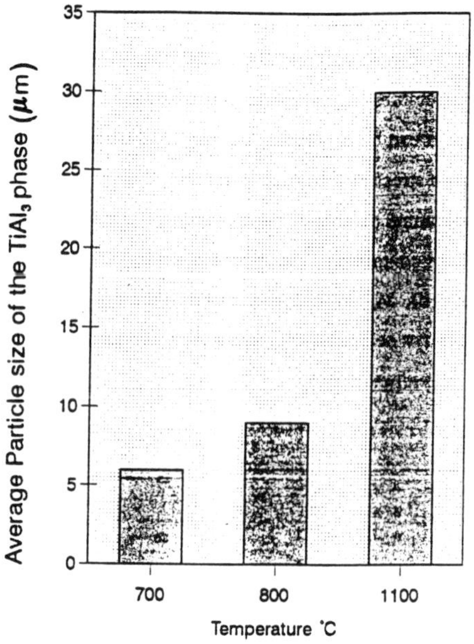

(b)

Figure 5. (a) Typical microstructure obtained from the PM process; and (b) effect of sintering temperature on the average $TiAl_3$ particle size. Note the cracking in the intermetallic phase.

VPS alloys. Both stainless steel and Ti substrates were used to collect the sprayed materials. The microstructures of the alloys deposited on these substrates are shown in Figure 6. The morphology of the $TiAl_3$ appeared to be similar to that of the powder metallurgy product i.e. irregular particles with a size range between 5 and 20 μm. The stainless steel substrates reacted with the deposited materials, forming complex intermetallics at the interface.

The temperature of the substrate was a key factor for producing the desired microstructures. When the substrate temperatures was below 800 °C, residual titanium was found in the sprayed material. Differential thermal analysis (DTA) performed at high heating rates on the elemental Al and Ti powders revealed that the reaction between Al and Ti proceeds at about 825°C, Figure 7. The alloys sprayed at temperatures higher than 800°C contained no residual Ti. The high substrate temperature required for this reaction creates problems in producing thick samples, as Al is in the liqid state at these temperatures. Since the alloys were produced on vertical substrates, the unreacted Al tended to flow downward, resulting in Ti enriched deposits, This explains the large volume fraction of $TiAl_3$ phase observed in alloys produced by VPS, as shown in Figure 6.

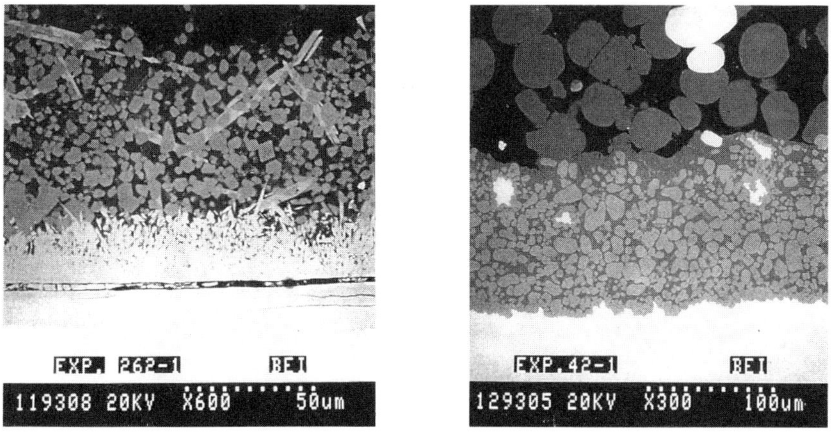

(a) (b)

Figure 6. Microstructures of spray deposited layers applied by VPS on (a) stainless steel and (b) Ti substrates. The substrate is at the bottom of each micrograph.

REFORM Alloys. The products of the REFORM process were collected as either a deposited layer or a powder. A typical microstructure of the reactively formed Al-$TiAl_3$ alloys is shown in Figure 8 for a deposited layer. The SEM backscattered image shown in this Figure reveals two major phases, a light and a dark gray phase. The results of quantitative EDS analysis showed that the bright phase contains 25 at.% Ti and 75 At.% Al, which corresponds to the composition of $TiAl_3$. The even distribution of the fine intermetallic particles (about 2-3 μm) in an Al-matrix is encouraging. On the basis of intermetallic particle size and distribution, the REFORM products could offer better properties than those obtained by other technologies and thus might be attractive for the fabrication of commercial products.

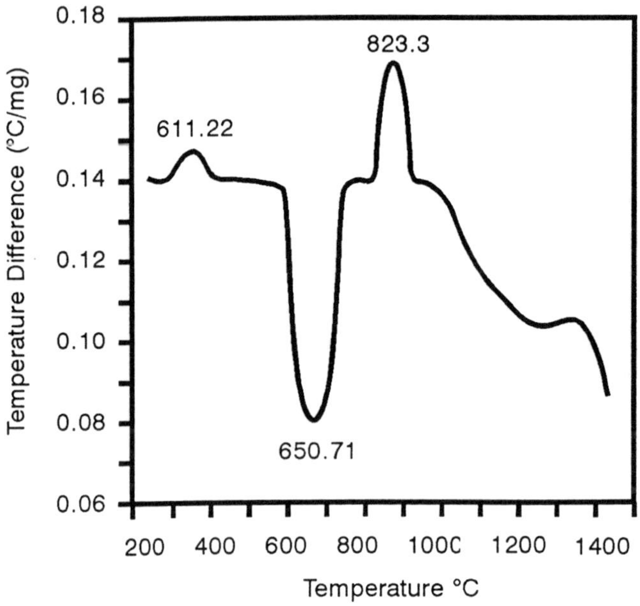

Figure 7. DTA analysis for Al-5 wt%Ti

The results of EDS analysis of REFORM powder also revealed the presence of chlorine on the external surface of the powder particles. In contrast, no chlorine was detected inside the particles. This implies that chlorine is a surface contamination in powder meterial. The chlorine content was also quantitatively analyzed by Inductively Coupled Plasma-Atomic Emission Spectrometry (ICP-AES) and found to be 5 at.%. Thermogravimetric analysis performed on the REFORM powders revealed that Al and Ti chlorides present on the surface evaporate in an inert atmosphere when heated to temperatures above 600 °C. It has not not been evaluated whether clorine contamination can be eliminated by such treatment in the case of deposited layers. This is being investigated as chlorides could be detrimental to mechanical properties.

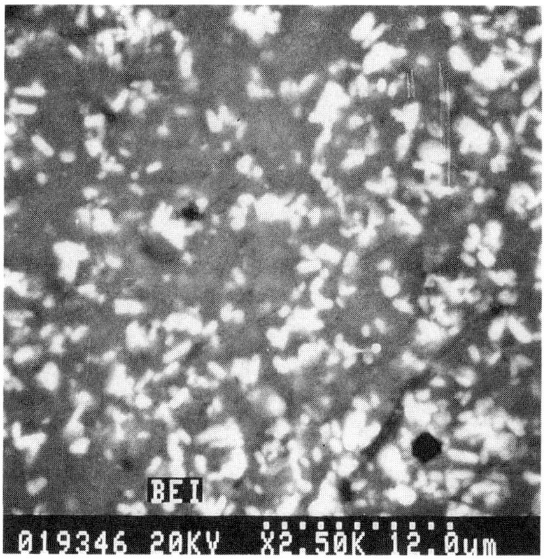

Figure 8. Microstructure of the Al-$TiAl_3$ alloy produced by the REFORM process.

CONCLUSIONS

1. By reacting $TiCl_4$ and Al powder in a plasma jet, the reactive plasma spraying process known as REFORM can be used to produce Al-$TiAl_3$ alloys.

2. By comparing the REFORM products to those produced by the other metallurgical processes considered in this study (casting, PM and VPS), it was found that the REFORM process is unique in its ability to produce alloys containing very fine dispersions of $TiAl_3$ particles (2 to 3 μm) in an Al matrix.

3. The $TiAl_3$ morphologies appeared to be needle-like in cast alloys, coarse equiaxed in PM and VPS alloys and very fine equiaxed for REFORM produced material.

REFERENCES

1.J.D. Whittenberger, "Solid State Processing for High Temperature Alloys and Composites", Solid State Powder Processing, eds; A.H. Clauer and J.J. deBadillo, The Minerals, Metals & Materials Society, p. 137, 1990.

2.G.S. Murty, M. J. Kocczak, "Superplastic Characteristics of Modified 7075 Al and Al-Ti Alloys Processes by Rapid Solidification Powder Metallurgy", Processing and Properties for Powder Metallurgy Composites, ed. P. Kumar, K. Vedula, and A. Ritter, Warrendale, PA: TMS, p. 139, 1988.

3.C.M Adam, "Structure/Property Relationships and Applications of Rapidly Solidified Amorphous and Crystalline Alloys", ed. B.H. Kear et al., Amsterdam: Elsevier Science Pub. Co., Inc. p. 411, 1982.

4.J. Weber and D. Chellman, " Keynotte Address Mechanically Alloyed Aluminum Alloys for aircraft Applications, Presented at the ASM International Conference On Structural Applications of Mechanical Alloying, Myrtle Beach, SC, March 27-29, 1990.

5.P. Ronsheim, A.Mazza, and A.N. Christensen, "Thermal Plasma Synthesis of Transition Metal Nitrides and Alloys", Plasma Chemistry and Plasma Processing, Vol. 1, No. 2, p.135, 1981.

6.Y. Tsunekawa, K. Gotoh, M. Okumyla, and N. Mohrl, "High Temperature Stability of Titanium Aluminide Matrix Composites Fabricated by Reactive Low Pressure Plasma Spraying", Proceedings of the International Thermal Spray Conference & Exposition, Orlando, Florida, USA,189-194. 8 May-5 June, 1992.

7.P. G. Tsantrizos. L.T. Mavropoulos, M.I. Boulos, J. Jurewicz, and K. Chen, ISPC-10 Conference Proceedings,Bochum , Paper 1.4, 1991.

8.M. Yamaguchi, "High Temperature Intermetallics-With Particular Emphasis on TiAl", Materials Science and Technology, 8, p. 299, April 1992.

CHARACTERIZATION OF SURFACE MODIFICATION INDUCED IN TEFLON BY UV LASER

M.T. Shehata[1], B.R. Casault[1], G. Williams[1], T. Malis[1], J. Montgomery[2], P. Eisele[2] and I. Miller[2]

ABSTRACT

A metallographic/Scanning Electron Microscope study was carried out to develop methodologies to characterize the nature and depth of UV laser- initiated colour changes in Teflon-coated wire. In order to obtain the depth of laser initiated process one has to examine cross sections of the Teflon coating. The preparation of such a section should minimize the risk of altering the wire's topology and sub-surface characteristics. In this context, three different methodologies in preparing wire cross-sections were investigated. The methodologies were: (i) metallographic polishing, (ii) liquid nitrogen fracture, and (iii) microtome sectioning. For each methodology, the extent of surface modification was studied and the feasibility of the methodology to characterize the interaction introduced by the UV laser was examined. This paper describes the experimental procedures and presents a complete summary of results obtained in this investigation.

INTRODUCTION

Teflon insulated wire has been selectively used within the aerospace industry due to its excellent dielectric properties, light weight and resistance to environmental conditions. Broad acceptance of this insulation has been inhibited by the lack of permanent marking or

----

[1]Metals Technology Laboratories, CANMET,Ottawa, Ontario, K1A 0G1
[2]Lumonics Inc., 105 Schneider Rd.,Kanata, Ontario K2K 1Y3, CANADA

labeling techniques which do not damage or alter the electrical or mechanical properties of the wire insulation. Active research in development of UV Excimer laser marking sources and photochemical additives to the Teflon insulation formulation has resulted in a laser marking process which allows for rapid, non-contact and non aggressive printing of aircraft wire (1). The technique relies on the photoinduced reduction of titanium dioxide pigment in the Teflon insulation. The resulting colour change yields a high contrast black mark against the white insulation layer.

To verify the depth of the laser interaction and determine if there is any material ablation or laser induced surface modification, a metallographic / Scanning Electron Microscopic study was carried out to develop methodologies to characterize the nature and depth of the marking process. In order to obtain the depth of any surface modification during the marking process one has to prepare and examine cross sections of the Teflon coating. The preparation of such a section should not alter the topology of the insulation layer. In this context three different methodologies in preparing wire cross-section were investigated. The methodologies are: (i) metallographic polishing, (ii) liquid nitrogen fracture, and (iii) microtome sectioning (2). For each methodology the extent of the marking process was studied and the feasibility of the methodology to prepare the cross sectional sample is examined.

EXPERIMENTAL PROCEDURES

As-marked sample: A Teflon coated wire sample was supplied by LUMONICS for this investigation. The Teflon was subjected to laser marking using laser energy density of 0.5 J/cm^2. Figure 1 shows a photograph of the as-marked wire where the marking reads "LUMONICS WIRE MARK".

SEM Examination: A Philips XL30 SEM was used in the preliminary examination of the laser-marked surface of the samples in the as-received conditions to characterize the type of surface modification induced by the laser marking. The SEM was also used to examine the depth of the marking process in cross sections prepared by: (1) conventional metallography (2) liquid nitrogen fracture surface (3) microtome sectioning.

Liquid nitrogen fracture: The Teflon coating was notched opposite to the middle of the letter W by a knife to insure that the fracture surface ran across the laser markings. After making the notch the

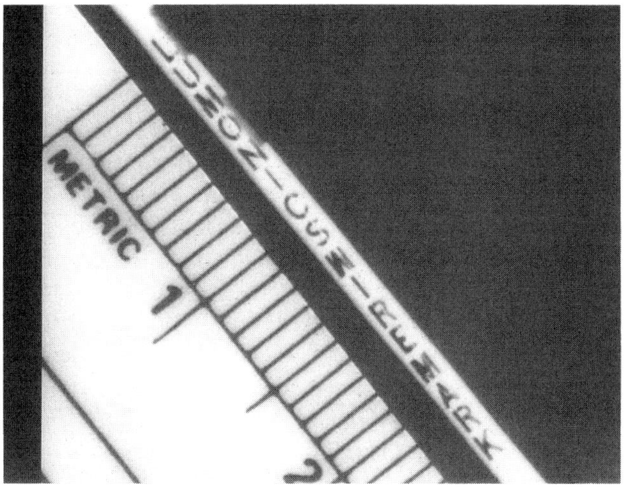

Figure 1. A photograph of the as-received Teflon coated wire which was marked LUMONICS WIRE MARK by the Excimer laser marker.

Teflon wire was dipped in liquid nitrogen for about five minutes to embrittle the Teflon. This was immediately followed by a snap fracture running through the letter W.

Microtomy: Wire samples were embedded in "L R White" hard grade resin. Polymerization was attained in a laboratory oven preset at 60 degree C and held at temperature for 24 hours. The wire and resin were contained in closed capsules. The resin blocks were rough trimmed with a scalpel and the final facetting was done with a glass knife. The final cut was made by a 45 degree angle Diatome diamond knife.

RESULTS AND DISCUSSION

EXAMINATION OF AS-RECEIVED SAMPLES:

To ensure that the integrity of the cable was not altered by the marking process, various tests were performed in another study to measure: shrinkage, fluid immersion, the dielectric strength and mark contrast. These tests showed that the electrical and mechanical integrity of the insulation was not altered by the marking process. The visible depth of the marking process by the optical microscope is approximately 10 μm. This means that there is no interaction of any significance of the laser light with the primary insulation of the wire, as the Teflon outer sheath of the cable is 200 µm. The only evidence of any subtle surface modification by the laser marking process is evident in the SEM micrographs in figures 2 and 3. Figures 2 shows at different magnifications from left to right part of the letter M and the letters A, R and K from the word MARK that was printed on the Teflon by the Excimer laser. Figures 3a & 3b show the top right portion of the letter K where both the marked area and the unmarked area are shown. Figure 3c shows the marked area at high magnification indicating that the surface modification is mainly in the form of micro porosity i.e. small voids (at most a few micrometers in size). Since Teflon is a fluorocarbon, the interaction of the laser with an absorbing medium such as TiO_2 could have resulted in localized heating and the evolution of gas from the Teflon which accumulated in the form of bubbles leading to the observed porosity.

It may be noted that the 5KV electron beam used in the SEM did not produce additional damage of any significance in the Teflon coating.

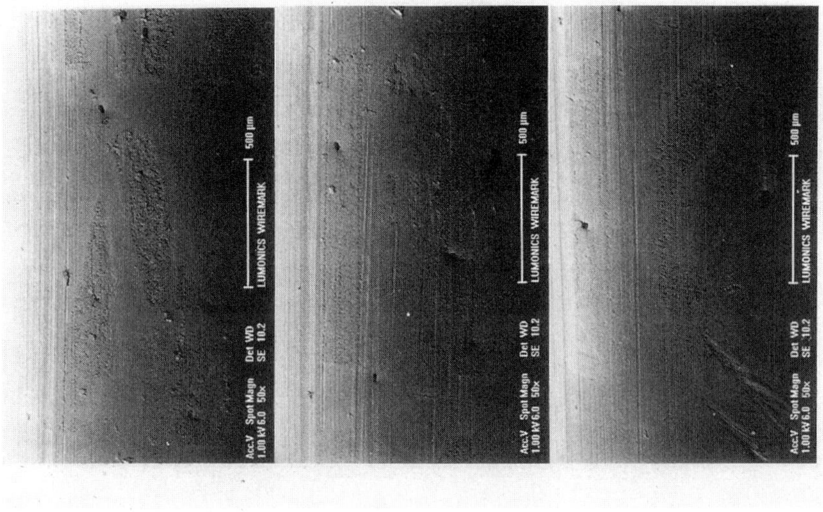

Figure 2. SEM micrographs at low magnification showing, from left to right, part of the letter M and the letters A, R and K from the word MARK that was printed on the Teflon by the laser marker.

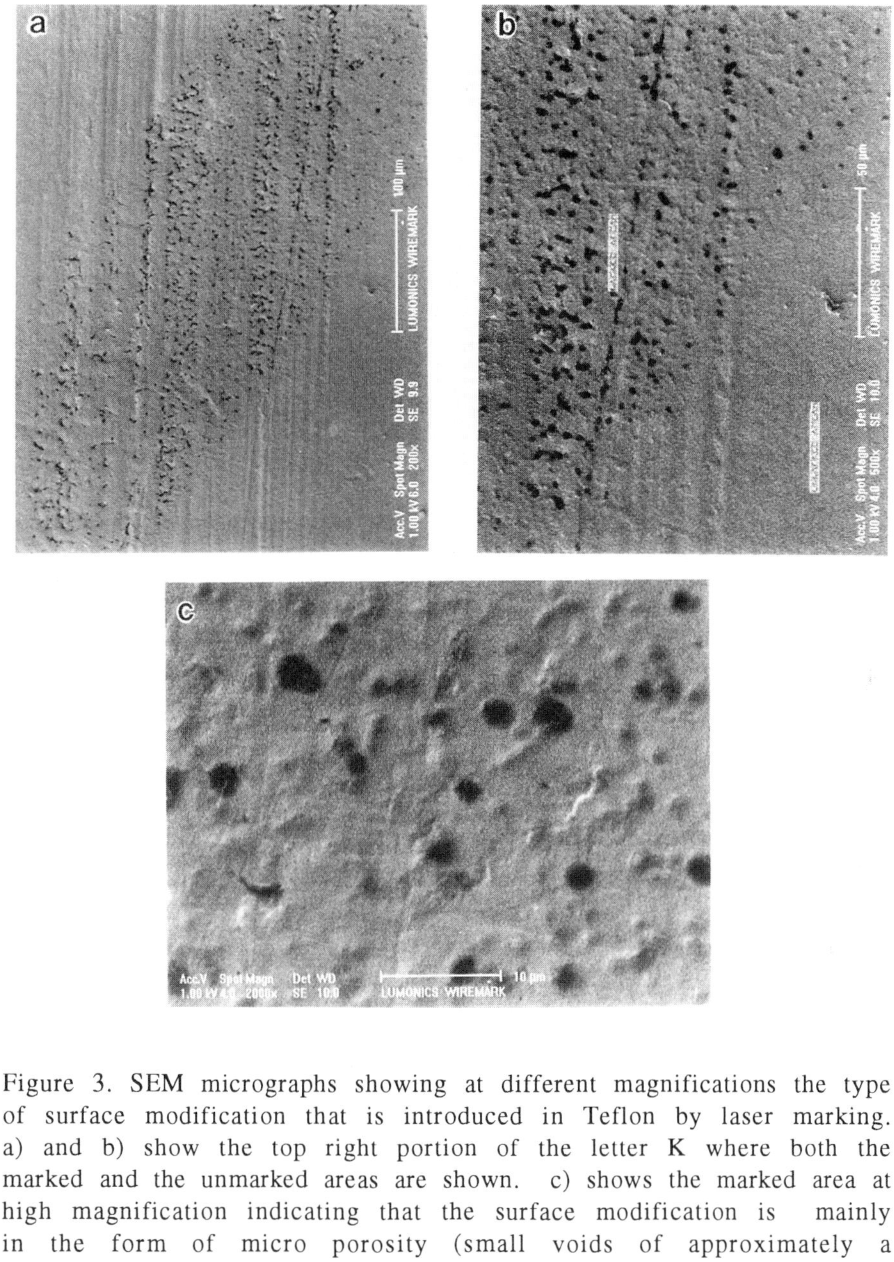

Figure 3. SEM micrographs showing at different magnifications the type of surface modification that is introduced in Teflon by laser marking. a) and b) show the top right portion of the letter K where both the marked and the unmarked areas are shown. c) shows the marked area at high magnification indicating that the surface modification is mainly in the form of micro porosity (small voids of approximately a micrometer in size)

EXAMINATION OF LIQUID NITROGEN FRACTURE CROSS SECTIONS:

Figure 4 shows a number of SEM photomicrographs showing at different magnification a liquid nitrogen fracture cross-section at a laser marked area. Figure 4a shows the general fracture cross-section which runs across the letter W on the side wall across the other side from the notch produced by the knife cut. Also one can observe that the Teflon coating consists of two layers of about 200 micrometers each. Figures 4b-d are closer views of the area containing the edge of the fracture cross section at the laser marked area. Figures 4b & 4c each show both marked and unmarked areas on the side wall whereas figures 4d & 4e each show a close up at the edge of the marked area. Some porosity is observed in the near-surface (10 μm) area in both figures 4d & 4e, but it can not be exclusively distinguished from other damage produced on the fracture surface during the fracture process. In other words, the liquid nitrogen fracture technique proved to be of very limited use in characterizing the damage induced in Teflon coating by laser marking.

EXAMINATION OF METALLOGRAPHIC CROSS SECTIONS:

Metallographic sections were examined by light optical microscopy as well as scanning electron microscopy (SEM). The results of the optical microscopy examination are shown in figure 5 whereas the results of the SEM examination are shown in figure 6. The optical microscope showed a faint shadowed area near the surface where the Teflon is marked by lasers. However, the optical microscope is interfaced with an image analysis system that has image enhancement capabilities which was utilized to provided greater visibility of the modified layer as shown in figure 5. The modified layer shown in the enhanced image (figure 5a) is painted in red by the image analysis system as shown in figure 5b and is about 10 micrometers in thickness.

The SEM examination of the metallographic section resulted in similar observations for the marked layer as shown in figure 6. Both the marked and unmarked areas are shown for comparisons in figure 6c. From the earlier SEM observations of the as-marked sample (Fig.3) the surface modification in the marked area is expected to appear in the form of micro voids. However, it seems that the metallographic preparation has affected the appearance of the marked layer. The edges of the micro voids may have been distorted during the grinding and polishing operations since Teflon is a soft material. Thus it is difficult to estimate the depth of damage, but 10 μm would be a reasonable estimate.

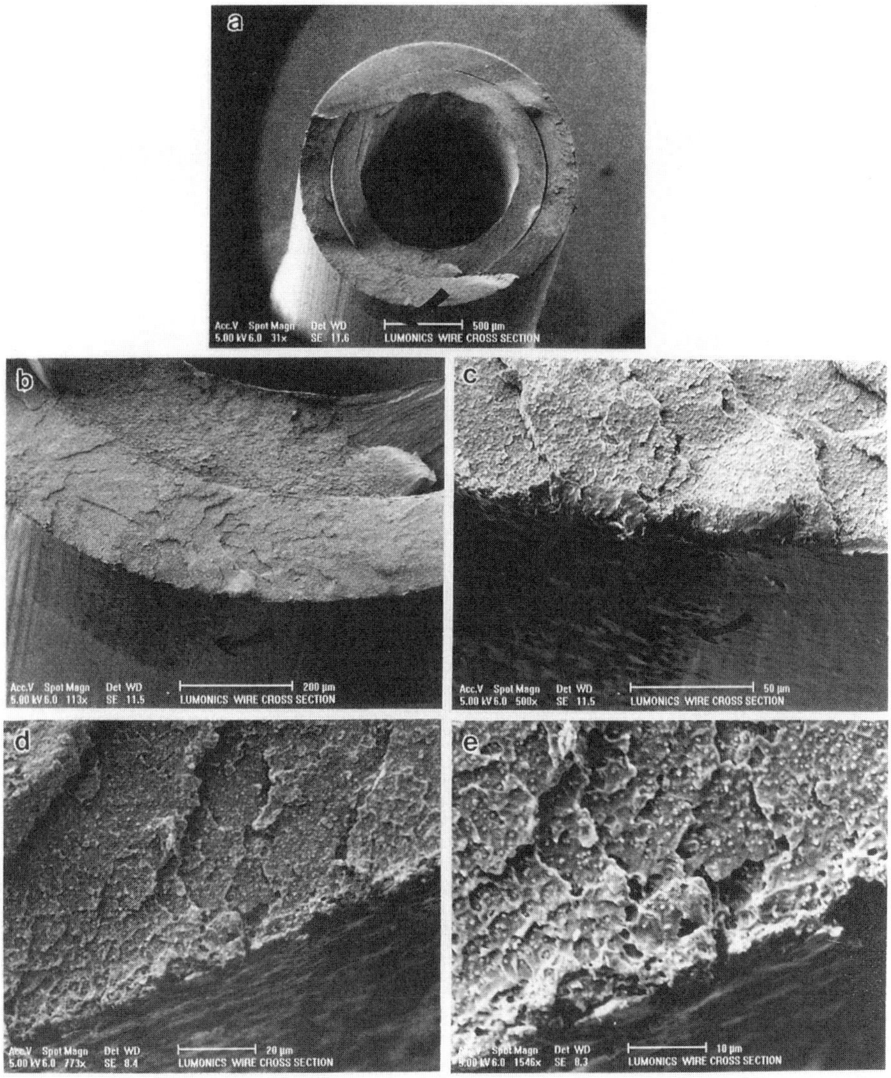

Figure 4. SEM photomicrographs showing at different magnification a liquid nitrogen fracture cross section at a laser marked area. a) shows that the general fracture cross section runs across the letter W on the side wall across the other side from the notch produced by the knife cut. Both marked and unmarked areas on the side wall are shown in (b) and (c) whereas the edge of the marked area is shown in (d) and (e).

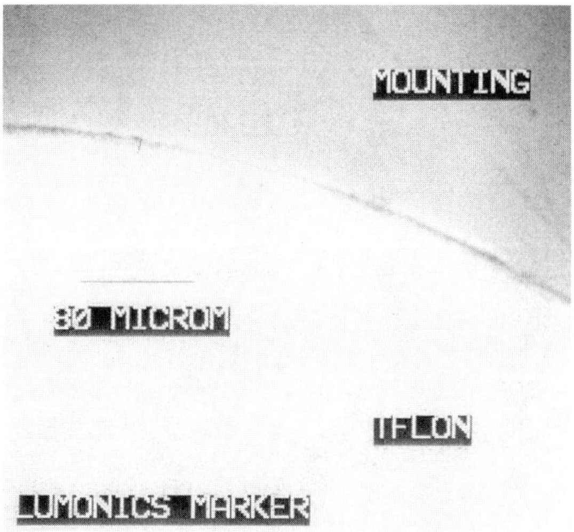

Figure 5. An image enhanced photomicrograph of a metallographic section where the Teflon is marked by lasers. The faint shadowed area near the surface represents the marked layer.

EXAMINATION OF MICROTOME CUT CROSS-SECTION:

The microtome cut cross section was examined by the SEM and the results are shown in Figure 7. Both marked and unmarked areas are shown for comparisons in figures 7b & 7c. Clearly the surface finish produced by the microtome diamondcut has the most desirable characteristics i.e. with the least distortion possible for the micro voids. The surface modification produced by the interaction with the laser beam is mainly in the form of micro porosity (submicron voids) that extend to at most 10 μm under the surface. Careful observation of figure7e indicates that the voids are formed preferentially around the small white particles (titanium dioxide) which are embedded in the Teflon to give the white color to the Teflon coating. It should be noted that a few ultra thin sections (about 50 nanometers thickness) were produced and examined in a Philips EM400 TEM at 120 KeV. While no attempt was made to examine the marked surface, the titanium dioxide particles were readily observed and analyzed. Moreover, the Teflon proved stable under the irradiation of 120 KeV electrons, an important factor if TEM examination should prove necessary.

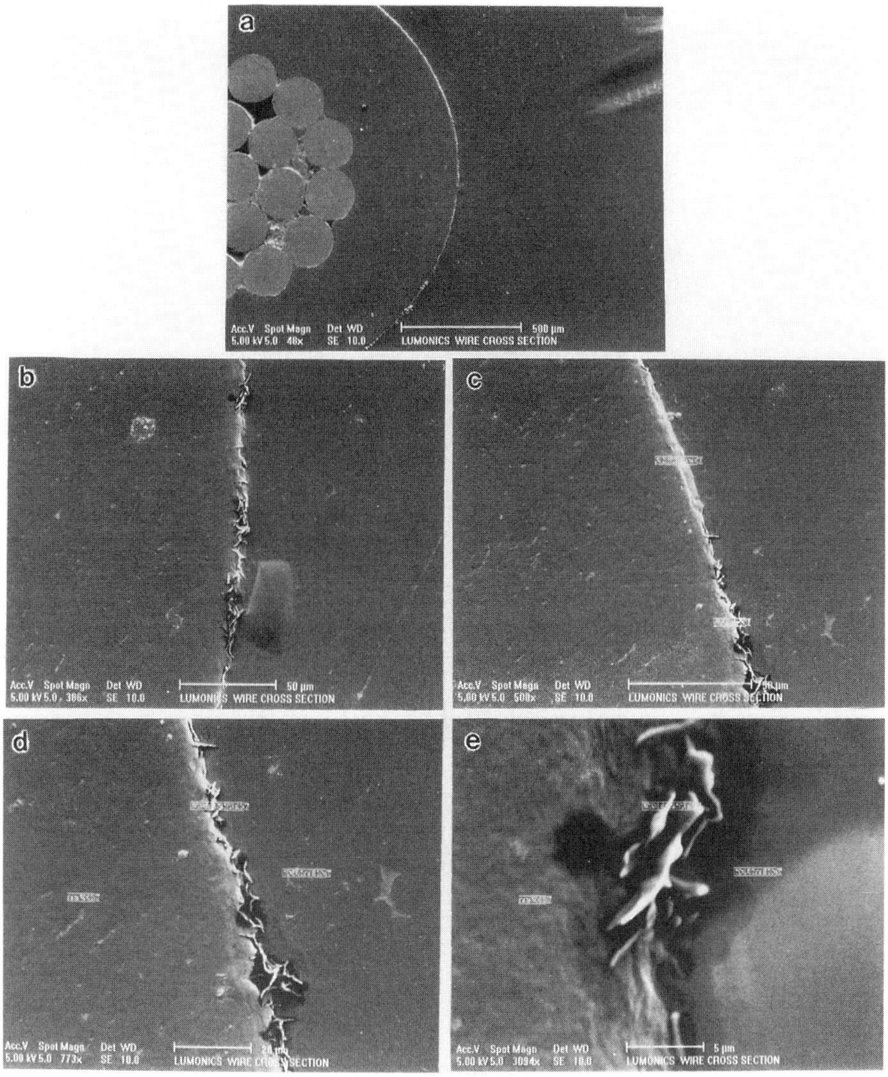

Figure 6. SEM photomicrographs showing at different magnification the metallographic section where the Teflon is marked by lasers. A general view of the metallographic cross section is shown in (a), both marked and unmarked areas are shown (c) whereas only the marked area is shown in (b), (d) and (e).

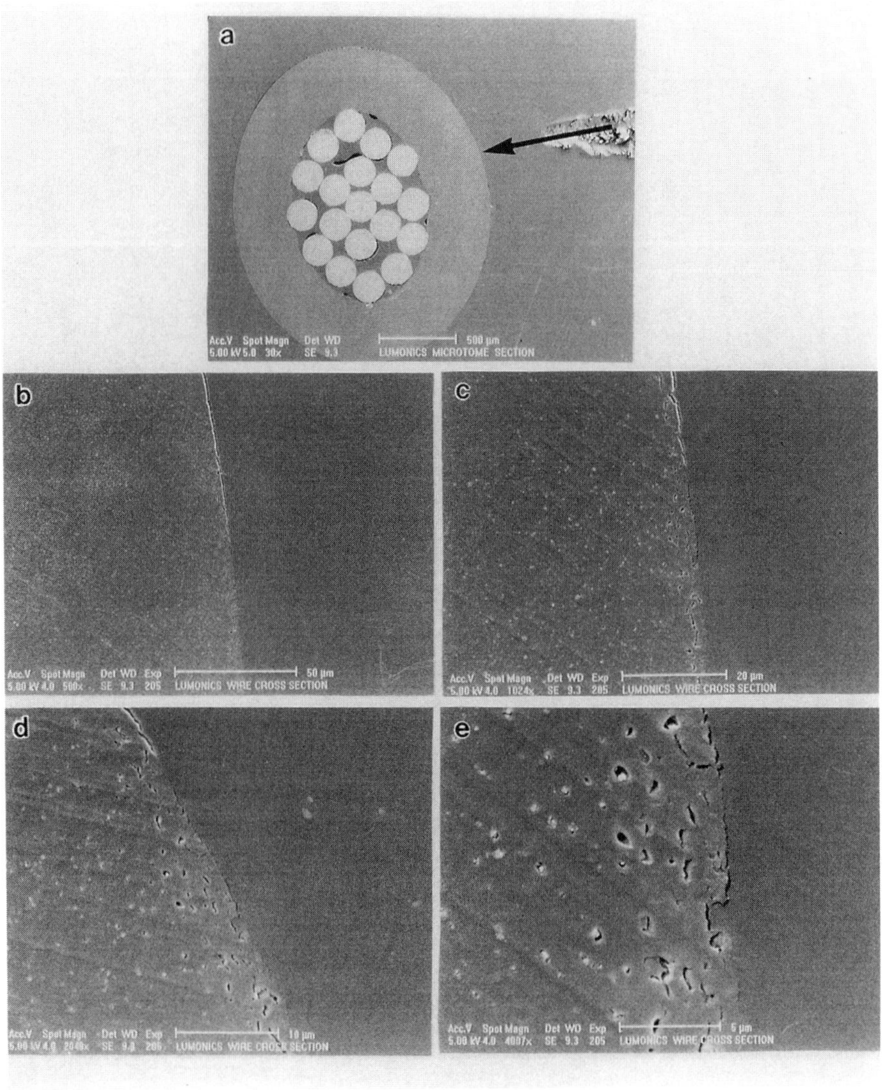

Figure 7. SEM photomicrographs showing at different magnification the microtome cut cross section where the Teflon is marked by lasers. A general view of the cross section is shown in (a), both marked and unmarked areas are shown in (b) and (c) whereas only the marked area is shown in (d) and (e).

CONCLUSIONS

1) Of the three methodologies used in preparation of wire cross-section, the microtome cut technique yielded the best results, with no discernible physical damage introduced to the surface or cross-section of the wire insulation.

2) The use of a 5 KeV electron beam in the SEM did not introduce additional damage of any significance in the Teflon coating. This indicates the feasibility of utilizing the SEM to characterize the damage introduced in Teflon by laser marking.

3) The examination of metallographic sections by optical microscopy indicated a faint shadowed area near the surface where the Teflon is marked by lasers. The image enhancement capabilities provided more visibility of the marked layer which is shown to be about 10 μm in thickness.

4) The penetration depth of the laser interaction was confirmed by the SEM analysis to be approximately 7-8 μm. The only observed surface modification was in the form of micro porosity i.e. small voids (at most a few micrometers in size). Since Teflon is a fluorocarbon, the interaction with the laser with an absorbing medium such as TiO_2 could have resulted in localized heating and the evolution of gas from the Teflon which accumulated in the form of bubbles leading to the observed porosity.

5) Further investigations are recommended to alter the grain size and concentration of TiO_2 to reduce the degree of localized heating and further improve micro-topology of the wire insulation.

REFERENCES

1. S.W. Williams and P.C. Morgan, Excimer Laser Printing of Aircraft Cables, Internal report, Sowerby Research Center, British Aerospace Plc., Filton, Bristol, England.
2. T. Malis and D. Steele, Specimen Preparation for TEM of Materials, Materials Research Society Symposium, Vol. 199, p.3, 1990.

TEM AND SEM INVESTIGATION OF DYNAMIC RECOVERY AND RECRYSTALLIZATION IN HOT DEFORMED METAL MATRIX COMPOSITES

Xiaoxin Xia and H.J. McQueen

ABSTRACT

Transmission and Scanning electron microscopies are used for microstructural investigation of 10, 15 and 20% volume fraction of either SiC or Al_2O_3 particle reinforced Al matrix composites after hot deformation tests. The TEM observations show that the dislocation density of the matrix increases, while the subgrain size decreases with rise of the reinforcing particle volume fraction and strain rate, and with decrease of the test temperature. The subgrains are smaller, less uniform and more highly misoriented than those in the bulk alloy. For higher volume fraction of particle reinforcement, more non-cellular structure is found in the MMCs at the same test conditions. The highly misoriented crystallites are confirmed as dynamic recrystallized grains by TEM and SEM. The research confirms that the hot deformation involves both recovery and recrystallization; this is different from lightly alloyed Al in which only dynamic recovery (DRV) occurs. Dynamic recrystallized (DRX) grains are found nucleated at the ceramic particle/matrix interfaces and in the Al matrix because of the distribution of highly localized strains.

Department of Mechanical Engineering, Concordia University, Montreal, Quebec, H3G 1M8, Canada.

INTRODUCTION

Particle reinforced metal matrix composites (MMCs) have been confirmed to exhibit improved stiffness, strength and creep resistance [1-8]. The discontinuously reinforced MMCs can be fabricated by conventional metal working technique, for example, rolling or extrusion; thus they have great potential to be used in industry to replace the conventional Al alloys and steels for special purposes. Previous research [7,9-12] has found that the hot working properties of MMCs are strongly dependent on their microstructural characteristics. Thus, TEM Selective Area Diffaction Pattern (SADP), SEM Back Scattered Image (BSI) and Back Scattered Diffraction Pattern (EBSDP) techniques are utilized for microanalysis of the hot worked specimens in this study.

EXPERIMENTAL PROCEDURES

10 v%, 20 v% Al_2O_3 and 15 v% SiC particle reinforced 6061 Al MMCs are supplied by Alcan Corporation, Kingston Laboratories and Duralcan, San Diego. The chemical compositions are given in Table 1.

Table 1. Chemical compositions of the MMCs

Reinforced materials with 6061 Al matrix	Mg	Si	Cu	Fe	Mn	Al	Particle size (μm)
10 v% Al_2O_3	1.0	0.6	0.3	0.7	0.15	Bal.	20
20 v% Al_2O_3	"	"	"	"	"	"	20
15 v% SiC	"	"	"	"	"	"	15

The hot torsion test temperature (*T*) range is from 200 to 500 °C with strain rates ($\dot{\varepsilon}$) of 0.1, 1.0 and 5 s^{-1}. The torsion test procedures and methods of analysis are described in earlier publications [4,13].

TEM and SEM specimen preparation

After hot torsion tests, the specimens are immediately quenched in a water spray to retain the hot worked microstructure. JEOL 2000CX. TEM and JEOL 300CX SEM with special beam set (1 μm) are used for microstructural investigations. TEM and SEM specimens are cut from the surface and parallel to the axis of the torsion specimen. TEM specimens are ground to 180 μm thick slices. To minimize mechanical deformation during preparation, 3 mm diameter discs are cut by a core drill with 50 μm diameter diamond cutting medium. Mechanical dimpling and ion-milling techniques are used to obtain thin areas from the discs for investigation. The angle between the ion-milling beam and specimen surface is 15 degrees. SEM specimens for EBSP and BSI observations are electro-chemically polished at 30V and -30 °C by 30% nitric acid and 70% methanol solution.

RESULTS

The stress-strain ($\sigma - \varepsilon$) curves of the three MMCs and the bulk alloy are shown in Figure 1. The curves display a much greater hardening behaviour with increase of the reinforcement volume fraction (f_V) and $\dot{\varepsilon}$ at low temperatures. As T increases to 500 °C, the hardening behaviour of the flow curves is diminished, thus the MMCs and bulk alloy present a similar softening behaviour. The size and distribution of the reinforced particles are investigated by SEM (BSI), as shown in Figure 2. The average size of the reinforced particles in Figure 2 is given in Table 1.

TEM observations show that the dislocation density (ρ) of all the MMCs increases with decrease of the T and increase of $\dot{\varepsilon}$ and f_V. The average subgrain sizes, given in Table 2, show an opposite behaviour to dislocation density with change of $\dot{\varepsilon}, f_V$ and T (Figures 3 and 4).

Highly misoriented crystallites, found by TEM (SADP), are presented in Figure 5. It reveals that near SiC particles, there are some DRX grains.

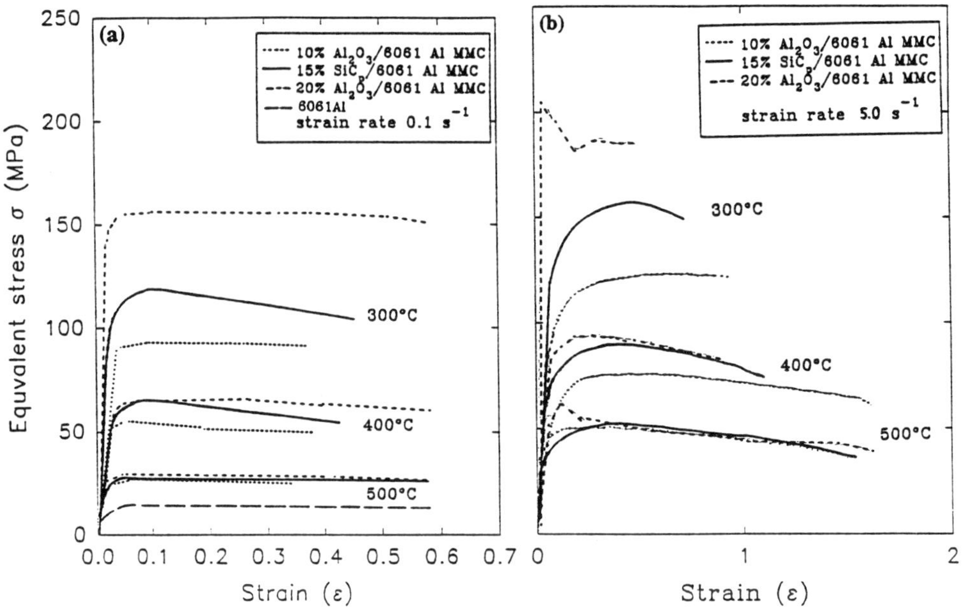

Figure 1. The stress versus strain curves of 10 v%, 20 v% Al_2O_3 and 15 v% SiC reinforced MMCs with temperatures from 200-500 °C and strain rates of (a) 0.1 and (b) 5.0 s^{-1}. In (a) the strain (ε) scale has been magnified and the torsion curves are cut off at $\varepsilon = 0.6$ when ε_f is greater.

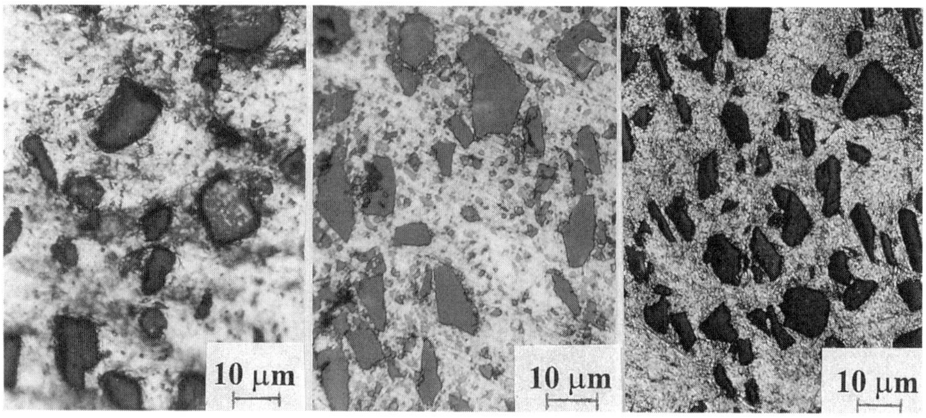

Figure 2. The particle distribution in Al_2O_3 and SiC reinforced MMCs is taken by optical microscopy (a) 10 v% Al_2O_3/6061Al MMC, (b) 15 v% SiC/6061Al MMC and (c) 20 v% Al_2O_3/6061Al MMC.

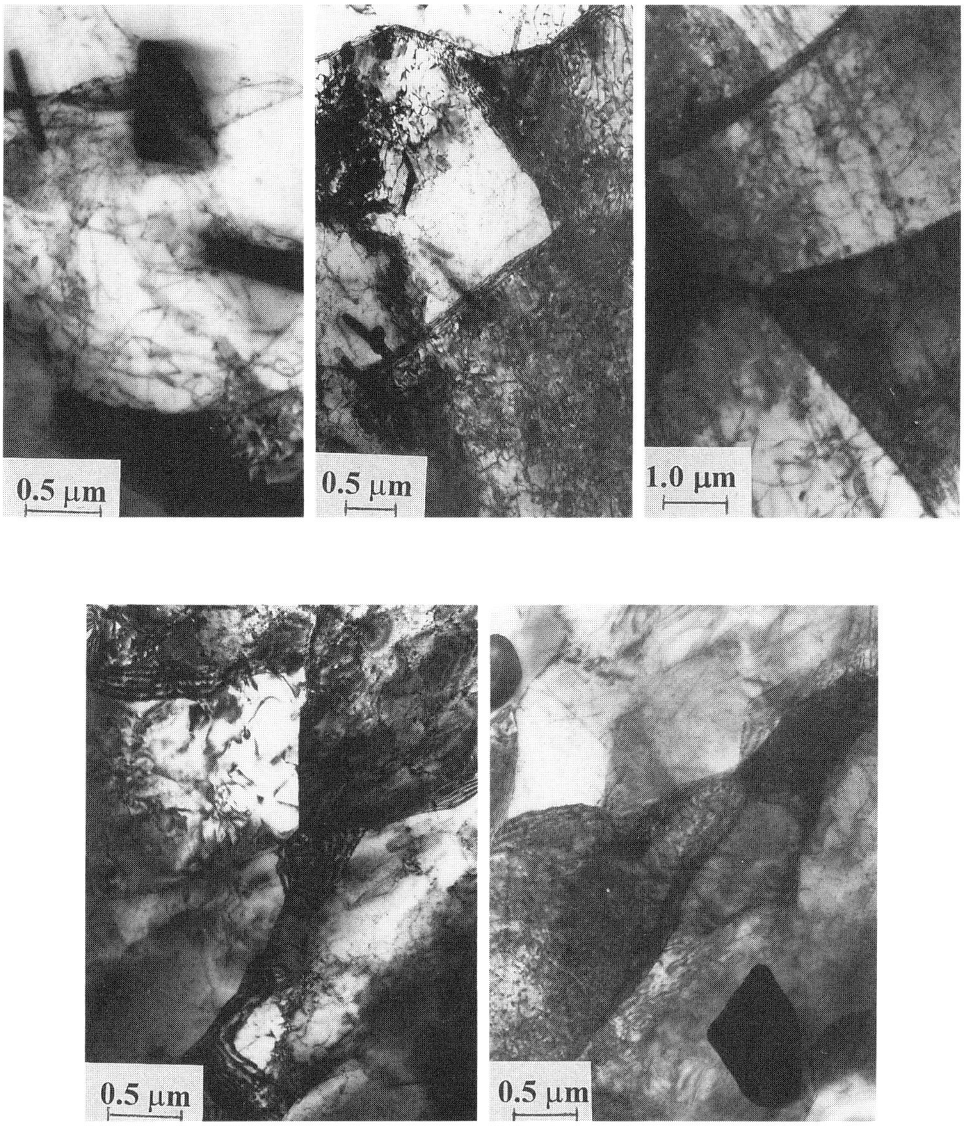

Figure 3. Dislocation density and subgrain size vary with the test temperatures and strain rates, 10 v% Al_2O_3/6061 Al MMC, at: (a) 300 °C, (b) 400 °C, (c) 500 °C, 1 s^{-1}; (d) 300 °C, 0.1 s^{-1}, (e) 300 °C, 5 s^{-1}.

Figure 4. Dislocation density varies with the reinforcing particle volume fraction: (a) the bulk alloy, (b) 10 v% Al_2O_3/6061Al, (c) 15 v% SiC/6061Al, (d) 20 v% Al_2O_3/6061Al MMC, 300 °C, 1 s^{-1}.

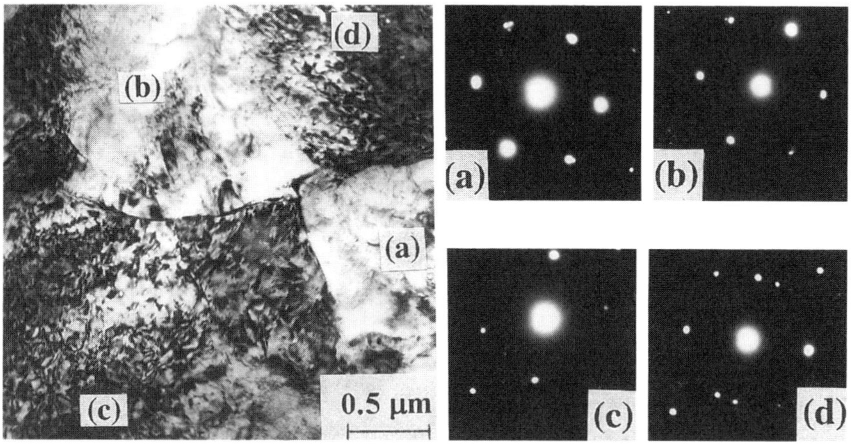

Figure 5. Highly misoriented crystallites are distinguished by TEM (SADP), 20 v% Al_2O_3/6061 Al MMC, 540 °C, 0.1 s^{-1}.

Table 2 Subgrain size variation with $\dot{\varepsilon}$ and *T* of particulate MMCs

Temp. (°C)	Strain rate (s^{-1})	Subgrain size *d* (μm) Bulk alloy *Compression*	10 v%Al_2O_3 *MMC*	15 v%SiC *MMC*	20 v%Al_2O_3 *MMC*
	0.1	-	-	0.7	-
200	1.0	1.4	-	0.6	-
	4.0	-	-	-	-
	0.1	-	1.5	0.9	-
300	1.0	-	1.0	1.0	0.4
	4.0	-	0.8	1.0	-
	0.1	3.8	-	1.2	0.5
400	1.0	-	2.0	0.9	0.7
	4.0	-	-	0.7	-
	0.1	-	-	1.8	2.0
500	1.0	-	3.0	1.6	2.0
	4.0	5.6	4.0	1.5	0.3

SEM (BSI) with small beam size (1 μm dia.), known as Dingley technique, also reveals the large misorientation between the two neighbouring grains (Figure 6) [14].

Thus the DRX grains are observed at the particle/matrix interfaces and in the Al matrix (Figures 5 and 7). It must be emphasized that the DRX grains do not completely replace the original matrix as they do in ordinary metals, such as Cu, Ni and stainless steels when steady state deformation is attained [15-17].

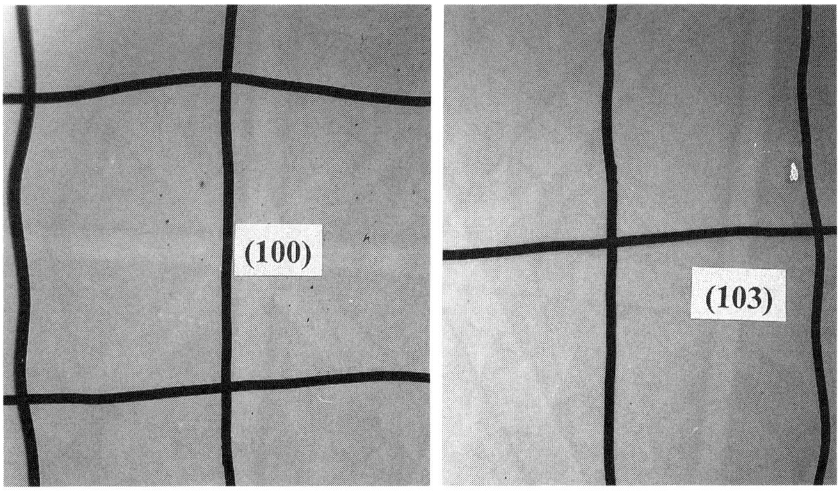

Figure 6. Dingley technique, Kikuchi patterns present the high angle boundary between two neibourgh grains (a and b), 10 v% Al_2O_3/6061 Al MMC, 400°C, 1 s^{-1}.

DISCUSSIONS

TEM observations show that the dislocation density (ρ) increases with increase of $\dot{\varepsilon}$ and decrease of T in MMCs. It has been found that ρ is much higher and subgrain size is much smaller than that of the bulk alloy in comparison at the same test conditions. This is the main cause of hardening during hot deformation. The dislocation density is observed also to rise in the MMCs' matrix with increase of the reinforcing-particle f_v, following:

$$\rho_{20\,v\%MMC} > \rho_{15\,v\%MMC} > \rho_{10\,v\%MMC} > \rho_{bulk\ alloy}$$

The much higher dislocation density forms during cooling and heating from the difference of thermal expansion coefficient (ΔCTE) between reinforcement and matrix forming.

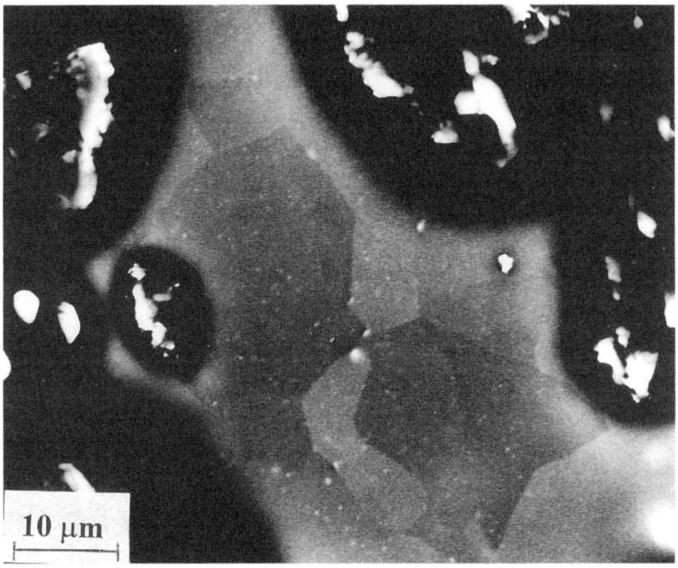

Figure 7. DRX grains are found at the interfaces of the reinforcing-particle/matrix and in the Al matrix, 10 v% Al_2O_3/6061 MMC, 400°C, $1s^{-1}$.

In addition, the rigid reinforcing particles restrict plastic flow in the surrounding matrix which results in fine substructures. These effects explain why the subgrain size has been found to decline with increase of $\dot{\varepsilon}$, f_v and decrease of T. Cellular substructures are generally found in the MMCs, but large amounts of the non-cellular structure are observed in 15~ 20 v% reinforced materials at 200 ~ 400 °C. It indicates that DRV is imperfect, particularly in the higher f_v MMCs.

When MMCs are hot-worked at higher temperature, DRV and DRX both contribute to softening as shown in Figure 8. Generally, cellular and non-cellular substructures resulting from DRV occupy about 70% of the entire matrix and DRX grains occupy 30% of the matrix.

The large volume fraction of substructure observed in the matrix confirms that DRV is still the main softening mechanism and is more important than DRX notably in lower volume fraction reinforced MMCs.

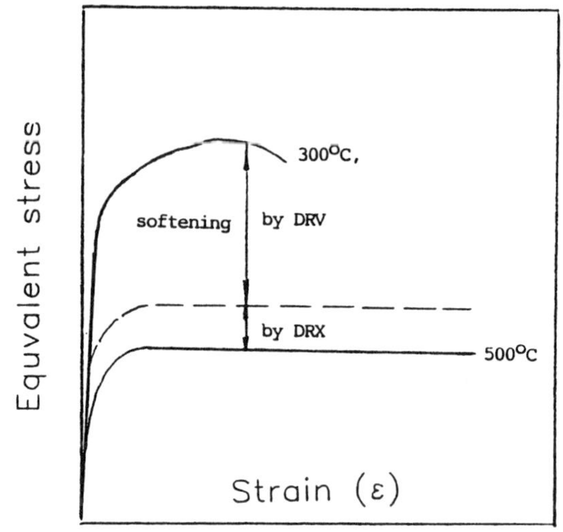

Figure 8. An illustration of the contribution of DRV and DRX on stress-strain curves during softening. The peak and work softening to the steady state regime of classical DRX is absent because the mechanism differs from the simple discontinuous mechanism which occurs in metals with relatively uniform dislocation substructure and highly mobile grain boundaries [15-17]. In the present high $\dot{\varepsilon}$ case, fine grains do not replace the large initial grains in a single wave, but rather form in dispersed clusters as a result of either particle stimulated nucleation or continuous DRX in relation to fine particle pinning. At 0.83 T_m (melting temperature, K) which is 500 °C in Al and 1200 °C in steels, the steady state regime with complete recrystallization would be established by $\varepsilon = 0.5$ in steels yet it has not been established in the MMCs at strains 4 to 5 times that value.

CONCLUSIONS

• In comparison to the bulk alloy, the initially high ΔCTE dislocation density and the particle constraint to flow give rise to smaller subgrain size in the MMCs, thus contributing to the hardening.

Increase in the percentage of the non-cellular structure in the higher particle volume fraction reinforced MMCs contributes additional hardening, especially at lower test temperature. The substructure is highly non-uniform compared to the bulk alloy.

• The dislocation density decreases and subgrain size increases with rising test temperature and decreasing strain rate and volume fraction. Dynamic recovery occurs during hot deformation and is the principal cause of softening behaviour in the flow curves as *T* rises; but DRV is imperfect in MMCs, especially for MMCs with higher particle fractions.

• The highly misoriented crystallites observed are DRX grains. They nucleate at the reinforcing-particle/matrix interfaces and in the Al alloy matrix as a result of the non-uniform distribution of strain. The DRX grains do not completely replace the original grains.

ACKNOWLEDGEMENT

The authors would like to thank Dr. David J. Lloyd, Alcan Research Laboratories, Kingston, Canada and Dr. William Dixon of Duralcan, San Diego for supplying materials. They acknowledge the NSERC funds which made this research project possible.

REFERENCES

1. I.A. Ibrahim, F.A. Mohamed and E.J. Lavernia, "Particulate Reinforced Metal Matrix Composites - A review", *J. Mat. Sci.*, Vol. 26, pp.1137-1156, 1991.
2. P. Sakaris and H.J. McQueen, "Hot Workability of SiCp/6061 Al Composite", ***Al Alloys and Their Physical and Mechanical Properties (ICAA3)***, ed by L. Arnberg et al, pub. by NTH-Sinteff, Tronheim, Norway, pp.554-559, 1992.
3. P.L. Munoz, C. Cerezo and P. Nogueroles, "Automation of processing of Composite Materials, An Example, The Airbus 330/340", ***Composite Materials (ICCM9)***, ed. by A. Miravete, pub. by Woodhead Publishing Limited, University of Zaragoza, Madrid, Spain,Vol. 1, pp.51-80, 1993.
4. P. Sakaris and H.J. McQueen, "Comparative Hot Workability of SiCp/A356 and SiCp/6061 Al Composites and Their Matrices", ***Advances in Production and Fabrication of Light Metals - CIMM***, ed. by M.M.Avedesian, et al., pub. by CIM Montreal,Canada, pp. 605-617, 1992.

5. H.J. McQueen, P. Sakaris and J.Bowles, "Hot Ductility and Strength of SiCp/A356 Al Composite and Matrix Alloy by Tirsion Testing", ***Advanced Composite Materials'93***, ed. by T. Chandra and A.K. Dhingra, pub. by TMS-AIME, Warrendale, PA, pp.1193-1198.1993.
6. Minoro Taya and R,J, Arsenault, ***Metal Matrix Composites - Thermo-mechanical Behaviour***, Pergamon Press, 1989.
7. R.J. Arsenault, "Strengthening and Deformation Mechanisms of Discontinuous Metal Matrix Composites", ***Strength of Metals and Alloys***, ed. by D.G. Brandon, et al, pub. by Freund Publishing Co., Ltd, London, U.K., Vol.2, pp.31-46, 1990.
8. D.Yu, S. Bandyopadhyay, A.P. Mouritz and P.R. Munroe, "Deformation and Recrystallization Behaviour of SiC Particulate Reinforced 6061 Aluminium Metal Matrix Composites" ***Composite Materials (ICCM9)***, ed. by A. Miravete, pub. by Woodhead Publishing Limited, University of Zaragoza, Madrid, Spain, Vol. 1, pp.363-368, 1993.
9. E. Evangelista, A.Forcellese, F. Gabrielli and P. Mengucci, "Hot Formability of AA6061 PM Al Alloy", ***J. Mat. Process. Tech.***, Vol. 24, pp.323-332, 1990.
10.J.R. Pickens, T.J. Langan, R.O. England and M. Liebson, "A Study of the Hot-Working Behaviour of SiC-Al Alloy Composites and Their Matrix Alloys by Hot Torsion Testing", ***Met. Trans.***, Vol. 18A, pp.303-312, 1987.
11. D. Yu and T. Chandra, "Effect of Temperature and Strain Rate on Flow Behaviour and Microstructure in 6061 Al Containing Al_2O_3 Particles", ***Advanced Composites'93***, ed. by T. Chandra and A.K. Dhingra, pub. by TMS-AIME, Warrendale, PA, pp.1073-1077, 1993.
12. X. Xia, P. Sakaris and H.J. McQueen, "Hot Torsion Behaviour and Recrystallization of an Al/SiCp Composite Material", ***Composite Materials (ICCM-9)***, ed. by A. Miravete, pub. by Woodhead Publishing Limited, University of Zaragoza, Madrid, Spain, pp.157-162, 1993.
13. S. Fulop, K. Cadien, M.J. Luton and H.J. McQueem "A Servo-Controlled Hydraulic Hot-Torsion Machine for Hot Working Studies", ***J. of Testing Evaluation***, Vol. 5, pp.419-425, 1977.
14. X. Xia, P. Sakaris and H.J. McQueen, "Deformation and Microstructure of Al_2O_3 Particle reinforced 6061 Al Composite", ***New Materials and Ceramic***, ed.by R.A.L.Drew and H. Mostaghaci, pub. by The Metallurgicial Society of the CIM, Montreal,Canada, pp.134-145, 1993.
15. H.J. McQueen, E. Evangelista and N.D. Ryan, "Dynamic Recrystallization and Recovery: Mechanical and Kinetic Behaviour; Nucleation and Growth Mechanisms", ***Recrystallization in Metallic Materials***, ed. by T. Chandra et. al, pub. by TMS-AIME, Warrendale, PA, pp.89-100, 1990.
16. N.D. Ryan and H.J. McQueen, "Comparison of Dynamic Softening in 301, 304, 316 and 317 Stainless Steels", ***High Temp. Tech.***, Vol. 8, pp.185-200, 1990.
17. N.D. Ryan and H.J. McQueen, "Flow Stress, Dynamic Restoration, Strain Hardening and Ductility in Hot Working of 316 Steel", ***J. Mat. process. and Tech.***, Vol.21, pp.177-199, 1990.

Chapter 4

Light Metals

FORMATION OF SERRATIONS IN AL AND AL-5MG DURING INITIAL HOT STRAINING

E.V. Konopleva*, H.J. McQueen** and W. Blum†

ABSTRACT

Serrations are formed in the grain boundaries of Al and Al-5Mg alloys during deformation in torsion to strains of 0.4 or 0.8 at 400 or 500°C and rates of 10^{-3} to 5 s^{-1}. The serration development rate is linked to the progress of polygonization and serration dimensions are proportional to the spacing of sub-boundaries. Thus, both the development rate and the serration spacing increase as deformation temperature rises and strain rate falls.

INTRODUCTION

When Al and Al-5Mg are deformed at 25°C, the grains gradually lengthen in the tensile direction, becoming narrower in the compression direction; within them, the dislocation density increases and a cellular structure develops. The grain boundaries remain smooth, at least for strain up to 0.8. At high temperatures, 400 - 500°C, the dislocation density (and strain hardening) remains much lower. Due to dynamic recovery (DRV) an equiaxed sub-grain structure develops, becoming fully formed at lower saturation

* Inst. Solid State Physics, RAS, 142432 Chernogolovka, Russia
** Mech. Eng., Concordia University, Montreal, Canada, H3G 1M8
† Mat. Sc., Univ. Erlangen-Nürnberg, D91058 Erlangen, Germany

strains ε_s as the temperature T rises and strain rate $\dot{\varepsilon}$ decreases [1-14]. The grain boundaries (GB) gradually become wavy as the subgrains form, reaching an amplitude and wavelength proportional to subgrain diameter d [1-5, 9-14]. The serrations are thought to form as a result of the GB migrating in order to absorb dislocations in subgrain boundaries (SGB) [15-19] and are observed in Al alloys subjected to industrial rolling or extrusion [1-3]. The effects have been mistaken for dynamic recrystallization (DRX). The formation of serrations will be followed in Al and Al-Mg alloys over strains of 0.4 and 0.8 in torsion at selected T and $\dot{\varepsilon}$.

In the torsion test, as considered theoretically and confirmed by many experiments, a line parallel to the axis on the surface is wound into a helix, which gradually increases in angle to the axis [9,14,20]. In the present specimens, some very long grains were retained from the prior extrusion and they form into such a helicoid which reaches about 35° to the axis at $\varepsilon = 0.4$ and about 50° at 0.8. The equiaxed grains have become slightly elliptical at 0.4 and the orientation is barely visible, whereas at 0.8, the elongation and re-alignment are both quite noticeable; this effect should be independent of temperature. If a section is developed parallel to the axis and close to the gage surface, it is called tangential; however, in the larger diameter shoulder, the same section is much deeper and is referred to as a chord [9,15,16]. It is possible to examine a range of strains in the shoulder, the transition (filet) and the gage to observe how the equiaxed grains gradually become helicoidal, spiraling around the axis. The present research will show how this grain flow is rendered more complex by the simultaneous formation of serrations whose size increases as the subgrains become larger as T rises and $\dot{\varepsilon}$ declines.

EXPERIMENTAL TECHNIQUES

Aluminum and Al-5Mg alloys (impurities 0.03 Fe, 0.02 Si, 0.01 Mn) were supplied by Alusuisse in the form of extrusions. Torsional specimens with a gauge length (L) of 25.4 mm and diameter (2r) of 6.32 mm were threaded into the stationary bar fixed to the torque cell at one end and inserted into a rectangular slot in the rotating grip at the other [9,14]. This arrangement allowed for easy removal without accidentally straining the specimen. Testing was conducted on a computer-directed, servo-controlled, hydraulic torsion machine, at strain rates between 10^{-3} and 5.0 s^{-1} and temperatures of 400 and 500°C. Heating was provided by a quadruple elliptical radiant furnace connected to a programmable controller, which automatically varied power input to maintain test temperature. The data was stored by the computer and displayed as torque Γ versus angle θ. The equivalent flow stress σ is obtained from the Backofen equation and von Mises criterion [20,21]:

$$\sigma = \sqrt{3}\ \Gamma\ (3+m+n')/2\pi r^3 \qquad (1)$$

where m, the strain rate sensitivity, is determined experimentally by the slope of log Γ versus log $\dot{\varepsilon}$ plot. The work hardening rate n′ ($= d\sigma/d\varepsilon$) is zero for the peak stress, since the slope is horizontal at that point. The equivalent surface strain (ε) is obtained from measured θ (in radians), using the following equation, which once again employs the von Mises criterion (18):

$$\varepsilon = r\theta/(\sqrt{3}\ L) \qquad (2)$$

All specimens were protected by argon at high temperatures and were quenched in water in less than ten seconds after deformation was completed. An entire half of the specimen was ground so that a small tangential flat on the gage section extended into the shoulder in order that one may see the grains in the transition region, as they are transformed from the as-received condition into a helicoid wrapped around the axis (horizontal in figures). The specimens were ground and electropolished for 20 s at 20 V, 20°C in 20 ml $HC\ell 0_4$, 80 ml ethyl alcohol. They were anodized for 30s at 45V, 20°C in 8.4 ml HF 45.4 ml ethyl alcohol and 45.4 ml H_2O for examination in polarized light. The subgrains referred to in the results are related to POM; however, in the discussion, comparisons are made between POM, SEM-BSI and TEM subgrains (referring to polarized optical, scanning back scattered image and transmission electron microscopy).

RESULTS

The flow curves for Al and Al-5Mg under the conditions tested are shown in Figure 1 [7, 8]. The Al exhibits a monotonic increase in stress to a steady state plateau which is characteristic of softening by dynamic recovery. The flow curves for Al-5Mg show a rapid rise to a slight yield stress, then rapid progress to a plateau with a gradually declining stress. The yielding and plateau stresses are the result of Mg solute atmospheres exerting a drag on dislocations. The flow stress for 500°, 0.1 s^{-1} of the alloy is much higher than that of Al. When the steady state stresses were subjected to a constitutive analysis [7, 8], the stress exponent ($\dot{\varepsilon}$ dependence on σ) was about 3 for Al-5Mg (for σ<80MPa), compared to 4.5 for Al, thus confirming solute drag alloying effects. Both activation energies were about 150 kJ/mol, indicating climb control for Al and Mg diffusion control for the alloy.

The general results of the hot torsion to ε = 0.8 at 400°C, 0.01 s^{-1} for Al are illustrated in Figure 2 (X44), which includes a chord section fairly deep in the shoulder at the right, a transition where the filet is located in the centre and a tangent section close to the surface of the gage. The grains can be seen to change from equiaxed to elongated and aligned at 45° to the torsion axis. It is also noticeable that the GB transform from smooth in the shoulder to serrated in the gage. Subgrains are observable in some of the grains as mottled regions, which could be better distinguished at higher magnification. At a lower strain in the shoulder, grain distortion and alignment could not be seen but serrations could be observed, although smaller and on fewer GB. The following microstructures are presented for increasing T and decreasing $\dot{\varepsilon}$; an increase by one order of magnitude in $\dot{\varepsilon}$ has an equivalent effect on lowering stress as a rise of 50°C [7, 8].

The grain and subgrain structures observed by POM in Al deformed at 500°C, 0.1 s^{-1} are shown in Figure 3. The gage section at ε = 0.4 (Fig. 3a, X56) shows the almost equiaxed grains with some evidence of subgrains within and of GB serrations. At ε = 0.8 in the gage (Fig. 3c, X70), the grains are quite elongated and aligned near 45° to the torsion axis. The subgrains have become more

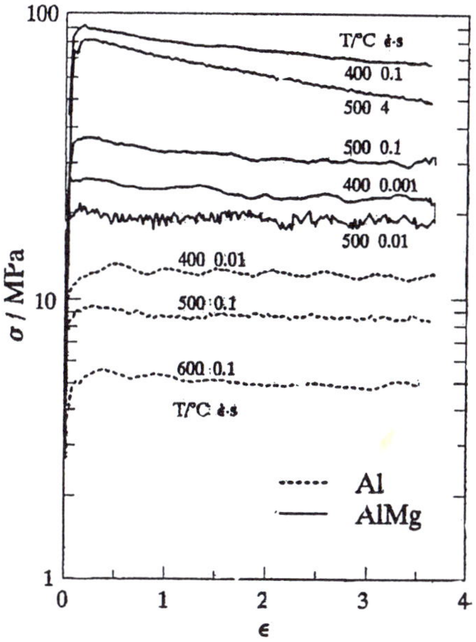

Figure 1: Stress strain curves for Al and Al-5Mg for the conditions examined. The alloy is much stronger than the pure metal under the same conditions [7,8].

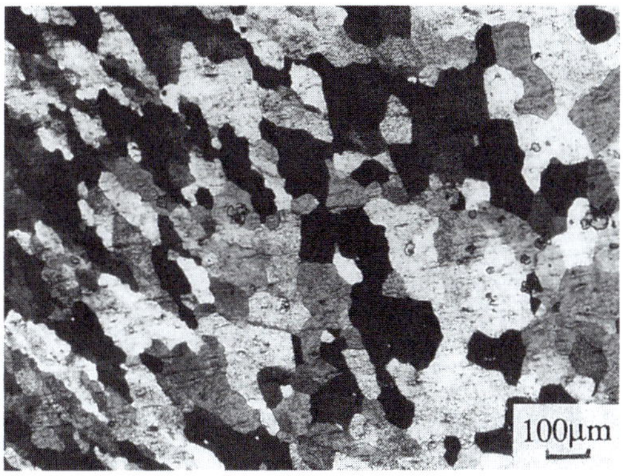

Figure 2: Polarized optical micrograph of Al deformed in torsion at 400°C, 0.01 s^{-1} to $\varepsilon = 0.8$, illustrating three regions: at right, shoulder chord section with equiaxed recrystallized grains; transition region at filet with some serrations and in left half, elongated serrated grains aligned at 45° to horizontal torsion axis.

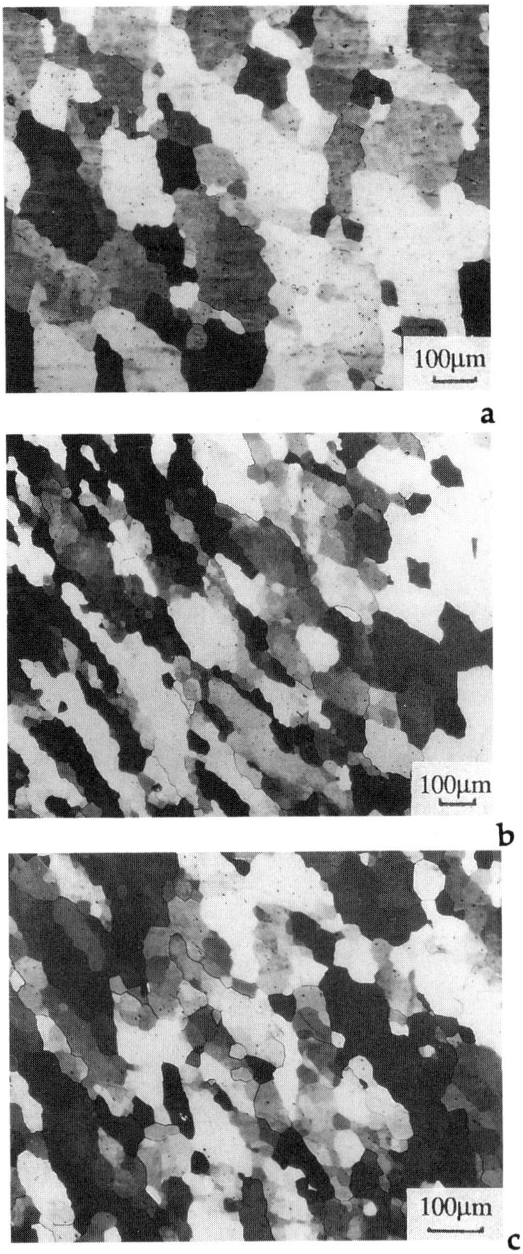

Figure 3: POM grain and subgrain structures developed in Al at 500°C, $0.1\ s^{-1}$: a) $\varepsilon = 0.4$, tangent section of gage; b) $\varepsilon = 0.8$, gage tangent at left and shoulder at right and c) $\varepsilon = 0.8$, gage tangent.

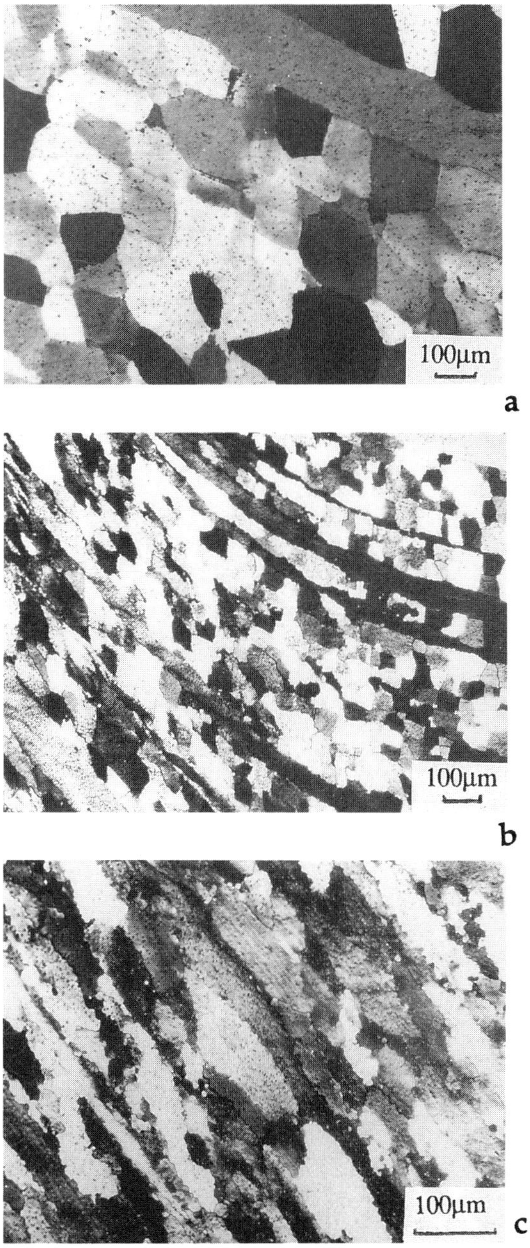

Figure 4: Polarized optical micrographs of Al-5Mg deformed at 400°C, 0.1 s^{-1} (much above the n=3 regime): a) ε = 0.4, gage tangent (long grain from extrusion indicates twist); b) ε = 0.8, gage tangent with shoulder at right and c) ε = 0.8, gage tangent. In comparison to Al in Figure 3, σ is ten times larger so subgrains are one-tenth the size [7,8], yet the serrations have similar amplitude but smaller half-wavelength.

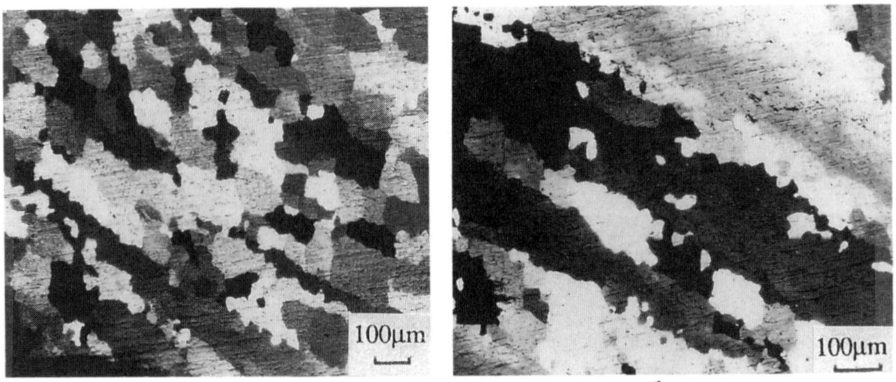

Figure 5: POM of Al-5Mg at 400°C, 0.001 s^{-1} (n=3) at ε = 0.4: a) gage tangent at left and shoulder at right and b) gage tangent.

prominent and seem to be about 0.2 - 0.4 of the grain width. The subgrains are nearly equiaxed and are clearly related to the serrations. The serrations can be seen to increase in size as the strain increases from shoulder to gage, (Fig. 3b, X44). Serrations and associated subgrains were also observed in Al deformed to both 0.4 and 0.8 at 600°, 0.1 s^{-1} in confirmation of lower T results.

In Al-5Mg deformed at 400°C, 0.1 s^{-1} to ε = 0.4 (Fig. 4a, X56), the slightly elongated grains (an unrecrystallized grain indicates rotation relative to horizontal torsion axis) exhibit only slight serrations on a few GB. When ε is raised to 0.8, the grain distortion and realignment are marked (Figs. 4 b, X56; c, X110) and the serrations are quite large on all grains. In some thin grains, the opposite serrations almost touch, causing pinching-off into isolated crystallites. The subgrains are exhibited within some grains as a mottling effect, thus indicating that they are incompletely formed. At 400°C, 10^{-3} s^{-1} to ε = 0.4, the development of serrations from shoulder to gage is clearly seen, especially in some retained elongated grains (Fig. 5a, X44). The extreme meandering of the serrations and the subgrains near the GB are clearly exhibited at a higher magnification (Fig. 5b, X70). Some pinched-off serrations are also present; they could be related to grains above or below the plane of section, since the helicoidal grains are initially cylindrical, with recrystallized grains randomly growing into them, as shown in Figs. 4a, b.

For 500°C, 5 s^{-1} at $\varepsilon = 0.4$, the grains are barely elongated and show slight serrations much less than 400°C, 0.1 s^{-1} (Fig. 4a). For 500°C, 0.1 s^{-1} (Fig. 6), the progressive formation of serrations was once again confirmed in the transition from equiaxed shoulder grains to slightly elongated ones in the gage; however, in a region of gage containing some initially elongated grains, the elaborate flourishing of serrations can be appreciated (Fig. 6a, X70). Moreover, there were good examples of pinching-off, notably at the ends of grains in the direction of grain alignment. When ε is increased to 0.8, the progressive development of strong serrations from shoulder to gage section can be clearly followed in Fig. 6b (X70). The serrations may be better appreciated in some prior elongated grains (Fig. 6c, X70). There are many examples of pinching-off, both complete and almost so. These results correspond to 400°C, 0.001 s^{-1}, as shown in Figs. 5a, b. When the strain rate is reduced to 0.01 s^{-1} (Fig. 7), serrations are seen to be much larger, relative to grain size than those at 0.1 s^{-1} (Fig. 6). The shoulder to gage transition clearly exhibits the increase in waviness of the boundaries as ε rises to 0.4 (Fig. 7a, X44). At 0.4, the initially equiaxed grains are quite distorted by the serrations, to the extent that partial and complete pinching-off have occurred (Fig. 7b, X56). At 0.8, there are both irregular thick grains and elongated grains that appear to be strings of subgrains (Fig. 7c, X56). When the strain is increased to 1.6, the pinching-off process has proceeded to such an extent and the subgrains have become so strongly emphasized that the elongated grains are indistinguishable (Fig. 8a, X63). When the same regions are simply etched instead of anodized, it can be seen that only a few of the boundaries are exhibited as high angle ones (Fig. 8b, X63).

DISCUSSION

The serrations clearly increase in frequency and in size as the strain increases over the range 0.4 or 0.8 (Figs. 3, 4, 7). This is also confirmed by examining the transition from shoulder to gage (Figs. 2, 3b, 4b, 5a, 6b, 7a, 8a). The serrations do reach a stable value as can

a

b

c

Figure 6: POM of Al-5Mg at 500°C, 0.1 s^{-1} (in the n=3 regime): a) ε = 0.4 gage tangent and at ε = 0.8: b) gage tangent with shoulder at right and c) gage tangent. In comparison to Al in Figure 3, σ is three times larger so subgrains are one-third the size [7,8], yet the serrations are about the same half-wavelength but larger in amplitude. The serrations are slightly less pronounced than in Figure 5, where the stress is somewhat lower and subgrains larger.

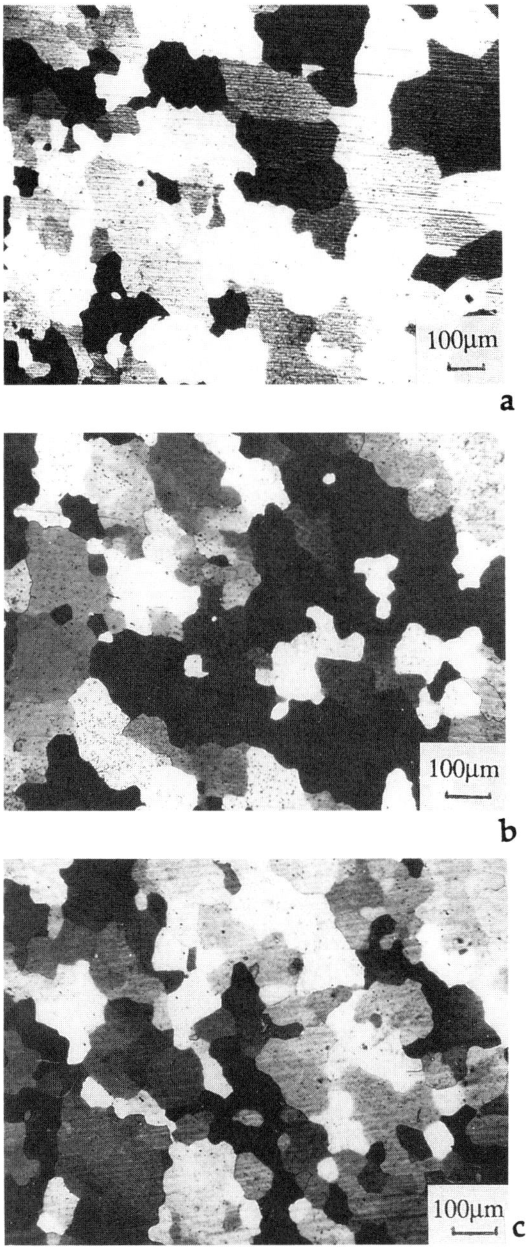

Figure 7: POM of Al-5Mg at 500°C, 0.01 s^{-1} (n=3) : at $\varepsilon = 0.4$ a) gage tangent with shoulder at right and b) gage tangent and c) $\varepsilon = 0.8$ gage tangent. In comparison to Al in Figure 3, σ is twice as large so subgrains are half the size [7,8], yet the serrations are much more pronounced and many have pinched-off.

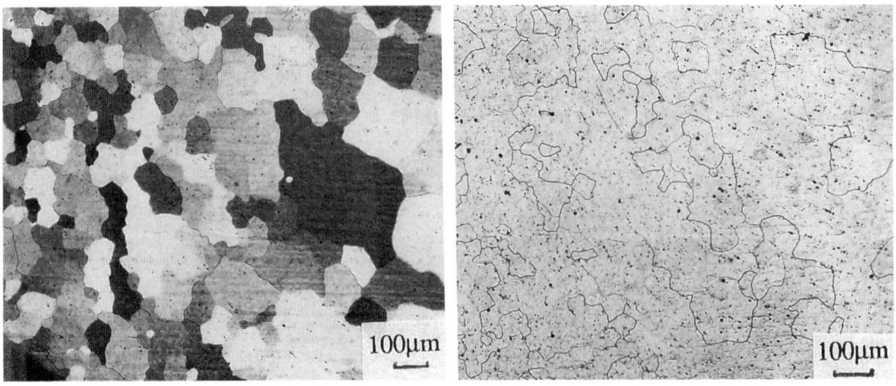

a b

Figure 8: Al-5Mg deformed at 500°C, 0.01 s^{-1} to ε = 1.6 gage tangent with shoulder at right: a) in POM and b) etched and in ordinary light.

be seen in the cases where they form more rapidly, such as at 400°C, 10^{-3} s^{-1} and 500°C, 0.1 or 0.01 s^{-1}. This has been confirmed by tests to high strains which are reported elsewhere [15].

The amplitude and wavelength of the serrations increase as T increases; from 400 to 500°C at 0.1 s^{-1} (Figs. 4, 6). The effect is similar when $\dot{\varepsilon}$ decreases, as seen both at 400°C, 0.1 to 10^{-3} s^{-1} (Figs 4, 5) and 500°C 0.1 to 0.01 s^{-1} (Figs. 6, 7). The equivalency of T and $\dot{\varepsilon}$, as estimated through their effects on flow stress is seen in Figs. 5 and 6 at 400°C, 10^{-3} s^{-1} and 500°C, 0.1 s^{-1}. This arises because of the relationship of serration dimensions to subgrain size, which is uniquely related to the flow stress. The serrations arise because the GB migrate as they absorb dislocations from SGB, the energy from the latter providing for the increased surface energy due to the GB undulation. This has been confirmed by TEM where SGB are clearly visible as dislocation arrays [7,8,16,17]. In POM, the subgrains always occupy concave sides of cusps, so that there are SGB on the convex sides. Finally, it is known that flow stress attains a steady state value at a strain ε_s, which is lower as T rises and $\dot{\varepsilon}$ falls. This explains why the serrations form more rapidly and reach a terminal size.

Serrations certainly form in both Al and Al-5Mg and appear to have many similarities, as can be seen at 500, 0.1 s^{-1} in Figs. 3 and 6. The serrations in Al seem to be about the same size; the comparison is affected by the initial grain size being smaller in Al than in Al-5Mg. The other difference is that the subgrains distinguished by differences in shade seem to be more readily visible in Al. The equality of metal and alloy becomes surprising when one notes from Fig. 1 that the stresses are much higher in Al-5Mg, yet the established data [7,8] show that the dependence of TEM subgrain sizes on stress are the same in both. This problem was cleared up by SEM back scattering image examination of the same region as POM [17]. In Al-5Mg, the TEM subgrains were about one third the size of the SEM, and only one sixth of the POM units [15-17]. In Al, the three techniques are similar, although the size does increase in the order TEM, SEM-BSI and POM, because of their decreasing sensitivity to very low angle SGB [7,8]. One factor in Al-5Mg is clearly the slow development of subgrains, which take approximately double the strain to form completely throughout the grains, as ε_s to reach steady state [7,8]. Apparently, the serrations in Al-5Mg begin to form as soon as the first SGB form near the GB, even though subgrains have not yet formed in the interior. This behaviour can also be related to the strong influence of the SGB on the flow stress of Al, whereas the Mg solute drag has the predominant influence in Al-5Mg [7,8].

The serrations also have another significant effect on the microstructure when they become large and meandering, so as to pinch-off. They apparently create new grains, which appear both between and inside original grains (Figs. 3, 5, 6, 7). However, these are quite different from nuclei formed by the bulging mechanism which develop and grow on a much broader front. Moreover, these crystallites do not grow, apparently because the strain energies within and without are balanced. As strain rises, these detached serrations (DS) may attain such a high volume fraction that they give an appearance of recrystallization; it has been termed geometric or rotational DRX [15-19]. However, the microstructure comes about from the undulations of the GB induced by DRV unit mechanisms [9-14], it can be referred to as DS-DRV.

The serrations have an especially marked effect when their amplitude becomes almost the same as the grain width, that is when it is only 2 or 3 times the subgrain diameter. The elongated grains begin to pinch-off, becoming shorter. This effect is apparent in Figure 8 at a strain of 1.6, and in tests to $\varepsilon = 4$ [15], where the POM micrograph exhibits a field of equiaxed crystallites, whereas the etched micrograph shows that only a few of the boundaries are very irregular high-angle GB and many must be SGB. This is in agreement with previous research, where the effect was called "geometric" G-DRX, since apparently small grains had formed during deformation [9,14]. Because the major mechanism has been DRV [9-14] in association with a limited degree of stress-induced boundary migration, an alternate name is GR-DRV (Grain Refining) [15,16,22].

CONCLUSIONS

The serrations form during high temperature deformation, increasing in rapidity of formation and in amplitude and wavelength as temperature is raised and as strain rate is reduced. These dependencies are similar to those for subgrains, confirming the relationship of the migrations to absorption of subgrain boundaries.

The serrations grow larger with strain until the steady state regime is attained; however, they can become more irregular, leading to pinching-off of crystallites which do not appear to grow.

The serrations in Al-5Mg are much larger, more meandering and more prone to pinching-off than in Al at the same stress and hence subgrain size; however, they are fairly similar for the same temperature and strain rate, where Al-5Mg has much finer subgrains.

When the temperature is very high and the strain rate very low, so that the serration amplitude approaches half the grain thickness, then the original grains can no longer be distinguished in POM. The microstructure appears recrystallized which has been named 'geometric', although the ordinary nucleation and growth mechanisms have not occurred.

ACKNOWLEDGEMENTS

The authors wish to acknowledge the support of the Humboldt Foundation, Germany and the Natural Science and Engineering Council of Canada. They are grateful to Q. Zhu and R. Merkel (University of Erlangen-Nürnberg) and E. Evangelista (University of Ancona) for discussion.

REFERENCES

1. H.J. McQueen, Metal Forum, (Aust)., 4, 1981, pp. 81-91.
2. H.J. McQueen, and J.J. Jonas, Plastic Deformation of Materials (Treatise on Mat. Sci. Tech., R.J. Arsenault, ed., Academic Press, New York, 6, 1975, pp. 393-493.
3. R.W. Evans and G.R. Dunstan, J. Inst. Metals, 9, 1971, pp. 4-14.
4. H.J. McQueen and N. Ryum, Scand. J. Met., 14, 1985, pp. 183-194.
5. H.J. McQueen and K. Conrod, Microstructural Control in Aluminum Alloy Processing, E.H. Chia and H.J. McQueen, eds. TMS-AIME, Warrendale, PA., 1986, pp. 197-219.
6. H.J. McQueen, Hot Deformation of Aluminum Alloys, T.G. Langdon et al., eds.,TMS-AIME, Warrendale,PA., 1991, pp. 31-54.
7. H.J. McQueen, W. Blum, Q. Zhu and V. Demuth, Advances in Hot Deformation and Textures, T.R. Bieler and J.J. Jonas, eds., TMS-AIME, Warrendale, PA., 1993, pp. 235-250.
8. Q. Zhu, W. Blum and H.J. McQueen, Strength of Materials, (ICSMA 10), H. Oikawa et al., eds., Japan Inst. Metals, Tokyo, 1994, pp. 803-806.
9. O. Knustad, H.J. McQueen, N. Ryum and J.K. Solberg, Practical Metal. 22, 1985, pp. 215-229.
10. H.J. McQueen, J.K. Solberg, N. Ryum and E. Nes, Aluminum Alloys, Physical and Mechanical Properties, E.A. Starke and T.H. Sanders, eds. EMAS, Worley, UK, 1986, pp. 513-528.
11. J.K. Solberg, H.J. McQueen, N. Ryum and E. Nes, Phil. Mag. 60, 1989, pp. 447-471, 473-485.
12. M.E. Kassner and M.E. McMahon, Metall. Trans. 18A, 1987, pp. 835-846.

14. G.A. Henshall, M.E. Kassner and H.J. McQueen, Metall. Trans., 23A, 1992, pp. 881-889.
15. E.V. Konopleva and H.J. McQueen, Mat. Characterization, 1995. (In press.)
16. H.J. McQueen, N.D. Ryan, E.V. Konopleva, and X. Xia, Can. Metal. Quart., 34, 1995. (In press.)
17. W. Blum, Q. Zhu, R. Merkel and H.J. McQueen, Z. Metallkde., 1995.
18. F.J. Humphreys and M.R. Drury, Aluminum Technology, T. Sheppard, ed., (London, UK: The Inst. Metals, 1986), pp. 191-196.
19. M.R. Drury and F.J. Humphreys, Acta Metal., 34, 1986, pp. 2259-2271.
20. E. Evangelista, N.D. Ryan and H.J. McQueen, Metal Sci. Tech., 9, 1991, pp. 75-92.
21. S. Fulop, K.C. Cadien, M.J. Luton and H.J. McQueen, J. Test Evaluation, 5, 1977, pp. 419-426.
22. T. Sakai (private discussion, University of Electro-Communications, Tokyo, 1994.

CHLORIDE INCLUSIONS IDENTIFICATION IN CAST ALUMINIUM INGOT

J. Boutin, Y. Raymond, M. Lebleu, H. Dufour and M. Belley

Alcan International Limited, Arvida Research and Development Centre, P.O.Box 1250, 1955 Mellon Boul., Jonquière, Québec, Canada, G7S 4K8

ABSTRACT

This paper describes the procedure for sampling metallographic specimen and evaluating the number, size and type of chloride inclusions in aluminium and its alloys. Liquid aluminium is sampled from the trough and horizontal discs are cast. The solidified discs are cut, mounted in bakelite and wet polished. Using an optical microscope, the potential chloride inclusions are recognized based on their shape and other morphological characteristics, and their number and size distribution are measured. Subsequently, the specimen are dry polished to prevent dissolution of the chloride inclusions and their exact chemical composition is further determined by Scanning Electron Microscope (SEM) using an Energy Dispersive X-ray (EDX) technique. Among the total number of declared potential chloride inclusions on the optical microscope, only a fraction of them are formally identified as real chloride inclusions.

INTRODUCTION

With the development of aluminium metal fluxing devices using chlorine gas as an inclusion floating agent [1,2,3,4,5], the presence of chloride inclusions in the final aluminium ingot is becoming a major concern, particularly for foil and lithographic products. The presence of chloride inclusions is detrimental for some rolling processes. This technique for chloride inclusions counting and identification will help evaluate the efficiency of the filtrating devices by determining their optimum operating conditions. Both cleaner rolling ingots and better filtrating conditions will result in substantial cost savings for aluminium producers.

The first occurrence of chloride inclusions was observed many years ago by Alcan's metallographers, in PoDFA (Porous Disc Filtration Apparatus) specimens. The PoDFA technique [6] is an Alcan technology that consists in filtrating liquid aluminium through a porous alumina disc to catch solid inclusions at the filtrating disc interface. The caught inclusions area is evaluated and divided by the weight of metal filtered through the porous disc, which gives the metal cleanliness value expressed as millimetre square of inclusions per kilogram of metal. Because the chloride inclusions are still liquid at the solidification temperature of most aluminium alloys, they are flushed out through the porous disc with the filtered liquid metal and only a few such chloride inclusions could be observed and measured in the PoDFA specimen.

Therefore, to evaluate all chloride inclusions present in the aluminium ingot, another sampling technique has to be used. The first sampling approach was to cut 50 mm thick slices at various length positions in the sheet ingot and to polish and perform a metallographic evaluation of the entire slice longitudinal surface on the optical microscope. This type of sampling was extremely costly and time consuming, but had the advantage of giving a lot of information about the chloride inclusions distribution throughout the ingot thickness. Further work showed good comparison between results of chloride inclusions measured in the ingot slices and those measured in cast horizontal discs (the same discs used for OES analysis). It is much easier to conduct a chloride inclusion survey of a cast without destroying the ingot with this sampling technique. Results presented in this work were obtained from HD specimens [7].

The analysis is performed first after a wet polishing by optical microscopy. Based on a few chloride inclusion morphological characteristics, the number of chloride inclusions is counted over a minimum of 5 centimetre square of surface, their size distribution and maximum diameter are measured and reported. Subsequently, the specimen is dry polished and examined by Scanning Electron Microscopy (SEM) for formal identification by Energy Dispersive X-Ray (EDX).

This paper deals with the metallographic techniques of specimen preparation, conventional wet and dry polishing, microscopic optical evaluation of the number and size of potential chloride inclusions and SEM-EDX formal identification of chloride inclusions.

METAL SAMPLING

The metallographic analysis was performed on horizontal discs [7] taken at various positions in the casting process. The rapid chill and the geometry of the HD specimen reduces the occurrence of gaseous and shrinkage porosity which eliminates any ambiguous situations where pores can be mistaken for potential chloride inclusions. The HD specimen can first be used for chemical analysis by Optical Emission Spectroscopy (OES) and then sent for metallographic analysis for chloride inclusions. This may serve to establish link between the element concentration in the aluminium alloys and the presence of inclusions. The dimension of the Alcan HD specimen is 58 mm in diameter and 10 mm thick. When the HD specimen is cut in two on its diameter, the surface available for microscopic evaluation is 5.8 cm^2. The analysis of the number of potential chloride inclusions performed on horizontal discs have shown great sampling versatility. But the reproducibility between two discs sampled side by side is not expected to be better than 50%. Nevertheless, this precision is good enough to express significant difference of the order of about ten times between two discs sampled before and after a filtrating device.

METALLOGRAPHIC SPECIMEN PREPARATION

Sectioning and mounting

The HD specimen riser is removed and the HD specimen is cut in two on its diameter. One of the two halves is again cut in two and the quarters are mounted on a 1½ inch (38 mm) diameter bakelite mount in the hot press (Figure 1). The second half is kept as a backup specimen in case of any further cross-checks.

Figure 1: Horizontal disc used for metallographic evaluation of the potential chloride and the mounted parts in a bakelite mount

Polishing Procedures

Two polishing techniques are used: the conventional wet polishing technique using water as a lubricant and the dry polishing technique using dodecane hydrocarbon as lubricant. When using the conventional water polishing, chloride inclusions partly dissolve upon contact with water, leaving holes filled with some remaining residue on the polished section.

The conventional wet polishing technique is performed on a multi-specimen automatic polisher and the steps are presented in Table 1. For intermediate polishing, either the diamond compound or slurry can be used.

A dry polishing technique was developed at ARDC to polish the specimen surface without dissolving the chloride inclusions. It uses dodecane hydrocarbon 99+% purity grade as lubricant instead of water. The dodecane is colourless as well as odourless. The most important characteristic is its high purity; it contains less than 1 ppm of S, Cl, N, Na, Ca and K. The dodecane is hygroscopic and to avoid having it being in contact with air, the bottle is filled with nitrogen and a syringe is used to pull out the amount of dodecane necessary to polish the specimen.

The abrasive used for intermediate polishing is a natural high concentration diamond compound that contains C, O and Na, as detected by the SEM-EDX technique. When sodium is detected in a chloride inclusion, part of it could be attributed to the polishing residue. There is no doubt on the nature of the sodium chloride when the inclusion shows a clear cubic shape.

The dry polishing procedure is presented in Table 2 for a multi-specimen automatic polisher. For this procedure, no final polishing is required, and an intermediate polishing down to 1 micron (diamond) in size is enough. The dry polishing technique is used to confirm the presence and the nature of the chloride inclusions only.

The dry polishing procedure should be considered as guide line only and may be adapted after the necessary precautions concerning possible contamination by the polishing lubricant and abrasive compound have been taken.

Care must be taken when using a dodecane hydrocarbon lubricant because it is a strong skin irritant and gloves must be worn when handling it. Furthermore polishing with a hydrocarbon lubricant should be performed in a fumehood or in a very well-ventilated room. The material safety sheet should be consulted for other precautions to be taken when using dodecane.

CHLORIDE INCLUSIONS EVALUATION

The chloride inclusion evaluation is carried out in two steps: First, the specimen is water polished and the potential chloride inclusions are evaluated by optical microscopy in the bright field mode illumination. These inclusions are declared "*potential chloride inclusions*" based on their shape and other morphological characteristics described below. Subsequently, the specimen are dry polished and examined by SEM to formally identify the type of "*chloride inclusions*" by EDX analysis.

Table 1: Water conventional polishing procedure to polish aluminium specimen with the Abramin multi-specimen polisher.

	ABRASIVE	CLOTH	LUBRICANT	PRESSURE (N-force)	SPEED (rpm)	TIME (min)
GRINDING	60 grit	SiC	water	150	150	until flat
	240 grit	SiC	water	100	200	1
	400 grit	SiC	water	100	200	1
POLISHING	Diamond 6 μm	Pellon	Blue oil	150	150	8
	Diamond 1 μm	Pellon	Blue Oil	100	150	7
	Colloidal Silica	OP-CHEM	In a bowl	100	150	3

Table 2: Dry polishing procedure for chloride identification on the Abramin multi-specimen polisher

STEP	ABRASIVE	CLOTH	LUBRICANT	PRESSURE (N-force)	SPEED (rpm)	TIME (min)
GRINDING	60 grit	SiC	Dodecane	150	150	until flat
	120 grit	SiC	Dodecane	150	150	2-5
	240 grit	SiC	Dodecane	150	150	2-5
	400 grit	SiC	Dodecane	150	150	2-5
POLISHING	Diamond Compound 6 µm	Pellon	Dodecane	150	150	9
	Diamond Compound 1 µm	Pellon	Dodecane	150	150	7

Main Morphology Characteristics of Chloride Inclusions

Using optical microscopy, it is quite easy to recognize the chloride inclusions on the specimen surface after they have been water polished, based on only a few morphological characteristics. These characteristics were determined after numerous observations on the optical microscope and formal cross-identification by SEM-EDX technique and are summarized as follows:

1. The chloride inclusions are liquid at the temperature of solidification of the aluminium and always appear spherical as they solidify in the metal. Therefore, they appear as circular holes on the plane of polish (Figure 2).

2. These circular holes appear black when observed at 100X under the optical microscope. At higher magnification (200X or 500X), the metallographer can detect some undissolved chloride residue in the holes or other smaller inclusions surrounding the holes. The chloride residue makes the hole difficult to focus. Sometimes, this residue bubbles out of the polished surface indicating that the chloride inclusions have expanded upon contact with water. Other times, the gaseous porosity show clean and empty circular holes (Figure 3) on the plane of polish. It is therefore easy to focus on the bottom of the holes, showing white and shiny aluminium.

3. The chloride inclusions can be differentiated from shrinkage porosity, because the latter are rather misshapen, showing cell contours (Figure 2). A quick etching with 0.5% HF will clearly show the cell contours which delimitate the shrinkage porosity.

4. In dirty aluminium, the chloride inclusions are usually surrounded by small inclusions such as TiB_2, $(Ti,V)B_2$ and Al_4C_3. Therefore, the chloride inclusions show up in the middle of a grain or a cell structure, because these small inclusions are preferential sites for the initiation and growth of the grains.

5. At the furnace outlet or before filtration devices, the chloride inclusions agglomerate around the oxide film stringers (Figure 4).

Optical Microscopic Evaluation Technique

Considering all the previous chloride inclusion characteristics, the metallographer scans an area of 5 cm^2 or more, at 100X magnification on the optical microscope. Once a small circular black hole is detected on the plane of polish, the metallographer uses the two other magnifications (200X and 500X) and checks by focusing up and down, to see if there is any remaining residue in the hole. The metallographer will detect the presence of the hole in the middle of a grain or a cell structure, the presence of small surrounding inclusions, the clean inside wall of a gaseous porosity and the misshapen dendritic-like shrinkage pore.

At each detected potential chloride, the metallographer measures the size using the square calibrated graticule or the micrometer installed in the microscope eyepiece. The number of potential chloride inclusions is counted and divided by the area of the specimen surface analyzed, to give a "Number of potential chlorides per centimetre square".

Chloride Inclusion Formal Identification by SEM-EDX technique

After counting and sizing the potential chloride inclusions, the specimen is dry polished and degassed in a vacuum chamber for 24 hours to minimize the SEM chamber contamination by hydrocarbon. Then, the specimen are examined by SEM and EDX analysis is performed on any circular inclusions that are charging and expanding under the electron beam. The presence of peaks of Cl, Ca, Na, K, Mg, Li will reveal the nature of the chloride inclusions detected. Figures 5-7 show typical chloride inclusions identified by the SEM-EDX technique.

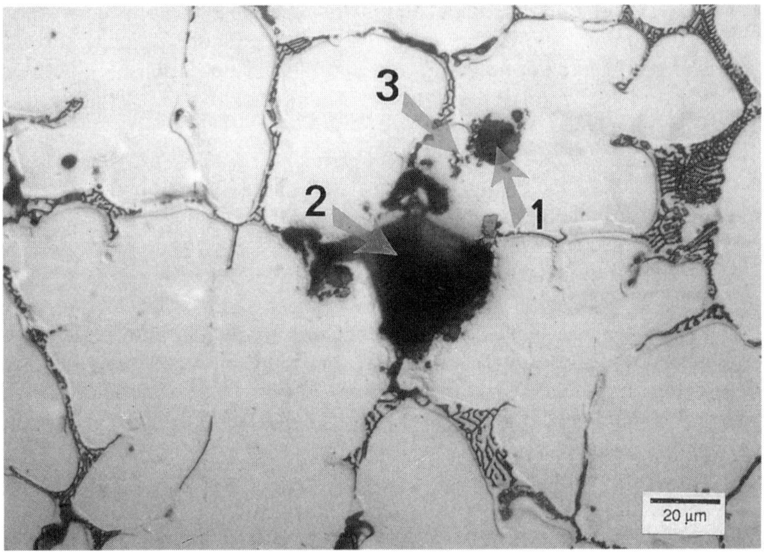

Light Microscope Wet Polishing 500X

Figure 2: 1) Black circular chloride inclusion, 2) misshapen shrinkage pore and 3) Al_4C_3 inclusions

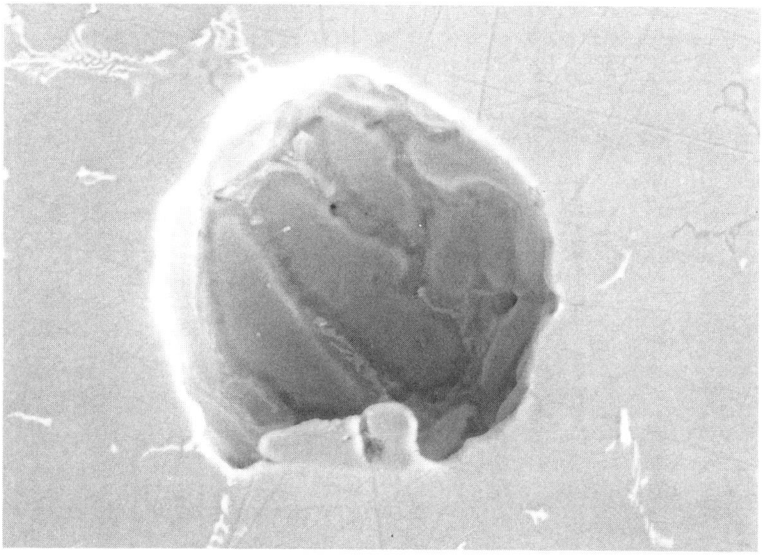

S.E.M. Dry Polishing 1000X

Figure 3: Empty circular hole of a gas pore showing bright aluminum, well focussed at the bottom.

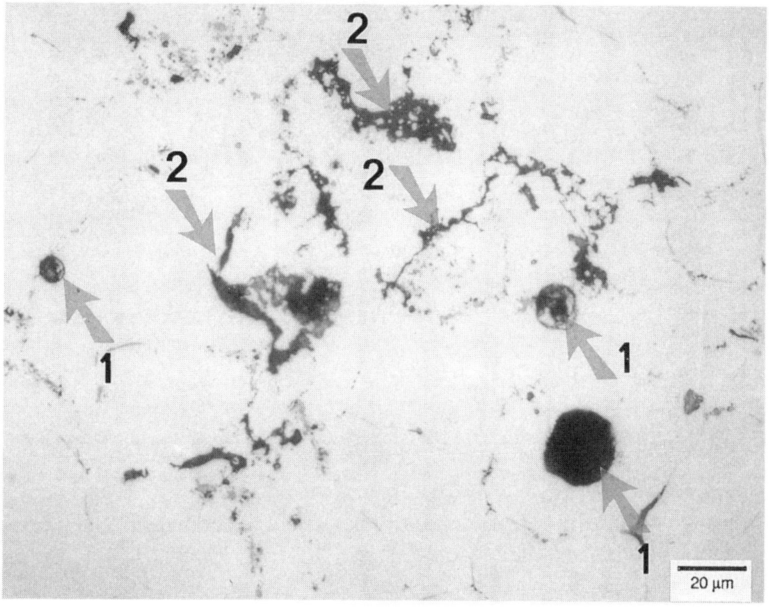

Light Microscope Wet Polishing 500X

Figure 4: Presence of chloride inclusions (1) agglomerated in oxide films (2). Specimen sampled before the filtrating device.

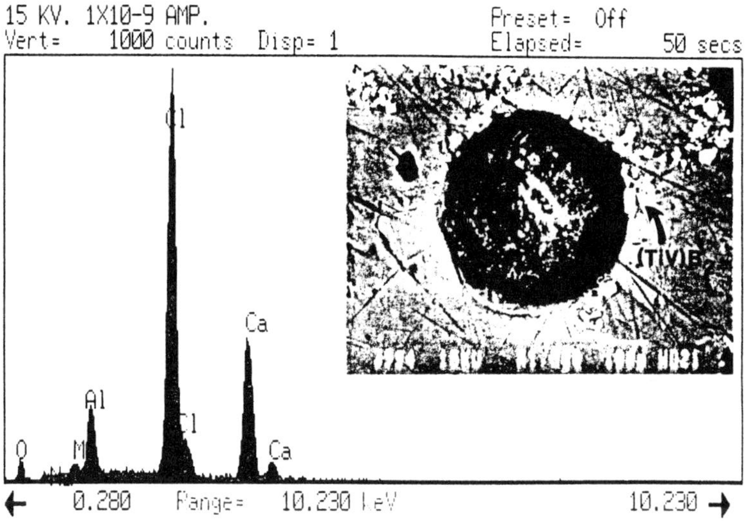

Figure 5: Calcium Chloride and the corresponding SEM-EDX spectrum.

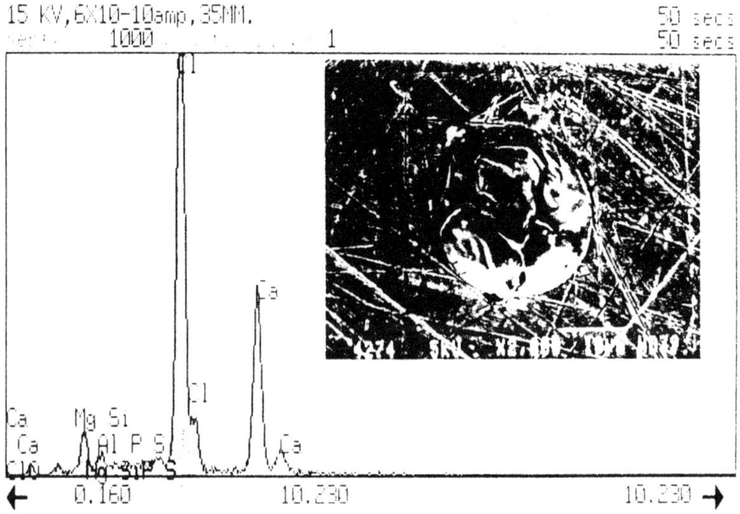

Figure 6 : A mixture of Calcium and Magnesium Chloride and their corresponding SEM-EDX spectrum.

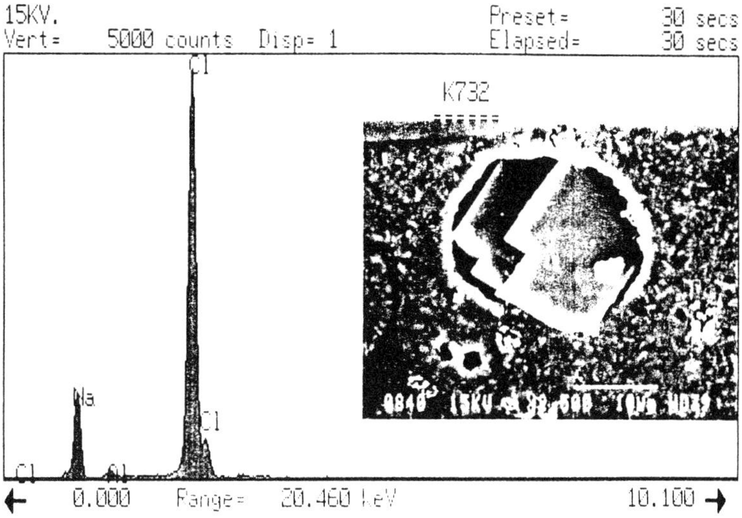

Figure 7: Sodium Chloride and its corresponding SEM-EDX spectrum.

The most often encountered chloride inclusions are sodium chloride (NaCl), calcium chloride ($CaCl_2$), magnesium chloride ($MgCl_2$) and potassium chloride (KCl) or mixture thereof. Lithium chloride (LiCl) also exists in lithium-based alloys, but we have never observed any at ARDC.

RESULTS AND DISCUSSION

Tables 3 and 4 present a few results [8,9] of potential chloride inclusions obtained at various positions in the metallurgical process. Table 3 compares the number of potential chloride per cm^2 with PoDFA results expressed in mm^2/kg of metal. While the PoDFA results are higher immediately at the furnace outlet, the number of potential chloride inclusions are much greater and their size bigger after the chlorine fluxing device. The reduction in the number and size of the potential chloride inclusions after the Alcan Bed Filter clearly indicates that the chloride inclusions are filtered very efficiently in the ABF. The mechanisms of filtration of chloride inclusions by the ABF is believed to be associated with the fact that the chloride inclusions agglomerate with the oxide films and other inclusions, making the overall agglomerates much bigger and easier to be trapped in the filtrating device. The evaluation of various filtrating devices efficiency was measured using both the PoDFA values and the potential chloride inclusions rating and are reported elsewhere [3,5]

Table 5 compares the number of potential chloride inclusions measured by optical microscopy after water polishing with chloride inclusions measured after dry polishing and formally identified by the SEM-EDX technique, for two specimen of

Table 4: Size Distribution of Potential Chloride Inclusions Measured at Various Positions in the Casting Process.

Sampling Positions	0-10 µm #/cm²	10-20 µm #/cm²	20-30 µm #/cm²	30-50 µm #/cm²	> 50 µm #/cm²	Total #/cm²	Maximum Diameter
Furnace Outlet	0.2	0	.2	0	0	0.4	22 µm
After Chlorine Fluxing	0.8	0.2	0	0.2	0	1.2	32 µm
After ABF	0.2	0	0	0	0	0.2	5 µm

Table 5: Comparison of the Number of Potential Chloride Measured by Optical Microscopy With Chloride Inclusions Formally Identified by SEM-EDX for Two Specimen Sampled After Chlorine Fluxing

Specimen I.D.	Analysis Technique	0-10µm #/cm²	10-20µm #/cm²	20-30µm #/cm²	30-50µm #/cm²	> 50µm #/cm²	Total #/cm²	Percent Difference
622A14 After Chlorine Fluxing	Optical	1.9	1.2	0.2	0	0	3.3	---
	SEM-EDX	0.8	0.8	0.2	0	0	1.8	55%
633A18 After Chlorine Fluxing	Optical	1.3	0.5	0	0	0	1.8	---
	SEM-EDX	0.5	0.5	0	0	0	1.0	56%

chlorine fluxed metal. The last column of Table 5 gives the relative percent of the number of real chloride over the potential chloride inclusions, which is about 55% in both cases. This difference can be explained mainly by the fact that the dry polishing procedure does not produce as good surface quality as the water polishing technique, which results in missing a few inclusions, particularly the smallest ones. This is clearly observed in the results presented in Table 5. This suggests that some further development of the dry polishing technique could improve the precision of the chloride inclusions evaluation technique.

Table 3: Number of Potential Chloride Inclusions Measured in Chlorine Fluxed Metal in Comparison with the PoDFA Metal Cleanliness Values

Sampling Positions	PoDFA All Inclusion Types mm^2/kg	Potential Chloride Inclusions $\#/cm^2$
Furnace Outlet	0.091	0.16
After Chlorine Fluxing Device	0.031	1.7
After ABF	0.007	0.3

REFERENCES

[1] Dubé, G and Waite, P., Processing of Molten Aluminum and Aluminium Alloys, Science and Engineering of Light Metals, RASELM'91, Tokyo, Japan, Oct. 1991, pp 1007-1014.

[2] Gariépy, B. and Dubé, G., Treating Aluminium With Chlorine, Alcan International Limited, Arvida Research & Development Centre, Patent No.:US5145514, 19 February 1991.

[3] Waite, P. and Bernard, B., Recent Experience With the Use of Sulphur Hexafluoride for Aluminium Fluxing, 119th TMS Annual Meeting of AIME, 1990, pp 775-784.

[4] Dumont, R., L'Italien, M. and Waite, P., The Alcan Filter-Degasser: A New In-Line Metal Treatment Technology, 121th TMS Annual Meeting of AIME, San-Diego 1992, pp 1077-1084.

[5] Waite, P., Applicability of Sulfur Hexafluoride for Furnace Fluxing, Light Metal Age, 49, (9-10), 1991, pp 16-17.

[6] Doutre, D., Gariepy, B., Martin, J.-P. and Dubé, G., Aluminium Cleanliness Monitoring: Methods and Applications in Process Development and Quality Control, The Metallurgical Society of AIME, Light Metal 1985, pp 1179-1195.

[7] ASTM Method E7-16, Methods for Emission Spectrochemical Analysis, American Society for Testing Materials, 7th Edition.

[8] Dumont, R., Dupuis, C., and Meslage, P., Private communication.

[9] Brochu, C., Private communication.

CHARACTERIZING PARTICLE DISTRIBUTIONS IN MODEL ALUMINUM ALLOY SYSTEMS

A.K. Pilkey[1] , J.P. Fowler[1], M.J. Worswick[1]
G. Burger[2] and D.J. Lloyd[2]

ABSTRACT

Quantitative image analysis is applied to study second phase particle populations in model Al-Si sheet. Very large scale, high resolution particle fields and their corresponding tessellations are constructed using image merging and matrix erosion algorithms. Measures of particle clustering are then extracted from these images. Inter-particle dilational spacing (IPDS) frequency plots were found to provide the most useful description of particle spatial distribution.

INTRODUCTION

Optimal processing of aluminum alloy sheet requires that the addition of alloying elements for increased strength be carefully controlled such that the resulting second phase particle population is able to delay fracture and maintain formability. It is well established [1] that the presence of particle populations within ductile metals can promote a mode of failure known as ductile rupture through void nucleation, growth and coalescence. The intrinsic effect of particle size, shape and distribution on formability, however, has not been quantified for most materials, including aluminum

[1] Department of Mechanical and Aerospace Engineering, Carleton University, Ottawa, Ontario, Canada, K1S 5B6

[2] ALCAN International Limited, Kingston Research and Development Centre, PO Box 8400, Kingston, Ontario, Canada, K7L 5L9

alloy sheet, due in large part to the difficulties involved in obtaining statistically significant measures of large fields of particles. To address this problem, a customized image acquisition and spatial analysis system has been developed and employed in the current research.

The formation of particle clusters in metals can often be attributed directly to specific stages during the fabrication process, such as solute segregation effects during casting and/or the production of stringers due to particle fracture and rearrangement during rolling. Generally, it is found that distributions of particles in particle-containing alloys are random in nature with a small degree of "clustering" (i.e. packets of closely spaced particles). It is this degree of clustering that needs to be measured. Several approaches have been proposed to spatially characterize particle clusters[2-10].

The present research examines the effect of particle populations on formability by considering three model aluminum alloy sheet systems: Al-Si, Al-Mg-Fe and Al-Mg-Fe-Mn. Within each system, the levels of second phase constituents have been varied parametrically to obtain a variety of particle sizes, shapes and distributions. In addition, several processing variations have been employed, including the addition of particle refining elements and the modification of rolling schedules in an attempt to produce distinctly different particle populations at nominally constant composition. This paper presents an assessment of the degree of clustering in comparable alloys by applying matrix erosion tessellations and a dilational counting technique to large-scale high-resolution particle fields. Formability tests are currently underway, facilitating future studies to assess the correlation between formability and particle population data.

MATERIAL

The primary alloys used in this study were cast as 95 X 225 X 380 mm DC ingots with nominal compositions of 2, 3 and 4% Si. After scalping, the ingots were heated to 450°C for one hour and then hot rolled to a thickness of 5 mm. Cold rolling passes were then applied until a final thickness of 1.6 mm was obtained, followed by a five minute 500°C anneal. In total, three sets of DC cast alloys were fabricated by: (i) adhering to regular casting procedures, (ii) adding 100 ppm Sr to act as a particle refining agent and (iii) applying a 48-hour particle coarsening treatment at 520°C to a regularly cast and rolled sheet. A secondary set of alloys, containing 2 and 4% Si, were produced from 50 X 100 X 150 mm book molds. After scalping, these castings were subjected to a prolonged heat treatment, with a linear change from room temperature to 460°C over a 12 hour period, followed by a steady temperature of 460°C for an additional six hours. The book molds were then hot rolled to 3.2 mm, cold rolled to 1.6 mm, and annealed for five minutes at 500°C.

Figure 1 shows typical microstructures from four of the Al-Si alloys considered. It is evident from these micrographs that the book mold cast alloy, 1(a), has much larger particles with a less homogeneous distribution (i.e. clustered) than the other three alloys, 1(b)-(d), which were cast into DC ingots.

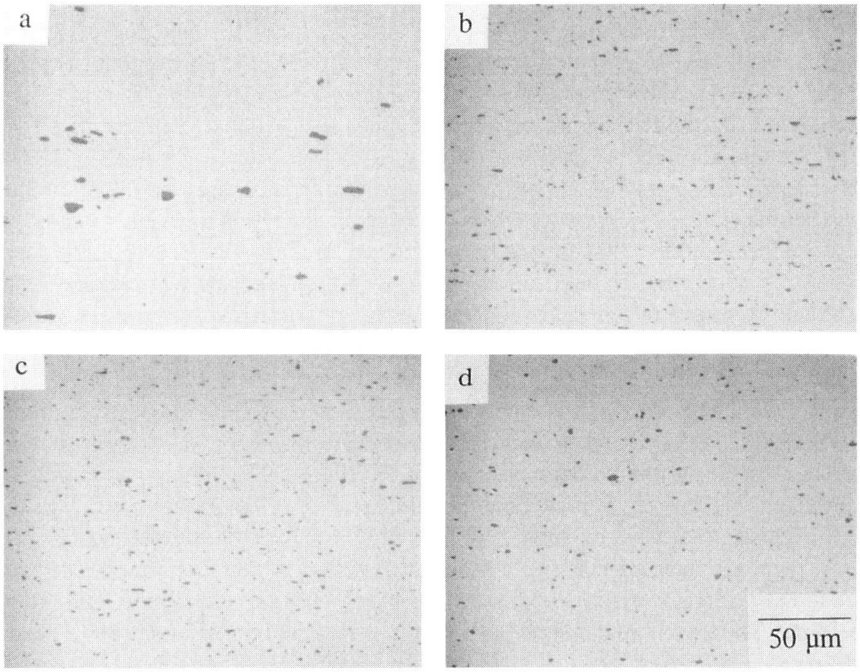

Figure 1. Optical micrographs showing side view of particle populations in: (a) book mold cast, Al-2%Si; (b) DC cast, Al-2%Si; (c) DC cast, Al-2%Si, Sr refined; (d) DC cast, Al-2%Si, coarsened. Side view contains rolling and through-thickness directions. All samples are unetched.

METALLOGRAPHY AND IMAGE ACQUISITION

Metallographic samples of the side view (i.e. containing rolling and through-thickness directions) of each alloy were prepared using a semi-automatic grinding/polishing system and the following protocol: (i) #320 SiC paper, water, 15 KPa, 150 rpm, 2 minutes; (ii) #600 SiC paper, water, 15 Kpa, 150 rpm, 2 minutes; (iii) 6 micron diamond suspension, napless cotton cloth, 15 KPa, 150 rpm, 5 minutes; (iv) 0.05 micron alumina, water, napped rayon cloth, 15 KPa, 150 rpm, 5 minutes;

(v) 0.04 micron colloidal silica, napless pad, 5 Kpa, 150 rpm, 90 seconds. All microstructures were left unetched.

Digital images of the prepared samples were acquired using a standard PC-based 480 X 640 pixel, 256 grey scale frame grabber connected to a 1/2 inch CCD camera mounted on an Olympus PMEB metallograph. The camera was mounted directly on an eyepiece port in order to avoid the inferior image quality obtained using the standard CCD camera port. Video noise was reduced by using multiple frame averages and field illumination corrections were applied. Binary encoding of particle fields was achieved by simple thresholding of the grey scale image. Threshold values were set at the midway point between the grey scale histogram peaks of the dark particles and the light Al matrix.

In order to obtain large field images while maintaining high resolution, multiple overlapping images were acquired and later merged to obtain a single image. For most of the samples, a 50X objective was utilized which resulted in a pixel resolution of 0.1625 microns after digitization. A 10 X 10 sample grid was acquired with a nominal overlap between fields of 10-20 microns. This approach resulted in a final image size of roughly 0.90 X 0.65 mm (5500 X 4000 pixels) after merging. For the sheet samples rolled from book molds, the average particle size was considerably larger than that in the DC cast samples. Therefore, the image magnification had to be reduced to a 20X objective in order to ensure that enough particles would be present in each of the overlapping portions of individual fields to enable successful merging of the images. Particle fields acquired with the 20X objective, corresponding to a pixel size of 0.4065 microns, required only a 4 X 4 grid to sample the same area.

The image merging task and subsequent tessellation operations are performed on a Silicon Graphics Iris Unix workstation using custom software written for this purpose. The algorithm employed to merge overlapping images minimizes a measure of overlap error based on the "sum of the exclusive OR of the pixels from the overlapping regions of the two images". Tessellations of these large images are then performed by a matrix erosion algorithm with alternating 4- and 8-connected erosion passes. The use of alternating passes has been found to generate tessellation cells with shapes that better reflect the geometry of their respective particles. Relevant statistics obtained from these tessellations include local area fractions, first, second and third nearest neighbour particle spacings, as well as the size, aspect ratio and primary axes orientation for both particles and tessellation cells.

RESULTS AND DISCUSSION

Matrix erosion tessellations were performed on the large scale merged images of each Al-Si alloy. A portion of one resulting tessellation (book mold cast, Al-2%Si) is illustrated in Figure 2. The presence of particle clustering is evident. Several measures of the particle population and its tessellated space were collected from these processed images and used to assess the degree of clustering in the model materials.

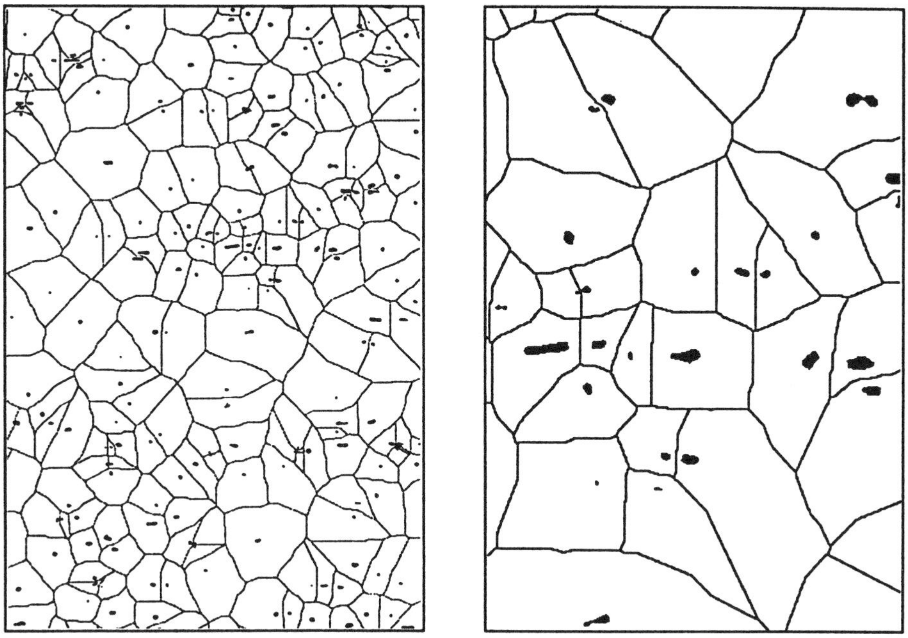

Figure 2. Portions of the large scale tessellated image constructed for the book mold cast Al-2%Si alloy.

It was demonstrated by Burger et al.[4] that the mean and standard deviation of tessellation cell aspect ratio, and the standard deviation of tessellation cell area are strongly sensitive to variations in clustering. High values for these parameters indicate a stronger degree of clustering. The coefficient of variance (C.V.) for tessellation cell area has been extracted from the tessellated images of each alloy and compiled in Table 1. The coefficient of variance (standard deviation normalized by mean) has been used here to allow for direct comparison between alloys. From the data in Table 1, it is evident that an increase in particle volume fraction causes a significant decrease in the degree of clustering. However, comparisons between alloys of differing compositions are of little interest here because the strong correlation between particle volume fraction and mechanical behaviour is certain to overwhelm any particle clustering effect. When comparing alloys of nominal constant silicon content, it is evident that the Sr addition and post-fabrication coarsening anneal had little effect on these cluster measures. It does, however,appear that the Al-2% alloy cast into a book mold show signs of being more clustered than its DC cast counterparts; a trend supported by the micrographs in Figure 1.

Inter-particle dilational spacing (IPDS) frequency data, employing the dilational counting technique of Shehata[2], was collected at regular intervals while each

Alloy Series	Si (wt%)		
	2	3	4
Book Mold Cast	0.67	---	0.32
DC Cast	0.62	0.57	0.31
DC Cast/100ppm Sr	0.60	0.58	0.30
DC Cast/Coarsened	0.60	0.59	0.29

Table 1. Coefficient of variance (C.V.) for tessellation cell area in Al-Si alloys.

particle field was being tessellated. This method involves counting the number of features which merge into a cluster during each successive erosion pass. The resulting frequency distributions for selected alloys are plotted in Figure 3. Peaks in these plots correspond to levels of clustering in the particle population, with the sharpness of the peak signifying the strength of clustering at that IPDS. In turn, frequency peaks at larger IPDS values represent successively higher order clusters (i.e. clusters of lower order clusters). Furthermore, at any given IPDS increment it is possible to calculate the average number of particles contained in the clusters which have formed, referred to here as the cluster density. Cluster density figures have been attached to each prominent peak in Figure 3. In all cases, the cluster density at the first peak is near three particles/cluster. Close inspection of the alloy microstructures reveals that a considerable number of particles have been broken into two, three, or four pieces during rolling. Due to the interdependence of IPDS and edge-edge nearest neighbour spacing, it is expected that the first order peak is located close to the average nearest neighbour spacing (denoted as NNS in Figure 3). The effect of the particle refining and coarsening treatments can also be established from these frequency plots. Figure 3(b) demonstrates that the particle refining did not have a considerable effect on the degree of clustering but did reduce the IPDS of first order clusters. On the other hand, the particle coarsening treatment (3(c)) removed some of the first order clustering while causing an increase in the IPDS. In all, the first order clustering exhibited by the DC cast alloys 3(a)-(c) is of a level that is to be expected in a random distribution. However, the frequency plot obtained from the book mold cast Al-2%Si alloy (3(d)) does show a significant indication of second order clustering, with a cluster density in excess of six particles/cluster. This outcome implies that, when compared to book mold cast sheets, the as-cast structure of DC ingots and/or the additional hot and cold rolling applied to them produces a more homogeneous distribution of particles.

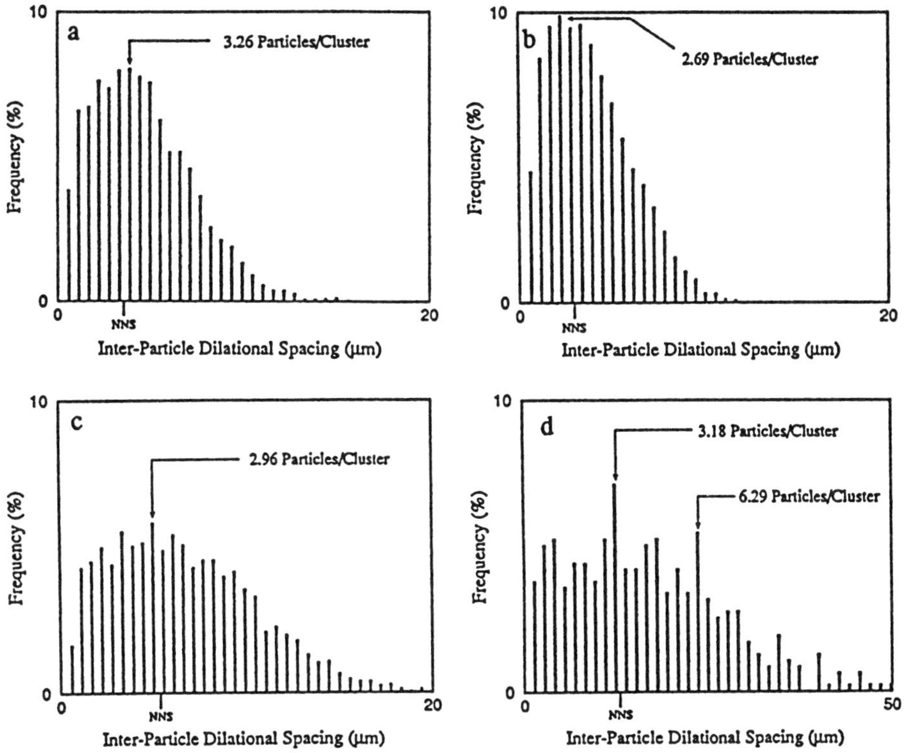

Figure 3. Inter-particle dilational spacing (IPDS) frequency plots for: (a) DC cast, Al-2%Si; (b) DC cast, Al-2%Si, refined; (c) DC cast, Al-2%Si, coarsened; (d) book mold cast, Al-2%Si. Mean edge-edge nearest neighbour spacing is denoted by NNS.

CONCLUSIONS

The degree of clustering present in several Al-Si sheets has been assessed using two established methods: (i) a comparison of cluster-sensitive tessellation parameters and (ii) an interpretation of IPDS frequency data obtained from a matrix erosion tessellation. It has been observed that the comparison technique yields only a measure of the relative degree of clustering. In contrast, IPDS plots reveal both the presence and degree of clustering at various orders (i.e. clusters, clusters of clusters), as well as the particle density and spatial scale of clusters. Furthermore, it is possible to use this data to locate a given set of clusters in the particle field by performing a partial tessellation up to the IPDS value of the appropriate frequency peak.

For the series of Al-Si sheets considered in this study, it has been determined that an observable degree of first order particle clustering, with a cluster density of roughly three particles/cluster, is present in each of the Al-2%Si alloys. It has also been established that this first order of clustering is associated with the average edge-edge nearest neighbour spacing. In one of these model sheets, a book mold cast Al-2%Si alloy, a second order of clustering has also been detected, having an average cluster density of six particles/cluster. Consequently, it follows that the nature of DC cast structures and/or the additional hot and cold rolling required for DC ingots produces a lower degree of clustering in the particle population.

REFERENCES

1) Ashby, M.F., Embury, J.D., Cooksley, S.H., and Teirlinck, D., "Fracture Maps With Pressure as a Variable", **Scripta Metallugica**, Vol. 19, p. 385, 1985.

2) Shehata, M.T., "Application of Image Analysis in Characterising Dispersion of Particles", in **Short Course on Image Analysis Applied to Mineral and Earth Sciences**, W. Petruk, Ed., Mineralogical Association of Canada, Ottawa, p. 119, 1989.

3) Vander Voort, G.F., "Evaluating Clustering of Second-Phase Particles", **Micon 90: Advances in Video Technology for Microstructural Control**, ASTM STP 1094, George F. Vander Voort, Ed., ASTM, Philadelphia, p. 242, 1990.

4) Burger, G., Koken, E., Wilkinson, D.S. and Embury, J.D., "The Influence of Spatial Distributions on Metallurgical Processes", in **Advances in Phase Transitions**, Pergamon Press plc, Headington Hill Hall, Oxford, p. 247, 1988.

5) Spitzig, W.A., Kelly, J.F. and Richmond, O., "Quantitative Characterization of Second-Phase Populations", **Metallography**, Vol. 18, p. 235, 1985.

6) Wray, P.J., Richmond, O. and Morrison, H.L., "Use of the Dirichlet Tessellation for Characterizing and Modelling Nonregular Dispersions of Second-Phase Particles", **Metallography**, Vol. 16, p.39, 1983.

7) Shehata, M.T. and Boyd, J.D., "Measurement of Spatial Distribution of Inclusions", in **Inclusions and Their Influence on Material Behaviour**, ASM International, Metals Park, OH, p. 123, 1988.

8) Vander Voort, G.F., "Characterization of Inclusions in a Laboratory Heat of AISI 33 Stainless Steel", in **Inclusions and Their Influence on Material Behaviour**, ASM International, Metals Park, OH, p. 49, 1988.

9) Embury, J.D. and Burger, G., "The Influence of Microstructure on Toughness", in **Proceedings of the 7th International Conference on the Strength of Metals and Alloys**, Pergamon Press 2, p. 1893, 1985.

10) Schwarz, H. and Exner, H.E., "The Characterization of the Arrangement of Feature Centroids in Planes and Volumes", **Journal of Microscopy**, Vol. 129, p. 155, 1983.

TORSIONAL TESTING OF COMMERCIAL GRADE BERYLLIUM TUBE

L. Vega[1], J. M. Herrera[2], C. S. Niou and S. W. Stafford

ABSTRACT

Advanced aerospace structures require lightweight stiffened materials. Beryllium in the cross-rolled, fine-grained, high purity state may serve as a candidate material. This test program attempted to characterize the mechanical behavior of extruded beryllium tube specimens in torsion. Each of the tests was monitored with acoustic emission in order to further identify the deformation/fracture behavior of the beryllium test coupons. Sample materials were also submitted for examination using optical, transmission and scanning electron microscopy to evaluate grain size, matrix and grain boundary dislocation structures, fractographic conditions and other pertinent features. One important consequence of this work is that beryllium's deformation and fracture behavior appears to be controlled by its grain boundaries. Localized slip in the grain boundaries also serves to nucleate microcracks along cleavage planes and along grain boundaries.

---

[1]Department of Metallurgical & Materials Engineering and [2]Mechanical and Industrial Engineering Department, The University of Texas at El Paso, El Paso, Texas 79968 USA

INTRODUCTION

Beryllium (Be) remains a candidate material for structural components in aircraft, missile and spacecraft vehicles. However, the propensity of this material to premature brittle fracture temporarily limits widespread application. Thus, the advantageous properties of Be make it a superior material for vehicle structures providing the properties which comprise vehicle integrity can be designed to efficiently use the limited tensile ductility inherent to beryllium. For the present time most uses of Be generally involve either compressive loading conditions or very low tensile loads because the required hot working produces texturing and unacceptable anisotropy in the achieved mechanical properties.

The work contained in this report represents the follow-up examination of the general deformation and fracture behavior of beryllium in tapered sheet specimens [1]. In order to further identify the mechanical behavior of beryllium, additional testing was performed to include torsion testing of extruded tube specimens 3.18cm OD X 20.32cm X 0.254cm (1.50 inch OD X 8 inch X 0.100 inch) of S-200 grade. Torsion testing was chosen to favorably assess the influence of pure shear stress on the deformation/fracture behavior of Be in another product state of high anisotropy. Each test was monitored with acoustic emission (AE) test sensors and equipment. As indicated in the previous study [1], AE is well suited for studies of deformation and fracture of brittle materials. For this test program, cumulative AE counts and events were recorded, along with various aspects of the event amplitude distributions. This work will be published in a future paper.

EXPERIMENTAL PROCEDURE

The S-200 grade tube specimens, represented in Figure 1 (a), were supplied by Nuclear Metals, Inc. (MA). An Instron tensile machine Model No. TTC with a 1382N (10,000 lb) maximum load cell capacity was modified to accommodate a test stand incorporating a lathe chuck, a rack and pinion gear mechanism for torsional load application and a rigid fixture to firmly clamp the other end of each tube specimen (Fig. 1 (b)). The rack/gear mechanism used a gear with a pitch diameter of 16.51cm (6.5 inches). Each end of the tube samples was fitted with an aluminum plug serving to reinforce the Be tube specimens during clamping. All of the test fixtures were designed and constructed for this project. Prior to testing the actual Be samples, steel and aluminum samples were machined to similar dimensions and loaded to simulate actual test conditions. The specimens were tested at different twist rates as shown in Table 1.

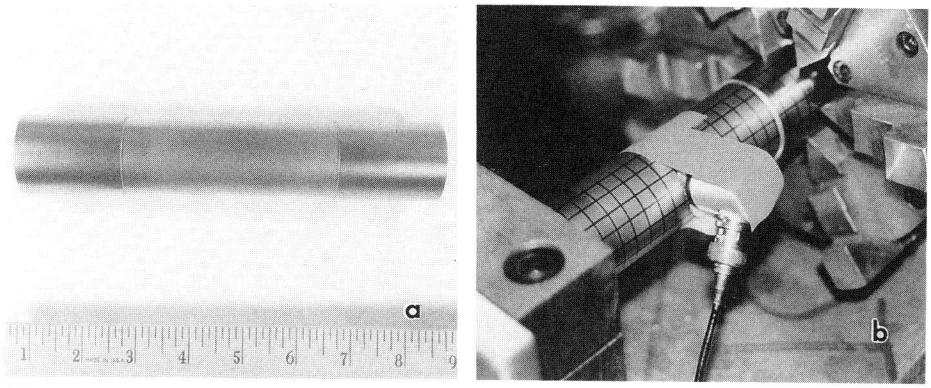

Figure 1. A representative as-received Be tube sample shown in (a) displaying a grid map and the general test setup shown in (b). Note the attachment of the AE sensor in (b).

Tube #1		0.352*		
Tube #2		1.76*		
Tube #3		1.76*		
Tube #4		8.81*		
*These values represent relative twist rates in degrees per minute.				

Table 1. Twist rates shown for the Be tube samples #1-4. Grid maps were also prepared on the surface of each tube specimen, shown in Fig. 1 (b), to facilitate the angle of twist measurement and to reference deformation behavior during testing.

Metallographic and fractographic sample preparation was performed in one of the two glove boxes designed and built expressly for this test program. The metallographic samples were mounted in KoldMount®and then prepared by standard metallographic procedures. The resulting samples were viewed under a bench metallograph using normal and polarized light illumination.

Fractographic samples were sectioned with the lubricant-cooled cut-off wheel contained in one of the glove boxes. Each sample was then mounted on specimen holders and observed in a scanning electron microscope.

Sample preparation procedures for thin film specimens used for TEM consisted of sectioning 3mm x 3mm x 1.4mm segments using a low speed diamond saw. These samples were subsequently ground to approximately 100-200μm thickness using 240 and 600 grit SiC paper. The specimens were prepared from three different locations in the plane orientation of tube #4; the selected areas were a) approximately 2mm from the fractured surface at a region of maximum shear stress, b) a region of maximum tensile stress also near the fracture, and c) from an undisturbed, remote location at the grips. An electrolyte consisting of 5% HNO_3, 1% H_2SO_4, 1% HCl, 31% ethylene glycol and 62% ethyl alcohol was used for electro-jet polishing with a bath temperature of -5 °C. The electropolishing potential was set at 100V. The electron-transparent thin sections were then observed using a Hitachi H-8000 analytical electron microscope operated at an accelerating voltage of 200kV.

RESULTS

Mechanical Properties

The results of the tube samples, summarized in Table 2, are quite consistent. The shear stress and shear strain characteristics of the torsionally tested tube specimens are shown in Figure 2.

As noted in Table 1, three different twist rates were attempted -- 0.352 degrees/minute for sample #1, 1.76 degrees/minute for samples #2 and 3, and 8.81 degrees/minute for sample #4. The behavior of samples #1-3, at the slower twist rates, are very similar while the higher twist rates slightly shift the shear stress vs. shear strain curves as a result of the higher strain rates. The shear yield and tensile strengths shifted upward in response to the higher strain rates. Clear indications of the presence of secondary yielding were not observed with the tube specimens[2,3].

Tube Sample No.	YS, Mpa (psi)	Shear Stress, Mpa (psi)	Shear Mod., Mpa (psi)	Twist Angle, degrees
#1	127 (18,400)	203.13 (29,462)	16,291.3 (2,362,850)	15.955
#2	149.50 (21,684)	253.51 (36,769)	16,776.54 (2,433,232)	40.549
#3	146.25 (21,213)	260.01 (37,712)	12,420.65 (1,801,463)	48.0147
#4	152.76 (22,156)	263.26 (38,183)	13,864.92 (2,010,936)	44.075

Table 2. Mechanical test results for Be tube samples #1-4.

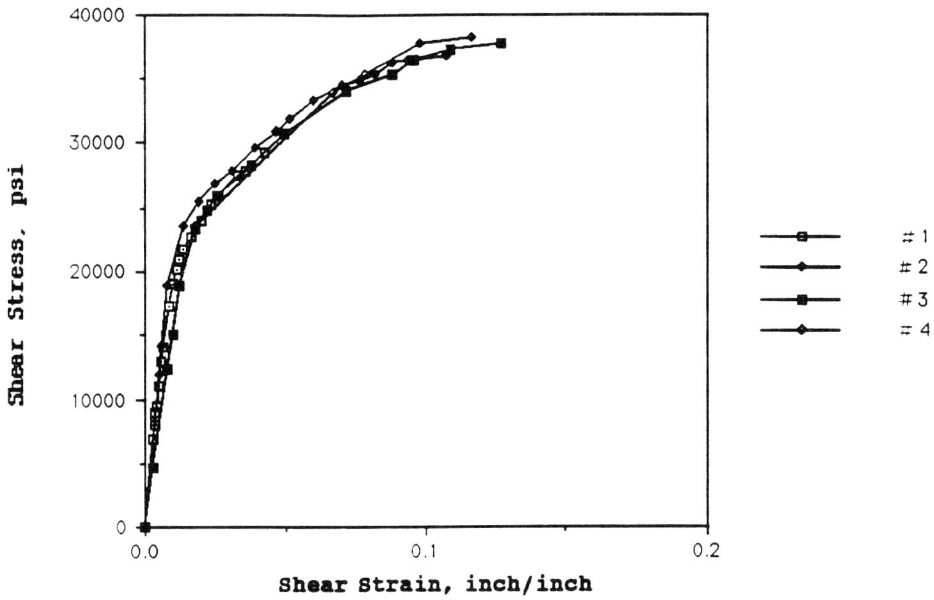

Figure 2. Shear stress vs. shear strain plot for Be tube samples #1-4.

Metallography

Samples were polished and observed under normal and polarized light conditions using magnifications between 100X and 1000X. A glycerol/HF/HNO_3 etchant which worked successfully on the beryllium sheet samples did not work on the tube specimens. Figure 3 (a) identifies the representative microstructural features. The average grain diameter for each sample group was approximately 30 microns using the ASTM E112 lineal intercept method. However, grain diameters were observed to range from approximately 3-70 microns. Twinning was not observed in any of these sections, however, secondary cracking was observed at the fracture surfaces as shown in Fig. 3 (b).

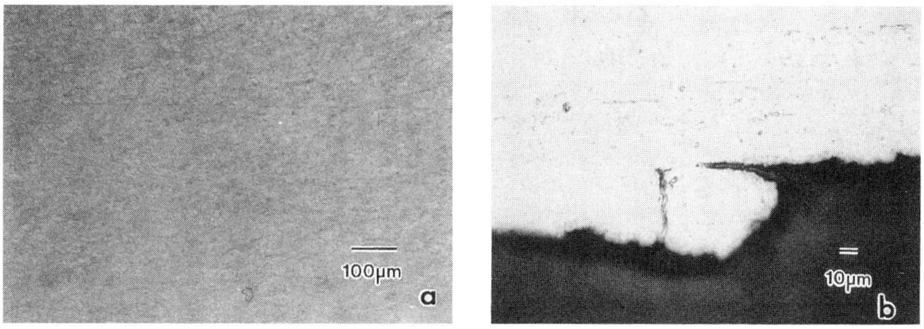

Figure 3. Representative microstructure viewed (a) under polarized lighting and (b) secondary cracking observed along the fracture surface as seen under normal lighting.

Fractography

Tube specimen #1 was not taken to rupture as with subsequent samples in torsional testing. Tube sample #2, documented in Figure 4 (a), illustrates the fracture conditions at the termination of the test. Interestingly enough, the macroscopic fracture paths follow both planes of maximum shear (axial and transverse rupture surfaces) and planes of maximum tensile stress (angled rupture surfaces) in torsion. A similar photographic documentation is shown for sample #4 in Fig. 4 (b).

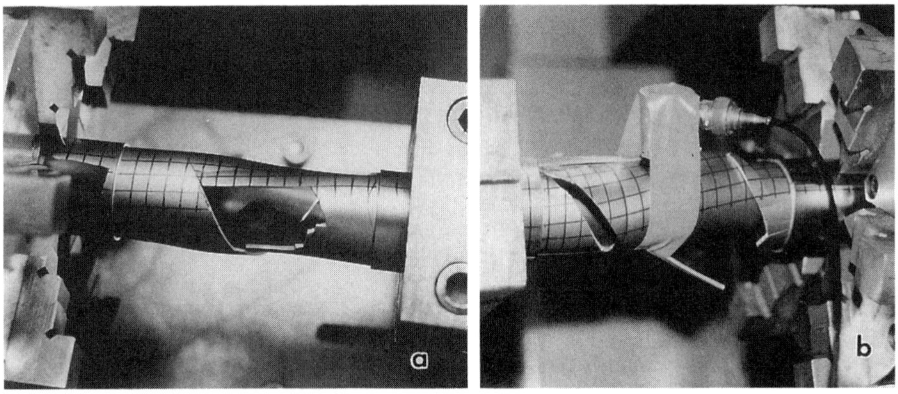

Figure 4. Fractured test specimens #2 (a) and #4 (b) following destructive torsional testing. Note the twist on both grid maps.

Microscopic evidence of the fracture surfaces coincidental to both planes of maximum shear stress and tensile stress were examined on tube sample #4. Fracture surface features again revealed evidence of brittle rupture. However, fracture evidence of transgranular cleavage and regions of intergranular rupture were detected. Figure 5 (a) represents a former plane of maximum shear. Figure 5 (b) serves to establish evidence of intergranular rupture showing apparent grain pull-out, intergranular facets and secondary intergranular cracking. Figure 5 (c) shows a very distinct river pattern in one region of transgranular cleavage. A former plane of maximum tensile stress or an angled fracture surfaces is shown in Fig. 6 (a). The microfractographic features shown in Fig. 6 (b) exhibit both transgranular cleavage an intergranular rupture.

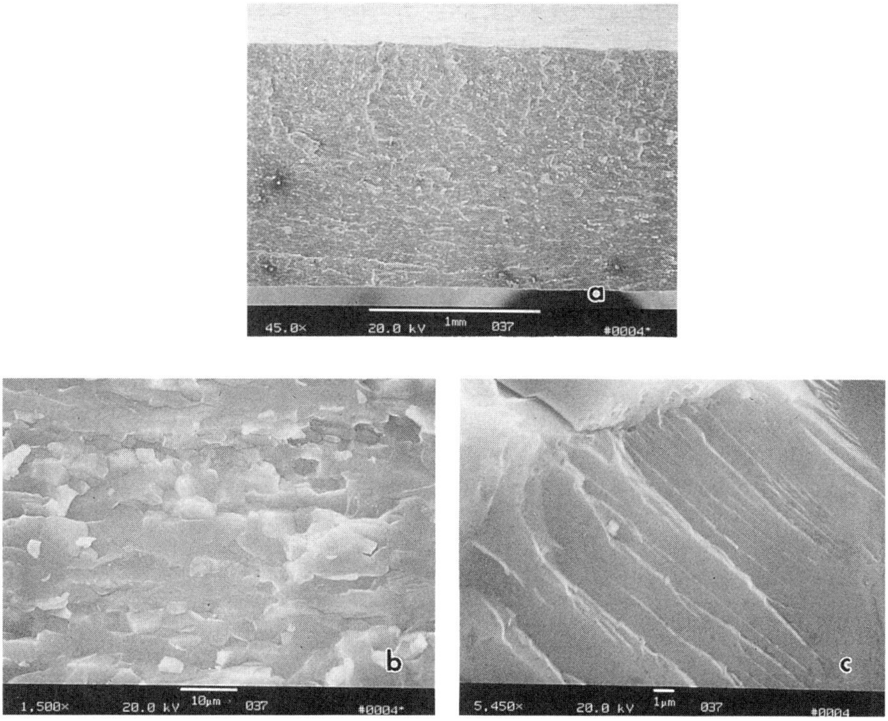

Figure 5. Beryllium tube wall thickness fracture at (a) former plane of maximum shear, (b) showing intergranular and transgranular rupture features. The river pattern displayed (c) in a region of transgranular cleavage.

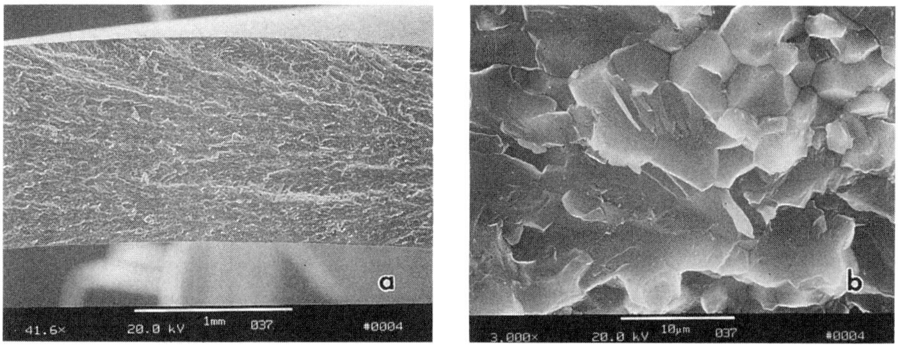

Figure 6. A former plane of maximum tensile stress (spiraling fracture) shown in (a) displaying transgranular cleavage and intergranular facets in (b).

Transmission Electron Microscopy

The microstructural features observed in the regions near the fracture surfaces reveal relatively higher dislocations densities, localized slip along the grain boundaries, and small ledge structures in the grain boundaries in contrast to the remote samples. The illustration in Figure 7 (a) shows the location of the selected areas for TEM. Figures (b) through (d) show the aforementioned features and are representative of the selected areas in the plane of tube #4. Evidence supportive of slip, twinning and dislocation breakaway from pins, was not observed in the TEM thin sections. The post-yield deformation region appears to be associated with strain localization or yielding in the grain boundaries leading to transgranular microcrack nucleation.

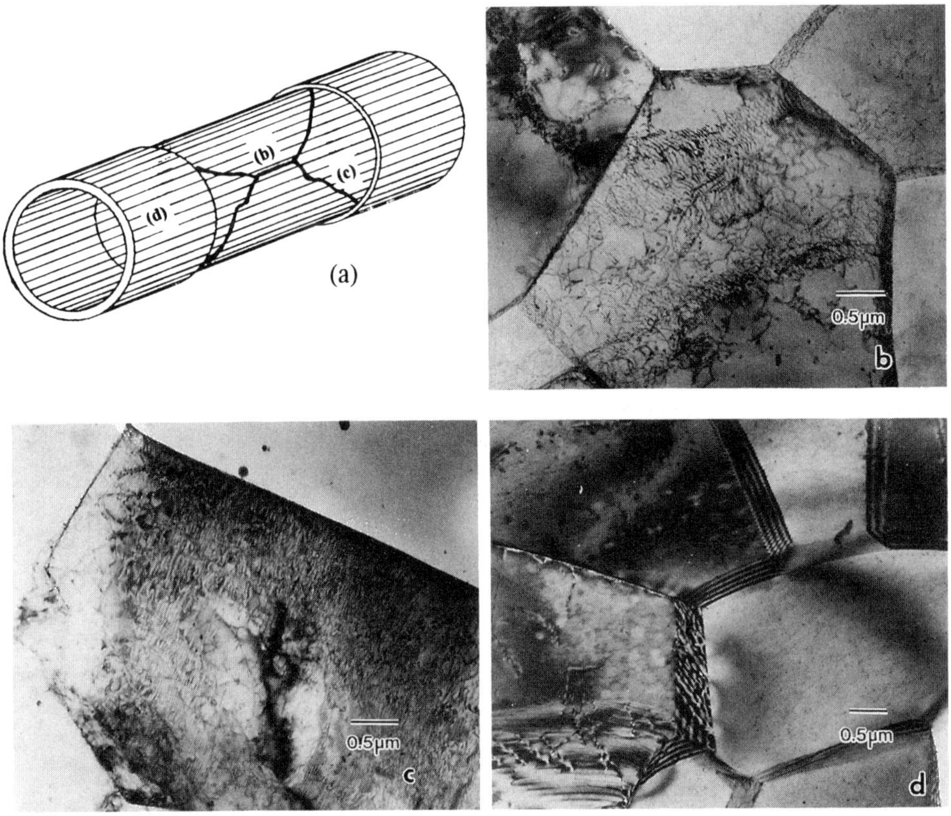

Figure 7. A schematic view (a) illustrating the location of the TEM samples in the in-plane orientation. The area shown in (b) and (c) revealed higher dislocations and small ledge structures when compared to the remote location (d) at the grip section of the tube.

Stress Analysis

The torsional stress was calculated using a standard mechanics of materials equation given in equation (1),

$$\text{Torsion: Shear Stress} = T\,r/J, \quad (1)$$

where T = torque, r = tube radius in the test section, and J = polar moment. For the torsion samples, the following relationships were used: $T = 3.25P$, where P is the load from the plots, r = the tube radius in the reduced test section, and $J = 0.5\,\pi\,(r^4_i - r^4_o)$.

DISCUSSION

The shear properties derived from torsional testing are also listed in Table 2. The average shear yield strength was 149.506 MPa (21,684 psi), the average shear tensile strength was 244.98 Mpa (35,532 psi), the average shear modulus was 14,838.50 Mpa (2,152,120 psi), and the average angle of twist was 37.15 degrees. We were unable to find references specifying torsional properties derived from actual testing to compare to these results. However, the ability of the tube specimens to deform to twist angles as high as 48.0147 degrees was unexpected.

The deformation and fracture characteristics of the test samples generally demonstrated minimal ductility [4,5] and a propensity for brittle rupture. From the observation of the residual microstructures, generally comparing before and after testing conditions, evidence of twinning and dislocation generation within the grains or at inclusions was absent. However, ample evidence supporting localized strain within the grain boundaries was observed., it appears that the deformation and fracture behavior are extremely localized. The torsion samples demonstrated complex grain boundary structures, including strain field contrast. These samples revealed greater grain interior dislocation densities than the tapered beam specimens [1]. This has been interpreted to be due to the nature of the manufacturing history of the S-200 grade hot rolled beryllium material utilized for the tube samples.

Observations made in the previous study determined that the predominant fracture mode was trangranular cleavage on basal and prism planes. Specific fracture surface features included river patterns, facets, and stepped cleavage surfaces. In beryllium, fractures of this type result from a high stress normal to the basal or prism planes. Tongues or transgranular cleavage deviating along an intersecting twin were not observed in either study.

CONCLUSIONS

1. The tube samples provided meaningful mechanical property data for beryllium in the grade tested. The torsional testing results should be unique, since the authors are unaware of previously reported results.

2. Transmission electron microscopy suggests that pre-yield deformation is largely due to localized slip along the grain boundaries. Evidence supportive of slip, twinning and dislocation breakaway from pins, was not observed in the TEM thin sections.

3. The post-yield deformation region appears to be associated with strain localization or yielding in the grain boundaries leading to transgranular microcrack nucleation. Apparently, intergranular microcracking also develops.

4. Beryllium's deformation behavior appears to be primarily controlled by its grain boundaries. This is consistent with the known role that grain boundaries play in significantly improving the strength and ductility of beryllium. Increasing the grain boundary surface area with decreasing grain size results in a greater number of grain ledges contributing to grain boundary deformation.

5. Fractographic evidence supports brittle rupture but with the appearance of both transgranular cleavage and intergranular rupture. This would indicate that grain boundary yielding serves to nucleate microcracks along cleavage planes and along grain boundaries.

ACKNOWLEDGMENTS

The authors would like to dedicate this effort to the memory of Dr. Ray W. Guard. We are also exceptionally grateful to Mr. William "Bud" Castner for his support and technical guidance.

REFERENCES

1. L. E. Murr, C. S. Niou, A. H. Advanti, L. Vega, S. W. Stafford and J. M. Herrera, Mater. Sci. & Engr. A151, 179 (1992).

2. J. M. Galbraith and L. E. Murr, J. Mater. Sci. 10, 2025 (1975).

3. L. E. Murr, Interfacial Phenomena in Materials and Alloys. Addison Wesley Publishing Co., Reading, MA. 1975; reprinted by TechBooks, Herndon, VA, 1990.

4. D. Webster and D. D. Crooks, Improved Beryllium Ductility Study. Lockheed Missiles and Space Co., Inc. Report No. LMSC - D507268, August 1976.

5. H. D. Hanes, S. W. Porenska, J. B. Melehan and P. J. Gripshover, Physical Metallurgy of Beryllium, Defense Metals Information Center, Battelle Memorial Institute, Columbus, OH, DMCC Report 230, June 1966.

PRECIPITATION IN A BETA-TITANIUM ALLOY UNDER DIFFERENT AGING TREATMENTS

Marcelo Henrique Prado da Silva[1]; Glória Dulce de Almeida Soares[1] and Luiz Henrique de Almeida[1]

ABSTRACT

Ti 15V-3Cr-3Al-3Sn (Ti 15-3) is a metastable β-Ti alloy that can reach high strength after aging where α phase precipitation occurs. Solution treated and pre-deformed samples were aged for a range of times at temperatures of 633 K, 723 K and 813 K and examined by no abstract TEM. Different shapes and distributions of α phase precipitates were observed. Mechanical tests were carried out to correlate the change in mechanical properties with phase transformations. The best combination of ductility and strength was attained by aging at 813 K.

INTRODUCTION

There is a strong trend towards the use of Ti alloys in the aerospace industry. These alloys have become prominent in recent years owing to their ease of fabrication and high strengths attained after aging. While older alloys have been re-examined, some alloys have attracted renewed interest and have seen extensive application [1].

One of the most important is Ti 15-3 which has a β transformation temperature of 1023 K-1041 K [2]. This alloy is used in the aerospace industry in structural components of nacelles at temperatures from 423 K to 573 K. In this temperature range, mechanical properties are influenced by strain aging [3], so, it is necessary to understand the mechanisms of phase transformation and the related mechanical property changes.

[1]COPPE/UFRJ, PO. Box 68505, 21945-970 - Rio de Janeiro, RJ, Brazil

EXPERIMENTAL

The starting material consisted of 10.2 mm thick plates of Ti-15-3 in two conditions: solution treated, and deformed 4% by stretching after solution treatment. In order to investigate the microstructure's dependence on cooling rate, re-solution heat treatments were made at 1073 for 30 min.. Three cooling rates were used: water quench, air and furnace cooling. The chemical composition of Ti 15-3 is 15.23 wt%V, 2.88 wt%Cr, 2.91 wt%Sn, 3.11 wt%Al and 0.12 wt%Fe. Air cooled samples were aged at 633 K, 723 K and 813 K for a range of times. Thin foils were prepared in TENUPOL equipment using a solution of 6% $HClO_4$, 59% CH_3OH and 35% $CH_3(CH_2)_2CH_2OH$ at 233 K and 12 V and observed in the TEM at 200 kV.

Mechanical properties were investigated by Vickers microhardness measurements and tensile tests. Tensile tests were made at a strain rate of 10^{-3} min.$^{-1}$ in the elastic regime and 10^{-2} min.$^{-1}$ in the plastic regime, as recommended in AMS-4914 [4].

RESULTS AND DISCUSSION

TEM examination of samples re-solution heat treated showed diffuse streaking characteristics indicating the presence of ω phase independent of cooling rate. This result disagrees with observations of Banerjee *et al* [5], who reported a strong dependence on cooling rate in a β Ti alloy.

Plots of microhardness data vs. aging time are shown in Figures 1 and 2. Figure 2 shows the effect of pre-deformation on aging response. As can be seen, pre-deformation of 4% had little influence, the only noticeable change being that hardness was slighthy greater at the longer aging times. Since the aging hardness can be related to the α phase distribution, the small difference observed in the 723 K curves could be attributed to the existence of more sites for nucleation in pre-deformed samples. As the pre-deformation increases, the α precipitates become more finely distributed [6].

At 633 K, the intermediate phases ω and β' coexist with α. The α precipitates showed a Burger's relation with the matrix classified as Type 1 α, according to Duerig and Williams [7]. Figure 3 shows a very fine precipitation of Type 1α with a $(2\bar{1}\bar{1}0)_\alpha$ // $(\bar{1}11)_\beta$. Type 1α is an hexagonal structure with a = 0.29 nm and c = 0.45 nm.

At 723 K, β' phase still co-exists with Type 1α, as observed on Figure 4. The intermediate phase β', dispersed throughout the matrix, promotes homogeneous α precipitation [8], explaining why the intergranular α phase was not observed in these conditions.

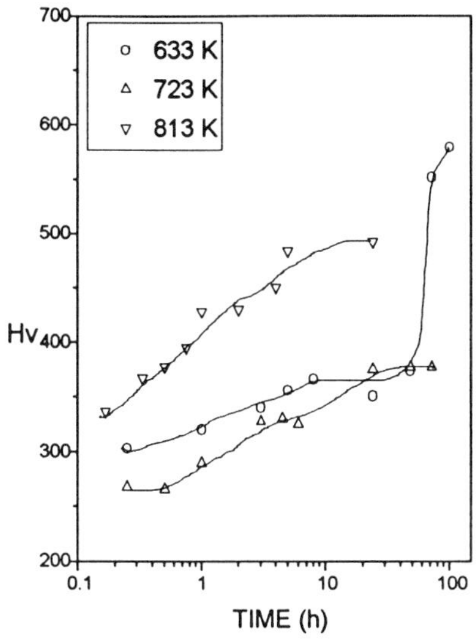

Figure 1. Aging curves (non-deformed samples)

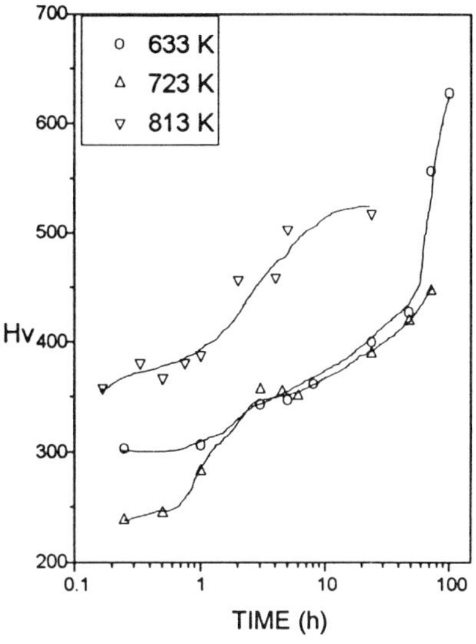

Figure 2. Aging curves (pre-deformed samples)

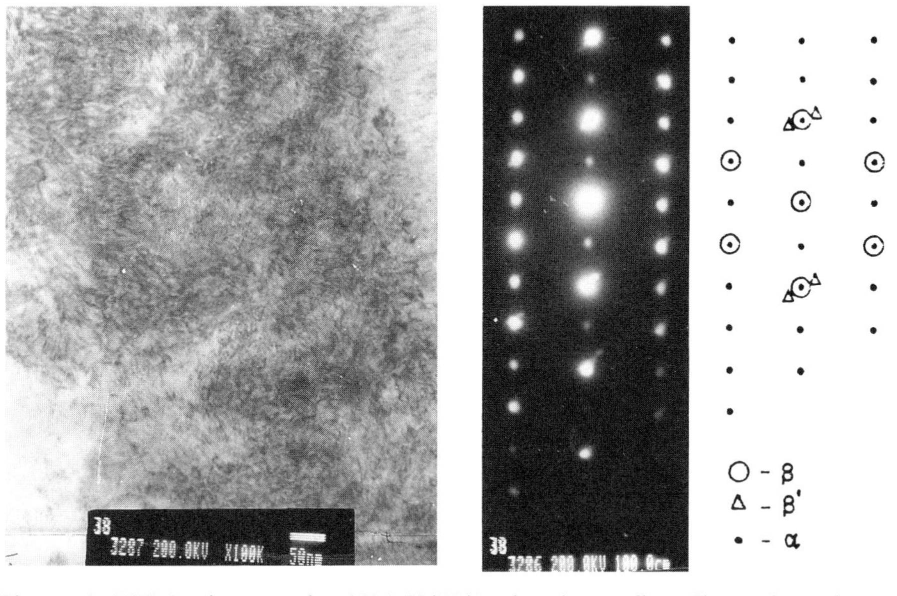

Figure 3. TEM micrography (633 K/72h) showing a fine dispersion of type 1 α (a) and its relation to the β matrix (b)

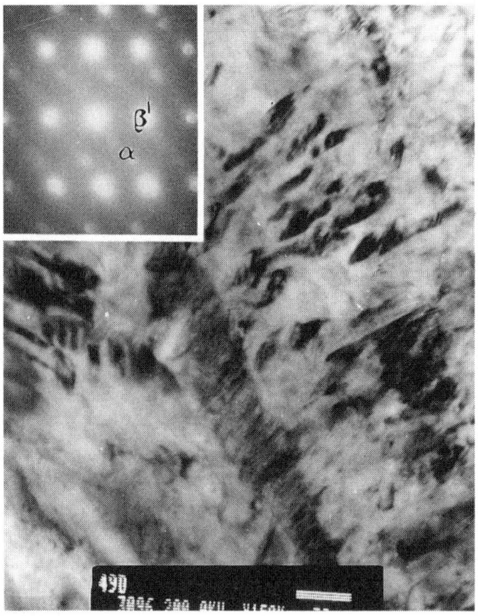

Figure 4. TEM image of a region with α and β' (723 K/72 h-4 %def.)

Aging at 723 K for 1 h produced a very fine precipitation of ω, β' and α. Both diffuse streaking and ω spots could be seen. Type 1α can assume a needle-like morphology, observed in Figure 5, especially at 633 K (for 72 h) and 723 K for shorter aging times. Observation of samples aged at 723 K for 72 h revealed the presence of both Types 1α and 2α. Type 2α appears for longer exposures as a consequence of losing Burgers orientation of Type 1α, exhibiting an almost acicular morphology. Microdiffraction of Type 2α indicates a lattice parameter of 0.30 nm for this phase. The lattice parameter was 0.30 nm for β', while that of the β matrix was 0.33 nm. The β' precipitates were observed together with the two types of α.

Heat treatment at 813 K inhibits ω and β', leading to a direct α phase precipitation [9], particularly of type 2α. Aging pre-deformed samples at 813 K for 45min. promoted α precipitation in deformation bands, as can be seen in Figure 6.

Tensile tests results are summarized in Table 1. The observed ductility was lower in samples aged at 633 K for 72h (deformed and undeformed), 723 K for 1h (deformed) and 723 K for 72h (deformed and undeformed) although these samples exhibited high strength. This effect can be associated with the needle-like Type 1α particles observed. Aging at 813 K produced the best combination of strength and ductility associated with large non-coherent Type 2α precipitates.

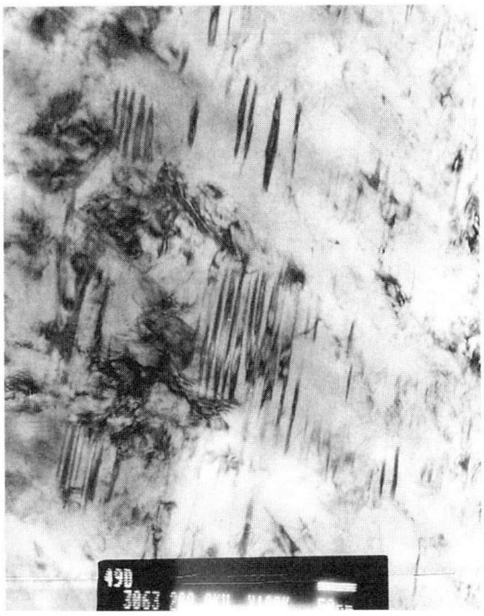

Figure 5. Type 1α with a needle-like morphology (723 K - 72 h - 4%)

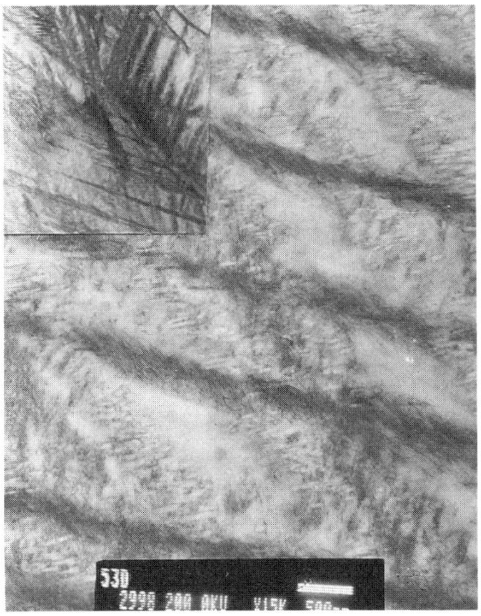

Figure 6. Alpha phase precipitation in deformation bands (813K - 45min-4%)

Table 1 - Tensile tests results

SAMPLE	ε%	σy (MPa)	σmax. (MPa)
633 K/8h	12.2	451.36	817.11
633 K/8h - def.	9.77	358.31	835.81
633 K/72h	7.36	500.961	1021.96
~~723~~ K/72h - def.	8.07	552.92	887.78
723 K/1h	10	621.67	813.24
723 K/1h - def.	6.96	560.62	848.71
723 K/72h	7.49	601.6	1554.91
723 K/72h - def.	8.84	1163.85	1399.68
813 K/45min.	12.9	710.53	827.65
813 K/45min - def.	11	654.68	882.25
813 K/24h	13	868.32	1206.04
813 K/24h - def.	12.4	819.56	1142.75

CONCLUSIONS

1 - Neither of the three cooling rates could avoid athermal omega phase formation after solution treating;

2 - Aging alloy Ti 15-3 at 633 K and 723 K promotes the precipitation of intermediate phases. These phases were found to be ω and β';

3 - Two kinds of alpha precipitates were detected: Types 1α and 2α. Type 1α is a Burgers oriented phase and can assume a needle-like morphology. As aging goes on, Type 2α appears;

4 - Needle like precipitates of Type 1α lead to high strength and low ductility, while coarser Type 2α leads to the best combination of strength and ductility.

REFERENCES

1) P. A. Blenkinsop, "Advanced alloys and Processes", Titanium '92 Science and Technology, Vol. I, Froes, F. H., and Caplan, I. L., eds., p. 1, TMS, 1991.

2) F. C. Ferraz, "Comportamento em Fadiga da Liga de Titânio Beta 15-3 em Condições de Utilização Aeronáutica", MSc. Thesis, UFRJ, Rio de Janeiro, 1991;

3) P. J. Bania, G. A. Lenning and J. A. Hall, "Development and Properties of Ti-15V-3Cr-3Sn-3Al (Ti-15-3)", Beta Titanium Alloys in the 1980's, p. 209, 1983.

4) AMS 4914 - "Aerospace Material Specification", SAE;

5) D. Banerjee, R. L. Mukherjee, R. L. Saha and K. Bose, "Microstructure and Tensile Ductility in a β Heat Treated Titanium Alloy", Met. Trans. A, Vol. 14A, p. 413, 1983.

6) Iman M. A., Poulose P. K., and Rath B. B., "Effect of Cold Work and Heat Treatment in Alpha Region on Mechanical Properties of Ti-15-3 alloy", Titanium '92 Science and Technology", Vol. I, Froes, F. H., and Caplan, I. L., eds., p. 177, TMS, 1993.

7) T. W. Duerig and J. C. Williams, "Microstructure and Properties of Beta Titanium Alloys", Beta Titaniium Alloys in the 1980's, p. 19, 1983.

8) C. G. Rhodes and N. E. Paton, "The Influence of Microstructure on Mechanical Properties in Ti-3Al-8V-6Cr-4Mo-4Zr- (Beta-C)", Met. Trans. A, Vol. 8A, p. 1749, 1977.

9) W. Shiong, S. Guigin and L. Youming, "Phase Transformation in ;Ti-15V-3Cr-3Sn-3Al Alloy", Titanium '92 Science and Technology, Vol. I, Froes, F. H., and Caplan, I. L., eds., p, 761, 1993.

Chapter 5

Stainless Steels

Recycle of Radiologically Contaminated Austenitic Stainless Steels

K. J. Imrich, D. R. Leader, N. C. Iyer and M. R. Louthan, Jr.*

ABSTRACT

The United States Department of Energy owns large quantities of radiologically contaminated austenitic stainless steel which could be recycled for reuse if appropriate release standards were in place. Unfortunately, current policy places the formulation of a release standard for U.S. A. industry years, if not decades, away. The Westinghouse Savannah River Company, Idaho National Engineering Laboratory and various university and industrial partners are participating in initiative to recycle previously contaminated austenitic stainless steels into containers for the storage and disposal of radioactive wastes. This paper describes laboratory scale experiments which demonstrated the decontamination and remelt of stainless steel which had been contaminated with radionuclides.

INTRODUCTION

Nuclear materials production facilities owned by the United States Department of Energy (DOE) use large quantities of austenitic stainless steel for structures, piping and components. Typical service conditions cause radiological contamination and/or neutron activation of much of this steel. Recent changes in the global political climate have lead to

---

* Westinghouse Savannah River Company, Aiken, SC 29808

significant transitions in mission within the DOE and to an increased emphasis on environmental restoration of many of the previously contaminated areas. Recycle, reuse, waste minimization and environmentally conscious disposal have become key issues in the materials utilization arena. This paper describes portions of an initiative to recycle previously contaminated austenitic stainless steels into containers for the storage and disposal of radioactive wastes. This initiative is part of a cooperative program involving the Savannah River Site, Idaho National Engineering Laboratory and university and industrial partenships.

BACKGROUND

Decontamination and recycle of radiologically contaminated metals and alloys could significantly reduce the waste volume, conserve national resources and improve the economics of the decommissioning of nuclear installations. However, socio-political considerations must be balanced with technical, safety and economic considerations. This is particularly true in the United States of America and has severely restricted the recycle process. Successful reintroduction of previously contaminated metals into the primary metals industry requires decontamination standards.

The European Union currently recommends a clearance level of 1 Bq/g for b and g emitters while Germany recommends a lower level of 0.1 Bq/g, the United Kingdom 0.4 Bq/g and Japan 0.5 Bq/g. Obviously, internationally accepted limits have not yet been developed; however, even with the lack of international standards the Siempelkamp plant in Germany, Studsvik of Sweden, and smaller plants in Japan and Slovakia have released recycled material into the free market (1). Worldwide trade, international movement of materials and the existing recycle market for non-contaminated metals and alloys assure that when any nation free releases previously contaminated metals and alloys, all nations eventually use and recycle products fabricated from previously contaminated materials. However, the ineffability of using previously contaminated metals in the U. S. A. places the formulation of a release standard for U.S. A. industry years, if not decades, away.

Evaluation of potential for recycling radioactive scrap metal (RCM) (2) in the U. S. A. has shown that:

a) "the perception of risk related to radioactive materials is more intense in the U. S. than in any other country", and

b) "not recycling RCM involves some trade-off of risk in industrialized countries for environmental and health impacts in developing countries" (presumely because of the environmental impact of mining operations to obtain the chromium necessary to replace that buried).

Apparently, current policies for recycle of radioactively contaminated scrap metal are such that, regardless of actions by most developed nations and the possible detrimental effects to developing nations, free release of RCM from U. S. nuclear facilities, including metals and alloys from DOE sites, is extremely unlikely.

REGULATED REUSE

The DOE owns large inventories of radioactive scrap metals (1.5 million tons), with much of that metal in a current state of storage that presents a future liability to DOE (3). For example, Savannah River Site has, from approximately 35 years of reactor operation, over 68 heat exchangers which are currently stored on pads outside a reactor building. Each heat exchanger is about the size of a railroad car, weights about 100 tons and is fabricated almost entirely from Type 304 stainless steel. Disposition and disposal of these radioactively contaminated heat exchangers can be accomplished through any of several pathways as illustrated in Figure 1.

The "leave on pad", or do nothing, option is not compatible with the environmental restoration initiatives at the Savannah River Site. Even if this long term alternative were environmentally appropriate the cost of continuous monitoring and physical protection of the storage area would become overwhelming. Transfer and burial of the heat exchangers would use a significant fraction of existing Savannah River Site burial ground capacity, reduce future options for reuse of the steel and waste critical materials (e.g. chromium). Therefore, burial may not represent a technically prudent disposal alternative. Decontamination and release of the clean steel into the commercial market would provide the best alternative if the regulatory instruments for such release were in place in the United States of America. Unfortunately, the steel in the heat exchanges contains both internal contamination from internal activation and tritium absorption and external contamination from the

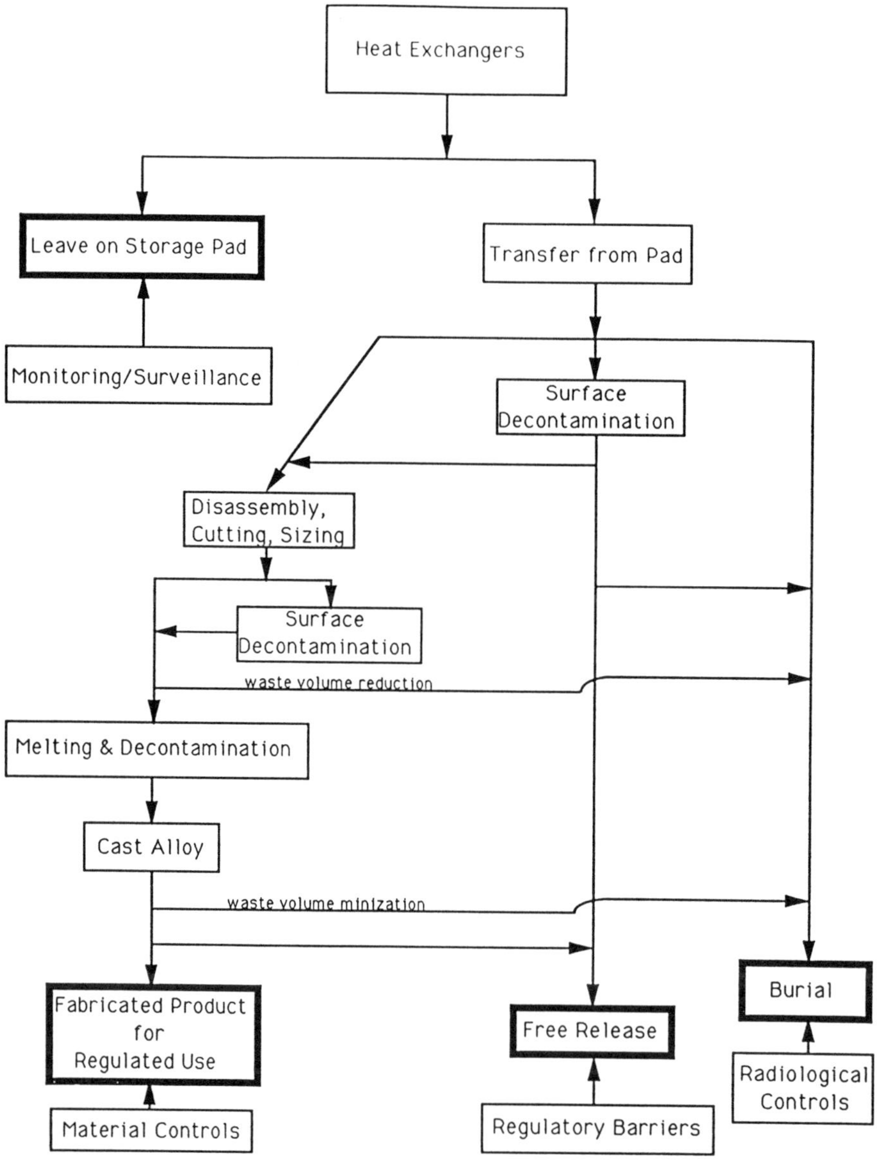

Figure 1. Flow Sheet Illustrating Possible Disposition Pathways for the Hundred Ton, Austenitic Stainless Steel Heat Exchangers Currently Stored at the Savannah River Site

accumulation of radionuclides on the exposed surfaces. The current regulatory framework provides for the removal of surface contamination but precludes free release of materials that have had any internal contamination. Regulated reuse therefore becomes the only effective alternative for disposal of the heat exchangers.

Regulated reuse involves decontamination and recycle of the steel such as that from the heat exchangers. However, distribution of the recycled steel is severely limited. Current plans are for the recycled steel to be fabricated into containers for nuclear waste. The waste containers will remain under DOE control, thus preventing free release of the steel into a non-regulated market. Advantages of regulated reuse include reducing the need for:

a) burial of contaminated material and

b) the introduction of additional steel into the DOE contaminated metals inventory.

The primary disadvantages of regulated reuse are:

a) the need to maintain continual control over the recycled materials, and

b) the potential difficulty in assuring a market for the components fabricated from the recycled stock.

The Savannah River Site recycle initiative is aimed at demonstrating that radiologically contaminated Type 304 stainless steel components, including heat exchangers from the reactors, can be successfully recycled. Up to two hundred tons of contaminated steel will be decontaminated, remelted, cast and fabricated into boxes and/or drums for the storage of contaminated waste. Although the technology required for stainless steel recycle is well established, recycle combined with decontamination and radiological hazard characterization have not been demonstrated. The Savannah River demonstration will validate the decontamination-recycle combination and provide metallurgical characterization of the recycled material. This paper describes the microstructural characterization of the first few laboratory heats of decontaminated-recycled Type 304 stainless steel.

MATERIALS AND PROCESSES

Feed stock for the initial decontamination-remelt demonstrations was small pieces or sections of the piping and heat exchangers which had been removed from the reactor process water systems and metallurgically evaluated (4,5). These sections, which had been decontaminated for metallographic examination, were radiologically characterized and remelted in an induction furnace. The initial melts provided approximately ten pound heats of steel which were sectioned, analyzed and cold rolled. Heats of steel were also prepared from non-contaminated feed stock to facilitate the evaluation processes.

Surface contaminantion was reduced by water rinsing and nitric acid cleaning. Nitric acid decontamination was used because of its compatibility with existing radioactive waste processing technologies at the Savannah River Site. Although nitric acid removes most of the smearable surface contamination, it does not effectively remove contamination that is trapped in surface flaws and cracks (Figure 2) From a practical standpoint the potential for such defects makes the level of effort required to assure complete decontamination of any component removed from reactor service difficult, if not impossible. Therefore, in-melt decontamination will be necessary for the recycle process and was accomplished in the laboratory studies by preferentially trapping radionuclides in a slag over the induction melt. The detailed slag-metal interaction studies necessary to optimize the melt refining/decontamination processes are underway (6).

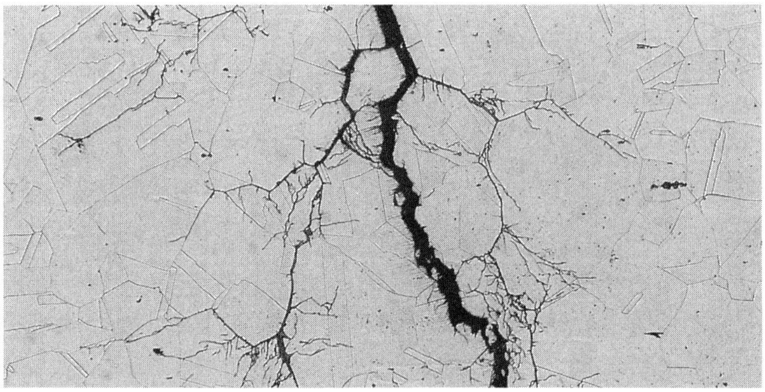

Figure 2. Stress Corrosion Crack in Process Water Pipe from SRS Reactor. (100X)

Selected pieces of reactor process piping and heat exchangers were melted to approximately 1500°C, then poured into either metal or sand molds. The ingot-mold system was air cooled to room temperature before the as-cast ingot was removed for evaluation and rolling. Melt offgas was monitored throughout the melt-pour process without indications of radionuclide release. The slag used for these initial studies was borosilicate glass. Transfer of radionuclides into the slag was noted. Two of the three ingots cast from previously contaminated steels showed contamination was undetected by common survey methods while one ingot was contaminated to higher levels than anticipated for regulated reuse. Heats of steels from the non-contaminated feed stock were prepared in the facilities used to process the contaminated feed and showed no indications of radionuclide uptake.

Elemental analysis demonstrated that the interior of ingots was an 18% Cr- 8% Ni alloy even though the exterior surfaces were depleted in chromium and perhaps in nickel. This demonstrated that neither selective oxidation nor slag refining preferentially removed significant quantities of the primary alloying elements. The induction furnace used for these studies was designed to melt metal in a graphite crucible, thus the initial recycle experiments were done in a graphite crucible. This led to significant carbon uptake during the melt-cast process for austenitic stainless steel. The carbon uptake, coupled with the slagging processes resulted in a decontaminated, high carbon "stainless steel". Subsequent tests were done by melting the stainless steel in an alumina or zirconia crucible that was placed inside the required graphite crucible. Use of the alumuna or zirconia crucibles minimized the potential for carbon uptake by direct contact with graphite and led to a successful melt-cast process for Type 304 stainless steel.

METALLOGRAPHIC OBSERVATIONS

Selected sections of the high carbon stainless steel ingot (Figure 3) were prepared for both optical and electron metallography. Energy dispersive analysis of the x-rays created during scanning electron microscopy (Figure 4) confirmed that the interior of the as-cast ingots was of an 18% Cr- 8% Ni composition. Optical metallography revealed the austenite-ferrite microstructure typical of the cast austenitic stainless steels (Figure 5). A gradient in microstructure from the ingot interior to the ingot exterior was observed. This gradient extended

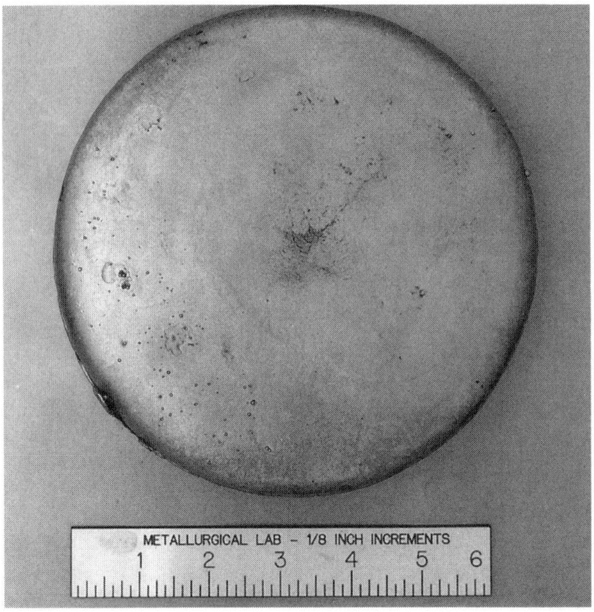

Figure 3. Ingot from Recycled Austenitic Stainless Steel

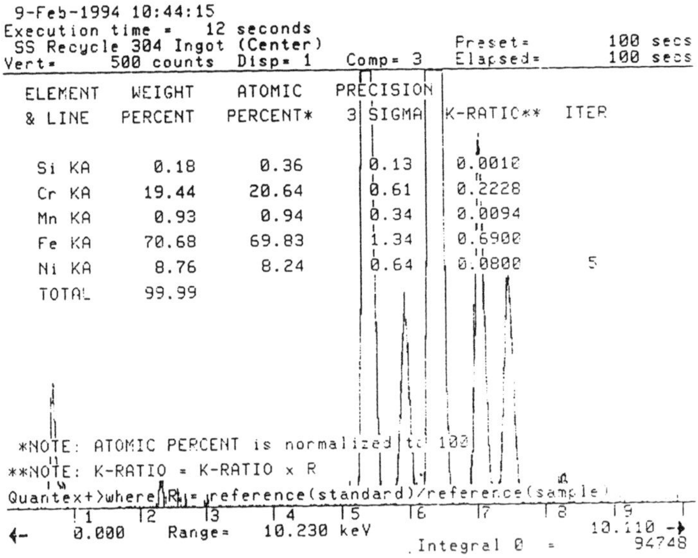

ELEMENT & LINE	WEIGHT PERCENT	ATOMIC PERCENT*	PRECISION 3 SIGMA	K-RATIO**	ITER
Si KA	0.18	0.36	0.13	0.0012	
Cr KA	19.44	20.64	0.61	0.2228	
Mn KA	0.93	0.94	0.34	0.0094	
Fe KA	70.68	69.83	1.34	0.6902	
Ni KA	8.76	8.24	0.64	0.0802	5
TOTAL	99.99				

Figure 4. Energy Dispersive Analysis of X-rays Created during SEM Examination of Center of Ingot

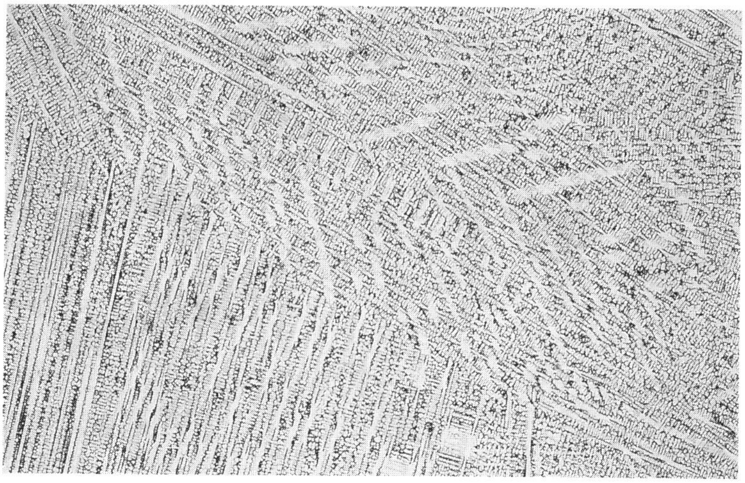

a) Macrostructure at 16X

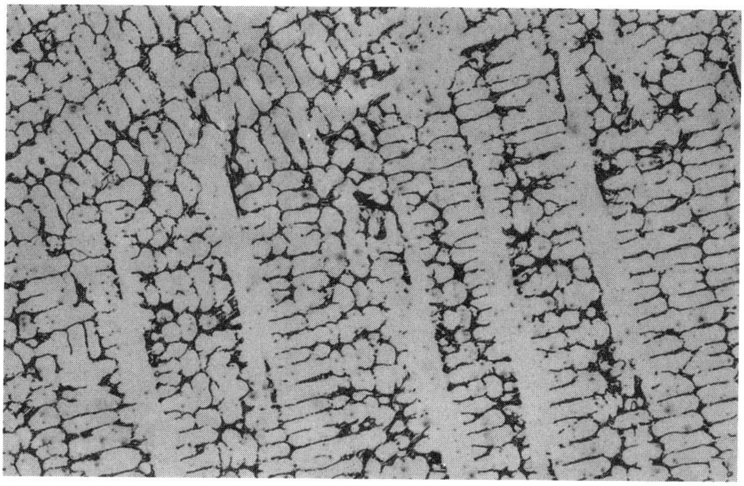

b) Microstructure at 100X

Figure 5. Structure of As-cast Austenitic Stainless Steel Ingot from Recycled Material approximately 1 mm into the ingot (Figure 6) and is associated with a near surface depletion of chromium and perhaps nickel (Figure 7). The location of the depleted region demonstrates preferential oxidation of the chromium during ingot solidification and is consistent with the observation that neither selective oxidation nor slag refining preferentially removed significant quantities of the primary alloying elements during the melting process. Additionally this observation demonstrates the benefit of the slag in preventing oxidation of the underlying metals.

SURFACE

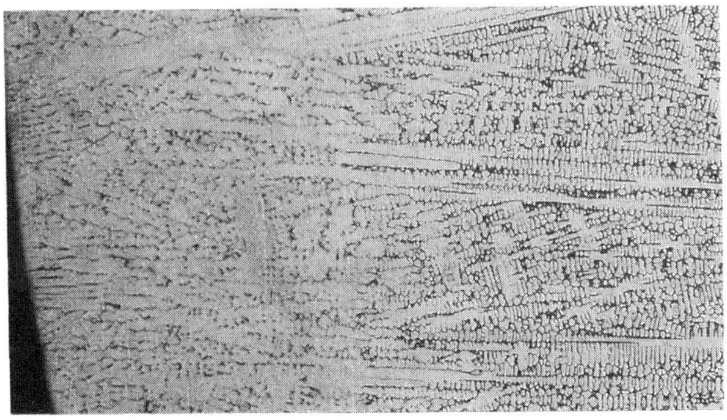

INTERIOR

Figure 6. Macrostructural Gradient Associated with Near Surface Chromium Depletion in As-cast Ingot, 50X

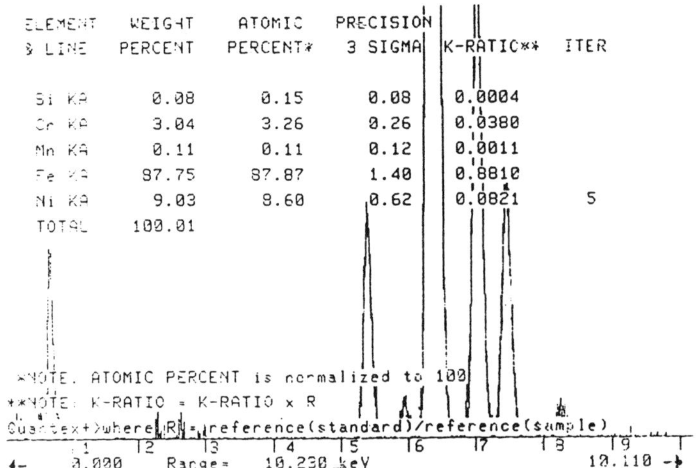

ELEMENT & LINE	WEIGHT PERCENT	ATOMIC PERCENT*	PRECISION 3 SIGMA	K-RATIO**	ITER
Si KA	0.08	0.15	0.08	0.0004	
Cr KA	3.04	3.26	0.26	0.0380	
Mn KA	0.11	0.11	0.12	0.0011	
Fe KA	87.75	87.87	1.40	0.8810	
Ni KA	9.03	8.60	0.62	0.0821	5
TOTAL	100.01				

Figure 7. Energy Dispersive Analysis of X-rays Created during SEM Examination of Surface of Ingot

The replacement of graphite molds with alumina crucibles eliminated, as expected, the unacceptable carbon uptake. This increased the ductility of the ingot and provided as-cast samples that were cold rolled to greater than ninety percent reduction in thickness. The rolling process, which was followed by a surface grinding operation to eliminate incipient edge cracks at several stages in the rolling process, elongated and broke-up the as-cast microstructure (Figure 8).

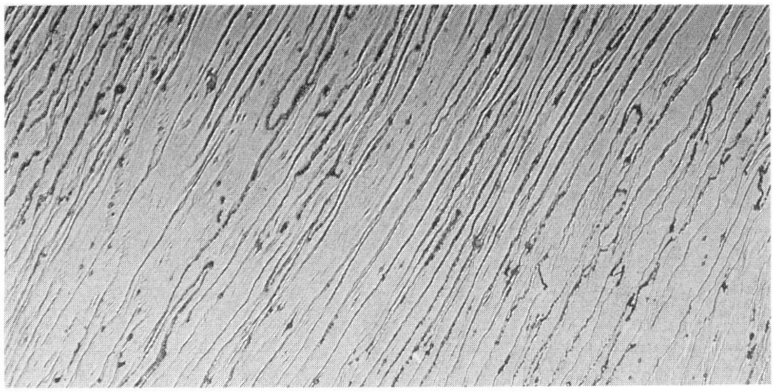

Figure 8. Microstructure of Ingot after Cold Rollong to 80% Reduction in Thickness. 250X

Annealing of the cold rolled ingot caused recrystallization and provided a microstructure typical of recrystallized, wrought austenitic stainless steel products (Figure 9).

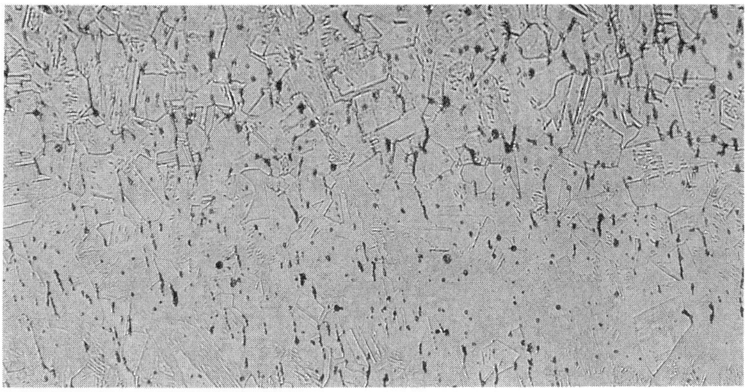

Figure 9. Recrystallized Microstructure After the Cold Rolled Section was Annealed at 1000°C for Two Hour. 250X

CONCLUSIONS

Laboratory scale experiments have demonstrated that Type 304 stainless steel which has been contaminated with radionuclides can be recycled for potential reuse. Decontamination, melting, casting, rolling and heat treatment experiments with previously contaminated feed stock and non-contaminated control samples have provided "recycled" steel which has the 18%Cr-8%Ni composition and microstructures similar to commercial grade alloys.

REFERENCES

1) M. Laser, "Melting and Recycling Waste from Decom" **Nuclear Europe Worldscan**, Vol. XIV, No. 1-2, page 58, 1994

2) L. A. Nieves of Argonne National Laboratory, "Evaluation of Radioactive Scrap Metal Recycling" May 19-20, 1993 Conference/Workshop on Recycle of Radioactively Contaminated Metals, Augusta, GA, Copies of slides from Presentation available from Westinghouse Savannah River Company

3) G. F. Kessinger of Westinghouse Idaho Nuclear Company, "WINCO Metal Recycle Program", ibid Reference 2

4) D. E. Rawl, Jr., S. L. West, D. A. Wheeler and M. R. Louthan, Jr., "Defect Cheracterization in Pipe-to-Pipe Welds in Large Diameter Stainless-Steel Piping" **Materials Characterization**, Vol. 33, p. 155, 1994

5) M. R. Louthan, Jr., S. L. West and D. Z. Nelson, "Inspection Indications, Stress Corrosion Cracks and Repair of Process Piping in Nuclear Materials Production Reactors" p. 314 in **Maintenance and Repair Welding in Power Plants**, American Welding Society, Miami, FL, 1992

6) S. A. Worcester, T. G. Twidwell, D. J. Paolini, T. A. Weldon and R. E. Mizia, "Decontamination of Metals by Melt Refining/Slagging" presented to American Vacumn Society Conference on Liquid Metal Processing, Santa Fe, MN, September 11-14, 1994

FUNDAMENTAL AND COMPARATIVE STUDIES OF $M_{23}C_6$ CARBIDE GROWTH ASSOCIATED WITH TWIN BOUNDARIES AND GRAIN BOUNDARIES IN 304 STAINLESS STEEL

R. J. Romero and L. E. Murr[1]

ABSTRACT

It is well known that $M_{23}C_6$ carbide precipitation occurs first on general grain boundaries, then on non-coherent twin boundaries, and finally, on coherent twin boundaries as evidenced by heat treatment of 304 stainless steel. This aging-related sequence for carbide growth kinetics is apparently related to the associated interfacial free energies: $\gamma_{tb} \cong 800$ mJ/m^2, $\gamma_{TB} \cong 200$ mJ/m^2, $\gamma_{tb} \cong 20$ mJ/m^2. We have observed some rather unique growth features for carbides associated with these distinct interfaces using transmission electron microscopy (TEM). Growth from regular grain boundaries, while often along {111} trace directions is confined to regions connected to the interface and carbides often form large, faceted precipitate blocks along the interface. Growth from non-coherent twin boundaries is usually lamellar with the long lamellar carbide interphase boundaries along specific {111} trace directions dependent upon the grain (or twin) surface orientation. Lamellar growth associated with the

[1]Department of Metallurgical and Materials Engineering, The University of Texas at El Paso, El Paso, TX 79968

coherent twin boundaries is observed to be either parallel to the {111} coherent boundary trace on non-coplanar, but along another {111} trace direction. In addition, carbides growing parallel or coincident with the coherent twin interface are often observed to be growing from non-coherent microsteps in the boundary plane. These growth-specific features for carbides in heat-treated 304 stainless steel can be rationalized in large part by examining the crystallographic/geometric features through a configurational equilibrium theory for relative interfacial torques associated phenomenologically with coherent twin boundary - grain boundary intersections. Numerous examples of these growth-specific carbide features in heat treated 304 stainless steel will be presented and the implications discussed in terms of theory and practice.

INTRODUCTION

Numerous TEM studies examining $M_{23}C_6$ carbide nucleation and growth and of corresponding precipitate morphologies have been conducted in 304 stainless steels [1-6]. At elevated temperature (>600°C), $M_{23}C_6$ carbides preferentially nucleate on grain boundaries and twins boundaries. For example, $M_{23}C_6$ carbides first precipitate on grain boundaries, then on non-coherent twin boundaries and finally, at long aging times, on coherent twin boundaries [7]. The precipitation sequence exhibited by $M_{23}C_6$ is possibly associated with the corresponding average boundary free energies: $\gamma_{gb} \cong 800$ mJ/m^2, $\gamma_{TB} \cong 200$ mJ/m^2, $\gamma_{tb} \cong 20$ mJ/m^2 [8, 9]. At short aging times, $M_{23}C_6$ precipitates first form on the grain boundaries. As aging continues, precipitates form on non-coherent boundaries. As the aging process continues, the $M_{23}C_6$ which formed on the non-coherent twin interfaces grow. At temperatures above 700°C for long aging times (>10 hours), $M_{23}C_6$ precipitates often become lamellar. The long lamellar interfaces coincide predominantly with the traces of [111] planes and have geometrical forms characterized by both [111] and occasionally [110]-type interfaces [4]. These carbide lamellae are observed to preferentially nucleate on non-coherent twin boundaries and on steps in coherent twin boundaries, and grow along [111] trace directions. Finally at long heat treatments, $M_{23}C_6$ precipitates and grow as lamellae along coherent twin boundaries or in intimate coincidence with the coherent twin boundaries or {111} twin planes.

Terao and Sasmal [4], suggested that lamellar growth of $M_{23}C_6$ appeared to be a consequence of planar distortion on {111} planes, and the morphology was found to depend strongly on a so-called "stress field". In effect, Terao and Sasmal suggested that the formation of precipitates was enhanced and necessitated by the presence of predominate directional stresses on {111} planes. However, the nature and origin of such directional stresses was not described in any detail. Recently, Romero and Murr [10, 11] utilizing a theory for configurational equilibrium for relative interfacial torques developed by Horylev et al [12], proposed that the driving force for lamellar $M_{23}C_6$ carbide growth is not attributed to some "directional stress" as previously suggested, buy may in fact be attributed to some resolved effective torque implied by the preponderance for lamellar carbide growth along the 90° inclined {111} planes at the twin boundaries. In the pursuit of their investigation of carbide growth features along twin boundaries, Romero and Murr observed that the precipitates are different on grain boundaries as compared to annealing twin boundaries. Consequently, the driving force for precipitation on grain boundaries is different compared to the annealing twins.

In this study, $M_{23}C_6$ carbide growth features associated uniquely with grain boundaries and twin boundaries are reported in 304 stainless steel. We examined $M_{23}C_6$ carbide growth and of corresponding morphologies coincident with traces of {111} planes. In addition, we investigated carbide interfaces relative to 90° inclined {111} planes. The carbide morphologies and growth features relative to the grain and coherent twin interfaces are discussed.

EXPERIMENT

Materials: In this experiment, four different heats of 304 stainless steel plates (nominal thickness = 6.25 mm) containing 0.01% - 0.07% carbon were obtained for this experiment. The nominal composition and nominal grain size of these heats in their as-received mill processed condition are illustrated in Table 1. The mill processed condition was used to avoid grain size variation during aging and eliminate any difficulties of twin observations during TEM (transmission electron microscopy).

Table 1. Nominal compositions and grain size in 304 stainless steels.

Heat #	wt%C	wt%Cr	wt%Ni	wt%Mo	wt%N	Grain Size, μm
1	0.01	18.13	8.92	0.34	.08	40
2	0.025	18.35	9.08	0.37	.08	35
3	0.05	18.29	8.22	0.19	.085	40
4	0.07	18.23	8.15	0.35	.08	35

Deformation and Aging Processing: The mill processed plates were cut into strips of approximately 125 mm in length and 12 mm in width. Undeformed strips were sectioned into approximately 12 mm square samples and aged at 775°C for 0.1 to 500 hours.

Additional strips of 0.05% C of as-received mill processed stainless steel were cold rolled to 70% and flash annealed at 1000°C for 1 minute to develop a detailed examination of precipitation on the twin boundaries and grain boundaries. The thermomechanical treatment only performed on 0.05% C heat, reduced the grain size to 15 μm. The recrystallized 304 stainless steel strips deformed to 20% engineering strain and then cut to 12 mm^2, were aged at 625°C for 0.1 to 500 hours.

Sensitization Response: The sensitization response was evaluated using the single loop version of the electrochemical potentiokinetic reactivation (EPR) test carried out in a modified flat corrosion cell [13, 14]. The EPR test measures the charge, Q, produced during the formation of Cr-depleted zones induced by precipitation of Cr-rich carbides on the interfaces, and normalizes the Q value with the grain boundary area to obtain the degree of sensitization (DOS). Optical examination of the EPR attacked surfaces was carried out to relate EPRDOS value to the microstructures. In this work, examination of the microstructures was crucial in the selection of strain and aging conditions where carbide precipitation was detected on the coherent boundaries in the 304 stainless steel samples. These samples were examined using conventional transmission microscopy (CTEM) to study the carbide features associated with all three boundaries (grain boundaries, non-coherent and coherent twin boundaries).

Carbide Precipitation Investigation: Carbide precipitation study was accomplished using transmission electron microscopy in a Hitachi H-8000 analytical TEM operated at 200 kV. Observations of precipitation, especially at grain and twin boundaries were normally accompanied by systematic, selected-area electron diffraction (SAD) patterns. SAD patterns record and identify the associated crystallography and geometry, where annealing twin boundaries intersect with grain boundaries and non-coherent steps on annealing twins associated with precipitate nucleation or growth.

RESULTS AND DISCUSSIONS

After numerous observations, a clear distinction between carbide growth at the grain boundaries and twin boundaries was obvious. The corresponding carbide morphologies were distinctively different at grain boundaries and twin boundaries. Carbides were observed to be lamellar in morphology at (of in association with) the twin boundaries. For some annealing twin configurations, lamellar carbides were observed to grow parallel to the trace of the coherent annealing twin interface or {111} twin interface, while for other configurations, the carbide lamellae grew out of the coherent annealing twin interface trace, but always in another {111} trace direction. Romero and Murr have suggested [5, 6] that lamellar growth occurring non-parallel to the coherent twin plane but coincident to another {111} plane results because of low effective interfacial torque or residual interfacial torque. Lamellar growth parallel to the 90° coherent {111} twin plane results because of a high effective interfacial torque acting on the twin plane. These twin configurations were noted for both (110) and (112) grain surface orientations. Very little is known about the morphology and growth features of $M_{23}C_6$ at the grain boundaries. In this investigation, the morphology of the precipitates are described as large and shapeless; while carbide growth was confined to the grain boundaries. Sometimes, the carbides had one faceted side corresponding with a {111} plane but rarely grew out of the boundary in a lamellar mode.

Shown in Fig. 1 are lamellar carbides growing out of micro-steps in an inclined coherent twin boundary for (110) and (112) grain surface orientations and growing along the trace direction of another {111} plane. These {111} planes are at 90° to the grain surface and the trace direction of {111} planes are denoted at T90 directions. In the (112) grain surface orientation shown in Fig. 1(a), lamellar carbides are observed growing from steps in the 61.5° inclined {111} coherent twin boundary, and growing in the T90 direction or in the [110]

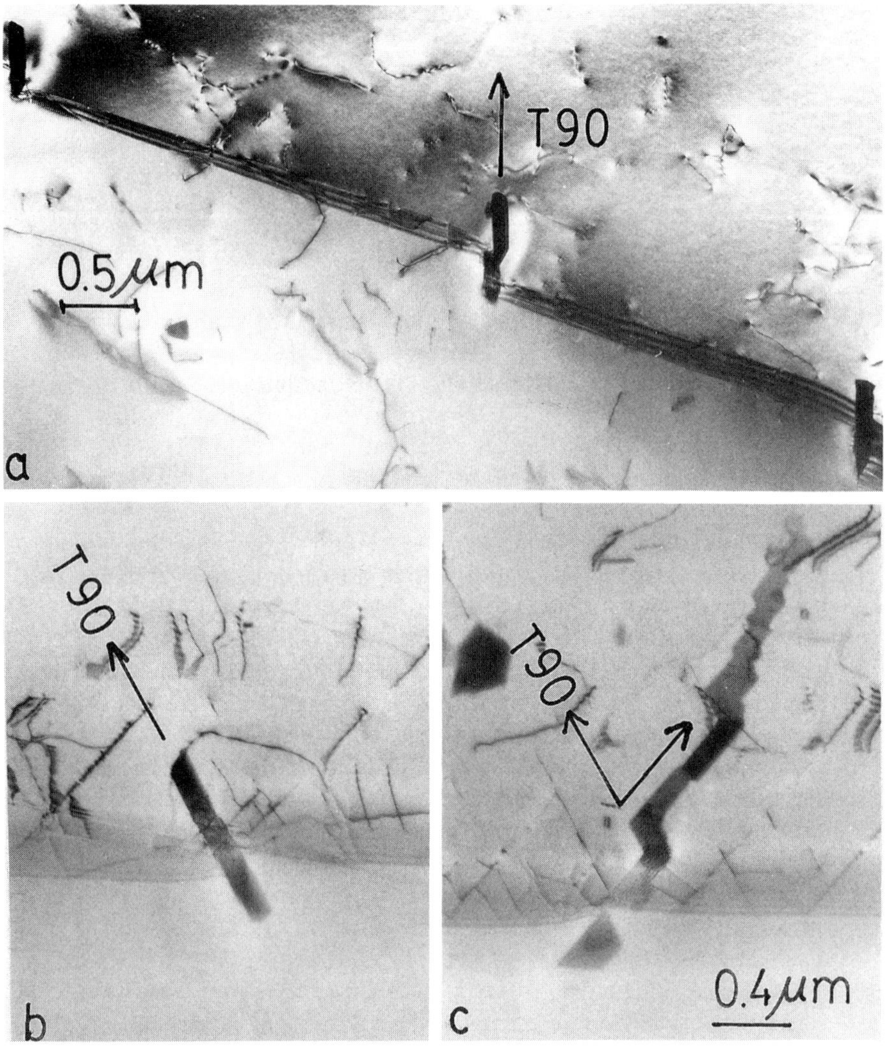

Figure 1. Lamellar growth of $M_{23}C_6$ from inclined (111) annealing twin boundary in (a) (112), (b) and (c) (110) grain orientations coincident with {111} planes 90° to the grain orientation. These directions are denoted T90 and coincides with <110> direction for (112) crystal orientation and with <112> for (110) crystal orientation. (c) Shows alternate growth coincident with both {111} and {111} planes at 90° to the (110) surface. Samples aged 10 hours at 775°C.

direction along (111). The interfaces of the carbide lamellae are faceted by (111). This sample contained 0.07% C and was aged at 775°C for 10 hours. In Fig. 1(b) and 1(c), a (110) grain surface is shown. Lamellar carbide is observed growing from steps in 35° inclined coherent twin boundaries. Figure 1(b) shows lamellar growth non-parallel to (111) coherent twin plane in the [112] direction or T90 direction. Figure 1(c), demonstrates preferential (lamellar) precipitate growth in a (110) grain surface orientation always follows the 90° {111} trace directions (the $(\bar{1}11)$ and $(1\bar{1}1)$ traces) or <112> direction (T90 directions). Again the carbide lamellae are faceted by the (111) and (111) planes. These samples, containing 0.05% C, were aged 10 hours at 775°C.

Figures 2 and 3 show a (112) grain surface orientation. Shown in Figs. 2 and 3 are lamellar carbides growing parallel to the coherent twin boundary which is at 90° to the crystal surface, and are representative of a high torque configuration [12]. As in Fig. 1(a), lamellar growth is always in the T90 direction corresponding to the $[1\bar{1}0]$ direction. The precipitate interfaces are faceted by {111} planes. The tendency for lamellar growth along the {111} coherent twin plane is illustrated in Fig. 3. In Fig. 3(a), we observe lamellar growth parallel to the coherent twin plane but in the regions where steps or non-coherent steps exist the precipitates are not lamellar and do not grow coincidental to {111}. In Fig. 3(b), we observe preferential lamellar growth along the coherent twin interface. The $M_{23}C_6$ carbide lamellae continue to grow along the {111} plane although the boundary changes direction noted by non-coherent steps in the twin boundary.

Numerous observations were conducted at grain interfaces which yielded some interesting results. We observed that the precipitate morphologies were distinctively different at grain boundaries compared to the coherent twin boundaries. For example, the precipitates are larger and do not grow as lamellae. Precipitate growth is usually confined to the immediate grain boundary region and rarely do they grow extensively out of the boundary. However, the precipitates are frequently faceted by the {111} planes.

Figures 4 and 5 show some representative examples of precipitate morphologies and growth features at the grain interfaces. Carbide growth out of the grain interface sometimes exhibits faceting along one side of the carbide interface. Figure 4(a) shows the $M_{23}C_6$

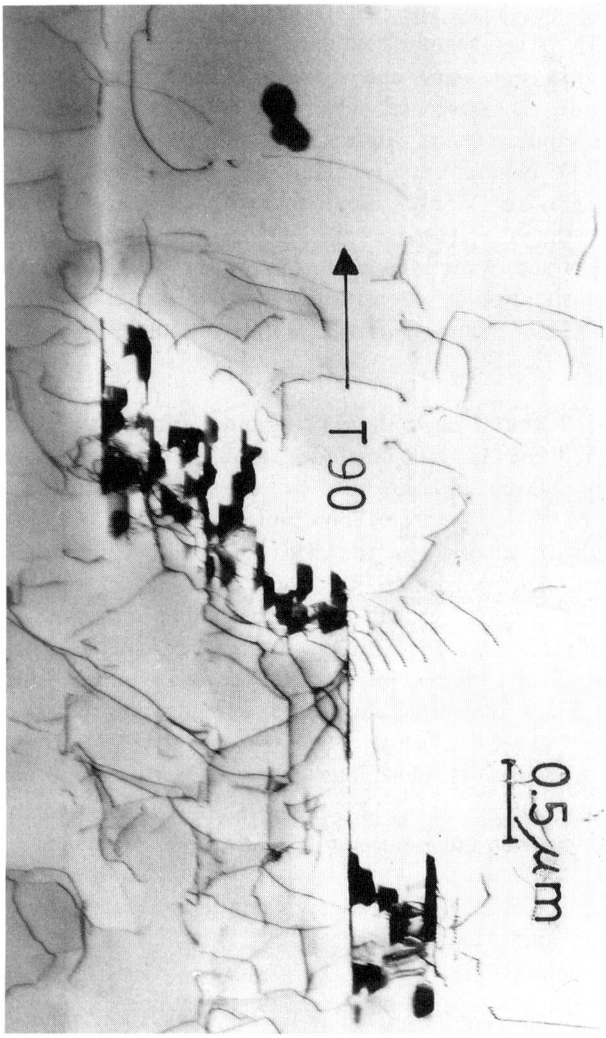

Figure 2. $M_{23}C_6$ precipitate growth in (112) crystal orientation growing parallel to 90° coherent twin boundaries in T90 direction. Sample aged 10 hours at 775°C.

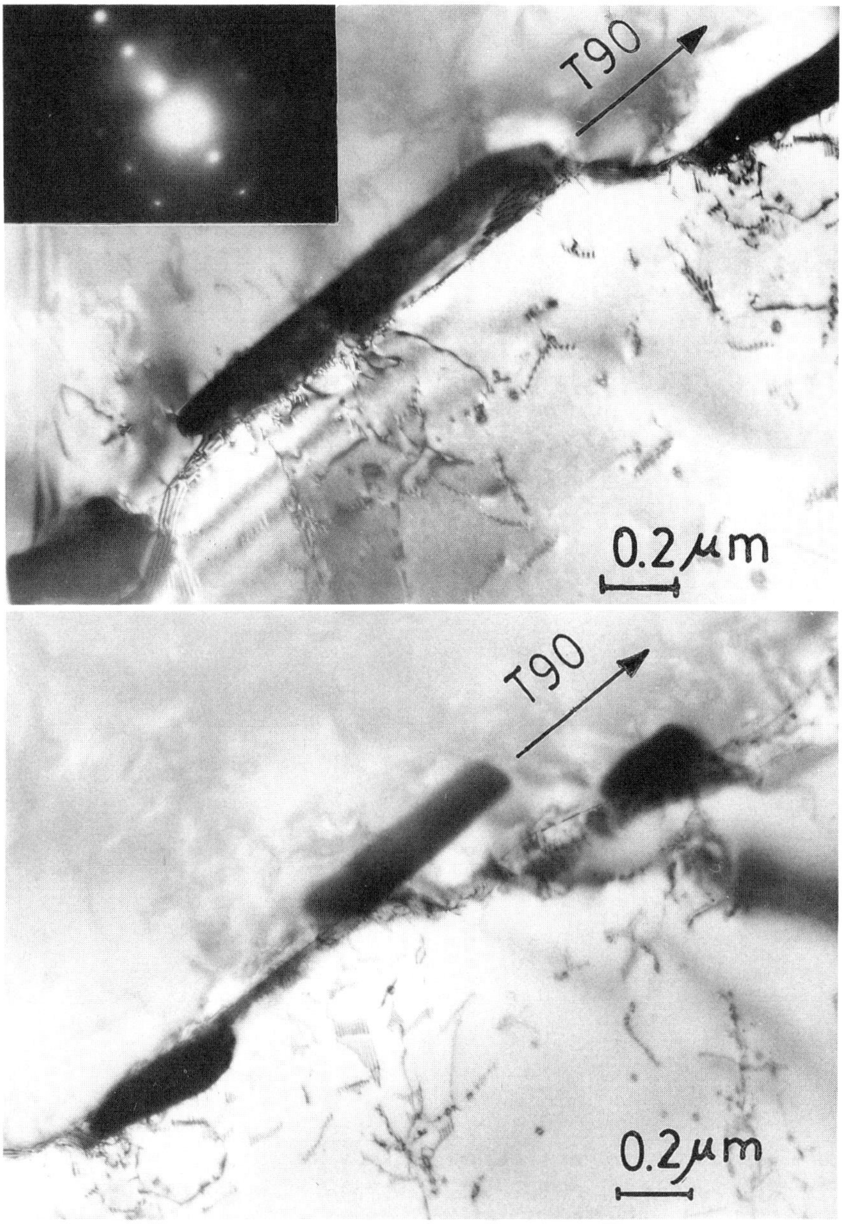

Figure 3. Examples of lamellar carbide growth parallel to the multi-faceted twin plane in T90 direction in the (112) crystal orientation. (b) Shows preferred growth along 90° inclined {111} plane.

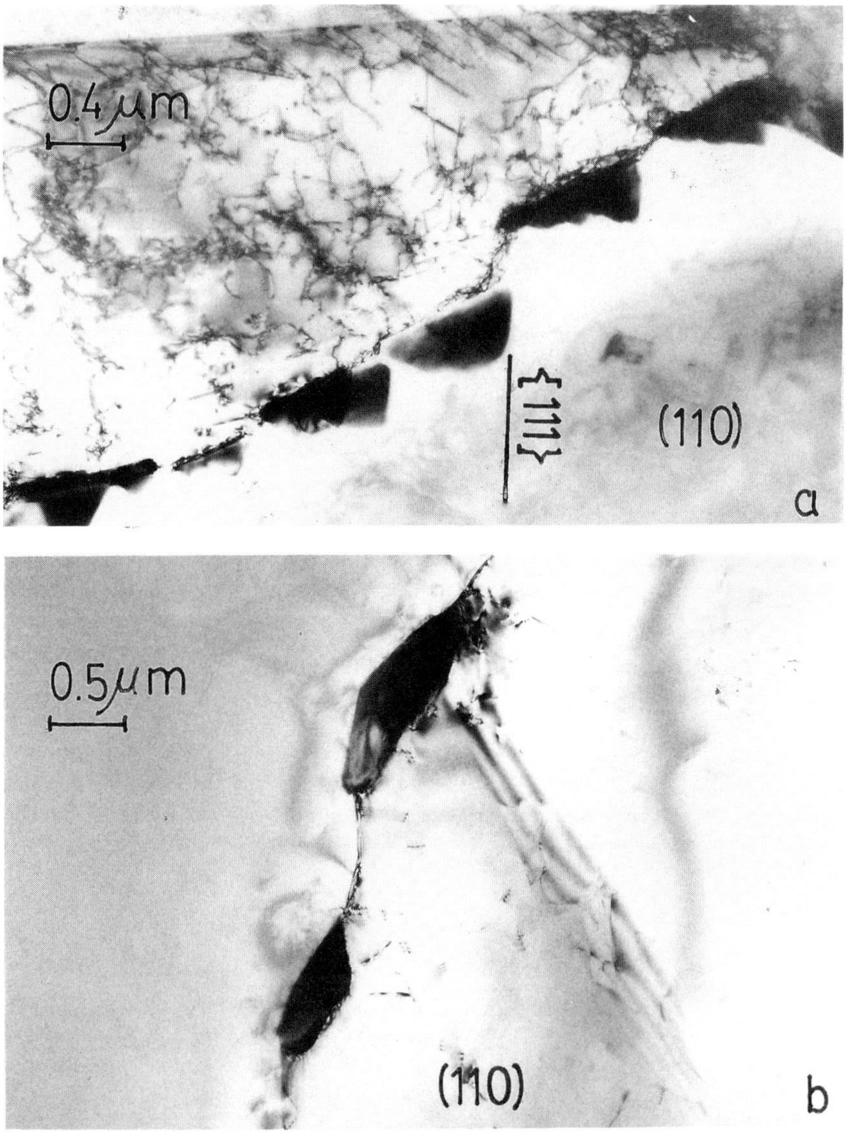

Figure 4. Examples of carbide growth at grain boundaries. (a) (013)/(110) neighbor grains. (b) (110)/(110) neighbor grains. Growth along {111} trace (faceting) is shown.

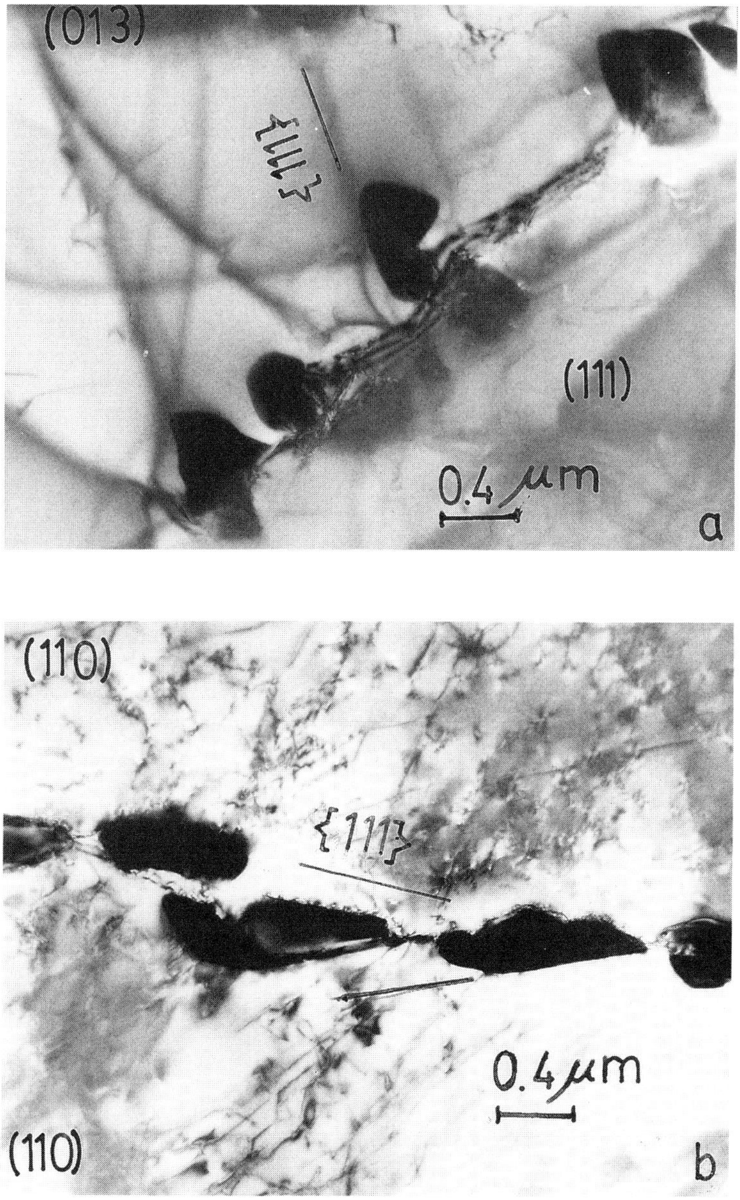

Figure 5. Additional examples of carbide growth from grain boundaries with grain orientations shown.

precipitates growing out of the interface into the (110) grain. One side of the precipitate interface apparently is faceted by a {111} plane inclined 35° to the (110) grain orientation. The other grain orientation is (013). In Fig. 4(b), the precipitates tend to be preferentially growing along instead of out of the grain interface. Apparently, two sides of the precipitate interface are faceted by the close packed planes or {111} plane. Grain orientation in grain B is (110).

CONCLUSIONS

At the twin boundaries, at 775°C, the carbides were lamellar in morphology and grew in the trace directions of 90° inclined {111} planes or T90 direction. The interfaces of the long lamellae were faceted or characterized by the {111}, specifically the {111} inclined 90° to the grain surface orientation., For (110) grain surface orientations, growth of lamellar $M_{23}C_6$ is always in the <112> directions and corresponding {111} traces. For (112) grain surface orientations, lamellar $M_{23}C_6$ growth was always in the <110> direction.

At grain boundaries, the grain boundary $M_{23}C_6$ precipitates were rarely lamellar. Sometimes, one faceted side of the precipitate corresponds with the trace of {111} planes. However, the grain boundary $M_{23}C_6$ carbide precipitates did not tend to grow out of the boundary in lamellar from along the trace directions of {111} planes with preference to T90 directions. The carbide morphologies at the grain boundaries are different compared to those at or near the twin boundaries. This suggests that the driving force for precipitation at the grain boundaries is different than at the twin boundaries and suggests kinetic differences which may be related to the energetics, or the consequence of grain boundary torque phenomena, which is not well understood.

ACKNOWLEDGMENTS

This research was supported in part by an NSF-RIMI Grant (HRD-9105065), GSA Grant PF90-018, and a Patricia Roberts Harris Fellowship (R.J.R.).

REFERENCES

1. H. Hatwell and A. Berghezan, "Nucleation and Growth of Chromium

Carbide Precipitates in Coherent Twin Boundaries in Austenitic Stainless Steels", *Iron & Steel Inst. Special report*, No. 64, p. 1603 (1959).
2. B. Hattersley and M. H. Lewis, *Philos. Mag.*, 10, 1075 (1964).
3. L. K. Singhal and J. W. Martin, *Acta Metall.*, 15, 1603 (1967).
4. M. Terao and B. Sasmal, *Metallography*, 13, 117 (1980).
5. R. J. Romero, E. A. Trillo, A. H. Advani, L. E. Murr, and W. W. Fisher, *Proc. 51st Annual Meeting of the Microscopy Society of America*, San Francisco Press, Inc., San Francisco, CA, 1166 (1993).
6. R. J. Romero and L. E. Murr, *Scripta Metall. et Materialia*, 30, 191 (1994).
7. R. Stickler, and A. Vinckier, *Trans. ASM*, 54, 362 (1961).
8. L. E. Murr, *Interfacial Phenomena in Metals and Alloys*, Addison-Wesley Publishing Co., Reading, MA (1975); reprinted by Tech Books, Fairfax.
9. A. H. Advani, R. J. Romero, L. E. Murr, D. J. Matlock, W. W. Fisher, P. M. Tarin, C. M. Cedillo, J. G. Maldonado, R. L. Miller, and E. A. Trillo, *Scripta Metall. et Materialia*, 27, 1759 (1992).
10. R. J. Romero and L. E. Murr, *Scripta Metallurgica et Materialia*, 30, 191 (1994).
11. R. J. Horylev and L. E. Murr, "An Electron Microscope Study of Configurational Equilibrium at Twin-Grain Boundary Intersections in Fcc Metals", in *The Nature and Behavior of Grain Boundaries*, H. Hu (Ed.), Plenum Press, New York, NY, 203 (1972).
12. R. J. Romero and L. E. Murr, *Acta Metallurgica et Materialia*, in press (1994).
13. A. H. Advani, L. E. Murr, D. G. Atteridge and R. Chelakara, *Metall. Trans. A*, 22A, 2917 (1991).
14. W. L. Clarke, R. L. Cowan, and W. L. Walker, "Comparative Methods for Measuring Degree of Sensitization in Stainless Steels", in *Intergranular Corrosion of Stainless"*, ASTM STP, 656, 99 (1978).

TYPE 304 STAINLESS STEEL SURFACE MICROSTRUCTURE:

PERFORMANCE IN HYDRIDE STROAGE AND ACID CLEANING

Elliot A. Clark

ABSTRACT

The performance of stainless steel containers in hydride storage bed systems have been evaluated, primarily using scanning electron microscopy. No adverse reaction between Type 304L stainless steel and either $LaNi_{5-x}Al_x$ (x=0.3, 0.75) or palladium supported on Kieselguhr granules (silica) during exposure in hydrogen was found in examination of retired prototype storage bed containers and special compatibility test samples. Intergranular surface ditching, observed on many of the stainless steel surfaces examined, was shown to result from air annealing and acid cleaning during fabrication of stainless steel. The ditched air annealed and acid cleaned stainless steel samples were more resistant to subsequent acid attack than vacuum annealed or polished samples without ditches.

INTRODUCTION

The DOE Savannah River Site uses metals and metallic compounds that form solid hydride phases to store, handle and purify hydrogen isotopes in the Tritium Facilities [1]. At lower temperatures these materials form hydrides in the presence of hydrogen gas, and absorb the gas. Hydrogen is liberated when the temperature of the hydride is increased. Hydride storage beds consist of the hydride compound contained in a stainless steel vessel, surrounded by a heating and cooling system to control the absorption and liberation of hydrogen gas. The bed is connected to a gas handling system for further gas processing.

Savannah River Technology Center, Westinghouse Savannah River Co, PO Box 616, Aiken, SC 29802

Metallurgical examination of a retired prototype hydride storage bed container revealed intergranular ditching on the inside surface that contacted the hydride *(Fig. 1)*. To determine whether this microstructure resulted from an unexpected reaction between the hydride and the stainless steel, an evaluation of the materials compatibility of hydrides with stainless steel was undertaken. The evaluation consisted of metallurgical analysis of stainless steel exposed to hydrides in both prototype storage and purification beds and samples exposed in small-scale experiments designed for this study. In addition, stainless steel samples were exposed to several types of acid cleaning and passivation solutions, to help explain observed stainless steel surface microstructures. The goal of the study was to determine whether any adverse metallurgical reaction occurs between austenitic stainless steels and hydrides used in process applications, under expected process conditions.

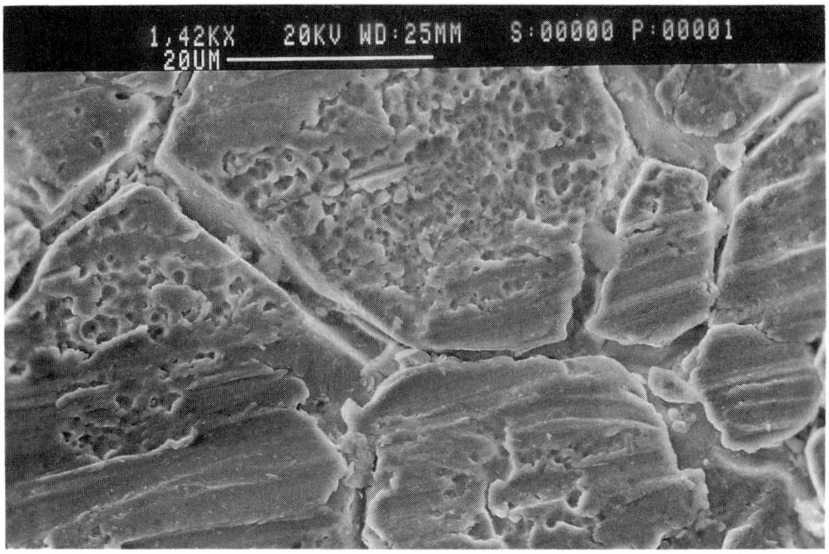

Figure 1. Interior surface of hydride bed container showing intergranular surface ditching

EXPERIMENTAL

General

Three types of stainless steel/hydride exposure experiments and one stainless steel/acid cleaning experiment were performed. In all cases, exposure effects were best characterized by scanning electron microscopy using secondary electron detection. Micrographs of exposed samples were compared with identically prepared but unexposed blanks where possible, to ensure that only changes in the surface microstructure resulting from exposure to hydrides or acid cleaning would be identified and not confused with the initial unexposed surfaces.

Hydride Compatibility Tests

The inside container wall surfaces of six retired prototype hydride storage beds were examined. Five of the beds contained $LaNi_{5-x}Al_x$ (x=0.3 or 0.75) hydride, and one contained palladium supported on Kieselguhr (silica) granules. In addition, an unused container identical to the one exposed to Kieselguhr supported palladium was examined. Three of the storage beds were used in tritium service, and three in deuterium or protium service.

Compatibility sample containers were machined from Type 304L stainless steel plate. $LaNi_{4.7}AlO_{.3}$ and $LaNi_3$ powder was placed inside the containers. Some of the containers were exposed to elevated temperatures in air, and some were encapsulated in quartz ampoules under either vacuum or in a helium/4% hydrogen mixture and subsequently exposed to temperatures up to 923 Kelvins (650° C.) for up to 3 months. Also, a special apparatus was used to expose one sample to 10.3 MPa (1500 psia) hydrogen at 573 Kelvins (300°C.) for one week.

Bench-top-scale hydrogen cycling systems were designed to simulate hydride storage systems. Each system had two Type 304 stainless steel tubes that contained $LaNi_{4.25}Al_{0.75}$ powder. Each tube was surrounded by a specially designed furnace, and a timed relay circuit enabled one tube in each system to be heated while the other was being cooled. At any given time one of the hydride containing tubes was heated to about 423 Kelvins (150°C.) to decompose the hydride and evolve hydrogen gas. At the same time, the other tube was cooled by forced ambient air and hydrogen gas was absorbed by hydride forming in the cooled tube. These systems provided continuous automatic cycling of hydrogen between the tubes, resulting in long-term exposure of the inside surface of the tube to the hydride environment while cycling both temperature and hydrogen. The cycling period was normally 40 minutes. One experiment had a total cycle times of 686 days, two had about 1 year cycle times, and others had cycle times of three months or less. Some tubes were used as supplied, some were solutionized at 1373 Kelvins (1100°C.) in vacuum for 1.5 hours, some were solutionized and then sensitized at 923 Kelvins (650°C.) in vacuum for 24 hours, and some were oxidized at 623 Kelvins (350°C.) in air for 40 minutes.

Acid Cleaning Experiments

The acid cleaning study was performed in two parts. First, coupons were heat treated in three groups (no heat treatment, solutionized at 1373 Kelvins (1100°C.) for 1 hour, and solutionized followed by a 923 Kelvins (650°C.) anneal for 24 hours) in vacuum and then acid cleaned and passivated [7]. Second, coupons from the same lot were polished to 600 grit, air annealed at 1373 Kelvins (1100°C.) and then acid cleaned and passivated [7]. The material used for both parts of the acid cleaning study was from one lot of Type 304L stainless steel, having a carbon content of 0.013 wt %, as reported by the supplier mill analysis and verified by an independent vacuum fusion analysis (Leco method).

Coupons from the three sample groups in the vacuum annealed part were exposed to two different cleaning solutions (2 vol. % HF +10 vol. % HNO_3 in water, and 14 vol. % Nitraad® + 25 vol. % HNO_3 in water) and two different passivation solutions (25 vol. % HNO_3 in water and the same with 3 weight % sodium dichromate), representative of those used in typical stainless steel manufacture [2, 3, 6]. The samples were cleaned for 20 minutes and passivated for 30 minutes at ambient temperature, and rinsed with water immediately after each cleaning or passivation step. Representatives of each type of vacuum annealed coupon were exposed to every combination of acid cleaning and passivation, and control samples served to isolate effects of each step and solution.

Air annealed coupons from the same lot were exposed sequentially to: 1) mechanical polish to 600 grit paper, 2) anneal in air at 1373 Kelvins (1100°C.) for five minutes, quench in water, 3) acid clean in 2 vol. % HF + 10 vol. % HNO_3 in water, for 20 minutes at ambient temperature, and 4) passivate in 25 vol. % HNO_3 in water for 30 minutes at ambient temperature. After each step, a sample was removed from the sequence for surface microstructure characterization.

RESULTS

Hydride Exposure

No evidence of a potentially adverse reaction between hydride-forming powder and stainless steel was observed on any exposed surfaces on the retired beds, machined compatibility samples, or cycling systems container tubes. The hydride container surface of the longest cycle system experiment, which lasted 686 days, exhibited physical impregnation of the finely divided hydride powder into the surface (*Fig. 2)*, as did some of the retired prototype beds. Even in these cases, no evidence of a potentially adverse chemical or metallurgical reaction was observed. No difference in surface microstructures was observed among the variously treated tube samples (solutionized, solutionized and annealed, oxidized, and as received) exposed in the cycling system experiments.

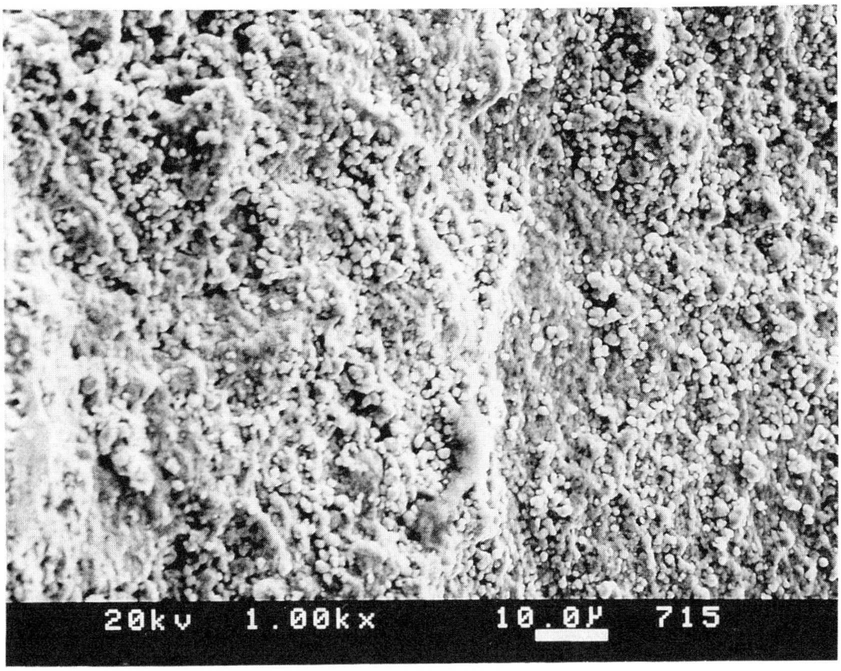

Figure 2. Micrograph of interior of tube exposed to $LaNi_{4.25}Al_{0.75}$ while cycling hydrogen for 686 days.

Some of the interior surface stainless steel microstructures had a morphology before exposure typical of the final machining process and others had intergranular ditching. No evidence exists from these observations that intergranular surface ditching forms during contact with hydrides. Results from the acid cleaning tests (below) show that intergranular ditching occurs during normal stainless steel fabrication.

Contact of both $LaNi_{4.7}Al_{0.3}$ and La_3Ni reduced or eliminated the darkening observed on the machined compatibility samples exposed to vacuum and helium/4% hydrogen mixture at elevated temperatures *(Fig. 3).* The darkening of the steel away from the compound likely occurs because of surface oxidation, and occurred even in the helium/ hydrogen mixture. The reduction in oxidation of the steel contacting the lanthanum compounds is attributable to the highly oxidizing nature of lanthanum. It is unknown whether the lanthanum compounds reduced existing oxide on the stainless steel surface or simply inhibited the growth of further oxide during exposure by local oxygen gettering. This phenomena is not expected to adversely affect the performance of the steel in hydride beds.

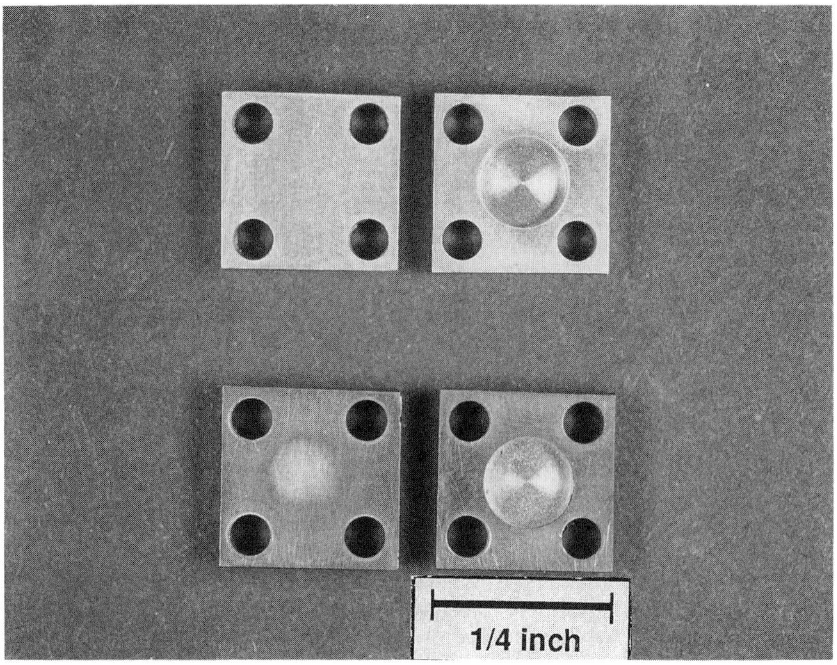

Figure 3. Photograph of compatibility sample exposed to $LaNi_{4.7}Al_{0.3}$ in vacuum at 923 Kelvins (650°C.). Top - Unexposed, Bottom - Exposed.

Acid Cleaning

Scanning electron microscopy images of the surface of the as-received coupons revealed extensive intergranular surface ditching. The as-received coupon surface microstructure did not change after exposure to any of the acid cleaning or passivation solutions, and was significantly more immune to acid attack than the vacuum annealed coupons.

Examination of air annealed samples after each step in the sequence polish/air anneal/acid clean/passivated revealed that intergranular surface ditching, identical to that found in the as-received coupons, formed during the acid cleaning step *(Fig. 4)*. An additional sample was identically treated except not air annealed, and no intergranular ditching or attack of any kind occurred. These results show that air annealing followed by acid cleaning is sufficient to form intergranular surface ditches on Type 304L stainless steel.

a. mechanical polish

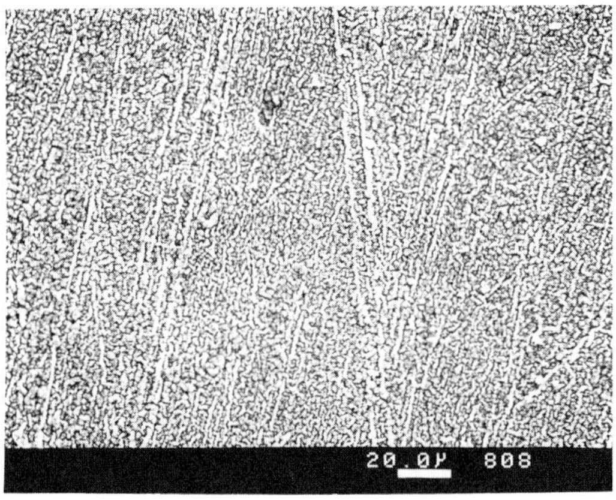

b. air anneal

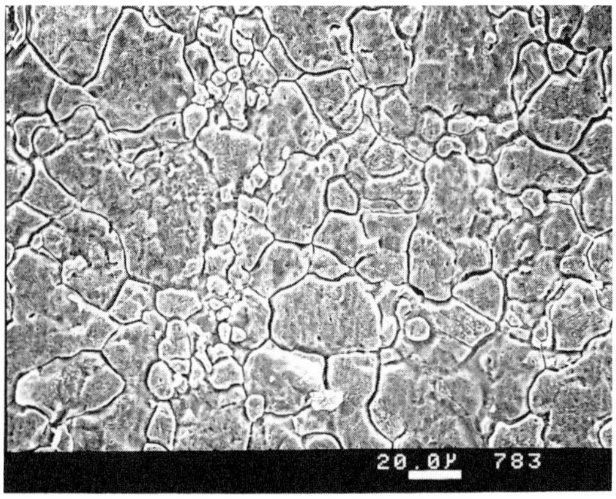

c. acid clean

Figure 4. Micrographs of coupons after each step in the sequence: a. mechanical polish, b. air anneal, c. acid clean.

Vacuum annealed coupons surfaces exposed to acid cleaning and passivation had varying degrees of two features: uniform attack of the surface grain faces and deepening of the grain boundary ditches *(Fig. 5).* The degree of attack varied with the sample, however in general the attack of solutionized and 923 Kelvins (650°C.) annealed samples was worse than that for the solutionized only coupons. The resistance of the vacuum annealed samples to acid attack was clearly inferior to that of the as-received samples.

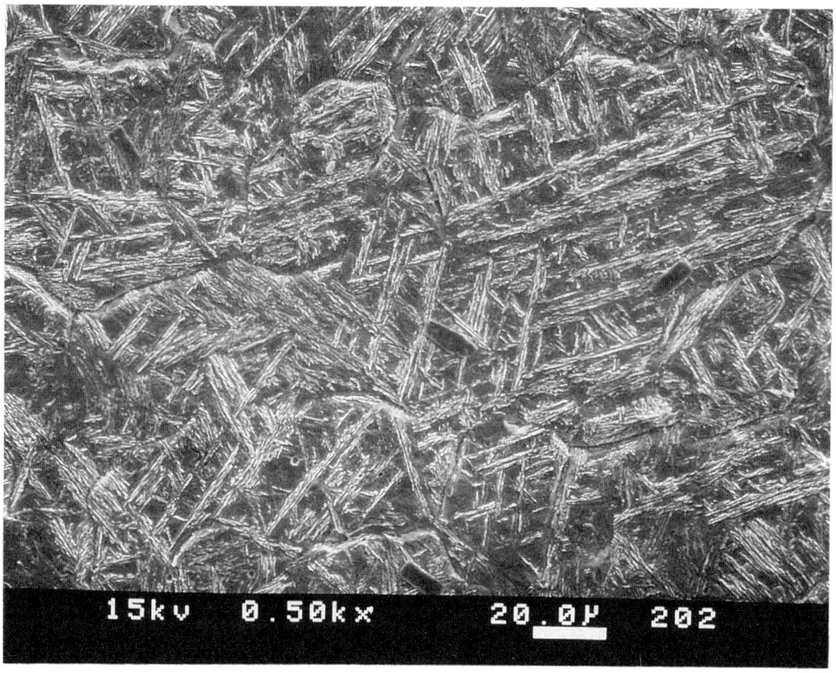

Figure 5. Micrograph of vacuum annealed sample exposed to acid cleaning and passivation.

Some vacuum annealed coupons were annealed at 923 Kelvins (650° C.) for one day to attempt to sensitize these samples. (The rate of sensitization is highest at approximately this temperature.) A normal electrochemical test for sensitization of stainless steels using oxalic acid (ASTM A-262) [4] revealed no indication of sensitization of any of these samples *(Fig. 6)*; sensitized samples would have formed intergranular surface ditches identical in morphology to that under study here. The carbon concentration in this material was so low (0.013 wt%, above) that even this anneal could not cause appreciable sensitization. Lower carbon concentrations have been shown to significantly reduce the rate of sensitization in Type 304 steels in another study [5].

Figure 6. Micrograph of coupon after oxalic test for sensitization (ASTM A-262).

DISCUSSION

The most likely cause of intergranular surface ditches on Type 304L stainless steel during air annealing and acid cleaning is impurity segregation to the line of intersection between grain boundaries and the surface, and reaction of these impurities with oxygen during the air anneal. Acid cleaning removes the compounds formed, leaving a void at the grain boundary/surface intersection (the observed ditched microstructure) and an attack resistant passive film on the stainless steel. The inability of the 923 Kelvins (650°C.) 24 hour vacuum anneal to sensitize the vacuum annealed samples, together with the ability to form intergranular ditches by air annealing and acid cleaning identically prepared samples, is evidence that chromium carbide precipitation does not play a role in this type of ditch formation. The thick visible black oxide surface film formed during air annealing prevented both X-ray analysis and Auger Electron Spectroscopy from identifying the compounds formed.

CONCLUSIONS

Microstructural observations of Type 304L stainless steel surfaces exposed to hydrides ($LaNi_{5-x}Al_x$ (x=0.3, 0.75) and Kieselguhr supported palladium), La_3Ni, or various acids (simulating cleaning and passivation) after different heat treatments (none, vacuum annealing, air annealing) allow the following conclusions to be drawn.

1) No compatibility problem exists between austenitic stainless steel containers and either $LaNi_{5-x}Al_x$ (x=0.3, 0.75) or Kieselguhr supported palladium in hydride storage systems environments. Although the fine hydride powder expanded and became impregnated into the stainless steel container wall, no reaction between the hydrides and steel was observed.

2) Contact of $LaNi_{5-x}Al_x$ and La_3Ni powders with stainless steel reduced or eliminated stainless steel oxidation at elevated temperature. This is not expected to adversely affect the performance of the hydride bed containers.

3) The intergranular ditches found on the interior surfaces of retired prototype storage beds and other as-received austenitic stainless steel coupons formed during normal fabrication of the original components of the bed vessel. These ditches are caused by acid cleaning after annealing in air.

4) The intergranular ditching is not related to typical stainless steel sensitization (formation of chromium carbides at grain boundaries).

5) Air annealed and acid cleaned stainless steel, having surface ditches, is more resistant to attack by nitric and nitric hydrofluoric acid mixtures than vacuum annealed stainless steel or freshly machined stainless steel.

ACKNOWLEDGMENTS

The help of D.P. Healy, A. S. Holsten, A. P. Watkins, C.B. Johnson, E.E. Kemper, D. Hopkins, D. Morgan, R.R. Lathrop, K.A. Dunn, C.L. Angerman, T. Motyka, L.K. Heung, and W.C. Mosley in performing this work is gratefully acknowledged.

This article was prepared in connection with work done under Contract No. DE-AC09-88SR18035 with the U.S. Department of Energy.

REFERENCES

1) M.S. Ortman, L.K. Heung, A. Nobile, and R.L. Rabun, III. "Tritium Processing at the Savannah River Site: Present and Future". *J. Vac. Sci. Technol. A.* 8 (3), pp 2881-2889 (May/June 1990).

2) *Standard Practice for Cleaning and Descaling Stainless Steel Parts, Equipment, and Systems.* Designation A 380-78, in "1987 Annual Book of ASTM Standards, vol. 1.03- Steel: Plate, Sheet, Strip, Wire". American Society for Testing and Materials, Philadelphia, PA , pp 201-217 (1987).

3) *Cleaning and Finishing of Stainless Steel.* Metals Handbook, 9th Edition, Vol. 5: Surface Cleaning, Finishing and Coating, American Society for Metals, Metals Park, OH, pp 553, 559-560 (1982).

4) *Standard Practice for Detecting Susceptibility to Intergranular Attack in Austenitic Stainless Steels.* Designation A 262-86, in "1987 Annual Book of ASTM Standards, Vol. 1.03- Steel: Plate, Sheet, Strip, Wire", American Society for Testing and Materials, Philadelphia, PA, pp 67-95 (1987).

5) A.J. Sedriks. *Corrosion of Stainless Steels.* John Wiley & Sons, New York, p 122 (1979).

6) Turco Products, Inc., Westminster, CA 92684.

7) *Heat Treating of Stainless Steel.* Metals Handbook, 9th Edition, Vol. 4: Heat Treating, pp 623-624, American Society for Metals, Metals Park, OH (1982).

RECRYSTALLIZATION PHENOMENA IN HIGHLY DEFORMED AND SENSITIZED 304 STAINLESS STEELS

J.G. Maldonado[1] and L.E. Murr

ABSTRACT

Austenitic 304 stainless steel uniaxially deformed to a 40% true strain level has shown some interesting characteristics in terms of the microstructural changes occurring during heat treatment at 670°C (for aging times up to 10h) to produce sensitization in the material. The primary objective of this study was to investigate the effects of strain-induced α'-martensite on transgranular (TG) carbide precipitation. Strain-induced α'-martensite (produced during deformation) showed a transformation/annihilation with thermal treatment. A different phase was observed to have formed in some of the previously strain-induced α'-martensite locations. Diffraction studies showed that the phase was also martensite, but its morphological characteristics were indicative of the lath-martensite phase and a product of aging-induced recrystallization and growth processes. After the shortest heat treatment (0.1h) the microstructure showed very significant changes in which fine-intermixed regions containing fine-grained austenite and lath martensite were observed. This apparent recrystallization phenomena, which continues to occur in longer heat treatments, appears to have a marked importance, and plays a major role in the sensitization behavior of austenitic 304 stainless steel. The highly deformed condition of the material provides a suitable mechanistic environment for the recrystallization process to occur. Also, the intermixing of the strain-induced α'-martensite phase in the austenitic matrix may have a significant bearing on the observed recrystallization phenomena.

---

[1]Department of Metallurgical and Materials Engineering, The University of Texas at El Paso, El Paso, TX 79968

INTRODUCTION

Austenitic 304 stainless steel (SS) is an ideal candidate for the containment, transport and disposal of waste materials, except for the problem of sensitization that these materials undergo when they are isothermally treated or continuously cooled through the 500-850°C temperature range [1]. This sensitization problem is further worsened and accelerated when a deformation process is involved. Strain-induced α'-martensite, produced during plastic deformation of the 304SS, nucleates at microshear band intersections in the material. These microshear band (composed of overlapping arrays of stacking faults and deformation twins) intersections define a precise set of microstructural conditions necessary for the nucleation of the α'-martensite as described previously by Olson and Cohen [2]. Also, these same types of intersections are known to be preferential sites for the nucleation of TG chromium-rich carbides in sensitized SS. Because of the similarities of the nucleation of α'-martensite and TG carbides, Briant and Ritter [3] suggested that α'-martensite served as a precursor to the nucleation of the carbides, and this was responsible for the much higher degree of sensitization (DOS) observed in deformed 304SS as compared to deformed 316SS (316SS does not form α'-martensite during uniaxial plastic deformation).

We have proven that α'-martensite is not a precursor to TG carbide precipitation [4], but it may contribute to the overall process by providing the necessary microstructural conditions for a localized recrystallization of the matrix material. A similar recrystallization phenomena has been observed by Fakhutdinov et al. [5] in austenitic 18-10 stainless steel, in which they observed submicron austenite grain colonies originating in volumes that were originally occupied by strain-induced martensite. This localized recrystallization phenomena appears to have a very definite effect on the sensitization behavior of 304SS in contrast to 316SS as will be shown in the results of the microstructural studies. The recrystallized microstructure, consisting of fine-grained austenite and lath martensite, represents a fundamental difference in the mechanisms of TG carbide precipitation operating in 304 and 316SS. We believe that establishing the effects of the α'-martensite phase on the observed localized recrystallization is essential to understanding the fundamental issues affecting the anti-corrosive properties of austenitic stainless steels.

The microstructural evolution with thermal treatment, after uniaxial deformation, was observed in a Hitachi H-8000 scanning transmission electron microscope (STEM). Corrosion studies were conducted to correlate the microstructural changes with the susceptibility to corrosion of the material, produced by the thermomechanical treatment. Parallel studies were also conducted on 316SS, which was used in some respects as a control material

EXPERIMENTAL

High carbon 304 and 316SS materials (0.05 - 0.06% C), of composition shown in table 1 were used in this study. Strips, measuring 125 mm in length and 12.5

mm wide, were cut from a 6.3 mm thick plate material. The strips were solution annealed at 1100°C for 1h to eliminate any carbides introduced during mill processing, and to assure that a uniform microstructure existed. The strips were then uniaxially deformed to a 40% true strain level. After deformation, the strips were cut into smaller specimens measuring 12.5 mm by 12.5 mm, and isothermally treated at 670°C from 0-10h. The samples were cold mounted, ground and polished to a 1 μm diamond paste finish. A commercially available ferritescope was utilized to quantify the amount of martensite since the martensitic phase is ferromagnetic in nature. Magnetic measurements have been discussed previously by other researchers [6].

	%C	%Mn	%P	%S	%Si	%Ni	%Cr	%Mo	%Co	%Cu	%N
304SS	.051	1.47	.025	.014	.38	8.22	18.29	.19	.11	.19	.085
316SS	.057	1.86	.024	.019	.58	13.48	17.25	2.34	.02	.10	.030

Table 1. Chemical composition of the 304 and 316SS used in this study (weight percent).

The electrochemical potentiokinetic reactivation (EPR) test was used to determine the degree of sensitization (EPRDOS) produced in the samples after the thermomechanical processing. For more information on the EPR testing technique, the reader is referred to Clark et al. [7]. The microstructural evolution and the TG carbide precipitation process were studied in a Hitachi H-8000 scanning transmission electron microscope (STEM) operated at 200 kV. Convergent beam electron diffraction (CBED), also known as microdiffraction, was used extensively to identify individual phases within the observed microstructure. Some limited x-ray dispersive spectroscopy was also employed for the identification of chromium-rich carbides.

RESULTS & DISCUSSION

MICROSTRUCTURAL EVOLUTION WITH THERMAL AGING TREATMENT

Figure 1 illustrates the microstructural features observed in the as-deformed 304SS material. The planar intersections (microshear band intersections) which are observable, composed of heterogeneous bundles of stacking faults and deformation twins, show the presence of the strain-induced α'-martensite phase. This martensite was observed to exist only in a fraction of the intersections observed. The total relative amount of α'-martensite is directly proportional to the strain level used in this study. Also, the total extent of martensite transformation is dependent on the strain state and strain rate, as previously discussed by Murr et al. [8]. The strain-induced martensite was identified with the aid of microdiffraction. The dark-field image shows distinctively the martensitic phase present (figure 1). The ferritescope studies showed that the as-deformed sample contained the maximum amount of the martensitic phase (10% bulk, [4]).

The 0.1h isothermally aged 304SS sample (figure 2 (a)) failed to show the strain-induced martensite, but instead an elongated structure was observed to occur in the previously occupied martensite volume intersections. Because of its morphological characteristics and extent of growth, even outside the intersection volume, we concluded that this phase was probably the product of an annihilation/transformation process of the strain-induced α'-martensite phase. This

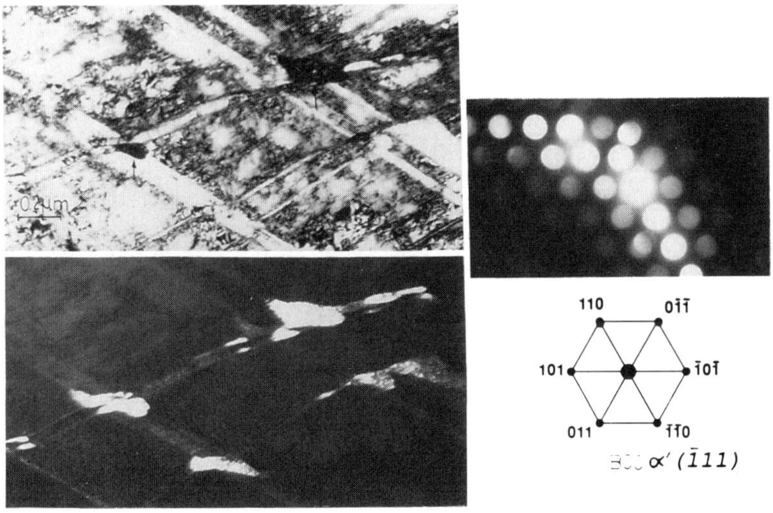

Figure 1. Microstructural features observed in the as-deformed 304SS sample. Strain-induced α'-martensite can be observed to have formed at twin-fault intersections in the material.

phase was also determined to be a martensitic (bcc) structure. Even this aging-induced, newly created or recrystallized, martensite was observed to decrease with an increase in the heat treatment time. This was also confirmed with the ferritescope studies which showed that the relative amount of the total volume of martensite falls quickly after the 0.1h treatment, and it is below the detection limit of the instrument in all subsequent heat treatments (> 0.1h, [4]).

After the shortest heat treatment (0.1h), the 304SS-0.4h-sample showed a dramatic change in its microstructure. Figure 2(b) shows the presence of clustered cell-like zones which seem to originate from the shear-band intersection regions. Some of the cellular phases encompassed entire sections of grains, and in some cases the phases seem to be oriented in specific directions and often appeared as bands in the microstructure. The diffraction pattern shows a complex pattern due to the intermixing of fine-grained austenite and the lath martensite (figure 2(b)). As a way of proving the previous point, figure 3 shows two individual phases which were observed (close-up view), and that were identified through microdiffraction as being fine-grained austenite and lath martensite respectively.

In contrast, the 316SS material did not show any significant microstructural changes with increasing aging treatment.

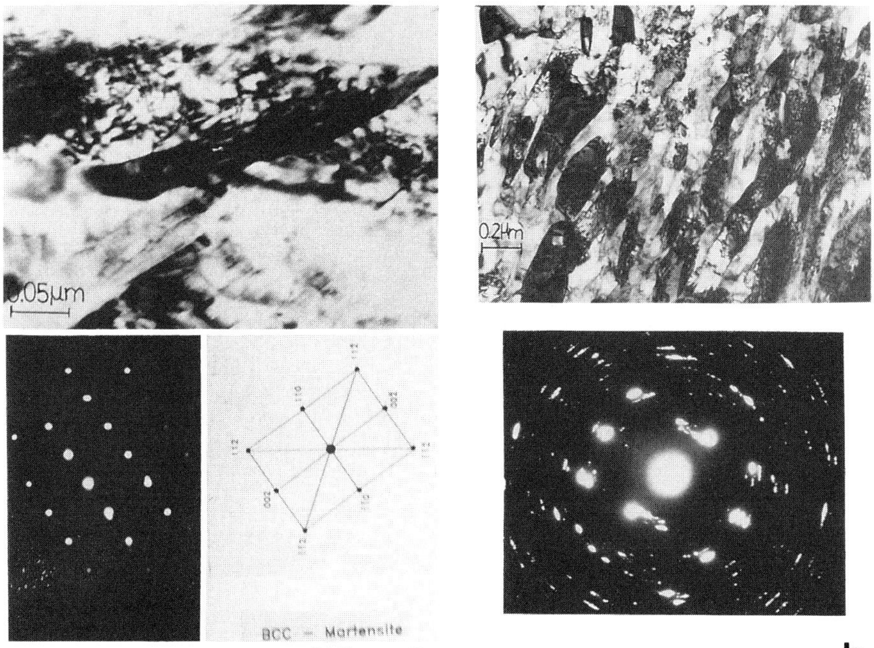

a b

Figure 2. (a) Elongated structure formed after the shortest heat treatment, 40%-670°C-0.1h 304SS sample. (b) Clustered regions found at twin-fault intersections which are composed of a fine-intermix of recrystallized austenite and lath martensite, 40%-670°C-0.4h 304SS sample.

RECRYSTALLIZATION EFFECT ON CARBIDE PRECIPITATION AND SENSITIZATION OF AUSTENITIC STAINLESS STEELS

Twin-fault intersections were observed to be preferred locations for TG carbide precipitation in 304 and 316SS [4] . This precipitation continued to occur even in those samples in which no martensite was observed, which proves that the martensitic phase is not a precursor to carbide precipitation. However, the martensite seems to provide requisite sites for the observed recrystallization. These localized recrystallization zones provided additional nucleation sites for chromium-rich carbides. Figure 4(a) shows one of these recrystallized zones (composed of fine-grained austenite and lath martensite) in which carbides are observed to have nucleated from the interphase boundary within the intermixed region or zone. The dark-field image gives a better view of the individual carbide precipitates. Carbide precipitation is also observed in samples in which no lath martensite is present, and in which the microstructure consists of fine-austenitic grains (figure 4(b)).

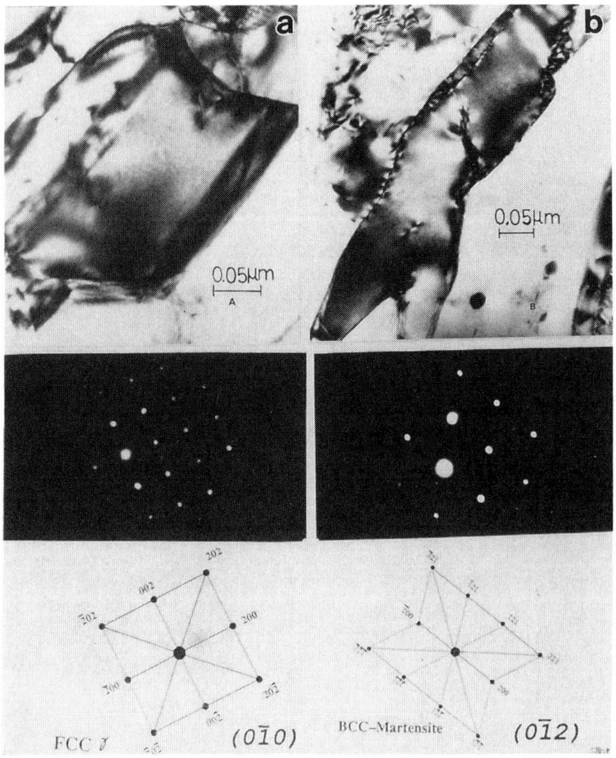

Figure 3. Close-up view of the recrystallized zone shown in figure 2(b), 40%-670°C-0.4h 304SS sample. Microdiffraction established the presence of fine grained austenite (a), and lath martensite (b).

The observed recrystallization seems to have a definite effect on the sensitization behavior of 304SS. Figure 5 illustrates the sensitization behavior of both 304 and 316SS. The much higher degree of sensitization observed in the deformed 304 in contrast to 316SS can be explained in terms of the additional precipitation sites provided by the recrystallized microstructure. If we consider the fact that 304SS has an increasing α'-martensite volume fraction at shear-band intersections initially, then the accelerated and larger EPRDOS may be a reflection of the recrystallization of α'-martensite which occurs in the 304SS material. This effectively reduces the "grain" size at local regions near shear-band intersections, some of which may coalesce to form larger regions. These refined interphase regions will become chromium-depleted like the grain boundaries (regular grain boundaries in the parent austenite microstructure), and the overall DOS will rise

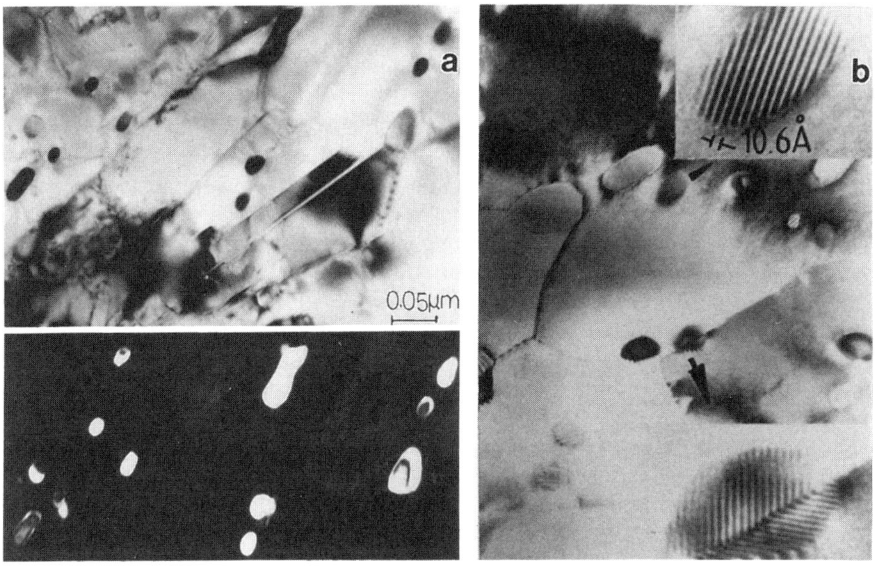

Figure 4. (a) Transgranular carbides on the boundaries of the recrystallized microstructure, 40%-670°C-0.4h 304SS sample. (b) TG carbides on fine-grained austenite boundaries with detailed views of the carbide lattices, 40%-670°C-1h 304SS sample.

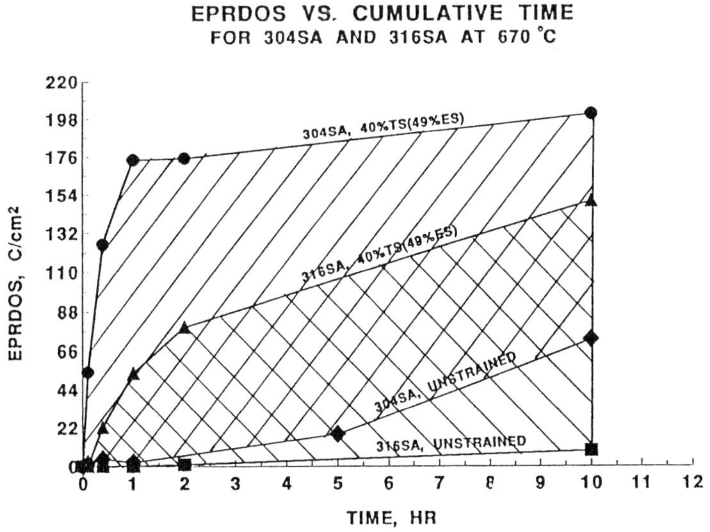

Figure 5. Degree of sensitization plotted as a function of aging time for 40%-670°C 304 and 316SS.

rapidly. While this may explain the accelerated and larger sensitization in 304SS, there is also the difference between the 304 and 316 stainless steels in the unstrained, aged condition. This may be due in part to the difference in chemistry which, to a larger extent, is responsible for the fact that the stacking fault free energy of the 316SS is roughly twice that for the 304SS (~ 21 mJ/m^2) [9]. The 316SS contains a larger amount of molybdenum than 304SS which is known to accelerate chromium-diffusion kinetics and also delays the start of sensitization, though we believe that this effect is minimal in comparison to the newly created nucleation sites for precipitation that occur with the microstructural breakdown (recrystallization).

The recrystallized microstructure, consisting of fine-grained austenite and martensite, represents a fundamental difference in the mechanisms of TG precipitation operating in 304 and 316 stainless steels. However, the apparent recrystallization phenomena could also be just a growth process, and not a recrystallization and growth process. This can be explained in terms of small volumes of α'-martensite formed during deformation (that originate from within the shear-band intersections) where the α'-martensite is surrounded by untransformed austenite. This inherent intermixing of α' and γ (described as a salt and pepper structure by Murr et al.[8]) when given the necessary thermal impetuous would grow.

Figure 6 is an attempt to more specifically illustrate the transgranular carbide precipitation differences in 304 and 316 stainless steel after deformation and aging. Energetics are key in the TG precipitation process. This fact can be better ascertained by observing that the recrystallized interphase interfaces have the same energy as the non-recrystallized austenitic grain boundaries [9].

CONCLUSIONS

While strain-induced α'-martensite is not a precursor to the precipitation/depletion in austenitic stainless steels, it seems to provide specific microstructural requirements for a localized recrystallization of the microstructure to occur. This apparent recrystallization, which is aging-induced and has been observed to occur in 304SS only, composed of fine-grained austenite and lath martensite, represents a fundamental difference in the carbide precipitation process between 304 and 316SS since the recrystallized interphase interfaces are favored locations for the precipitation process. The energetics of these recrystallized interphase interfaces is very similar to the energetics of the austenitic non-recrystallized grain boundaries.

ACKNOWLEDGEMENTS

This research was supported by the National Science Foundation (RIMI) HRD 9105065, and a Patricia Roberts Harris Fellowship (J.G.M.).

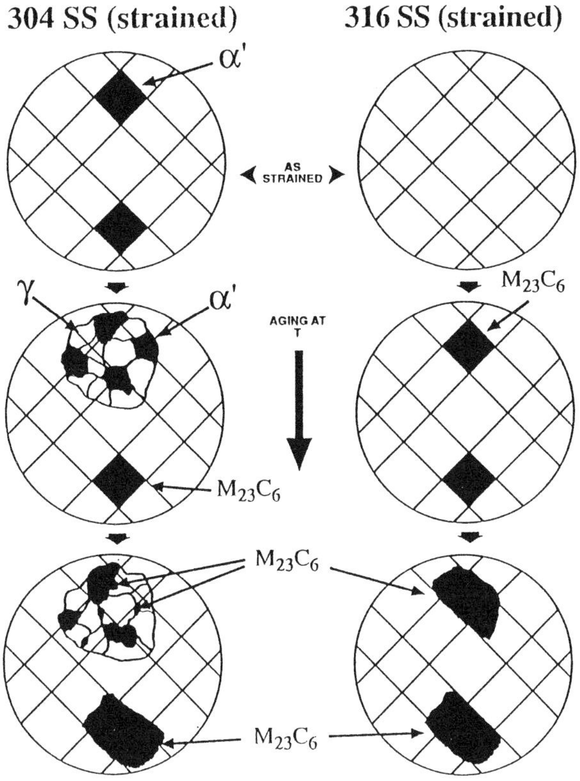

Figure 6. Schematic representation of TG carbide precipitation differences during aging of plastically deformed (uniaxial) 304 and 316SS.

REFERENCES

1. E.C. Bain, R.H. Aborn, and J.J.B. Rutherford, " The Nature and Prevention of Intergranular Corrosion in Austenitic Stainless Steels," **Transaction American Society Steel Treatise, Vol 21**, p. 481, 1933.

2. G.B. Olson, and M. Cohen, "Kinetics of Strain-Induced Martensitic Nucleation," **Metall. Trans. 6A**, p. 791, 1975.

3. C.L. Briant and A.M. Ritter, "The Effects of Deformation Induced Martensite on the Sensitization of Austenitic Stainless Steels," **Metall. Trans. 11A**, p.2009, 1980.

4. J.G. Maldonado, A.H. Advani, L.E. Murr, and W.W. Fisher "Strain-Induced Martensite Effects on Transgranular Carbide Precipitation and Chromium-Depletion in Austenitic Stainless Steels," **Proceedings of the International Metallographic Society 26th Annual Technical Meeting in Charleston, S.C.**, July 1993.

5. K.G. Farkhutdinov, R.G. Zaripova, and N.A. Breikina, "Submicrocrystalline 18-10 Stainless Steel: Structure Formation, Mechanical and Corrosion Properties," **Mater. Sci. and Eng., A174**, p. 217, 1994.

6. S.S. Hecker, M.G. Stout, K.P. Staudhammer, and J.L. Smith, " Effects of Strain State and Strain Rate on Deformation Induced Transformation in 304 Stainless Steel: Part I. Magnetic Measurement and Mechanical Behavior," **Metall. Trans. 13A**, p. 619, 1982.

7. W.L. Clark, R.L. Cowan, and W.L. Walker, **"Comparative Methods for Measuring Degree of Sensitization in Stainless Steels," ASTM STP 656**, p. 99, 1978.

8. L.E. Murr, K.P. Staudhammer, and S.S. Hecker, "Effects of Strain State and Strain Rate on Deformation-Induced Transformation in 304 Stainless Steel: Part II. Microstructural Study," **Metall. Trans. 13A**, p. 627, 1982.

9. L.E. Murr, **Interfacial Phenomena in Metals and Alloys**, Addison Wesley, Reading, Mass., 1975, (reprinted 1991).

CORRELATION BETWEEN SENSITIZATION AND CR-DEPLETION IN HIGH-NITROGEN AUSTENITIC STAINLESS STEELS

J.W. Simmons, B.S. Covino, Jr., S.D. Cramer, and J.S. Dunning (1)

ABSTRACT

In a U.S. Bureau of Mines study, sensitization of a high-nitrogen stainless steel (nominally Fe-19Cr-5Ni-5Mn-3Mo-0.024C-0.69N) due to Cr_2N precipitation at 800 and 900°C was measured using the single-loop electrochemical potentiokinetic reactivation (EPR) test. The EPR test results were subsequently correlated with direct measurement of Cr-depletion using analytical electron microscopy (AEM). The high-nitrogen stainless steel was susceptible to matrix Cr-depletion and sensitization caused by grain boundary, cellular, and transgranular Cr_2N precipitation. The degree of sensitization increased with increasing aging time at 800°C up to a maximum value at 30 hours. Longer aging times resulted in healing. Although aging at 900°C produced much higher volume fractions of Cr_2N than aging at 800°C, samples aged at 900°C were not sensitized. AEM measurements showed that samples containing Cr-depletion zones with minimum Cr-levels below about 14 wt% were subject to reactivation (sensitization) in the EPR test. Healing of specimens aged at 800°C occurred because the Cr-levels within the depleted matrix, in thermodynamic equilibrium with the nitrides, increased as nitride volume fractions increased and lowered the nitrogen activity. Samples aged at 900°C suffered from Cr-depletion, but the Cr-minimums produced were above the threshold necessary for sensitization.

---

(1) U.S. Bureau of Mines, Albany Research Center
1450 Queen Ave. S.W., Albany, OR 97321

INTRODUCTION

High-nitrogen stainless steels (generally > 0.4 wt% N) are emerging as a new class of engineering materials since they exhibit equal or better corrosion resistance than the 300 series alloys, and have yield and tensile strengths which are 300-400 percent higher without sacrificing ductility.[1,2] Although interstitial levels of nitrogen approaching 1 wt% can be retained in high-nitrogen austenitic alloys by proper thermomechanical processing, the resulting structures are thermally unstable and susceptible to the intergranular precipitation of Cr-rich nitrides (Cr_2N), much the same as the 300 series alloys are susceptible to Cr-rich carbide ($M_{23}C_6$) precipitation.[3-6] It is well-known that carbide precipitation leads to matrix Cr-depletion and sensitization, but the possible detrimental effects of nitride precipitation on the corrosion properties of high-nitrogen alloys have yet to be determined. This paper describes the relationship between the degree of sensitization, measured indirectly by the electrochemical potentiokinetic reactivation (EPR) test method (ASTM G 108-92),[7] and direct measurement of Cr-depletion by analytical electron microscopy (AEM).

EXPERIMENTAL METHODS

The high-nitrogen austenitic stainless steel used in this study had a nominal bulk composition (wt%) of Fe-19.3Cr-5.2Mn-5.2Ni-2.9Mo-0.03V, a carbon content of 0.024, and a nitrogen level of 0.69. This high-nitrogen material was produced using the pressurized electroslag remelting process, described elsewhere.[8] The material was forged at a temperature of about 1050°C and subsequently solution annealed at 1150°C for 2 hours, followed by water quenching. This annealing treatment resulted in a single phase structure with all of the nitrogen (0.69 wt%) dissolved interstitially in solid-solution and a grain size of 70 μm. Annealed materials were isothermally aged at 800 and 900°C for times ranging from 0.1 to 300 hours.

The single-loop EPR test was used to determine the degree of sensitization of the aged specimens. The test method consists of developing potentiokinetic curves of a polarized specimen by the use of a controlled potential sweep from the passive to the active region (reactivation). During the reactivation scan, non-sensitized stainless steels exhibit a very low current density resulting from the stability of the passive film. However, sensitized materials show a high current density due to the breakdown of the passive film in Cr-depleted regions near Cr-rich precipitates.[9,10] Sensitization is quantified by measuring the activation charge, Q, expressed in coulombs (given by the integrated area below the reactivation peak of the curve). The integral charge value, Q, is subsequently normalized to the exposed grain boundary area of each test specimen to obtain the degree of sensitization, Pa (coul/cm^2). Generally, the higher the degree of sensitization, the more susceptible the material is to intergranular corrosion and intergranular stress corrosion cracking. The EPR test uses a metallographically prepared specimen polished to a 1μm diamond finish and an electrolyte which is an aqueous solution of 0.5 M H_2SO_4 + 0.01 M KSCN, held at 30°C.

Direct measurement of Cr-depletion was accomplished by AEM utilizing an accelerating voltage of 200 kV and a focused electron probe of approximately 10 nm. Specimens were twin-jet electropolished at ambient temperature using a solution consisting of 144 ml methanol, 36 ml H_2SO_4, and 20 ml glycerin.

RESULTS AND DISCUSSION

Aged microstructures

Previous work on this material showed that although all of the nitrogen can be dissolved interstitially by solution annealing at 1150°C, the resulting structure is thermally unstable and susceptible to nitride precipitation within the temperature range of about 600-1075°C.[6] At lower aging temperatures (i.e., below 800°C), the precipitation of Cr_2N occurs almost exclusively at grain boundaries. At temperatures between 800 and 1000°C, Cr_2N precipitation occurs sequentially at grain boundaries, by cellular decomposition, and finally, by transgranular precipitation. Precipitation kinetics in this alloy are most rapid at temperatures between 900 and 950°C.

Cellular precipitation of Cr_2N is a discontinuous precipitation reaction which occurs by the decomposition of the nitrogen supersaturated austenite matrix (γ_1) to an austenite of a lower nitrogen and chromium concentration (γ_2), and Cr_2N. The reaction can be described as $\gamma_1 \Rightarrow \gamma_2 + Cr_2N$. Cellular precipitation products are lamellar and have a morphology which is very similar to pearlite formed in carbon steels. This discontinuous precipitation reaction has been studied extensively by Kikuchi and co-workers for Fe-Cr-Ni-N stainless steels and the reader is referred to the literature for more details.[3,11]

For the purposes of this study, samples were aged at 800 and 900°C which resulted in a wide variety of microstructures. Optical micrographs illustrating the Cr_2N precipitation morphologies produced after aging for 1, 10, and 100 hours at 800°C, and 1 hour at 900°C, are shown in Figure 1. Aging at 800°C produced primarily grain boundary precipitation at shorter aging times (1 hour or less) followed by increasing amounts of cellular precipitation with increasing aging time, and at times greater than 10 hours, also included transgranular Cr_2N. Precipitation kinetics were much more rapid at 900°C than at 800°C and continuous grain boundary nitride coverage occurred after less than about 0.1 hour (6 minutes) at 900°C. Aging at 900°C for 1 hour resulted in 29 vol% cellular precipitates and 43 vol% after 10 hours (total volume fraction of γ_2 + Cr_2N).

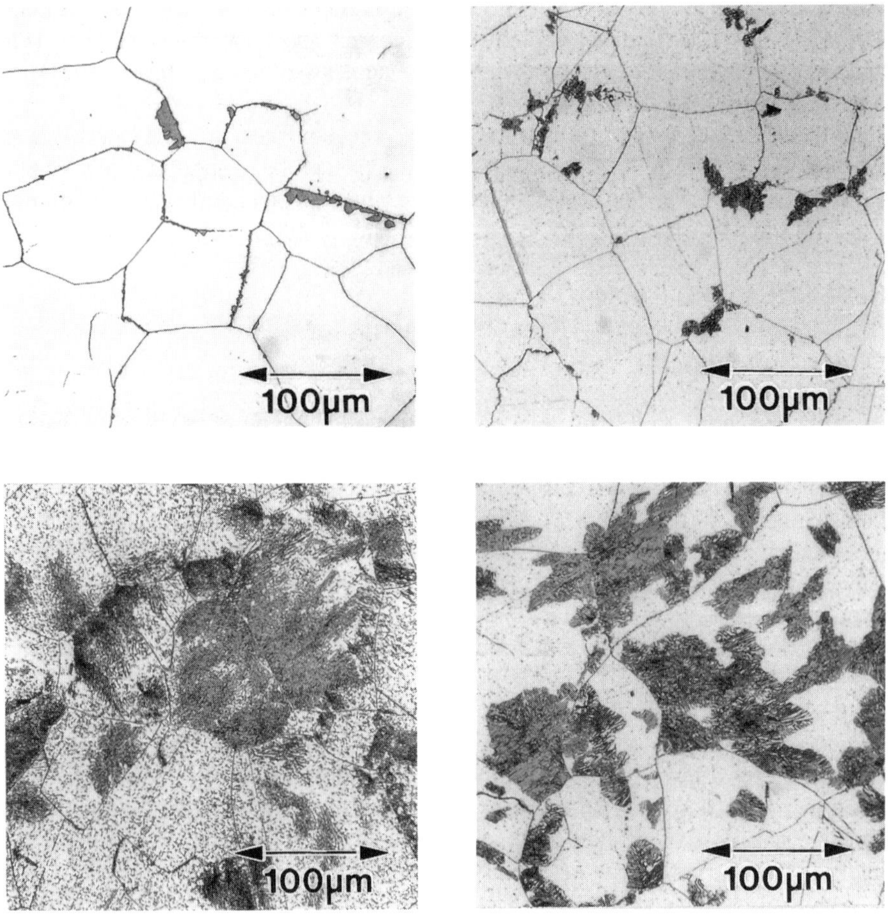

Figure 1. Optical micrographs of samples aged at 800°C for (a) 1 hour, (b) 10 hours, (c) 100 hours, and (d) sample aged at 900°C for 1 hour.

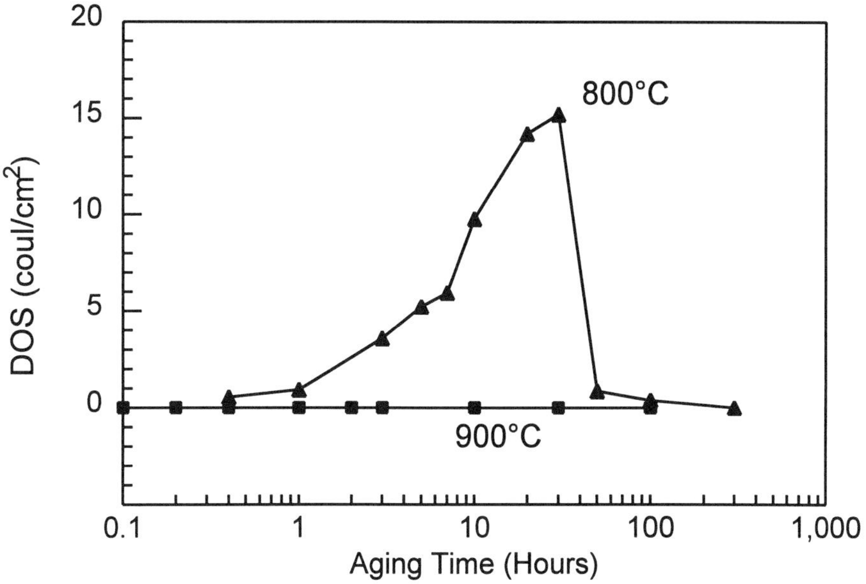

Figure 2. Degree of sensitization (DOS) determined by EPR test as a function of aging time.

Sensitization Susceptibility Measured by EPR Testing

The degree of sensitization of samples aged at 800 and 900°C, as a function of aging time, is shown in the graph of Figure 2. Aging at 800°C resulted in material sensitization with the degree of sensitization increasing with increasing aging time up to a maximum value at 30 hours of aging. Aging at 800°C for times longer than 30 hours resulted in decreasing sensitization values which decreased continuously with further aging up to 300 hours. The reduction of sensitization with increased aging time after a peak degree of sensitization has been reached is known as "healing." Aging for 300 hours at 800°C resulted in essentially complete healing of the material. Healing is a well known phenomena in carbon-alloyed stainless steels.[12-14] It should be pointed out that although samples aged for 100 and 300 hours at 800°C did reactivate during EPR testing, the integrated charge values for each specimen were extremely low, less than 0.4 coul/cm^2 for the 100 hour specimen and less than 0.1 coul/cm^2 for the 300 hour specimen. Therefore, in practical terms, these samples were completely healed.

Samples aged at 900°C for times ranging from 0.1 hours (6 minutes) to 100 hours did not reactivate during EPR testing and, therefore, were not sensitized. Aging at 900°C did not result in material sensitization in spite of the much greater total volume of Cr_2N precipitation which occurred at 900°C compared to 800°C. The explanation for this behavior will be discussed later.

Direct Measurement of Cr-Depletion and Correlation to Sensitization

Chromium-depletion profiles were taken primarily in two regions of the austenite adjoining Cr_2N precipitates: (1) normal to the grain boundaries, between nitride precipitates and (2) in samples containing cellular precipitates, Cr-profiles were measured across the boundary between the untransformed austenite and the austenite product of the cellular precipitate reaction, between Cr_2N lamellae. Grain boundary and cellular precipitate regions typical of those analyzed are shown in the TEM micrographs of Figure 3.

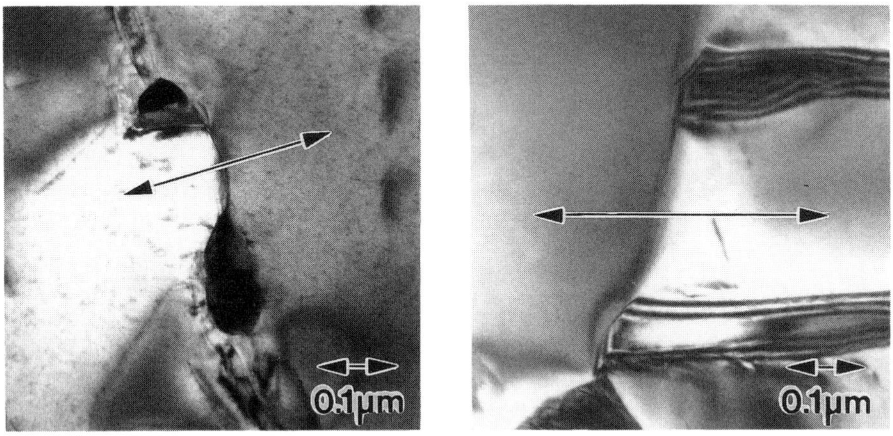

Figure 3. TEM micrographs of (a) grain boundary and (b) cellular precipitate interface regions across which Cr-depletion profiles were taken.

The average minimum Cr-concentrations measured at grain boundaries and at matrix/cell interface boundaries in samples isothermally aged at 800 and 900°C are shown in Figure 4. Sensitization, defined by reactivation and corrosive attack during EPR testing, occurred in samples with minimum chromium levels less than about 14wt% Cr. Samples which contained large volumes of Cr_2N, but had minimum Cr-levels above 14 wt%, were not susceptible to attack in the EPR test. These results correlate well with other work in which it was reported that, for 304 and 316 type stainless steels, the EPR test attacks regions where the Cr concentration is less than about 13.5 wt%.[14]

Although the measured values for Cr-depletion in samples aged at 100 and 300 hours at 800°C were similar to those measured in the samples aged for 1 hour at

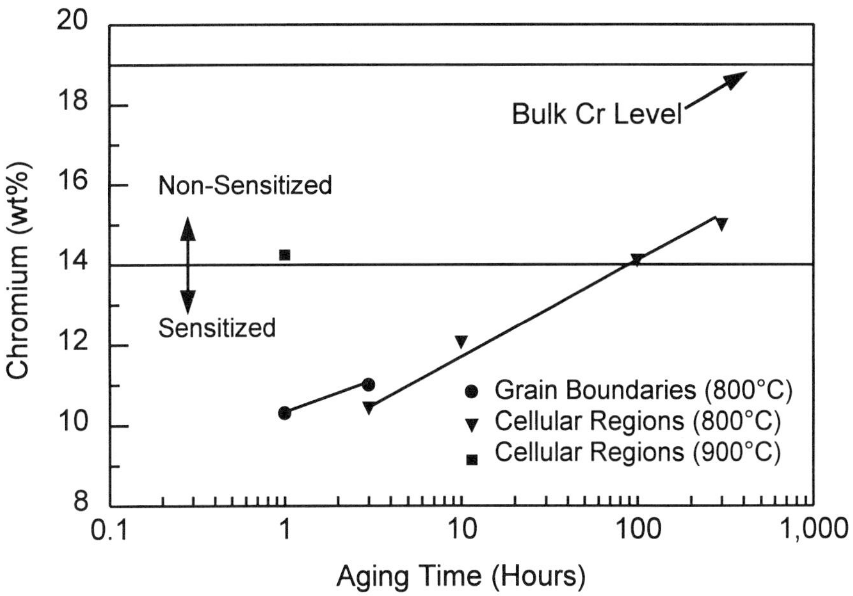

Figure 4. Minimum Cr-depletion values measured within austenite matrix adjacent to grain boundary and cellular nitride precipitates in samples aged at 800 and 900°C.

900°C, the former samples reactivated during EPR testing. The integrated charge values associated with these "healed" specimens, were, however, extremely small and their overall sensitization values below 0.5 coul/cm^2, as mentioned previously. The reactivation which occurred in the samples aged at 800°C may have had something to do with the fact that materials aged at 900°C never suffered from sensitization, while those aged at 800°C for long times, were recovering from sensitization. The long diffusion paths associated with the cellular regions in samples aged for long times at 100 and 300 hours at 800°C may have caused the rate of healing to vary from one region of the sample to another. This would not have been a factor for samples aged at 900°C since the initial Cr-minimums were above the threshold required for sensitization.

The increasing Cr-minimums measured with increasing aging time at 800°C are evidence of the healing process mentioned previously. In carbon alloyed steels Strawstrom and Hillert[12] found that the healing process could be accounted for by the increase in matrix chromium content at the carbide-matrix interface caused by changes in the carbon activity in the steel. The healing process occurring in nitrogen-alloyed stainless steels can be explained using an analogous argument. As precipitation progresses in the material, the bulk level of the interstitial element involved in the sensitization process, either carbon or nitrogen, decreases, thus decreasing the activity of the interstitial element. It has been predicted that the chromium concentration in the matrix in thermodynamic equilibrium with the nitride, increases with decreasing nitrogen activity.[6] Thus, healing of the sensitized region initiates from the very beginning of precipitation and eventually a Cr-minimum concentration is reached which is above the threshold value required for attack in the EPR test. Additionally, the EPR test results and Cr-depletion measurements of samples aged at 800 and 900°C indicate that the thermodynamically controlled Cr-concentration in the matrix, adjacent to the nitrides, increases with increasing temperature. This is analogous to what has been observed for carbide precipitation in 304 and 316 stainless steels.[14]

Measurements from specimens heat treated at 800°C for 3 hours indicated that Cr-minimums at grain boundaries and matrix/cell interface boundaries are equal. This is expected since the boundaries are of similar type and the Cr-diffusivities within these regions are much faster than that within the matrix. The Cr-depletion is therefore controlled by the bulk Cr-diffusivity and not the interface. Values measured at these interfaces should represent the thermodynamic equilibrium Cr-value for the matrix/nitride interface and should be equal regardless of precipitate morphology.

A typical Cr-depletion profile taken normal to the grain boundary in a sample aged for 1 hour at 800°C is shown in Figure 5. In this case, the total width of the depletion zone was about 400 nm. The Cr-depletion profile shown in Figure 5 is not symmetric with the grain boundary. This is typical of samples in which grain boundary migration occurs concurrently with precipitate growth.[13] Many of the Cr-depletion profiles measured in samples aged at 800°C were asymmetric. In the material studied, significant grain boundary migration occurred in samples heat treated at temperatures of 800°C and higher. In some areas, the minimum chromium value was actually measured up to several hundred nanometers away from the grain boundary. Additionally, for a given thermal treatment, variations were measured in the minimum chromium values present at different grain boundaries. This has been reported by other researchers as well, for Cr-depletion associated with carbide precipitation.[13] Cr-minimums measured in specimens aged for 1 hour at 800°C ranged from about 10-11.5 wt% Cr and had an average value of 10.3 wt%. This range of values shows very little scatter considering that the standard error in the measurements due to counting statistics alone is about 1 wt%.

A chromium profile taken across an untransformed matrix/cellular precipitate boundary interface, typical of specimens aged for 10 hours at 800°C, is shown in Figure 6. In the example shown in Figure 6, the minimum Cr-level measured was 11.7 wt%. Overall, the minimum chromium values measured within cellular precipitate regions varied from 10.8 to 12.8 wt%. No measurable depletion zone was detected ahead of the growth front and the Cr-content measured near the interface, within the cell, was consistent with those measured within the cell, away

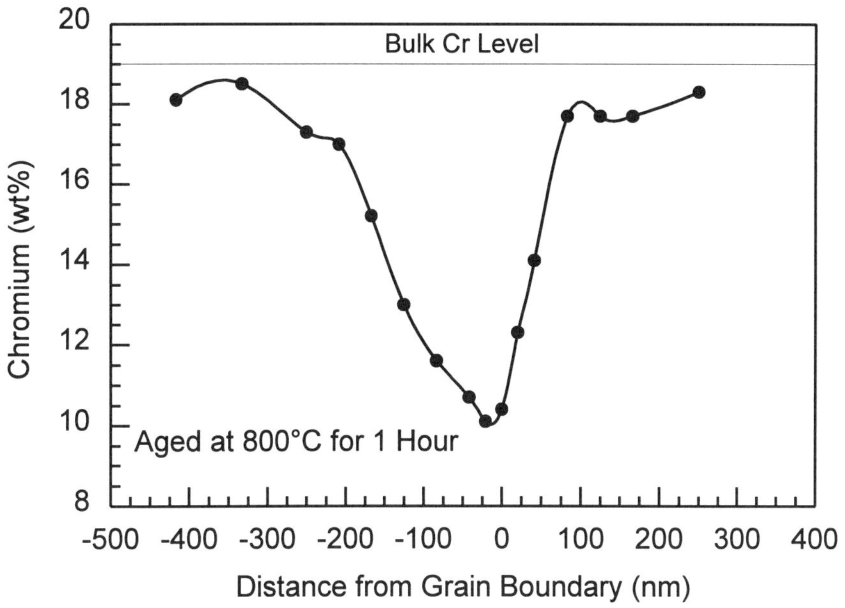

Figure 5. Cr-depletion profile taken across a grain boundary, typical of those measured in samples aged at 800°C for 1 hour.

from the interface. Although the overall degree of sensitization in samples aged for 10 hours at 800°C was much greater then after aging for 1 and 3 hours, the Cr-minimums were higher after the longer aging treatment, consistent with the healing phenomena. As long as the Cr-level is below the threshold required for reactivation in the EPR test, the overall degree of sensitization is primarily controlled by the volume of depleted material present within the sample. This area increases with increasing Cr_2N precipitation (e.g., increasing time at temperature).

The healing effect is further illustrated by the profile shown in Figure 7, taken across a matrix/cellular precipitate boundary interface in a specimen aged for 100 hours at 800°C. The level of chromium measured within cellular regions of this sample varied from a minimum of about 13.4 to a high of 15.2 wt%. In specimens aged for 100 and 300 hours at 800°C, there was virtually no untransformed matrix material left in the sample. Regions of the sample which did not transform by discontinuous precipitation to form the cellular precipitation products, contained significant transgranular precipitation. Measurements taken within the matrix of the transgranular precipitate regions indicated that the entire matrix of the material within transgranular precipitate regions contained chromium levels from about 14.5 to 15.5 wt% Cr. This explains why the "matrix" region in Figure 7 contained only 14 wt% Cr. One side of the matrix cell interface consisted of the cellular precipitate region,

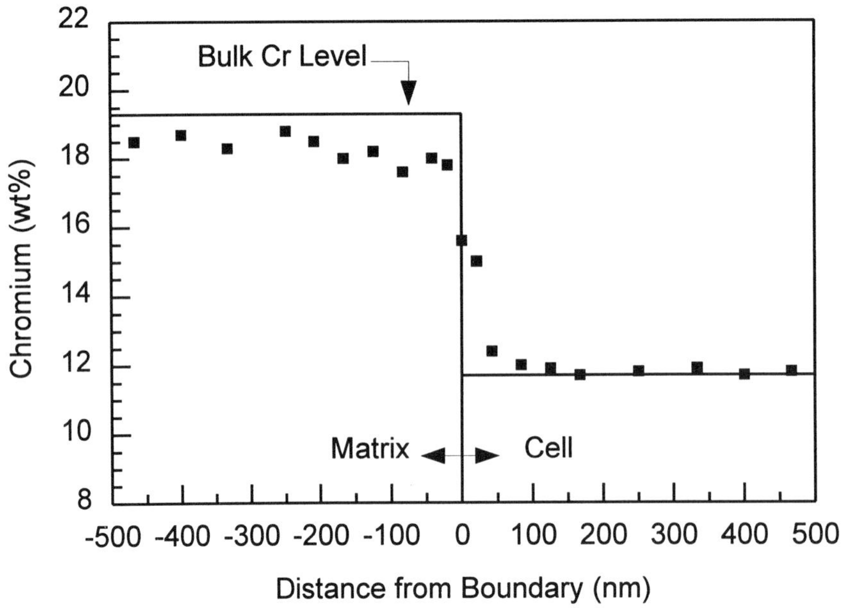

Figure 6. Cr-depletion profile taken across a matrix/cellular precipitate interface in a sample aged for 10 hours at 800°C.

and the other side, the "untransformed matrix," contained transgranular Cr_2N which depleted the overall matrix of chromium.

Specimens aged for 1 hour at 900°C had chromium depletion profiles similar to the specimen shown in Figure 6 (aged at 800°C for 10 hours) in that the matrix outside of the cellular precipitate region contained a chromium level equal to that of the bulk level of the alloy. However, samples aged at 900°C had minimum chromium levels which were about 2.5 wt% higher (at 14.2 wt% Cr) than the samples aged for 10 hours at 800°C.

SUMMARY

The present work correlated sensitization measured by the single-loop EPR test with direct measurement of Cr-depletion by AEM. Sensitization of low-carbon, high-nitrogen stainless steels is caused by the Cr-depletion of the matrix adjacent to Cr-rich Cr_2N precipitates. Reactivation, and the simultaneous corrosive attack, occurred during EPR testing when the minimum level of chromium within a given sample was below about 14 wt% Cr. Generally, sensitization increased with increased aging time at 800°C. However, a reduction in sensitization occurred with

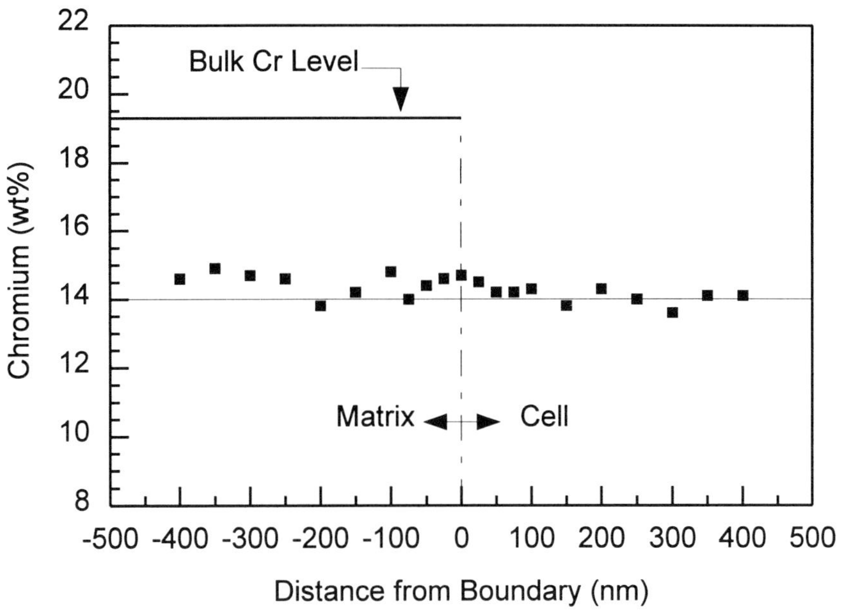

Figure 7. Cr-depletion profile taken across a matrix/cellular precipitate interface in a sample aged for 100 hours at 800°C.

increased aging time at 800°C after a peak sensitization value was reached at 30 hours of aging. This phenomena, known as healing, was caused by increasing Cr-minimums with increasing Cr_2N precipitation which lowers the chemical activity of nitrogen and increases the chromium level in equilibrium with the nitrides. Although aging at 900°C produced much higher volumes of Cr_2N than aging at 800°C for a given time, sensitization did not occur at 900°C since the minimum level of Cr within depleted regions was above the threshold for attack in the EPR test.

REFERENCES

1. M.O. Speidel, "Properties and Applications of High-Nitrogen Steels," Proceedings of the International Conference on High Nitrogen Steels, HNS 88, Lille, France, May 1988, edited by J. Foct and A. Hendry, The Institute of Metals, London (92) 1989.
2. M.L.G. Byrnes, M. Grujicic, and W.S. Owen, "Nitrogen Strengthening of a Stable Austenitic Stainless Steel," Acta Metall., Vol. 35, No. 7, (1853) 1987.
3. M. Kikuchi, M. Kajihara, and S.K. Choi, "Cellular Precipitation Involving both Substitutional and Interstitial Solutes: Cellular Precipitation of Cr2N in Cr-Ni Austenitic Steels," Mater. Sci. and Eng., A146 (131) 1991.

4. D.B. Rayaprolu and A. Hendry, "Cellular Precipitation in a Nitrogen Alloyed Stainless Steel," Mat. Sci. and Tech., 5, April (328) 1989.
5. R. Presser and J.M. Silcock, "Aging Behavior of 18Mn-18Cr High-Nitrogen Austenitic Steels for End Rings," Metal Science, 17, May (241) 1983.
6. J.W. Simmons, PhD Dissertation, "Microstructural Development in High-Nitrogen Stainless Steels," Oregon Graduate Institute of Science & Technology, Beaverton, Oregon (1993).
7. ASTM Book of Standards, G 108-92, ASTM, Philadelphia, Pennsylvania, Vol. 3.02 (449) 1992.
8. J. Menzel, G. Stein, and P. Dahlmann, "Manufacture of N-Alloyed Steels in a 20-ton PESR-Furnace," Proceedings of the 2nd International Conference on High Nitrogen Steels, HNS 90, Aachen, Germany, October 1990, edited by G. Stein and H. Witulski, Verlag Stahleisen mbH, Dusseldorf, Germany (365) 1990.
9. W.L. Clarke, R.L. Cowan, W.L. Walker, "Comparative Methods for Measuring Degree of Sensitization in Stainless Steel," Intergranular Corrosion of Stainless Alloys, ASTM STP 656, R.F. Steigerwald, Ed., ASTM (99) 1978.
10. J.B. Lee, "A New Electrochemical Potentiokinetic Reactivation Test for Determining Degree of Sensitization in Ferritic Stainless Steels," Corrosion, Vol. 42, No. 2, (106) 1986.
11. M. Kajihara, M. Kikuchi, and S.K. Choi, "Cellular Precipitation of Cr2N in Chromium-nickel Austenitic Steels," Phase Transformations '87, G.W. Lorimer, ed., The Institute of Metals, London (265) 1988.
12. C. Strawstrom and M. Hillert, "An Improved Depleted-Zone Theory of Intergranular Corrosion of 18-8 Stainless Steel," JISI, January (77) 1969.
13. E.L. Hall and C.L. Briant, "Chromium Depletion in the Vicinity of Carbides in Sensitized Austenitic Stainless Steels," Metall. Trans. A., Vol. 15A, May (793) 1984.
14. S.M. Bruemmer, L.A. Charlot, and A.W. Arey, "Sensitization Development in Austenitic Stainless Steel: Correlation Between STEM-EDS and EPR Measurements," Corrosion, Vol. 44, No. 6 (328) 1988.

THERMAL EMBRITTLEMENT TEST METHOD

FOR TYPE 410 STAINLESS STEEL

C. Czajkowski[1], E. Edwards[2], K. Warburton[1], and M. Dascoli[3]

Environmental and Waste Technology Center
Brookhaven National Laboratory
Upton, New York 11973-5000 U.S.A.

ABSTRACT

Investigations performed at Brookhaven National Laboratory (BNL) for the U.S. Nuclear Regulatory Commission (NRC) into thermal embrittlement and subsequent stress corrosion cracking of 410 stainless steel components led to the development of a nondestructive etch. Various thermal treatments on 410 stainless steels were evaluated using microhardness measurements and charpy impact testing. The results were then compared to the revealed microstructure after etching. Scanning Auger was performed on a embrittled specimen to evaluate the etched grain boundary constituents. The program's outcome showed that the ethereal picral etchant is an effective determinant of temper embrittlement in 410 stainless steel. The etch is non-destructive in nature and lends itself to field applications and receiving inspection of critical components.

---

[1] Brookhaven National Laboratory, Upton, NY 11973 - Research funded by U.S. Nuclear Regulatory Commission, Office of Nuclear Regulatory Regulation.

[2] Rensselaer Polytechnic Institute, Troy, NY 12180 - Students efforts supported by U.S. Department of Energy, Office of University and Science Education Programs.

[3] Riverhead High School, Riverhead, NY 11901 - Student efforts supported by U.S. Department of Energy, Office of University and Science Programs.

INTRODUCTION

Brookhaven National Laboratory (BNL) has performed for the U.S. Nuclear Regulatory Commission (NRC) three investigations into thermal embrittlement and subsequent stress corrosion cracking of 410 stainless steel components [1,2]. These involved components from the Joseph M. Farley Unit 1 Nuclear Power Station, the R.E. Ginna Nuclear Power Plant and the Beaver Valley Nuclear Power Plant.

Both the Farley plant and the Ginna plant concerned main steam isolation valve shafts (MSIV) that showed linear indications of cracking when examined during refueling outages. Subsequent analyses of the cracked 410 stainless steel components at BNL produced similar findings [1]:

1. The chemical analyses of the materials in all cases were consistent with that expected of a Type 410 stainless steel.
2. The microstructure observed was martensitic in nature.
3. The hardness measurements were high (~R_c 40).
4. The cracking was intergranular and occurred along prior austenite grain boundaries.
5. There was no evidence of fatigue interaction and no corrodent species were identified.
6. The cracking was considered to be an intergranular stress corrosion cracking phenomenon resulting from a high hardenss/susceptible material. The observation of temper colors on one fracture face (from the Ginna plant) after oxide removal suggested the possibility of pre-existing heat treatment cracks.

The Beaver Valley incident involved a river water pump shaft coupling which failed during operation. The coupling was used to join two shafts of a Byron Jackson Vertical Circulator river water pump. The pump rotates at 1170 RPM in river water of 0°-33° C (32°-90° F) and produces a discharge pressure of 469 Pa (68psi).

A preliminary root cause analysis performed by Lehigh University (for the utility) concluded that the failure occurred due to an improper heat treatment being applied to the 410 stainless steel couplings which resulted in a temper embrittled condition (making the component susceptible to cracking). A subsequent independent metallurgical investigation at BNL reached the following conclusions [2]:

1. The chemical analysis of the coupling verified that it was made from Type 410 stainless steel.
2. The observed cracking was intergranular and appeared to follow prior austenite grain boundaries in the tempered martensitic alloy matrix.
3. There was no evidence of heat tinting and no high temperature oxides present on the fractures.

4. The mechanical properties were reasonably consistent for a 410 stainless steel tempered at 538° C (1000° F). There was a reduction in ductility and yield strength from expected values. The impact energy of the "as received" coupling material was an extremely low 10.8 to 20.3 J (8 to 15 ft. lbs.) and was markedly increased 135.6 to 162.7 J (+100 to 120 ft. lbs.) after a corrective heat treatment was performed.
5. The root cause of the failure was considered to be temper embrittlement of the 410 stainless steel coupling brought about by improper heat treatment of the coupling during manufacture.

The above investigations prompted a literature search which disclosed that there have been various incidents of cracking of Type 410 stainless steels in nuclear plant applications. These included valve stems in bonnetless globe valves [3], a valve stem in a suppression pool suction supply valve [4] and a valve stem in a gate valve [5]. In each case, improper heat treatment was cited as a contributory cause to the failure.

Because all these cases involved failures apparently related to the 410 stainless heat treatment, the NRC decided it would be advantageous to be able to nondestructively determine the metallurgical condition of 410 stainless steel prior to in-service use. Subsequently, a contract was placed with BNL for the development of a non-destructive etch for the determination of possible temper embrittlement.

TESTING OF AN ETHEREAL PICRAL ETCH FOR THE DETERMINATION OF TEMPER EMBRITTLEMENT

Temper brittleness has been defined in the *Metals Handbook*, 8th Edition [6], as "Brittleness that results when certain steels are held within or are cooled slowly through a certain range of temperatures below the transformation range. The brittleness is revealed by notched-bar impact tests at or below room temperature." The only definitive way to prove temper embrittlement seems to be from the notched-bar impact tests, which is precisely the problem. "In view of the lack of a satisfactory method of metallographic discrimination between the tough and brittle conditions of steel susceptible to temper embrittlement, the impact test has remained the sole practical tool for investigation of the embrittlement phenomenon. It is conceivable that were the metallographer enabled to perceive the affects of temper embrittlement under the microscope, much could be added to the understanding of the nature of embrittlement." [7]

In 1947 the Ethereal Picral Etch [7] was tested by Cohen, Hurlich, and Jacobson on Mn-Ni-Cr-Mo, Ni-Cr, Ni-Mo, and Ni-Cr-Mo steels to detect embrittlement. It was believed that temper brittleness was due to specific impurities, such as antimony, phosphorous, tin and arsenic, which segregated to the prior austenite grain boundaries when the steel is held within a certain range of temperatures dependent on its composition. The ethereal picral etch is extremely sensitive to phosphorous and in the 1947 report produced satisfactory results that it could help detect conditions in the steels which made them susceptible to temper embrittlement. BNL personnel used the etchant for a report [2] in 1992 on 410 stainless steel in an attempt to prove the existence of temper embrittlement. Its satisfactory results prompted them to examine the ability of the ethereal picral etch to be an indicator of temper embrittlement in this type of stainless steel.

Materials and Procedures

Type 410 stainless steel was taken from two previously completed programs, each differing in composition and prior treatments. Those taken from a program for the Beaver Valley Power Plant were believed to be in the temper embrittled condition (as received), while those from Farley were apparently affected by stress corrosion cracking. Eleven samples were taken from each program. They were cut to similar dimensions (approximately 7.62 cm (3 in.) in diameter and 1.27 cm (1/2 inch)) although the samples from Farley were slightly wider. All heating had a soaking time of one hour. After austenitization (Table 1) samples were oil quenched, and after tempering samples were water quenched with the exception of sample #10, which was furnace cooled.

Table 1 Heat Treatment

Sample #	Treatment
1	Samples left as received
2	Austenitized 926° C (1700° F)
3	Austenitized 926° C (1700° F), tempered 93° C (200° F)
4	Austenitized 926° C (1700° F), tempered 204° C (400° F)
5	Austenitized 926° C (1700° F), tempered 315° C (600° F)
6	Austenitized 926° C (1700° F), tempered 426° C (800° F)
7	Austenitized 926° C (1700° F), tempered 537° C (1000° F)
8	Austenitized 926° C (1700° F), tempered 593° C (1100° F)
9	Austenitized 926° C (1700° F), tempered 648° C (1200° F)
10	Austenitized 926° C (1700° F), tempered 648° C (1200° F) - furnace cooled
11	Austenitized 926° C (1700° F), tempered 759° C (1400° F)

Each sample was mounted and polished down to 0.3 μm, then they were all etched using ethereal picral. The etchant solution (Table 2) [7] was prepared by dissolving the picric acid in ether and then adding 250 milliliters of Zephiren-water solution. The resulting mixture was thoroughly shaken. The solution was kept in a tightly stoppered bottle to prevent excessive evaporation of the ether. With time, the mixture separates into two layers, the top layer consisting essentially of a saturated solution of picric acid in ethyl ether containing Zephiren, and the bottom layer consisting of an aqueous Zephiren-picric acid solution. The solution is allowed to stand overnight prior to use. After the first shaking, no further agitation is necessary. A portion of the top layer was decanted into a small beaker and diluted with approximately one-third volume of ether. The samples were set in one layer in a glass dish and soaked in the etchant for approximately fourteen minutes. The samples were then carefully rinsed off with alcohol and blown dry. Specimens were examined at 250x, then were photographed at 320x after a light two minute polishing by pant 0.3μm alpha alumina. In order to verify the characteristics of each heat treatment, a microhardness test was performed (five locations) on each sample. The average result was interpolated into a Rockwell "C" hardness.

In order to determine a closer approximation of the mechanical property dependence of 410 stainless steel on the embrittled condition, additional samples were taken from Beaver Valley and Farley and heat treated as shown in Table 3, following the same procedure as the prior heat treatments. Photographs were taken of their structure at 320x, then the microhardness tests performed.

Table 2 The Etchant

Type	Amount
Picric Acid	50 grams
Purified Ethyl Ether	250 milliliters
Zephiren Chloride (12.8% solution)	10 milliliters
Water	240 milliliters

Table 3 Additional Heat Treatments

Sample #	Treatment
12	Austenitized 1700°F, tempered 850°F
13	Austenitized 1700°F, tempered 900°F
14	Austenitized 1700°F, tempered 950°F

RESULTS AND DISCUSSION

It is believed that for each steel there is some range of temperatures at which, if the steel is tempered or cooled slowly enough, it will be susceptible to temper embrittlement. The specific temperature ranges for the tested 410 stainless steels were unknown, so a variety of tempering temperatures were tested. There was a vast difference in the effects of the etchant on the separately treated specimens.

For those specimens simply austenitized and those tempered in the range of 93° C-426° C (200-800° F), and at 759° C (1400° F) there appeared to be no effect on the specimens from the etchant. The only observation was that small pits appeared on the surface (Figures 1 and 2). Those specimens tempered within the range of 537° C-648° C (1000°-1200° F) appeared to have their microstructure highly defined by the etchant (Figures 3 and 4). The prior austenite grain boundaries were defined in addition to the martensitic structure. After a slight repolishing, the prior austenite grain boundaries became more evident. Comparison of the two sets of samples tempered at 648° C (1200° F) showed that the specimens furnace cooled (effects of temper embrittlement became more serious as they were additionally slowly cooled through the vital temperature range) had darker highlights created by the etchant than those water quenched. It appears that the etchant can determine the degree of temper embrittlement on a specimen by the severity of its effects.

The hardness remained relatively constant until it reached a specific temperature at the beginning of its critical range. Figure 5 and 6, in which hardness is plotted as a function of tempering temperature, demonstrate this sudden drop which corresponds with the lowest impact energy (condition of temper brittleness)observable in the charpy impact test results in Figure 7. The hardness found in the "as received" specimen from Beaver Valley was fairly close to the hardness calculated from the Beaver Valley specimen tempered at 537° C (1000° F), its recorded treatment. The hardness calculated in the Farley "as received" specimen fell slightly lower than that from the specimen tempered at 426° C (800° F). The recorded tempering temperature of 509° C (950° F) of the specimen appeared to be close to the critical tempering temperature of the specimen at the beginning of the critical range. The hardness found was similar to the hardness calculated in those specimens that showed no signs of temper embrittlement by using the ethereal picral etch.

Figure 8 shows a typical scan of the fracture surface of the furnace cooled 648° C (1200° F) specimen seen in the Scanning Auger Multiprobe. It is evident that phosphorous was detected on the fracture face. The sample was then sputtered cleaned using an Argon sputtering gun operating at 3kV and 25 ma for 2 minutes. This typically will remove between 20-40 nm of material. After cleaning the sample was reanalyzed and no phosphorus was present (Figure 9). This was taken as a good indication that the ethereal picral etch highlighted the phosphorous segregation in the prior austenite grain boundaries.

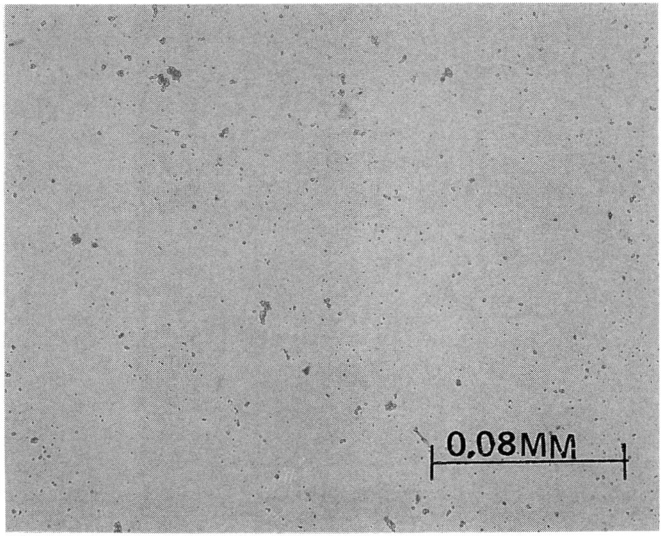

Figure 1. No observable grain boundary etching after austenitizing the Farley specimen at 926° C (1700° F).

Figure 2. The specimen tempered at 454° C (850° F) showed no microstructure after etching.

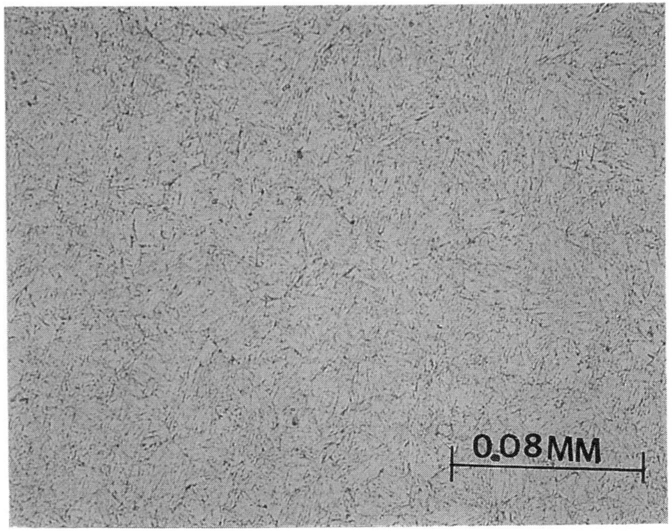

Figure 3. Microstructure is quite evident in etched specimen tempered at 482° C (900° F).

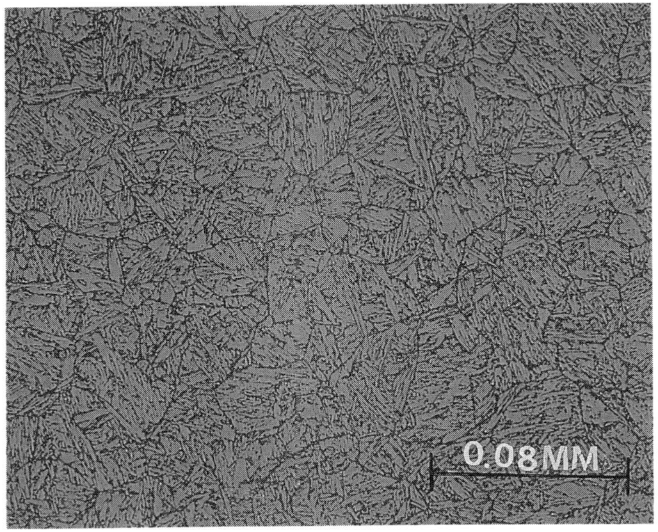

Figure 4. The microstructure of the specimen tempered at 648° C (1200° F) was also evident after etching.

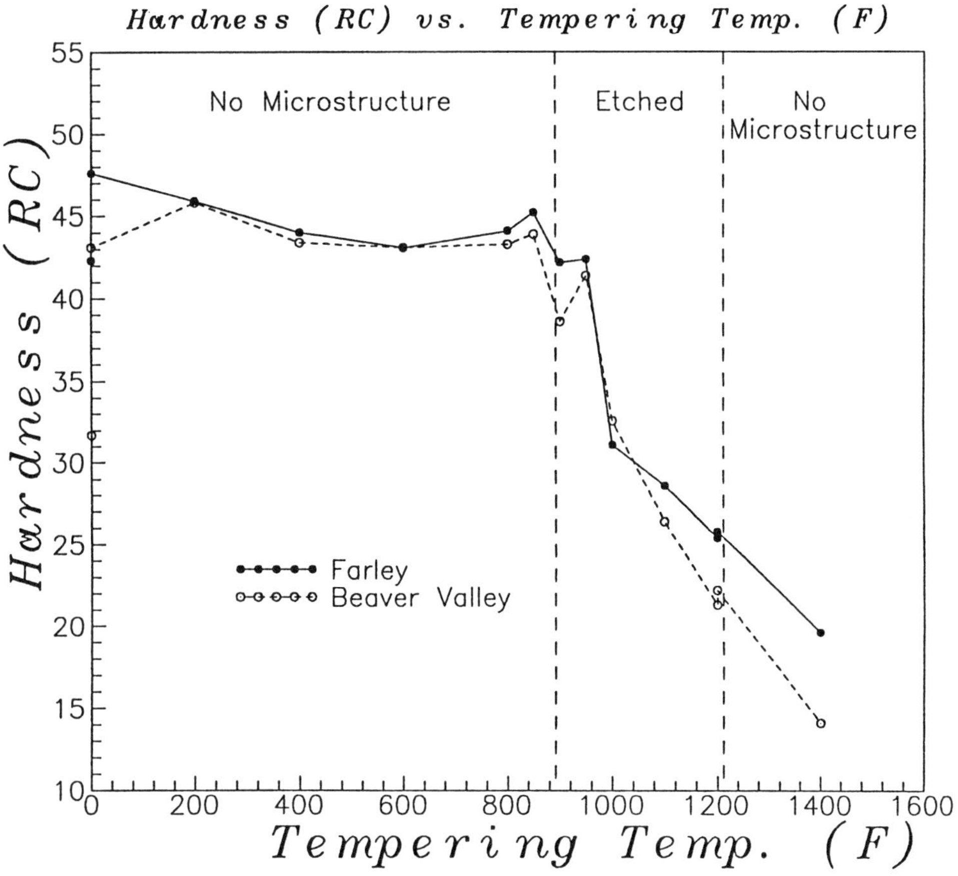

Figure 5. comparison of hardness values to tempering temperature (Farley and Beaver Valley specimens).

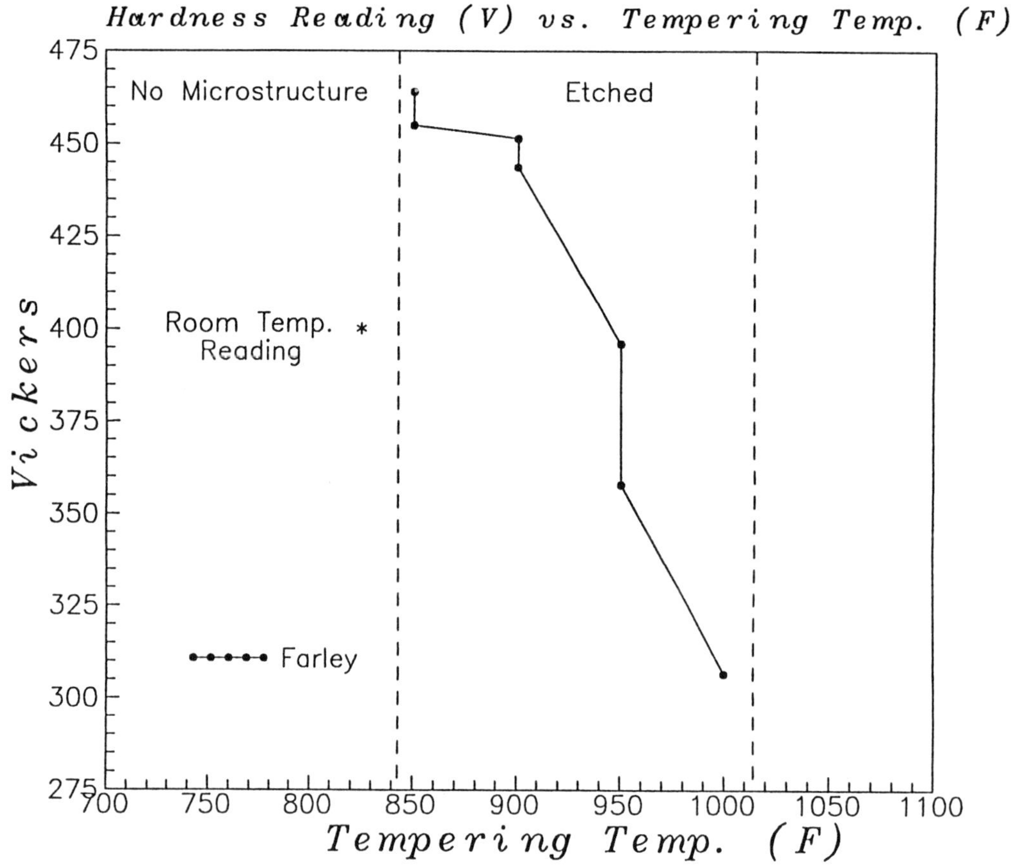

Figure 6. Comparison of hardness values to tempering temperatures for Farley specimens.

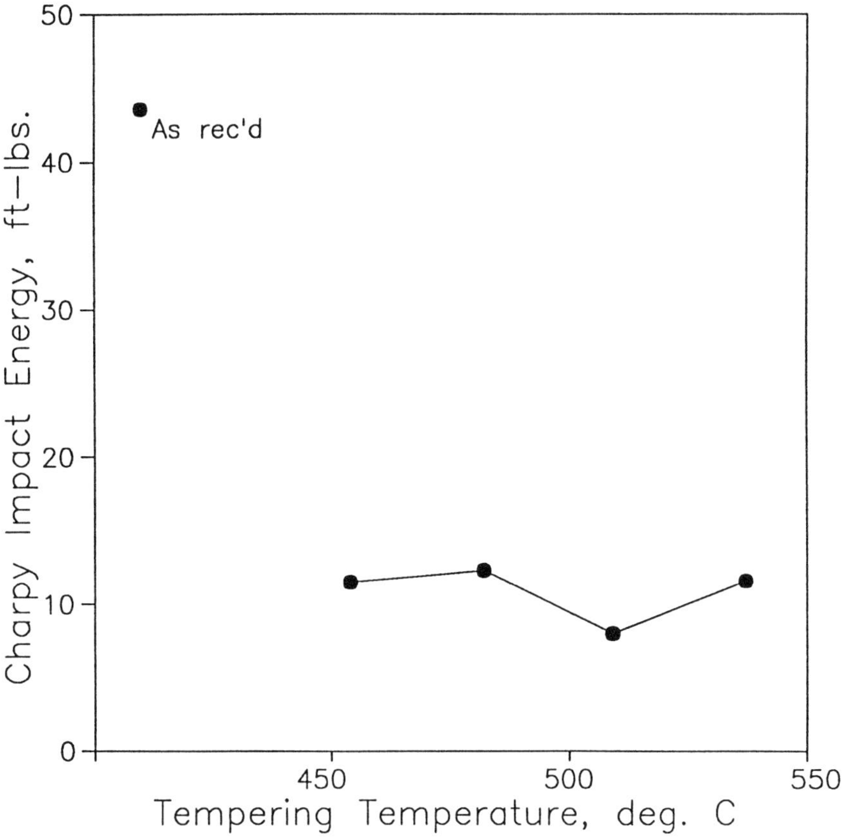

Figure 7.Impact energy as a function of temporary temperature for Farley specimens.

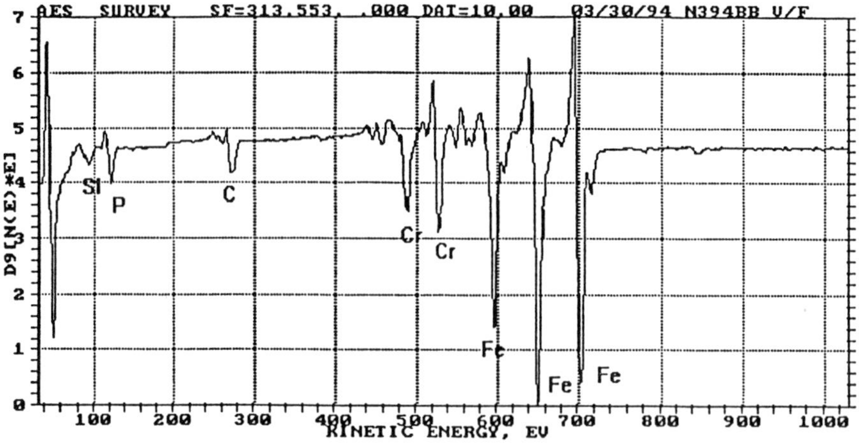

Figure 8. auger scan of sensitized type 410 SS (648° C/1200° F temper-furnace cooled).

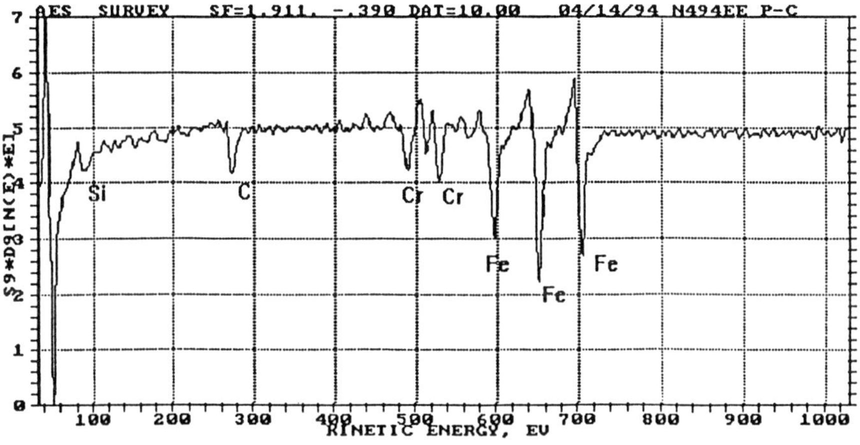

Figure 9. Auger scan of previous specimen (Figure 7) after sputter cleaning.
Note lack of P peak.

SUMMARY AND CONCLUSIONS

Twenty-two samples of 410 stainless steel with two separate compositions were heat treated in eleven different treatments prior to being introduced to the ethereal picral etch. The following conclusions were reached:

1. Specimens tempered between 482°C-648°C (900°-1200°F) showed a significant amount of etching, whereas the others showed no apparent etching.
2. A comparison of the measured hardness of these specimens to that in the *Metals Handbook*, shows that the heat treated specimens fall in the correct range for their tempering temperatures.
3. A specimen taken from the Beaver Valley Power Plant that had supposedly been temper embrittled showed significant etching in the "as received" condition by the etchant and its measured hardness falls within the drop caused by the reaction of stainless steel to temper embrittlement.
4. A specimen taken from Farley affected by stress corrosion cracking had no recorded effects from the etchant "as received". Its calculated hardness was similar to those found in 410 stainless steel prior to the drop caused by its reaction to temper embrittlement.

The ethereal picral etchant is considered to be an effective determinant of temper embrittlement in 410 stainless steel. The etch delineates microstructure of temper embrittled specimens of Type 410 stainless steel, while performing no apparent etching on Type 410 stainless steel not embrittled. The etch is non-destructive in nature and lends itself to field applications and receiving inspection of critical components.

REFERENCES

1) Czajkowski, C.J., Investigation of Failed Type 410 Stainless Steel Shafts from Main steam isolation Valves at Two Operating Plants, Paper 97, Corrosion 87, NACE, San Francisco, CA March 9-13, 1987.

2) Czajkowski, C.J., Brookhaven National Laboratory Technical Report L-1529-6-92, June 1992.

3) S.M. Stoller Corp., "Nuclear Power Experience," Volume PWR2, XV. Miscellaneous Systems, Sec. 2, "Valve Stem Cracks - Failure to Stress Relieve and Valve Seat Cracks."

4) U.S. Nuclear Regulatory Commission Licensing Event Report, 82-088/3L-3, May 24, 1984, Brunswick Unit 2.

5) U.S. Nuclear Regulatory Commission Licensing Event Report, 84-04/01, June 15, 1984, Browns Ferry Unit 3.

6) *Metals Handbook*, 8th Edition, Volume 1, American Society for Metals, p.38.

7) Cohen, J.B., Hurlich, A., Jacobson,M.M.., "A Metallographic Etchant to Reveal Temper Brittleness in Steel, *Transactions of the ASM,* Volume 39, pp. 109-138, 1947.

A TEM COMPARISON OF $M_{23}C_6$ CARBIDES AT LOW AND HIGH STRAINED 304 STAINLESS STEEL: SIZE, MODE, MORPHOLOGY

E. A. Trillo, R. J. Romero, J. G. Maldonado, and L. E. Murr

ABSTRACT

$M_{23}C_6$ precipitate sizes and morphologies of four different types of 304 stainless steel was investigated using the scanning transmission electron microscope (STEM). The materials composition ranged from .011 to .07 wt% C and was subjected to three different temperatures in the sensitization range: 775°C, 670°C, and 625°C. The 625°C experiments included samples deformed 10 and 20% true strain. The 670°C experiments included both low strain (10 and 20%) and high strain (40%) samples. The two lower temperatures yielded an increase in the size of carbides when increasing the amount of strain (<20%), and the precipitates nucleated and grew on grain boundaries and twin boundaries. However, when the steel has been strained to such an extent (>20%), a type of nucleation is preferred on microshear bands, and when heat treated the proper requisites are met for recrystallized zones that prompt nucleation on those boundaries. When increasing the temperature to 775°C, a morphological change occurs and the carbides take on specific and often lamellar growth patterns.

---

[1]Department of Metallurgical and Materials Engineering, The University of Texas at El Paso, El Paso, TX 79968

INTRODUCTION

The behavior of a material will be a direct result of the chemical composition and any subsequent heat treatment and/or mechanical processing. Stainless steel, when subjected through isothermal heat treatment or a continuous cooling through the range of 850-450°C, will be subjected to a phenomenon known as sensitization (1). Sensitization is the depletion of Cr at specific sites on the grain boundaries, twin boundaries, etc., due to the precipitation of $M_{23}C_6$ carbides. This leaves the material susceptible to stress corrosion cracking (SCC) and intergranular stress corrosion cracking (IGSCC). Although carbide formation has been studied as a precursor and indication of the threat of sensitization and SCC (2-5), few studies have produced a comparison of differences in material composition and thermomechanical processes with relation to carbide sizes, modes, and morphologies.

At elevated temperatures (>700°C), a specific pattern in the growth of $M_{23}C_6$ precipitates occurs on certain portions of the twin boundary. The carbide takes on a lamellar formation and on {111} coherent twin boundaries will grow in a {111} trace direction, while in other instances, lamellar precipitates grow out of the coherent twin boundary, coincident with another {111} trace direction. It has also been found that carbides grow coincidental to traces of the {111} 90° planes. As demonstrated by Romero and Murr (6), the resulting geometries are attributed to an effective interfacial torque (EIT) acting on the coherent annealing twin interface.

Carbide formation is also affected by the chemistry of the microsystem. In stainless steels, carbon content is the key variable concerning sensitization and carbide precipitation. Increasing the amount of carbon decreases the time that starts the sensitization process, yields higher sensitization levels, accelerates healing kinetics, and increases the size and density of carbide precipitation (7).

Strain also has a marked effect on carbide precipitation and sensitization development. Increasing the amount of strain has been shown to increase the degree of sensitization (DOS), and increase the amount of precipitation. When increasing the strain to levels of 40% true strain, an increased amount of overlapping arrays of stacking faults and deformation twins (called microshear band intersections) will provide the required conditions for the nucleation of α'-martensite. It will also provide the necessary energetics for the formation of Cr-rich carbides to nucleate on the same type of intersection site. However, the martensite produced by means of strain, has been found not to be a precursor to the carbide precipitation process (8).

There are obviously many variables that effect the precipitation of Cr-rich carbides in stainless steels. This research will attempt to give an overview of the main factors influencing carbide precipitation (temperature, carbon content, and strain) and relate these factors to the size of carbides, their morphology, and their growth locating preference.

EXPERIMENTAL PROCEDURES

The material utilized for this research consisted of five different grades of 304 stainless steel, received as a 6.25 mm thick plate in the mill-processed condition. The exact compositions and grain sizes are listed i Table 1. Three experimental temperatures were chosen in the sensitization heat treatment range: 775°, 670°C, and 625°C.

Table 1
Chemical Compositions in Weight Percent

Heat #	C%	Cr%	Ni%	Mo%	N%	P%
1	.011	18.13	8.92	.34	.08	.022
2	.025	18.35	9.08	.37	.08	.026
3	.051	18.29	8.22	.19	.085	.025
4	.07	18.23	8.15	.35	.08	.035
5	.05	18.31	8.08	.35	.084	.031

The first set of experiments included heats 1-4. Each of the carbon contents was sectioned into 12 mm x 12mm sample sizes and furnaced treated at 775°C for .1-500 hours. The samples were then polished down to a one micron diamond finish and tested using the single loop variation of the EPR test, and a modified cell. This electrochemical potentiokenitic reactivation test measures the current resulting from the chromium depleted areas of the corroded sample.

Following testing, several samples were observed in the Hitachi 8000 scanning transmission electron microscope (STEM) at 200 kV.

The remaining samples were deformed using a universal testing machine. Samples furnace treated at 670°C utilized heat #5. They were strained 10, and 20% true strain. Heat number 3 was strained to 40% true strain and furnace treated at 670°C for up to 10 hours. Heat treatments at 625°C encompassed heat #3 that was strained 10 and 20% true strain. All samples were then EPR tested, polished, and a select group of those were observed in the STEM.

RESULTS AND DISCUSSION

Carbide Morphology: Figure 1 illustrates the directional growth characteristics that $M_{23}C_6$ precipitates are able to take on at the elevated temperature of 775°C. In Fig. 1(a) precipitation occurs on a non-coherent portion of a twin boundary. Its lamallae formation and growth follows the (111) twin plane in the [112] direction. Figure 1(b) displays similar lamellar growth in a (110) grain orientation. Precipitation is occurring along alternate {111} and {111} planes at 90° to the (110) surface. These directions are denoted T90 directions.

The lamellar morphology and preferential growth is a result of the isothermal conditions and, more importantly, the effective interfacial torque (EIT). Figure 2 demonstrates low and high torque conditions for twin boundaries intersecting grain boundaries. The high torque case (right side) describes the resolved interfacial torques (arrows) that follow the T90 direction. Note the high torque directions correspond to twins having boundary lines which are perpendicular to the (110) surface seen in Figure 1(a), and the low torque configuration is seen in Figure 1(b).

An important aspect to note is that directional growth was examined in strained samples (less than 20%) at 625°C, although no TEM micrographs have been given. However, directional growth has not been observed in samples that contain over 20% strain.

Carbide Size Measurements: Figure 3 displays the results for the EPR test at 775°C. The data clearly illustrate an increase in DOS with increasing carbon content. A fact that is corroborated with electron microscopy. In Figure 3(b) the carbon content effect is illustrated by

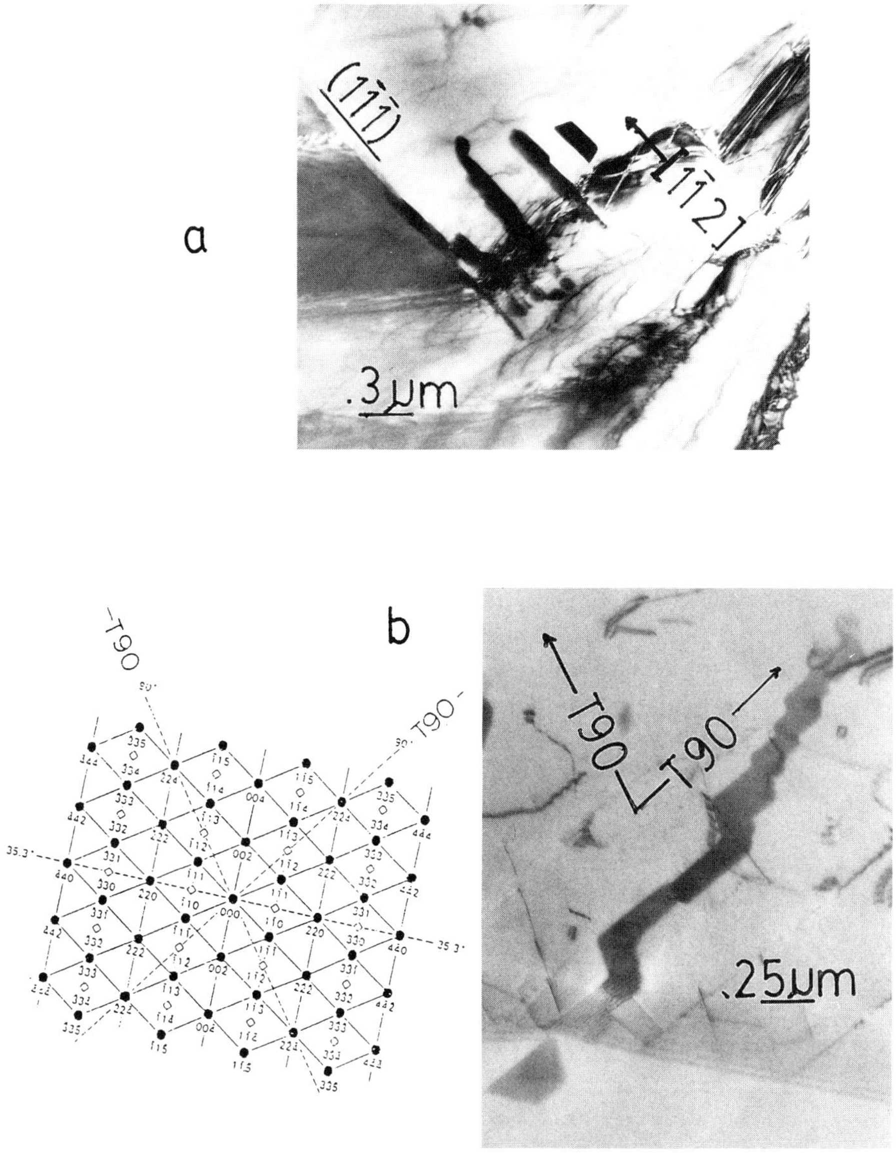

Figure 1. Lamellar carbide growth on (a) carbides growing out of a non-coherent portion of the twin boundary in the {112} direction. (b) carbide along alternate {111} planes in T90 directions.

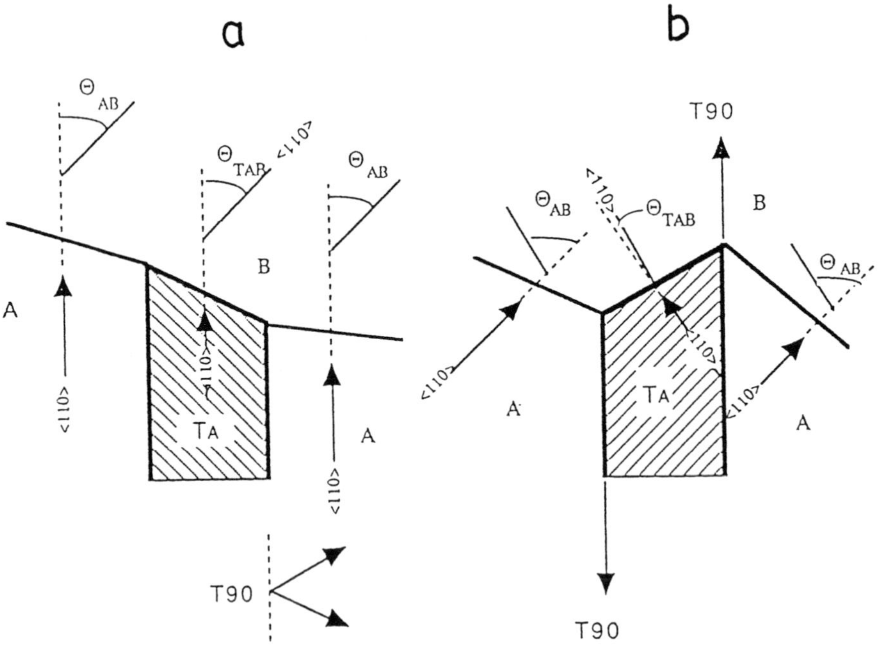

Figure 2. (a) Low and (b) high torque conditions for twin boundaries intersecting with (110) grain boundaries.

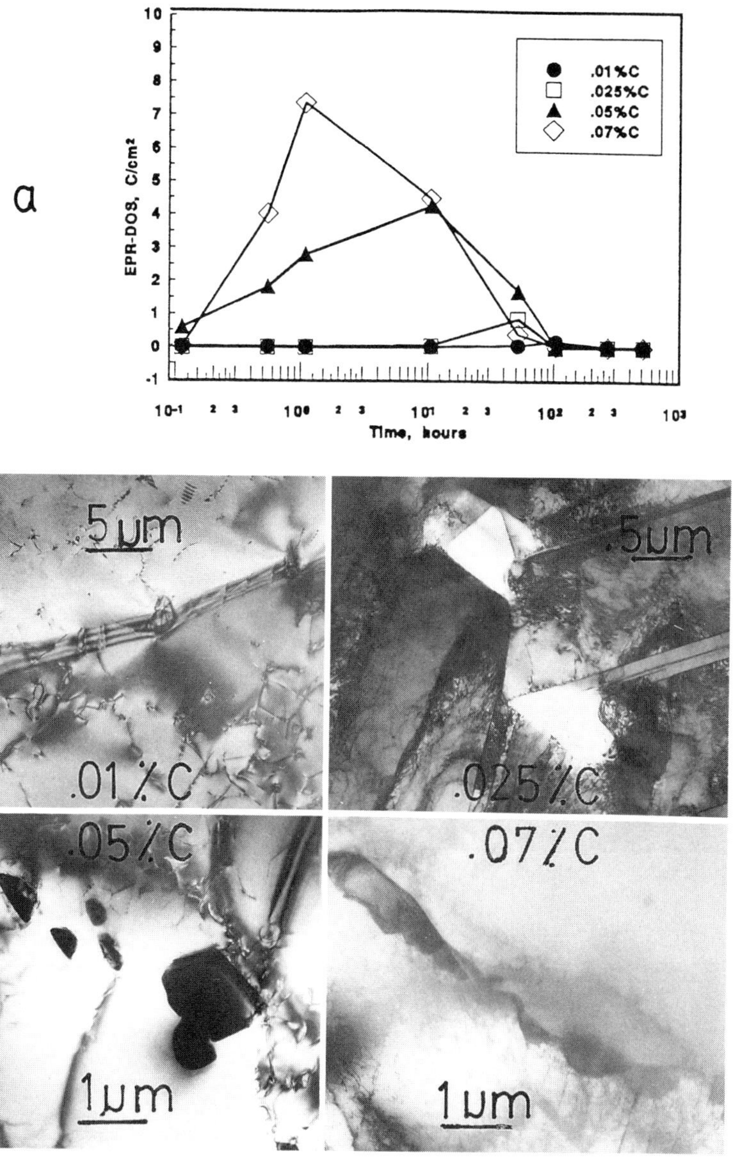

Figure 3. Results for the 775°C experiments. (a) EPR-DOS vs. heat treatment time graph for increasing carbon content. (b) TEM micrographs of increasing carbon content. Increasing the amount of carbon in the material increases the DOS and the size of the precipitates.

an increase in the size of the carbides at 50 hours furnace time. Carbide size measurements were taken on several separate representative areas and their average is recorded on Table 2. These actual measurements confirm visual inspections of a gradual increase in carbide size with higher levels of carbon content (thermodynamic effect).

Carbide size measurements were also obtained for the two other temperatures (670° and 625°C). However, in these experiments, the element of strain was added. Table 3 lists the results for the three temperatures for 0 and 20% strain, for a carbon content of approximately .05%, and a heat treatment time of 50 hours. The effect of strain on carbon content is clearly seen with this data. Increasing the amount of strain will increase the size of the carbide as was similarly seen with increasing carbon content. A TEM comparison of carbide sizes at the two temperatures and straining conditions are observed in Fig. 4. Strain produces an increase in the amount of dislocations which provide a pathway for the slower diffusive constituent (Cr) to gain access to the carbide. Also, a temperature effect may also be seen with respect to carbide size. The more elevated temperatures provide an accelerated kinetic effect that will incorporate additional Cr and C at the precipitate site for carbide enlargement.

Table 2
Carbide Size Measurements

Carbon Content, wt%	Carbide Size, μm
.01	.195
.025	.45
.051	1.08
.07	1.09

Table 3
Carbide Size Measurements for .05%C

	Carbide Size, μm	
Temperature, °C	0% ε	20% ε
775	1.08	--
670	.22	.34
625	.11	.15

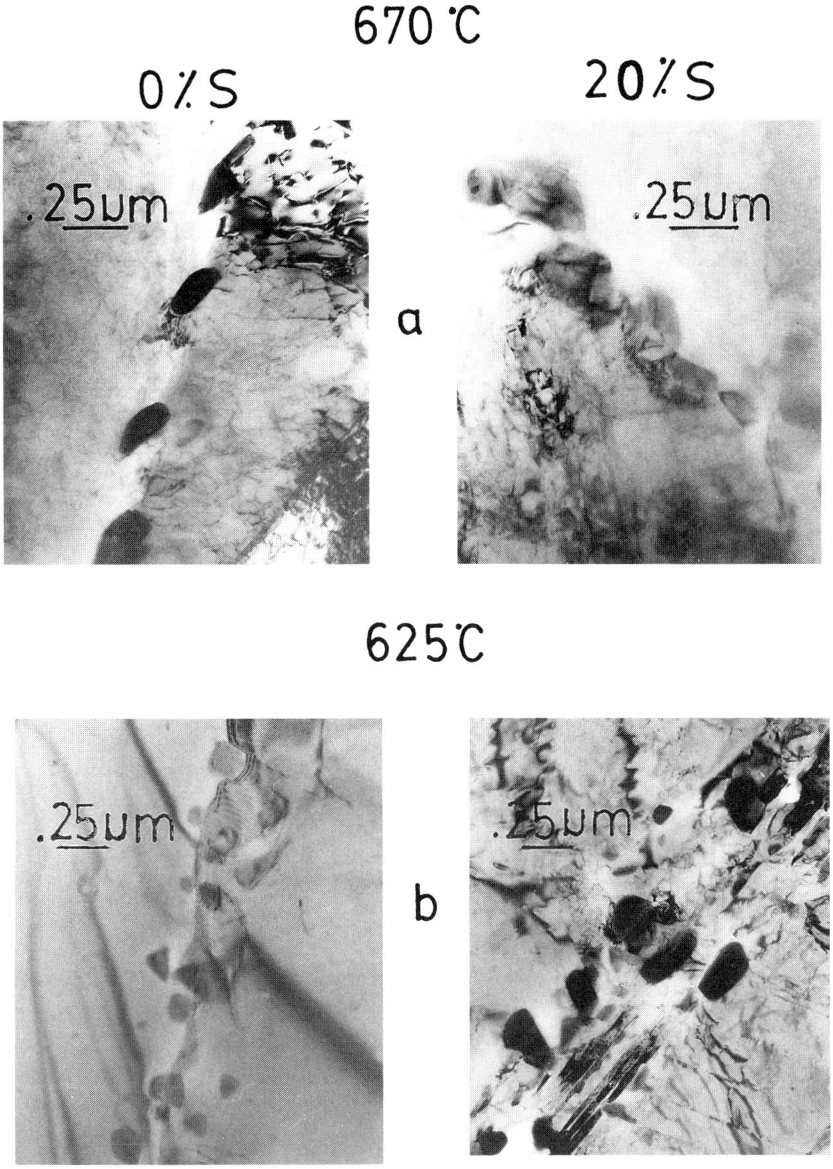

Figure 4. TEM micrographs of 0.05% 0 and 20% strained at (a) 670°C and (b) 625°C. Increasing the strain and temperature increases the size of the precipitates.

Effect of High Strains on Carbide Precipitation:
Increasing the strain to higher levels (40% true strain) also provides some unique carbide characteristics. The added levels of strain incorporate larger amounts of microshear intersections, as explained earlier. The energy of these bands is sufficient to nucleate and grow $M_{23}C_6$ precipitates. Figure 5(a) is such an example. The precipitate on this 40% strained sample is growing out of a twin-fault intersection. This seems to be the preferential location for transgranular carbide precipitation. Therefore strains below 20% have grain boundaries and twin boundaries that represent sites for nucleation and growth of precipitates, and strains above 20% contain twin fault intersections which become the more probable nucleation site.

Another interesting feature of the highly deformed material is that along with carbide precipitates an α'-martensite formation appears. The martensite formed seems to provide requisite sties for a recrystallized zone within the microshear sections. The recrystallized zones in turn provide added sites for carbide nucleation, as seen in Fig. 5(b). The interphase boundaries in the recrystallized zone displays carbide precipitates. A more clear image of the individual precipitates is seen in the accompanying dark-field micrograph.

CONCLUSIONS

Definite effects of temperature, strain, and carbon content are seen in 304 stainless steels. The highest temperature (775°C) yielded a preferential growth direction and morphology of $M_{23}C_6$ precipitates. Directional growth was not observed for samples that contained greater than 20% true strain. Incorporating larger amounts of carbon in the material at the highest temperature will induce larger carbide sizes, whose effects can be explained thermodynamically. Larger carbide sizes can also be found with increasing amounts of strain, which is an effect that was observed for the two temperatures of 670° and 625°C. Temperature plays a key role in increasing the size of the Cr-rich carbides by accelerating the kinetics of carbide formation. High strained samples (40%) will yield the requisites for microshear bands and a recrystallization zone effects, which become the more probable precipitate sites as compared to grain boundaries and twin boundaries of the lower strained material.

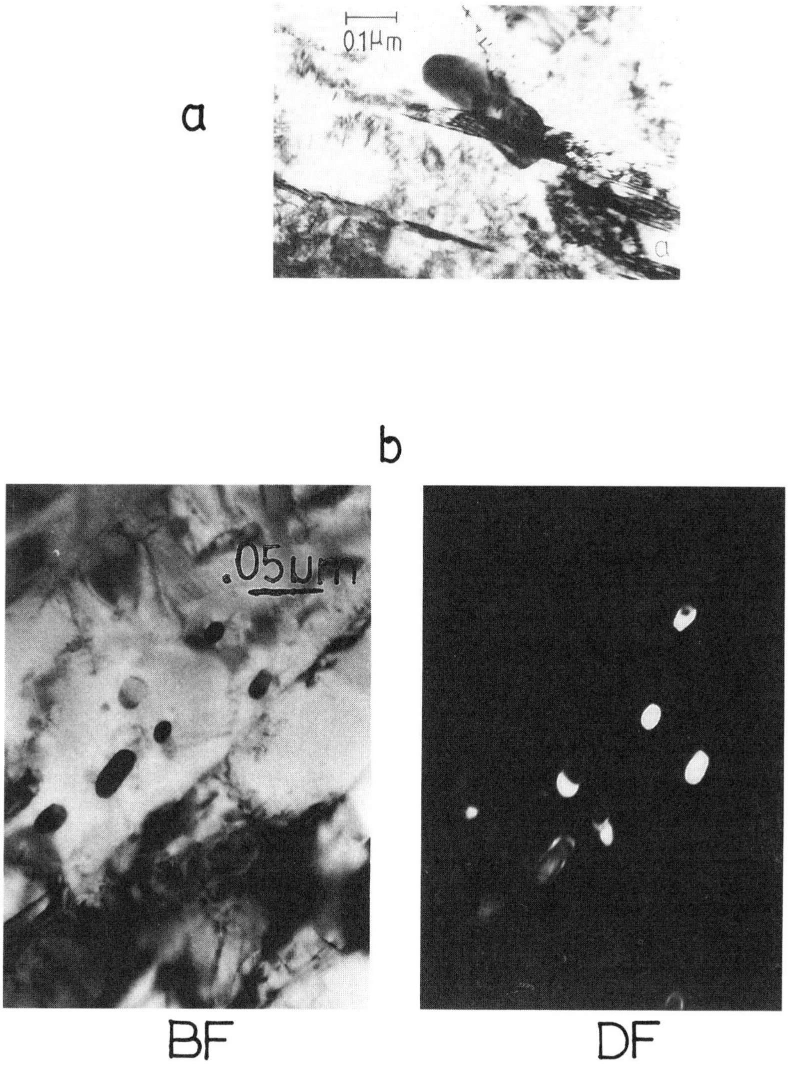

Figure 5. Precipitation in a high strained material. (a) Cr-rich carbide growing out of twin fault intersection . (b) Bright field and dark field image of carbides forming on recrystallized area of the grain.

ACKNOWLEDGMENTS

This research was supported by the National Science Foundation (RIMI) HRD 9105065 and the Patricia Roberts Harris Fellowship Program (E.A.T., R.J.R, and J.G.M.). We would also like to thank Reynalda Beltran and Daniella Martinez for their contributions to this work.

REFERENCES

1. E. C. Bain, R. H. Aborn, and J. J. B. Rutherford, Transaction American Society Steel Treatise, vol. 21, p. 481, 1933.
2. H. Hatwell, and A. Berghezan, Iron Steel Inst., Special Report, No. 64, p. 67, 1959.
3. B. Hattersley and M. H. Lewis, Philos. Mag., Vol. 10, p. 1075, 1969.
4. L. K. Sinhal, and J. W. Martin, Acta Metall., Vol. 15, p. 1603, 1967.
5. M. Terao, and B. Sasmal, Metallography, Vol. 13, p. 117, 1980.
6. R. Romero, and L. E. Murr, Scripta Metallurgica et Materialia, Vol. 30, pp. 191, 1994.
7. E. A. Trillo, A. H. Advani, L. E. Murr, and W. W. Fisher, Proc. 51st Annual Meeting of the Microscopy Society of America, p. 1164, 1993.
8. J. G. Maldonado, M.S. Thesis, "Strain-Induced Martensite Effects on Transgranular Carbide Precipitation and Sensitization in Austenitic Stainless Steels", the University of Texas at El Paso, El Paso, TX, July 1994.

A STUDY OF PRECIPITATION HARDENING IN PH 13-8 Mo STAINLESS STEEL

Luiz Henrique de Almeida[1], Sylvio José G. Magrani[1], André Freitas Ribeiro[1] and Iain LeMay[1,2]

ABSTRACT

The relationship between mechanical properties and microstructure for PH 13-8 Mo precipitation hardening stainless steel has been studied. Tensile, Charpy and CTOD tests were made on solution treated material and after aging at different temperatures. Special techniques were used for sample preparation for light, scanning and transmission electron microscopy. The nature of the precipitation was characterized as well as its influence on mechanical properties.

INTRODUCTION

The first precipitation hardening stainless steel was developed in the 1940s in order to fulfill the need for a high strength steel that posessed good fracture and corrosion resistance and, since then, many different types of such steels have been developed. Among these, the martensitic ones are preferred owing to the good response of the matrix to aging treatment. Although several extensive studies of the behavior of maraging stainless steels such as 17-4 PH have been made, very few papers deal with the PH 13-8 Mo Steel.

The first paper known to refer to this steel dates from 1981, Seetharaman et al. [1] making a basic approach to characterize its microstructure. Other studies were published from the late 80s, dealing mainly with its resistance to stress corrosion cracking (SCC) [2-4]. These

[1] COPPE/UFRJ, C.P. 68505, 21945-970 - Rio de Janeiro, RJ, Brazil
[2] Metallurgical Consulting Services Ltd, Box 5006, Saskatoon, S7K4E3, Canada

studies show that PH 13-8 Mo has good SCC resistance when compared with other similar steels.

The good mechanical behavior shown by the PH 13-8 Mo steel is attributed to the precipitation of a β-NiAl intermetallic compound finely distributed in the martensitic matrix during aging between 723 K and 823 K. Except for the paper by Munn and Andersson [3], no other work mentions the presence of carbides in the aged material and no relation between their presence and the mechanical behavior has been established. With regard to chemical composition, the work of Munn [4] has shown that a small variation in the alloy content has a considerable effect on the fracture resistance. In spite of the studies made to date, no clear relationship between the microstructure resulting from different heat treatments and the mechanical behavior has been developed.

In the present work, the alloy was submitted to various heat treatments and its mechanical properties evaluated by means of tensile, Charpy and CTOD tests. From TEM and SEM observation of the microstructure, a relationship between this and the mechanical properties has been proposed.

EXPERIMENTAL PROCEDURE

The material used in this work was a PH 13-8 Mo stainless steel in the form of hot forged bars of 38 mm in diameter. The chemical composition is given in Table 1.

Table 1 - Chemical Composition of the PH 13-8 Mo (wt%)

C	Si	Mn	Cr	Ni	Mo	Al	P	S	Ti	N
0.044	0.021	0.11	12.3	7.94	2.14	1.17	0.001	0.005	0.011	0.002

The material was solution treated at temperatures between 1148 K and 1373 K for 40 min. and quenched in water with temperature lower than 288 K for 2 h. The samples annealed at 1198 K and 1323 K were aged at 783 K, 823 K and 868 K for various times from 20 min. to 6 h and air cooled to room temperature.

Vilella's reagent was used as the etchant for light and SEM examination. The prior austenitic grain boundaries were revealed by electrolytic etching with 60 mL of HNO_3 and 40 mL of H_2O with a stainless steel cathode and current density of 0.14 A/cm^2. For TEM examination the samples were initially cut with a diamond saw and ground to a thickness of

50 μm followed by electropolishing at 258 K with a current density of 1.3 A/cm^2 in a solution of 5% $HClO_4$ and 95% CH_3COOH. Replicas were prepared with an evaporated carbon layer after the samples had been etched using Vilella's reagent.

The mechanical tests were performed at room temperature on samples that were machined after aging. The specimens used in the tensile tests were cylindrical with 4 mm diameter and a gage length of 24 mm. The strain rate used was 7.10^{-4} s^{-1}. The specimens used for the CTOD tests were of the bending type with the following dimensions: 16.75 x 34 x 156 mm.

RESULTS

The microstructure of the solution treated samples consisted of large and irregular blocks of massive martensite that grew with increase in annealing temperature. When observed by SEM the microstructure revealed using diluted Vilella's reagent shows marks within the blocks. TEM observation of this structure showed the presence of parallel laths with high dislocation density. EDS analysis performed on the lath boundaries showed the presence of carbides, identified by Cr and Ti peaks.

It was observed that, although the prior austenite grain size increased with increase in annealing temperature, the hardness remained constant.

The variation in hardness with aging time as a function of aging temperature and of prior annealing condition is shown in Figure 1. The hardness of the annealed samples is indicated by a dashed line. The hardness upon aging was unaffected by prior annealing temperature; on the other hand it was affected markedly by aging temperature. There was a considerable increase in hardness at lower aging temperatures, this remaining constant for longer times. At the highest aging temperature (868 K), after the hardness reached a maximum it decreased with longer times to values close to that of the annealed material.

Table 2 shows the mechanical test results for two annealing temperatures after aging for 1 and 4 h at 783 K and 868 K. The toughness was greater at the higher aging temperature and for longer aging time. It was also dependent on annealing temperature, being greater at the lower temperature. There was significant reduction in toughness of the samples aged at 783 K from the values for the annealed material, which were 196 J and 205 J for the 1198 K and 1323 K annealing temperatures respectively.

The aged samples have a lath pattern microsctructure when examined by SEM at high magnification, as shown in Figure 2, this becoming more dense at longer times and higher temperatures. Figure 3 shows aligned carbide particles observed in extraction replicas for the sample annealed at

1198 K a aged at 868 K for 4h. The EDS analysis showed peaks of Cr, Ti and Mo. Figure 4 shows an overaged sample (100 h at 868 K) previously annealed at 1198 K the β-NiAl precipitates in which were uniformly distributed and with spherical shape. The difraction patterns showed coherent particles with a lattice parameter of 0.287 nm.

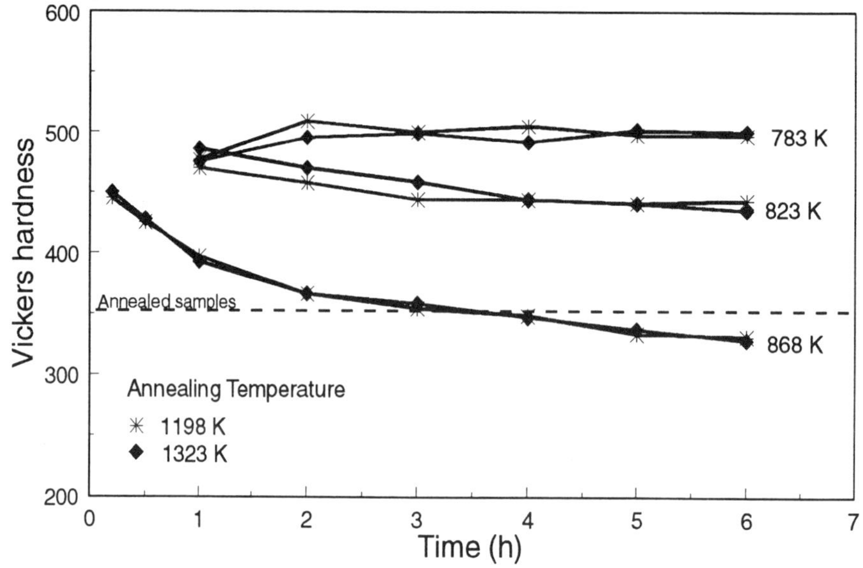

Figure 1. Hardness variation with aging time as a function of annealing and aging temperatures.

Tabela 2 - Mechanical properties of the steel PH 13-8 Mo at room temperature

Annealing temperature	1198K				1323K			
Aging temperature	783K		868K		783K		868K	
Aging time	1h	4h	1h	4h	1h	4h	1h	4h
Yield stress (Mpa)	1383	1478	1159	803	1360	1465	1080	908
UTS (MPa)	1588	1570	1303	1149	1570	1604	1252	1161
Unif. elongation (%)	6.0	3.4	6.6	8.9	5.7	4.1	6.6	7.6
Charpy Energy (J)	28	31	175	220	12	13	147	192
CTOD (mm)	-	0.004	-	0.41	-	-	-	-

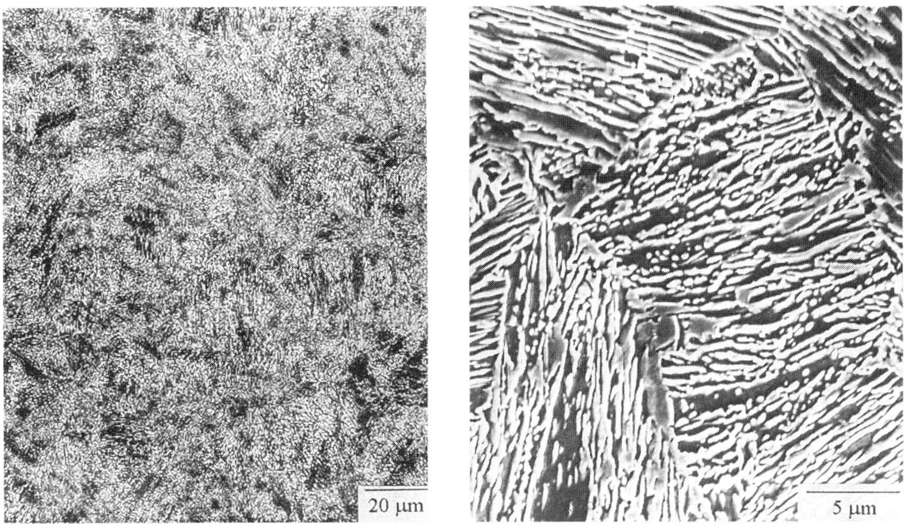

Figure 2. Microstructure observed by SEM, annealing temperature 1198 K, aging temperature 868 K, Vilella's etchant.

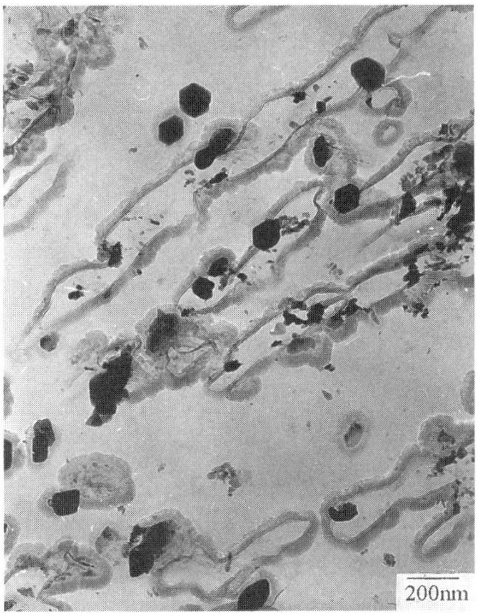

Figure 3. TEM replica examination, annealing temperature 1198 K, aging temperature 868 K, Vilella's etchant.

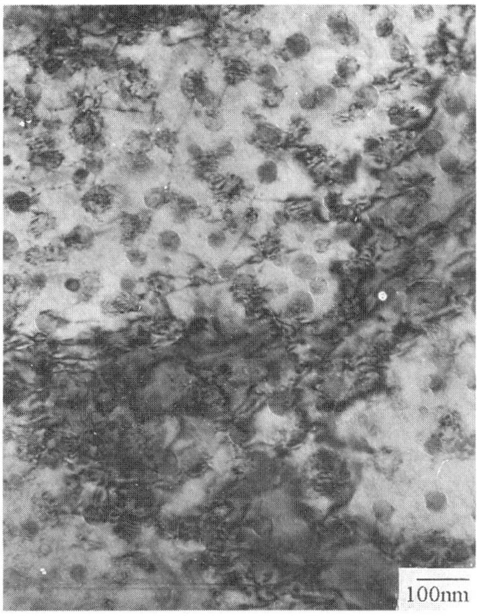

Figure 4. TEM examination, annealing temperature 1198K, aging temperature 868 K (100 h), 5% $HClO_4$ and 95% CH_3COO_4 etchant.

Figure 5. Type 1α with a needle-like morphology (723 K - 72 h - 4%)

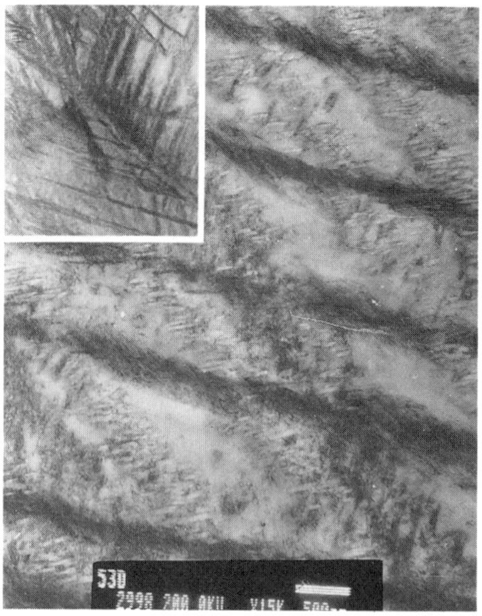

Figure 6. Alpha phase precipitation in deformation bands(813K-45min-4%)

Table 1 - Tensile tests results

SAMPLE	ε%	σy (MPa)	σmax. (MPa)
633 K/8h	12.2	451.36	817.11
633 K/8h - def.	9.77	358.31	835.81
633 K/72h	7.36	500.961	1021.96
633 K/72h - def.	8.07	552.92	887.78
723 K/1h	10	621.67	813.24
723 K/1h - def.	6.96	560.62	848.71
723 K/72h	7.49	601.6	1554.91
723 K/72h - def.	8.84	1163.85	1399.68
813 K/45min.	12.9	710.53	827.65
813 K/45min - def.	11	654.68	882.25
813 K/24h	13	868.32	1206.04
813 K/24h - def.	12.4	819.56	1142.75

CONCLUSIONS

Although it is not reported in the literature the presence of carbides can be indentified in both annealed and aged conditions. The visualization of carbides in the TEM in the annealed condition is very difficult due to the substructure of parallel laths with high dislocation density. A careful EDS analysis in the lath boundaries shows peaks of carbide forming elements, especially Ti, which is the one with a higher solutionizing temperature then those of Cr and Mo. The marks relating to the substructure within blocks of the massive martensite, as observed by SEM using diluted Vilella's reagent, are also an indication of carbide precipitation in this region.

In the aged conditions the precipitation of carbides increases and the SEM observations show a pearlite-like microstructure. Replica analysis shows aligned Cr, Ti and Mo precipitates.

The carbide precipitation characteristcs affect the toughness of this material. The absorbed Charpy energy for all test conditions is lower for the samples annealed at 1323 K than those annealed at 1198 K, although the tensile properties are not affected. For the higher annealing temperature, larger blocks of massive martensite and the presence of larger parallel laths can be seen. In this condition there are fewer lath boundaries, which promotes a more intensive aligned carbide precipitation.

The low values of the absorbed Charpy energy for the lower aging temperature can be interpreted as a tempering enbritlement-like phenomenon related to the fine carbide precipitation in a parallel lath subestructure with a high dislocation density.

The substructure recovering during the aging treatment is compensated for by the βNi-Al precipitation which maintains the high values of the yield and ultimate tensile strengths in a high toughness microstructure condition.

REFERENCES

1) V. Seetharaman, M. Sundararaman and R. Krishnan, "Precipitation Hardening in a PH 13-8 Mo Stainless Steel", Mat. Sc. Eng., Vol. 47, 1, p. 1, 1981.
2) R. R. Gaugh, "Stress Corrosion Cracking of Precipitation-Hardening Stainless Steels", Mat. Performance, Vol. 26, 2, p. 29, 1987.
3) P. Munn and B. Anderson, "Hydrogen Embrittlement of PH 13-8 Mo in Simulated Real-Life Tests and Slow Strain Rate Tests", Corrosion, Vol. 46, p. 286, 1990.
4) P. Munn, "Suscesptibility of PH 13-8 Mo to Hydrogen Assisted Cracking in Slow Strain Rate Tests", Scand. J. of Metallurgy, Vol. 18, p. 137, 1989.

CHARACTERIZATION OF TYPE 304L STAINLESS STEEL SOLID-STATE CLOSURE WELDS FOR RADIOACTIVE WASTE CONTAINMENT

S. L. WEST [1]

ABSTRACT

Radioactive wastes from nuclear materials production will be vitrified in glass and sealed in canisters. These cylindrical canisters are fabricated from Type 304L stainless steel plate and have a forged nozzle. The canisters will be sealed by resistance upset welding a slightly oversized plug into the nozzle. Automation and remote operation will minimize personnel radiation exposure. Production welding variables were recommended based on mechanical tests and metallography. Microstructures were characterized by optical metallography and scanning electron microscopy (SEM). Intentionally "cold" welds produced with low currents and short times exhibited insufficient interface length and incomplete bonding. At very high currents, long weld times and low force, maximum heating occurred with significant melting at the top. All welds made between these extremes exhibited predominantly solid-state bonding and little respective variation in microstructure along much of the interface with changes in current, force and time. Grain sizes and microhardness consistently varied along the interface from differences in dynamic recrystallization and grain growth. The degree of melting at the top was the most significant difference among welds made within the recommended parameter range.

INTRODUCTION

Wastes generated at the Savannah River Site (SRS), a Department of Energy (DOE) facility, during nearly 40 years of nuclear materials production are to be processed for long term storage. These wastes consist of fission products and other inert constituents in liquid and sludge form. They will be vitrified in borosilicate glass in the Defense Waste Processing Facility (DWPF) . Radioactive waste glass will be poured into canisters that

---

[1] Materials Technology Section, Savannah River Technology Center, Westinghouse Savannah River Company, Aiken, SC

will be sealed by resistance upset welding a plug into the canister nozzle. These canisters are cylinders approximately 61 cm (24 in) in diameter and 3 m (118 in) tall fabricated from 9.5 mm (0.375 in) thick plate with a forged bottom and nozzle joined to the cylinder by girth welds. All components are fabricated from Type 304L stainless steel. The plug is 1.27 cm (0.5 in) thick and approximately 12.7 cm (5 in) in diameter, slightly oversize for the nozzle opening (interference fit).

The filled canisters will be stored at SRS until transfer to a federal repository, where they will be placed into an additional container (overpack). Since no federal repository will be available in the near future, long term storage of the filled canisters at SRS is likely; therefore, it is essential to assure the integrity of the nozzle seal weld. From a structural standpoint, very little load will be placed on the joint, but the potential for corrosion or other environmental assisted degradation is of greater concern, and will be addressed in future studies. A high integrity joint will provide the greatest margin against degradation.

Resistance upset welding provides a reliable means for sealing the canisters and is well suited for use in the DWPF environment where its simplicity facilitates automation and remote operation, which will minimize personnel radiation exposure. It is generally a solid-state process (no melting) using high temperature and force to bond materials over a very short time period (< 2.5 sec in this case) [1]. These short times at elevated temperature also minimize the formation of a heat-affected zone in the material surrounding the weld. Grain growth across the interface subsequent to dynamic recrystallization during high temperature deformation appears to be an important mechanism by which the bonding occurs. Plastic deformation is, therefore, a key element in the formation of these solid-state welds in that it promotes recrystallization. The role which solid-state diffusion plays in the process is uncertain due to the short times at elevated temperature. The role of diffusion is also difficult to determine with materials having essentially the same composition, as in this application. Little composition gradient exists across the interface to indicate a diffusion distance. Regardless of the specific mechanism(s) by which joining is achieved, solid-state welds in an austenitic stainless steel exhibit a single phase structure (austenite) compared to duplex structures (austenite and retained delta ferrite) and associated local variations in composition and properties exhibited by fusion welds. Solid-state welds are also less susceptible to volumetric defects such as porosity from solubility changes during solidification of fusion welds. Good strength and ductility are demonstrated by properly made solid-state welds.

A set of welding parameters was developed and demonstrated to provide sound welds using full-scale mock-ups of the plug to nozzle weld (Figure 1). Production welding equipment was used. The variables involved were welding current, time (cycles of 60 Hz signal) and electrode force. During the experimental portion of this work, welds were made with two sets of parameters which varied around different central points. Some welds were intentionally made either very "cold" or "hot" to study the effects of extremes in parameters. Final welding parameters will be confirmed by a future weld procedure qualification at DWPF during cold chemical runs, prior to radioactive waste processing.

The parametric study demonstrated a range of acceptable parameters through leak tests, tensile tests, burst tests, fracture tests and measurements of bond length. Minimum requirements were a parameter range which allowed for normal process variability in a production environment, the production of a gas-tight seal weld with tensile and burst strengths equal to the canister material with reasonable ductility. Metallography was required to verify a minimum weld interface length of 8.5 mm (0.335 in), with zero flaws (discontinuities) along the interface.

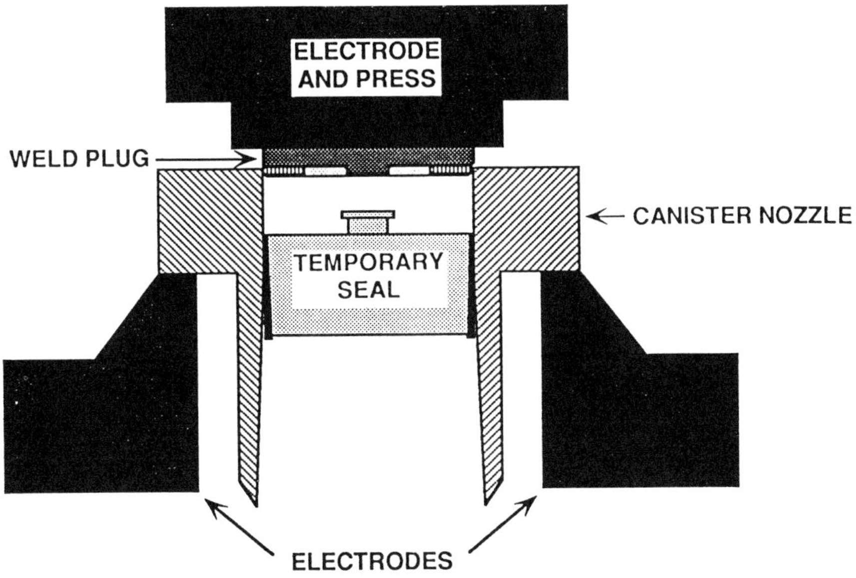

Figure 1. Schematic of resistance upset welding of canister or mock-up nozzle. Temporary seal is pressed into nozzle immediately after filling to preclude oxidation of weld area, and then is pressed in farther just prior to final welding.

OBJECTIVES

The major objective of this work was to analyze and document the microstructures of welds made during the parametric study. Other objectives included the identification of any relationships among the microstructure, welding variables and mechanical properties. Also, it was desirable to develop an understanding of the effect on the microstructure of a parameter shift from a set-point.

PROCEDURE

Metallographic specimens were cut from welds made both within the range of acceptable parameters and the initially broader parameter range outside this acceptable range. Figure 2 is a plot of the parameter spaces, showing a larger outer cube used for initial screening tests and the smaller inner rectangle of "acceptable" weld parameters. For this study, only corner points and center points within the parameter spaces were examined.

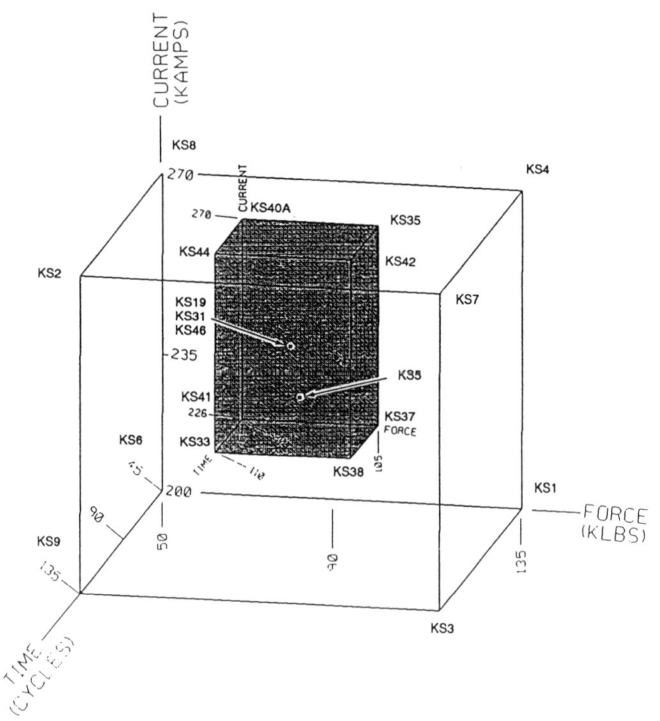

Figure 2. Plot of the welding parameter spaces. Initial screening tests (samples KS1 to KS9) utilized 222 to 600 kN (50 to 135 klbs) force, 200 to 270 kAmps current, and 45 to 135 60 Hz cycles of time. Final weld tests defined "acceptable" parameters as 334 to 467 kN (75 to 105 klbs) force, 226 to 270 kAmps current, and 80 to 110 60 Hz cycles of time, represented by inner shaded box (samples KS19 to KS44). Sample numbers shown are welds selected for characterization.

Welds were sectioned at two locations (180° apart), and each piece mounted in thermosetting epoxy resin. Mounted samples were then ground on emery paper to 600 grit, and polished with 15, 9 and 6 μm diamond paste, and 0.05 μm alumina powder. The microstructure was revealed by electrolytic etching in 10% Oxalic acid, using a stainless steel cathode. Resultant samples were then visually examined to verify that the weld interface and profile were equivalent in both samples, and one was selected for further examination. Samples analyzed and photographed at approximately 7X to 500X via optical microscopy. Several samples were subsequently examined via Scanning Electron Microscopy (SEM) at 500X and 1000X. The length of the bond (weld) interface was documented by taking photomicrographs near the top, bottom and middle of the weld. This was sufficient to cover the range of microstructures observed. Knoop microhardness (KHN) tests were also performed. Grain sizes were determined qualitatively using ASTM E 112 [2]. Grain sizes were not quantitatively determined since there was no obvious difference in grain size among welds made using differing welding parameters.

RESULTS

In general, little evidence of poor bond quality was found except in those welds which were intentionally very "cold" and resulted in a short interface length and unbonded regions. Starting microstructures were significantly different than those of the weld and surrounding region of plastic flow. The nozzle was a forged part while the plug was machined from wrought plate. Both exhibited a fairly equiaxed grain structure, but the plug material had a finer grain size (ASTM no. 5 vs. ASTM no. 3 for the nozzle).

Figure 3 shows a schematic of the plug-to-nozzle weld cross section. As stated previously, the plug was slightly larger than the nozzle opening, resulting in the extrusion of material from the joint. The shaded region in Figure 3 shows the approximate location and concentration of plastic flow in the material. Crevice formation is inherent to the process since metal is extruded from the joint interface area, creating the nascent surfaces that bond. Since there is no constraint in the metal extruded beyond the parts (termed "flash"), the unbonded portion of the interface separates. The more substantial crevice at the bottom is not associated with the bond line, but is located between the canister nozzle and material extruded from the joint. In the region between these upper and lower crevices, a typical "good" weld exhibits a bond which exceeds the minimum required length and appears "solid" without voids or interruptions. Grain size varied from top to bottom, with the largest grains at the top and smallest grains at the bottom, a combined effect of temperature, constraint and a concentration of the plastic upset.

The weld made with minimum initial parameters, low force, low current and short time was obviously "cold" and did not meet the minimum bond length requirement. Such welds would fail a post-weld plug height inspection and be rejected. However, regions of the weld actually appeared to be well bonded. When weld force was increased from this set point with current and time held constant, the quality of the weld actually decreased (Figure 4). This weld had the poorest quality of any weld in the test matrix. Intermittent areas of poor bonding, as indicated by a continuous dark feature (discontinuity) on the photomicrographs, were evident across the entire bond length and, therefore, this weld did not meet the acceptance criteria. However, this weld did demonstrate a reasonable tensile strength, so bonding of some regions did occur, which was confirmed by SEM (Figure 5).

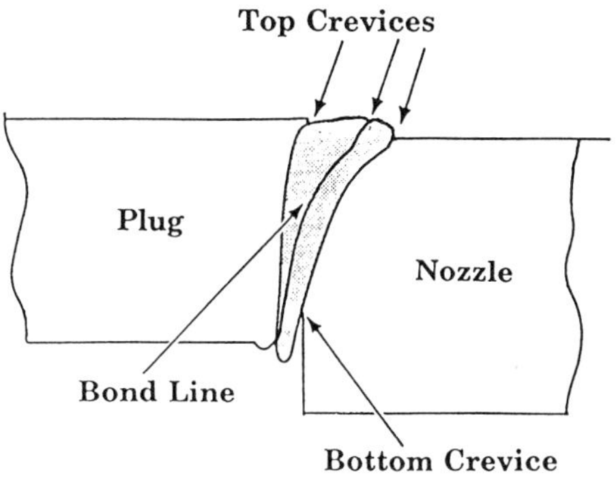

Figure 3. Schematic of a typical plug-to-nozzle resistance upset weld. Note that plug is oversize, resulting in extrusion of material from the joint, and formation of crevices. Shaded region is approximate location of plastic deformation.

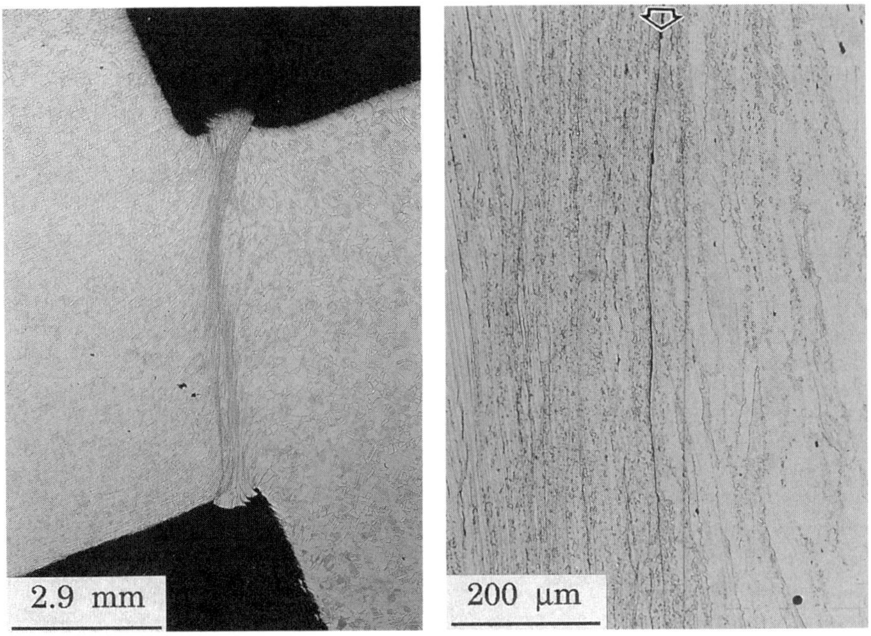

Figure 4. Optical macrograph and micrograph of initial test weld (KS1 in Figure 2) produced with maximum force (600 kN {135 klbs}), minimum current (200 kAmps) and time (45 cycles). Note short weld interface length and flow lines (left) and poor bonding at interface (right). Arrow indicates weld interface.

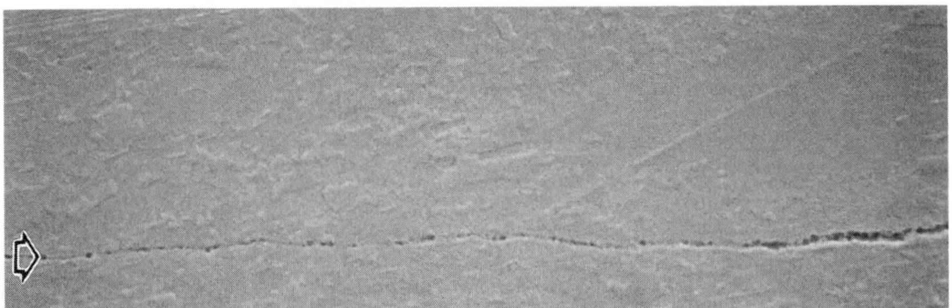

Figure 5. SEM micrograph of sample KS1 weld interface (Figure 4). Note that some localized bonding was apparent at higher magnification. Arrow indicates weld interface.

At the other extreme in parameters, the most melting was observed in the sample welded at peak current and time, but minimum force (Figure 6). This sample exhibited significant melting at the top. A potential concern of operating in this range of parameters is the stability of the process, which will be discussed later. This excessive melting was reduced for other welds made with maximum current by using either higher force or shorter weld time.

All other welds within the initial test matrix exceeded the minimum bond length and exhibited good bonding. However, the parameter space was modified to increase the minimum current to avoid cold welds, and the ranges of time and force were reduced for better control. The resultant parameter space is the shaded box shown in Figure 2. Welds made with these parameters all met or exceeded minimum requirements, and exhibited weld interfaces without voids or other discontinuities. The most significant microstructural difference between these welds was the degree of melting at the top. Some melting was evident at the top of all samples except those welded at the lowest current setting (226000 Amps). This melting is not foreseen as a problem from a strength or integrity standpoint. Corrosion tests will be performed on welds removed from glass filled canisters to address any potential impact of melting or the inherent crevices on the life of the canisters.

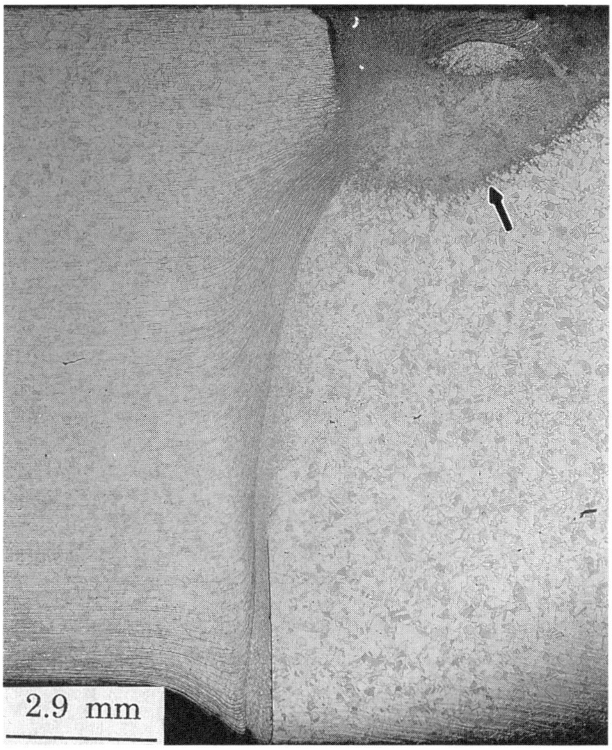

Figure 6. Optical macrograph of initial test weld (KS2 in Figure 2) produced with minimum force (222 kN {50 klbs}), maximum current (270 kAmps) and time (135 cycles). Note large melted region at top of weld, indicated by arrow.

Three welds were examined from the central point of the final, or recommended, parameter space (400 kN {90000 lbs}, 248000 Amps, 95 cycles), all of which appeared to be fairly consistent from a microstructural standpoint. Figure 7 presents one of these welds, which exhibited some melting at the top in the nozzle where material extruded from the joint area and tended to fold over on itself. At low magnification (Figure 7(a)), obvious flow lines in the plug material versus more subtle flow lines in the nozzle material make the bond line readily apparent. A hot worked structure is evident, and grain size varied along the length of the weld interface (ASTM no. 7 at the top, no. 9 in the center and no. 10 or finer at the bottom). Microhardness values were consistent with grain size in that the most severely hot worked structures with the finest grains were also the hardest (Figure 8). Note the formation of crevices as described previously (Figure 3). At higher magnification, the photomicrographs show a solid bond line with no voids. A very small crevice was apparent at the top, but the bond became solid just a short distance into the weld (Figure 7(b) and (c)). The bond in the center was solid, with grain growth evident across the interface (Figure 7(e)). The bond at the bottom was also solid well beyond where the crevice began (Figure 7(f)).

DISCUSSION

In general, the weld microstructures were not significantly affected by parameter variations within the range of recommended parameters. Weld interface quality was satisfactory throughout the recommended welding parameter ranges. No obvious correlations between parameters and interface length or grain size were apparent. Welds made with the initial range of parameters indicated that increasing current or time, or decreasing force tend to produce hotter welds. In particular, force influences the contact resistance and the resultant heating at the interface.

The grain size variability from top to bottom indicates variations in plastic flow and temperature responsible for recrystallization, but this was consistent from weld to weld. The extremely fine grains near the bottom of the weld indicate the high degree of constraint at that part of the joint and the occurrence of dynamic recrystallization. The larger grains at the top indicate less constraint and higher temperature, also manifested as melting. Much of the plastic flow was concentrated in the plug material, which is probably due to constraint and greater heat sink on the canister nozzle side of the joint.

The degree of melting was the most obvious difference among welds made with recommended parameters, with an increase in melting produced by high current, long time, and low force. This melting begins at the outer edge of the upset area. Melting was not necessarily found at the weld interface. No deleterious effect of this melting on properties has been identified. However, large amounts of melting may result in deleterious operating conditions such as electrode sticking or wear, or ejection of molten metal (expulsion).

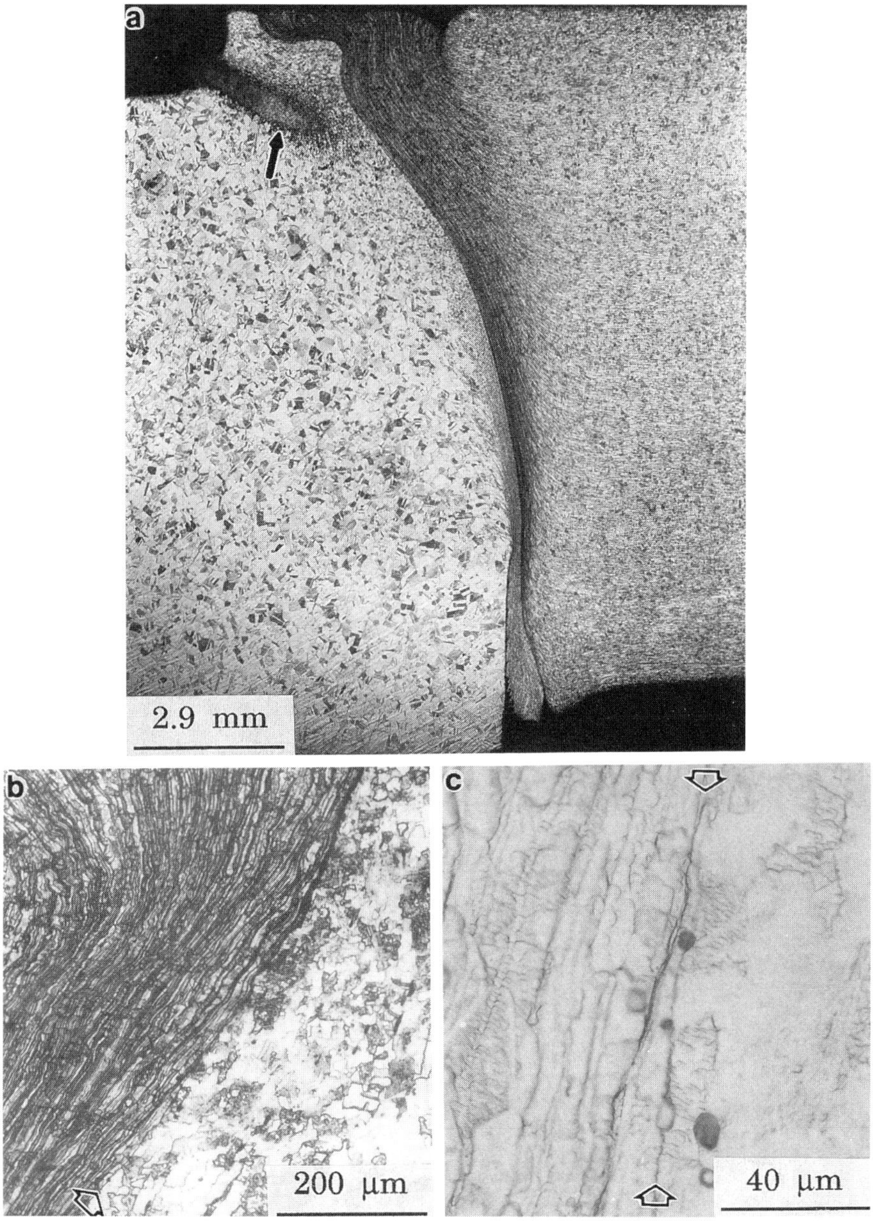

Figure 7. Optical macrograph and micrographs of weld (KS19 in Figure 2) produced with parameters at center of recommended parameter space (400 kN {90 klbs} force, 248 kAmps current, and 95 cycles time). Macrograph in (a) shows slight melting at the top and concentration of flow lines toward bottom. Micrographs show bond line at top (b,c).

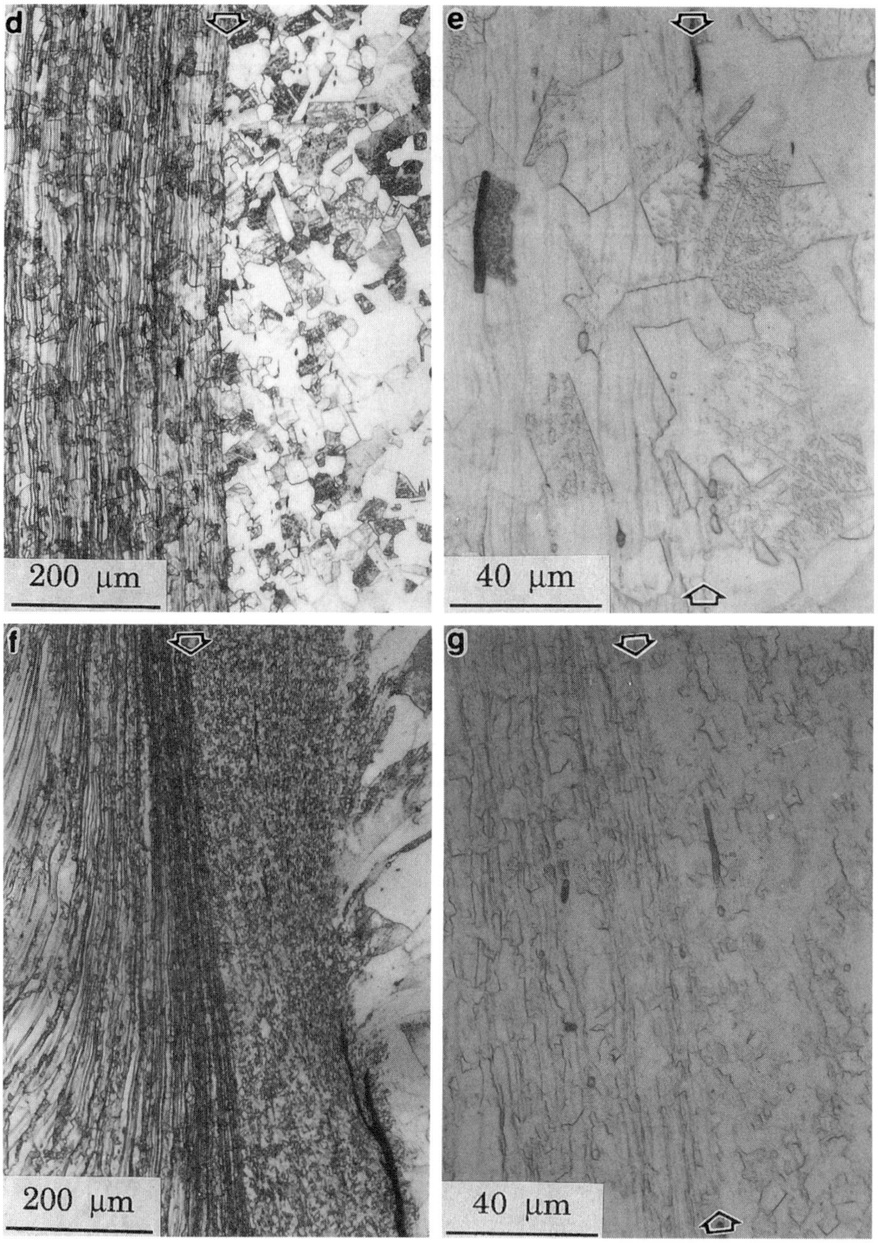

Figure 7. Con't. Micrographs show bond line at center (d,e) and bottom (f,g). Note that all regions exhibited excellent bonding, and that grain size decreases from top to bottom. Arrow in (a) indicates melted region and arrows in (b - g) indicate weld interface.

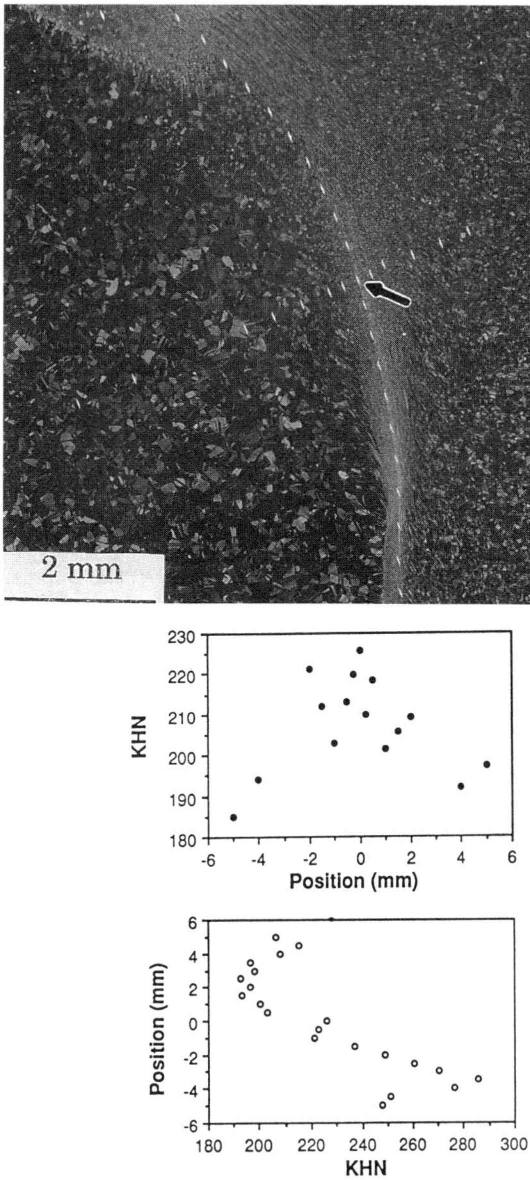

Figure 8. Microhardness (KHN) as a function of position from the weld interface at the center of weld KS19 (Figure 7). A 500 g load was used with a 10 s load time. Note that the nozzle material (left) was slightly softer than the plug material (right), hardness increased across the interface and down the interface, indicating how heavily worked the structure was. The zero position point is at intersection of the two hardness traverses (arrow).

The tensile and burst strengths were not specifically reported here, but all of the welds from the recommended parameter space exhibited good tensile strengths. This result is a reflection of the good bond quality and bond length exhibited by these welds. However, there was some variation in weld burst strengths. "Hotter" welds made at high current and low force exhibited higher burst pressures than those at high current and high force. This is undoubtedly due to the higher heat at the weld interface for lower force welds. The microstructural difference in observed bond quality for the high and low force welds was too subtle to explain the difference in these properties. The results of this work indicate that any combination of parameters within the recommended range will produce acceptable welds. However, there is a general trend for better mechanical properties with forces lower than maximum.

CONCLUSIONS

This work confirmed that all of the welds produced using parameters within the recommended ranges of the parametric study meet the acceptance criteria of minimum interface length and no discontinuities along that section of weld interface. Weld microstructures varied significantly from those of both base metals and along the length of the weld interface, but the most obvious microstructural difference observed for these welds is the degree of melting at the top, especially where material is extruded from the nozzle material. Slight variability in mechanical properties of the welds can not be easily explained by the microstructures. Therefore, a final welding parameter set-point near the center of the recommended parameter space seems appropriate to produce reliable, predominantly solid-state welds.

ACKNOWLEDGEMENTS

B. J. Eberhard and P. A. Kestin coordinated the initial work during the parametric study. T. B. Curtis performed the optical metallography and D. L. Williams operated the SEM. J. R. Harbour and W. R. Kanne, Jr. provided insight through many useful discussions.

REFERENCES

1) W. R. Kanne, Jr., "Upset Welding," ASM Handbook, Vol. 6, p 249, ASM International , Materials Park, OH, 1993

2) Metals Handbook, Vol. 7, 8th Edition., p 4, American Society for Metals, Metals Park, OH, 1972.

Chapter 6

Fatigue, Fracture and Failure Analysis

A FRACTOGRAPHIC INVESTIGATION OF THE ORIENTATION DEPENDENCE OF FATIGUE CRACKING MECHANISMS IN ALUMINUM-LITHIUM PLATE.

D.A. Hull[1], Z. Zhai[2], J. Morrison[2]

ABSTRACT

A fractographic investigation of Alumimun-Lithium 8090-T8771 fatigue crack growth mechanisms at selected plate orientations was carried out using optical and scanning electron microscopy. In an earlier study [1], constant amplitude fatigue crack growth rate data (FCGR) was obtained using compact specimens machined from 13 mm thick 8090 plate with the crack planes oriented at 30° intervals with respect to the rolling direction. In this study, fracture features observed on the broken fracture surfaces have been correlated with the corresponding crack extension behaviour observed in the da/dN versus ΔK plots. Reductions in FCGR behaviour resulted from the onset of crystallographic facets on the fracture surface. Facet location on the fracture surface was dependent on plate orientation and stress intensity. The changing importance of 8090 grain boundaries in assisting crack propagation was also related to plate orientation. Metallographic and crystallographic texture analyses were used to aid interpretation of the fractographic results.

INTRODUCTION

Interest in applying Aluminum-Lithium alloys in aerospace structures derives from the significant airframe structural weight reductions that can be achieved relative to conventional damage tolerant alloys. However, Al-Li fatigue crack growth and fracture properties need to be comparable to the conventional alloys that they are to replace. An understanding of Al-Li fatigue crack growth mechanisms will aid in

[1] Engineering Material Research, Toronto, Ontario, M3J 2S1

[2] Defence Research Establishment Pacific, Victoria, B.C., V0S 1B0

establishing the material boundary conditions necessary for damage tolerant design.

The Defence Research Establishment Pacific (DREP) has investigated the effect of Aluminum-Lithium 8090-T8771 plate orientation on constant amplitude fatigue crack growth rate (FCGR) [1]. Compact specimens (W = 51.0 mm and B = 12.5 mm) were machined from 8090 plate with the crack growth direction at 0, 30, 60 and 90 degrees relative to the transverse axis (Figure 1). The long transverse axis, was selected as a reference direction to which other off-axis plate orientations were referred because frequently observed macroscopic crack deflection did not occur in the L-T specimen.

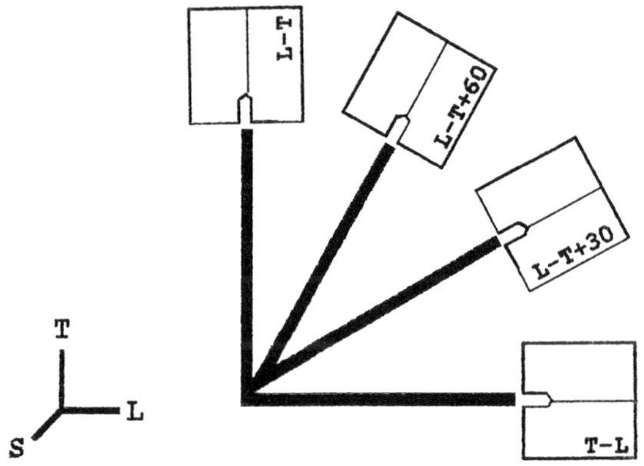

Figure 1. Compact Specimen Orientation.

Specimens were side grooved 10% on each side along the intended crack growth direction in an attempt to constrain crack growth to the plane normal to the direction of applied loading. Tests were conducted in a 100 kN capacity servo-hydraulically controlled load frame with a 10 Hz constant amplitude triangular wave form. The laboratory air temperature and humidity were ~ 23 C and 35% respectively. This paper describes a follow-on fractographic study of broken compact specimens evaluated at an R ratio of 0.5 [2].

EXPERIMENTAL PROCEDURES

Three dimensional optical micrographs were prepared from the 8090-T8771 plate and compared to a conventional aerospace 2024-T351 alloy. Samples were removed from the plate mid-thickness, polished and etched with Keller's reagent. The texture of the 8090 plate was analyzed at the plate surface and mid-thickness locations using a Schulz back-reflection X-ray diffraction method.

The fracture surfaces of the DREP specimens were inspected using a scanning electron microscope. Fractographic features were correlated with crack length using the calibrated micrometers of the SEM x-y stage.

RESULTS

Metallography and Texture.

The 8090 and 2024 metallographic samples are shown in Figure 2. The unrecrystallized peak-aged microstructure of the 8090 alloy exhibits planar "pancaked" shaped grains in the L-T plane. The addition of Zr for grain refinement results in the formation of coherent Al_3Zr precipitates [3], which retards recrystallization during processing and leads to the observed grain shape. The 8090 T-S and L-S planes illustrate the grain boundary structure present on two of the fracture surfaces that were studied.

The texture analysis showed that the degree of preferred crystal orientation was higher at the mid-thickness of the alloy. The dominant textures were {110}<211> and {123}<634>. Volume fraction calculations indicated that approximately 40% of the grains have one of these textures.

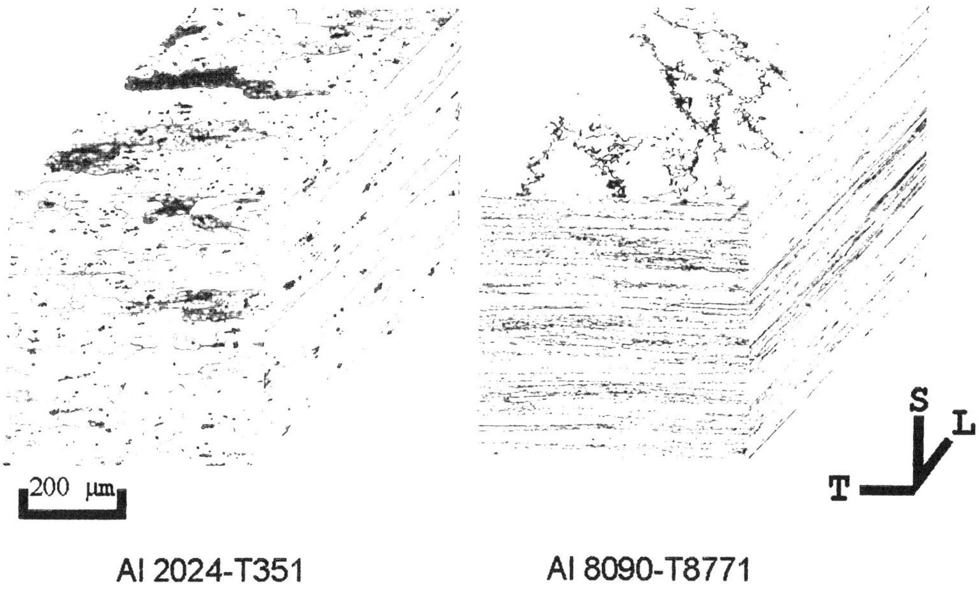

Figure 2. Alloy Al 2024-T351 and 8090-T8771 microstructures.

Fractography: L-T Orientation.

The 8090 L-T specimen was evaluated over the ΔK range 6 - 19 MPa√m. The da/dN versus ΔK plot (Figure 3a) indicates a gradual reduction in FCGR occurs as ΔK increases from 8 to 12 MPa√m, followed by an increase at higher applied stress intensity levels. The surface in Figure 3b exhibits increasing roughness, initially more noticeable at the specimen mid-thickness, then spreading to the sides with increasing crack length. A view of the "A" location at ΔK ~ 6 MPa√m is shown at

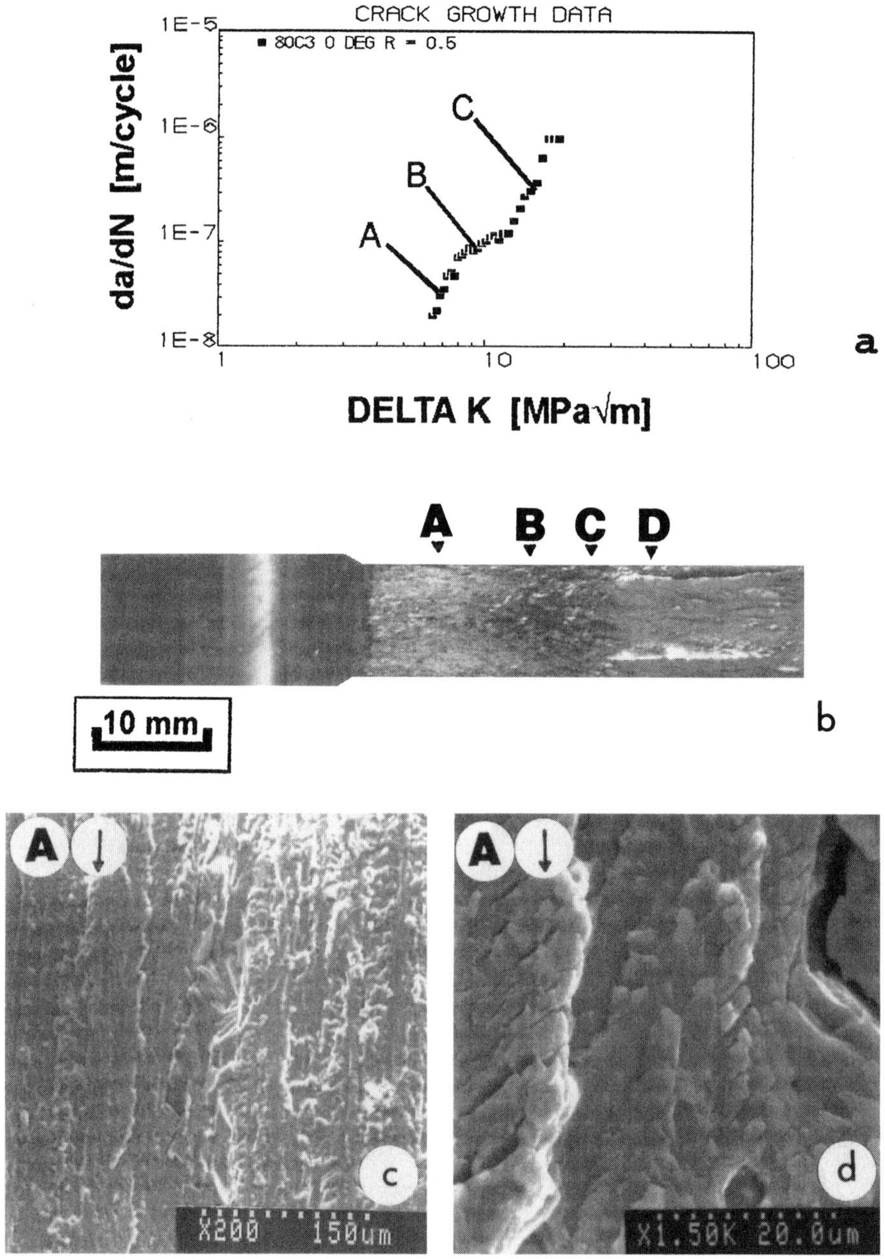

Figure 3. 8090-T8771 L-T Fracture Surface

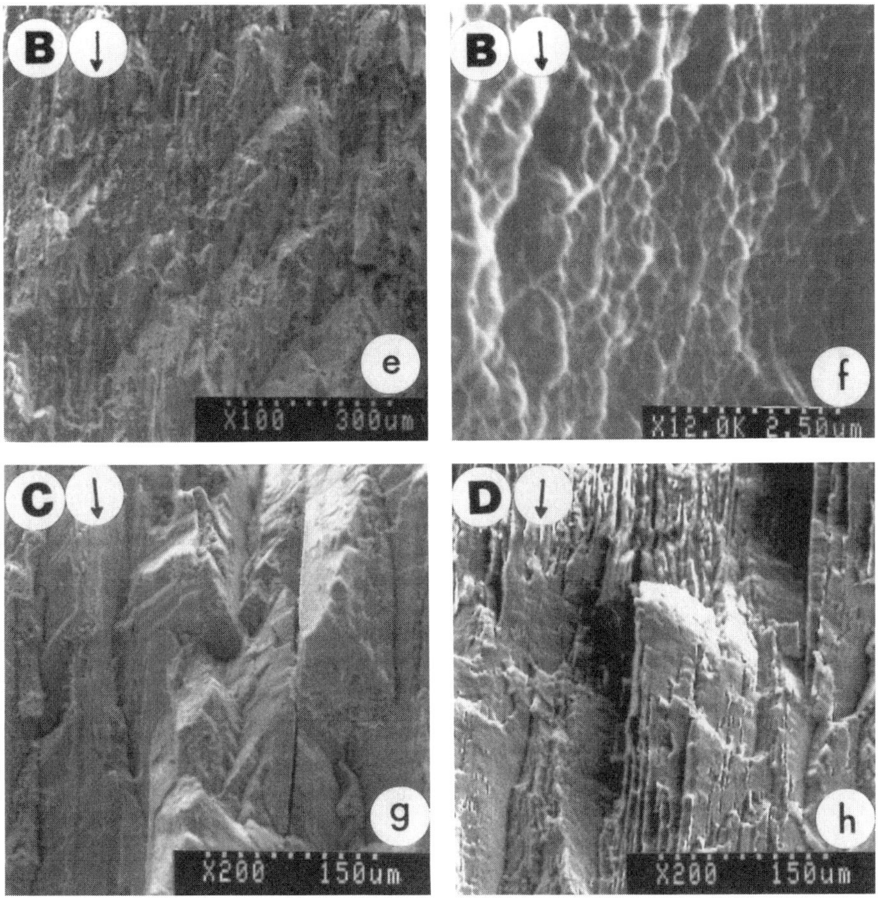

Figure 3 continued. 8090-T8771 L-T fracture surfaces.

low and high magnification in Figures 3c and 3d. In Figure 3c, the surface appears relatively flat and crack growth has followed the elongated grain boundaries. Figure 3d reveals extensive crack tip branching that occurs as a result of slip band cracking. In addition, secondary cracking along grain boundaries was occasionally noted. Figures 3e and 3f were taken at "B" where a flattening of the FCGR curve occurred, ($\Delta K \sim 10$ MPa√m). Figure 3e shows the center thickness where larger facets were found, while Figure 3f reveals that a typical facet surface is covered with uniform shallow dimples. The facets produced an increase in surface roughness as the crack front deflected from the primary crack plane. Spreading of the facets to the specimen sides produced a wedge shaped region of increasing roughness with the apex of the wedge pointing to the machined starter notch. At "C" in Figure 3g ($\Delta K \sim 12$ MPa√m) facet size can span several grains but changes in the facet inclination angle frequently occur at grain boundaries. Finally, a combination of grain boundary and transgranular shear fracture is apparent at the "D" fast fracture location in Figure 3h.

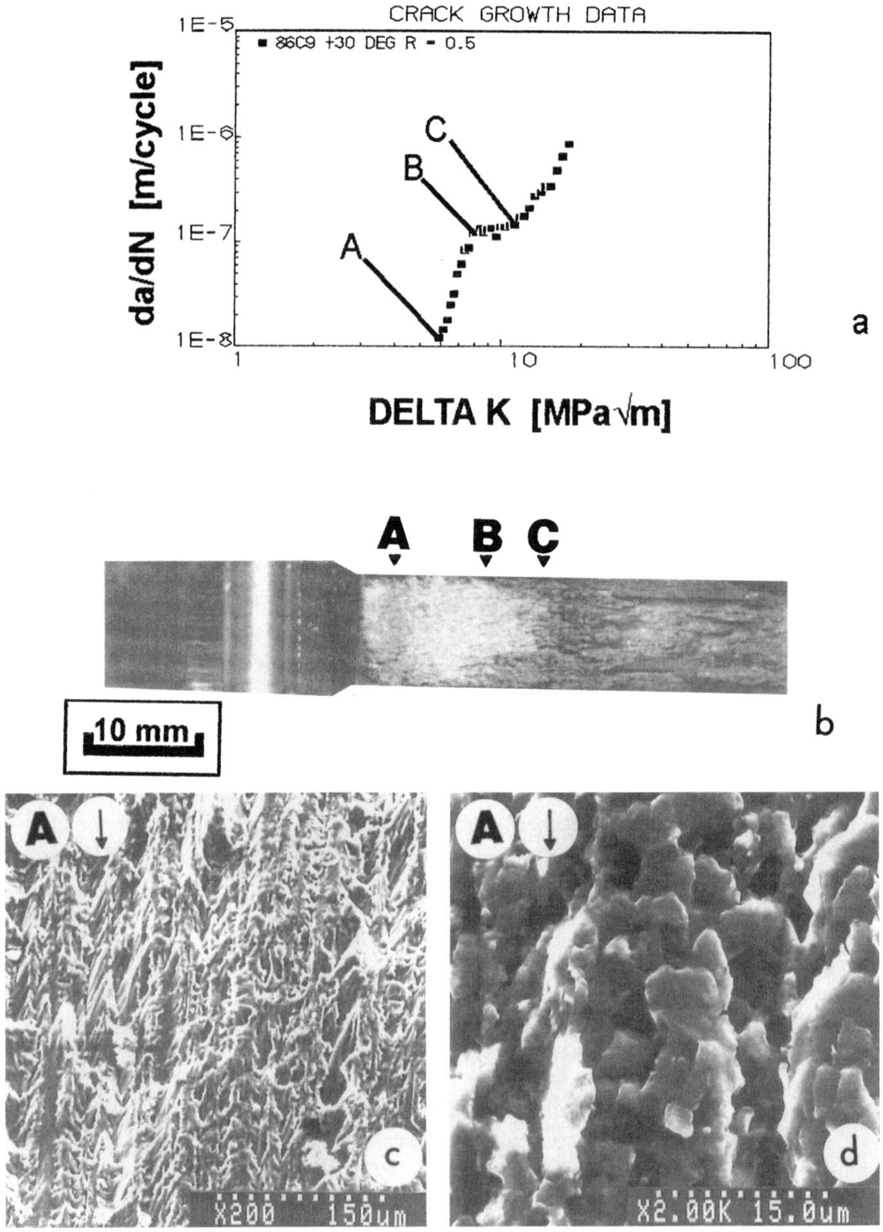

Figure 4. L-T+30 8090-T8771 fracture surfaces.

L-T+30 Orientation

The da/dN versus ΔK plot for the L-T+30 specimen in Figure 9a is similar to the L-T specimen plot in Figure 4a with a flattening in FCGR at intermediate ΔK.

Figure 4b shows three distinct regions of increasing surface roughness with increasing crack length. The first has a macroscopically smooth appearance. At $\Delta K \sim 8\ MPa\sqrt{m}$, the second region begins with a change in surface roughness at the specimen edges that progressively extends into the specimen interior. The third region has a further increase in surface roughness 2 - 3 mm prior to the onset of unstable crack growth. Regions 2 and 3 of Figure 4b are similar in appearance to the two found in Figure 4b. The "A" fracture surface at $\Delta K \sim 6\ MPa\sqrt{m}$ is shown in Figure 4c and has a series of bright ridges and dark inclined planes within a particular grain. Local crack advance occurred on a particular inclined plane for approximately 10 μm before changing to the next inclined plane (Figure 4d). Figure 4e shows the "B" location ($\Delta K \sim 8\ MPa\sqrt{m}$) where crystallographic facet first occurred at the edges of the specimen fracture surface. At this ΔK the facets are bracketed by the grain boundaries visible in Figure 4f. The fracture surface at "C" shows that transgranular crystallographic facets were numerous at $\Delta K \sim 14\ MPa\sqrt{m}$. Dimples on one facet are presented at increasing magnifications in Figures 4g and 4h.

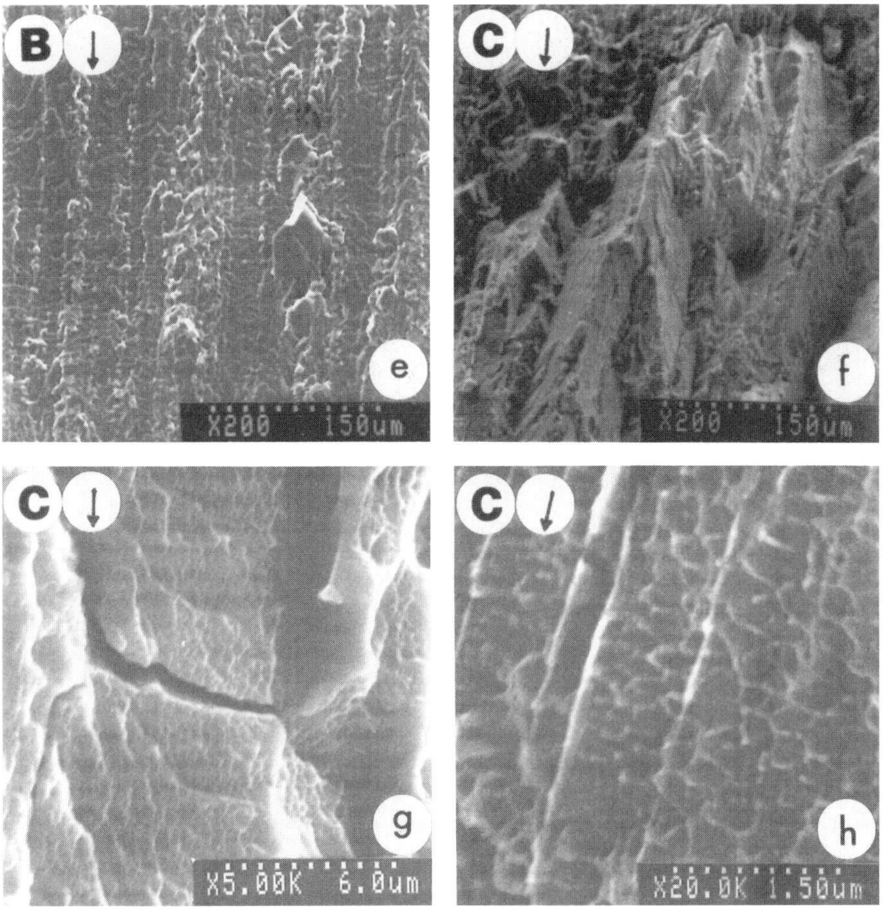

Figure 4 continued. 8090-T8771 L-T+30 fracture surfaces.

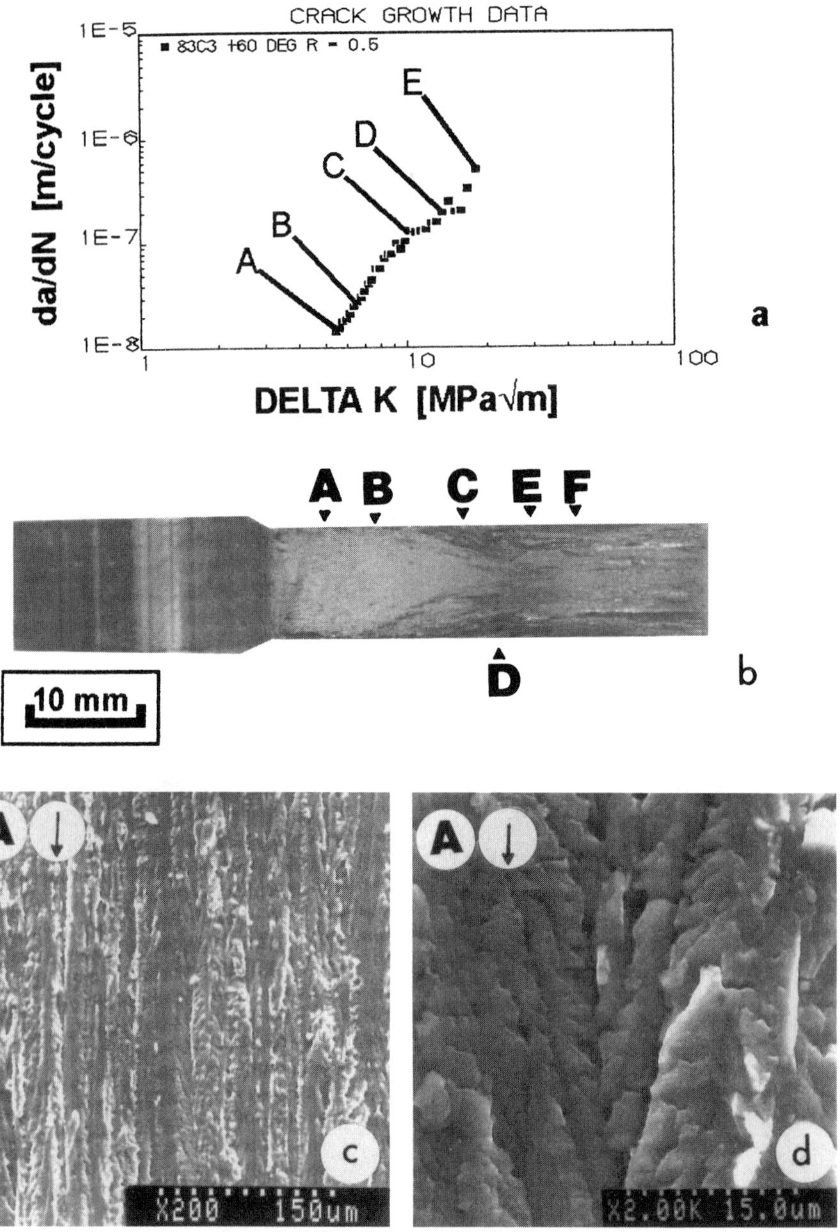

Figure 5. 8090-T8771 L-T+60 fracture surfaces.

L-T+60 Orientation

The L-T+60 da/dN versus ΔK data extends over a ΔK range of 6 to 18 MPa√m in Figure 5a, with no retardation in FCGR, although an increase in scatter was noted at $\Delta K > 10$ MPa$\sqrt{m}$. Figure 5b shows increasing surface roughness started at the sides and extended inward. However, this region did not reach the mid-thickness prior to fast fracture. Consequently a macroscopically smooth region extended along the centerline from the tip of the pre-crack to the fast fracture region. The "A" location ($\Delta K \sim 6$ MPa$\sqrt{m}$) is shown in Figures 5c and 5d. At low magnification, grain boundaries are delineated by changes in features along adjacent grains. Some features that resemble cleavage river patterns at low magnification consist of extensive secondary cracking (Figure 5d). Figure 5d also indicates a large amount of crack tip deflection. At "B" ($\Delta K \sim 7$ MPa$\sqrt{m}$) in Figure 5e, the cracking that occurred along grain boundaries is shown.

Secondary cracking along grain boundaries was more frequent on this specimen relative to the two preceding specimens. At $\Delta K \sim 11$ MPa$\sqrt{m}$, individual grain boundaries show a laminated morphology (location "C" in Figure 5f). In Figures 5f and 5g, local crack tip deflection was reduced and numerous open secondary cracks were noted on the fracture surface. In Figure 5h at "D" ($\Delta K \sim 14$ MPa$\sqrt{m}$) near the specimen sides the crystallographic facets were produced as the crack

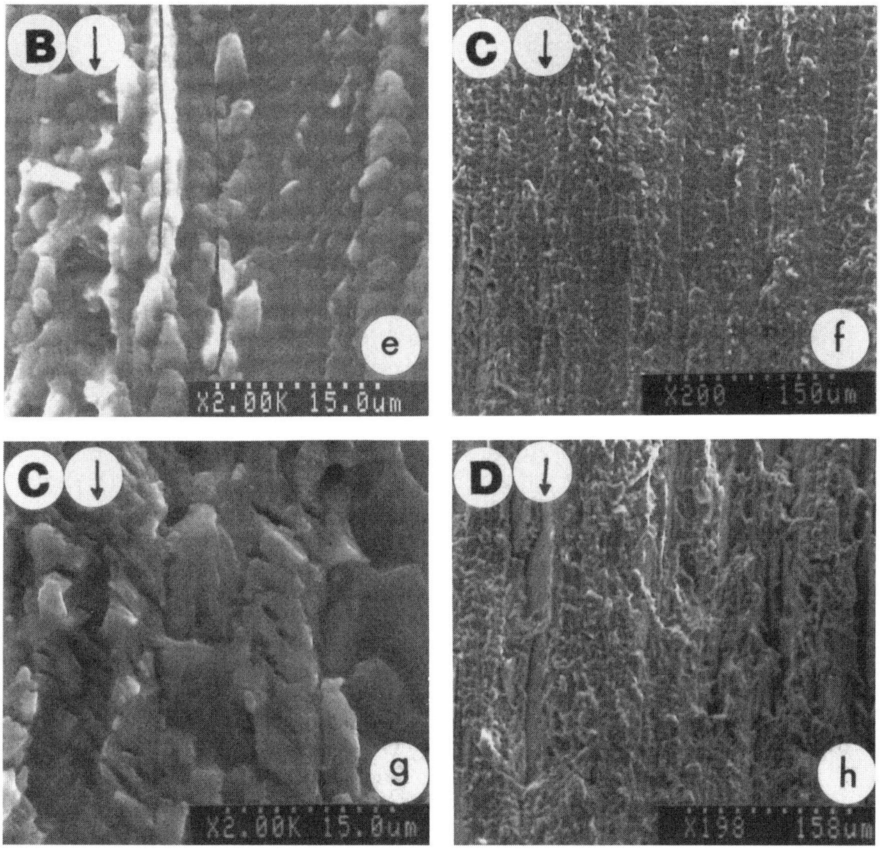

Figure 5 continued. 8090-T8771 L-T+60 fracture surfaces.

followed grain boundaries. At "E" (ΔK ~ 18 MPa√m) secondary cracking parallel and normal to grain boundaries occurred prior to fast fracture (Figure 5j). At "F" in Figure 5k fast fracture followed grain boundaries with secondary cracks occurring by grain boundary delamination. It should be noted for this specimen that facets were not numerous on the fracture surface and a reduction in the FCGR was not observed.

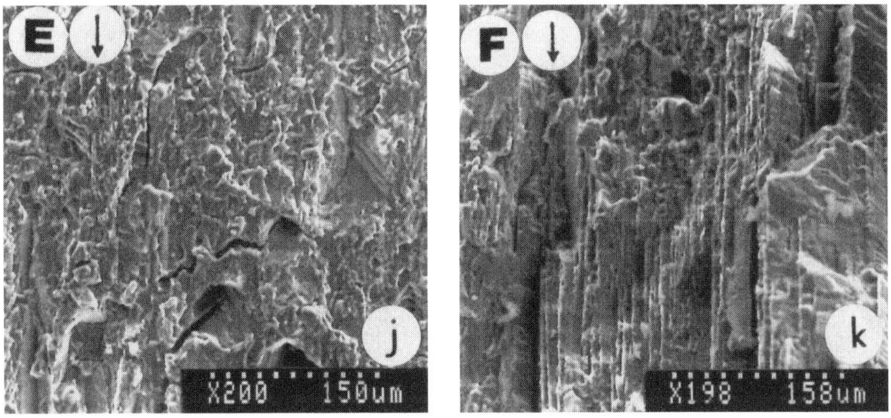

Figure 5 continued. 8090-T8771 L-T+60 fracture surfaces.

T-L Orientation

Figure 6a shows that the da/dN versus ΔK plot has a noticeable FCGR reduction over the ΔK range 12 - 16 MPa√m. Figure 6b illustrates the large amount of macroscopic fatigue crack deflection exhibited by this specimen (up to 45° from the T-L plane). Conversely unstable crack growth occurred along the T-L plane. The specimen fracture surface contained a half cylindrical groove in the center that reached a depth of 4.0 mm relative to the edges. Both the sides and bottom of this groove were coated with the black wear and debris particles usually only found at R ratios of 0.1. This observation suggested significant abrasive contact between mating cylindrical fracture surfaces had produced wear particles. The surface at "A" (ΔK ~ 8 MPa√m) in Figure 6c is significantly rougher than on the other specimens at this ΔK value. This surface was produced before significant macroscopic crack deflection occurred. A detailed view is provided in Figure 6d. Figure 6e is from the "B" location where crystallographic facets were initially formed (ΔK ~ 10 MPa√m). Macroscopic crack deflection from the T-L plane had already begun. The center thickness "C" location (ΔK ~ 12 MPa√m) in Figure 6f shows that crack propagation was predominantly crystallographic, which may explain the observed macroscopic crack deflection. Fracture along grain boundaries produced grain delaminations that enabled the crack to change elevation between crystallographic facets to form the groove/hill geometry in the specimen interior. The grain boundary structure is illustrated metallographically on the L-S plane of Figure 1. Near the specimen edges grain boundary fracture was rare, but it occurred more frequently 4 mm from the free surface and resulted in a rapid change in crack front elevation. At "D" (ΔK ~ 18 MPa√m), Figure 6g again shows fracture along the grain boundaries, and at "E" fast

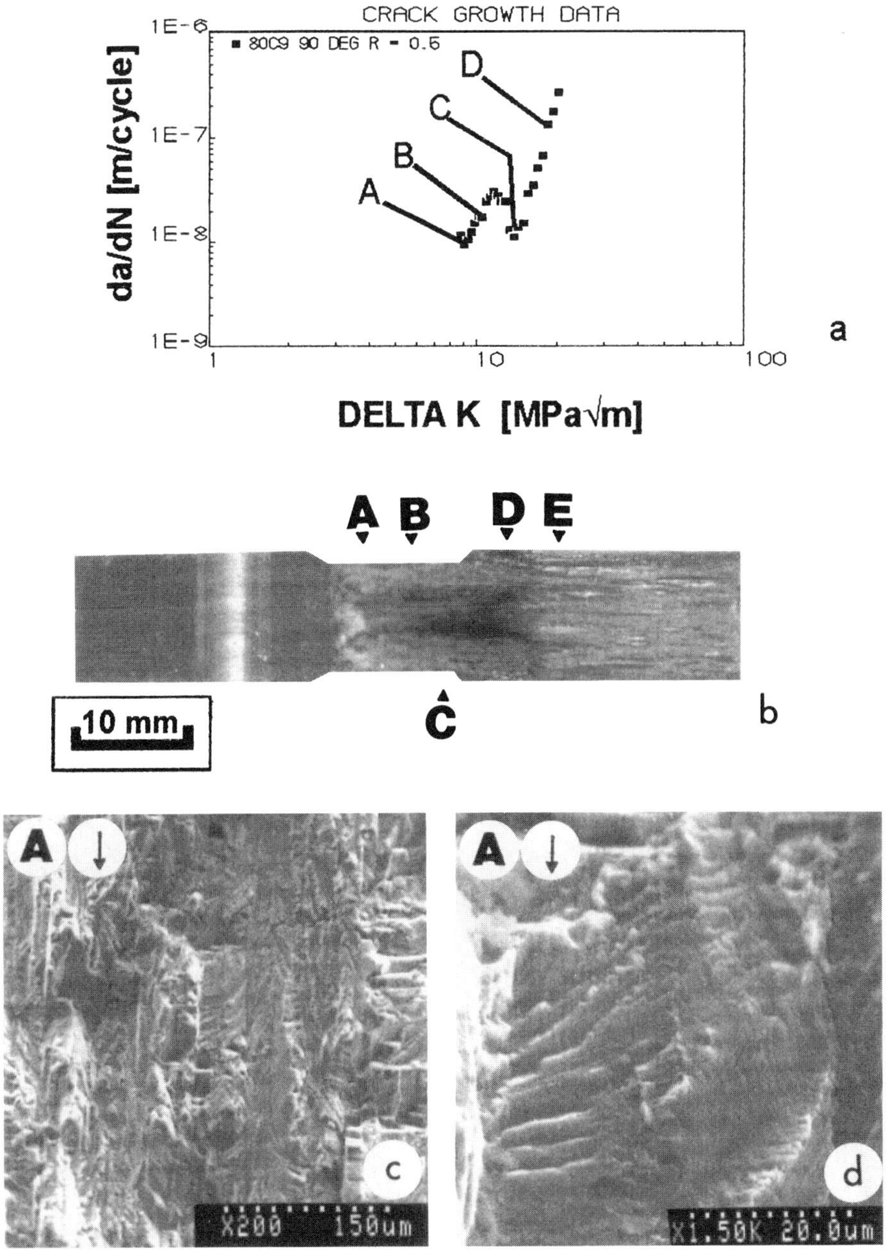

Figure 6. 8090-T8771 T-L Fracture Surface.

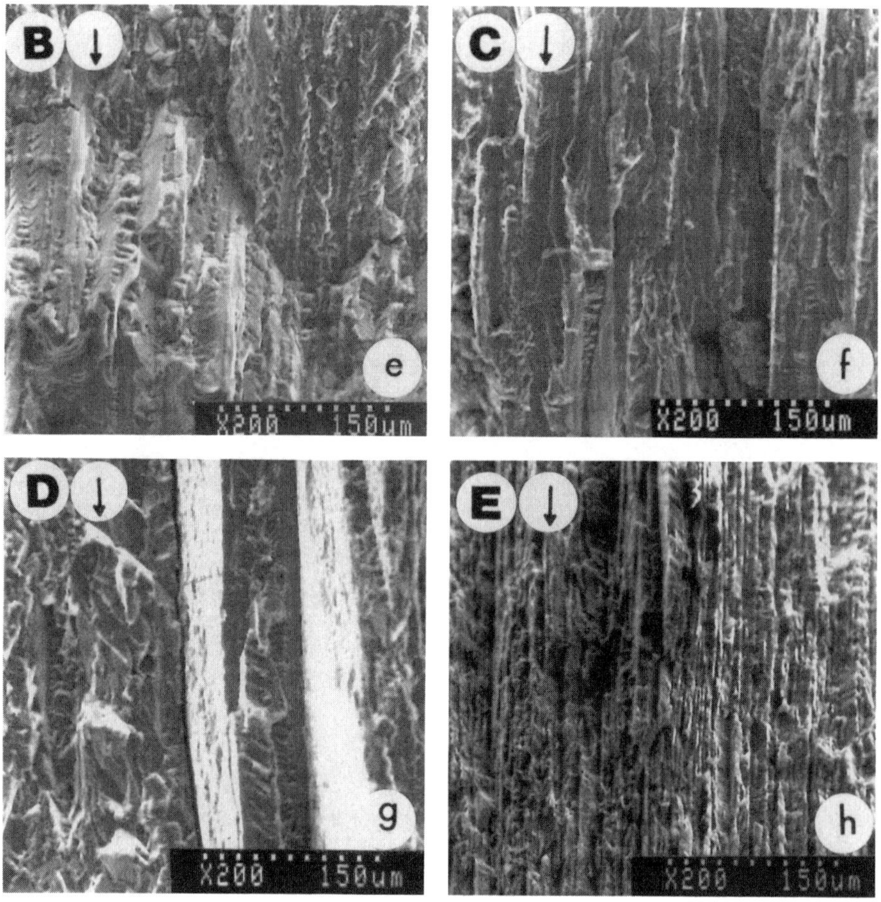

Figure 6 continued. 8090-T8771 T-L fracture surfaces.

fracture occurred almost entirely in this manner (Figure 6h). These latter two figures demonstrate the larger role played by grain boundaries during T-L crack propagation relative to the other specimen orientations.

DISCUSSION AND CONCLUSIONS

The experiment boundary conditions i.e., waveform, environment, etc., used to evaluate the DREP specimens have revealed specific 8090-T8771 fatigue and fracture mechanisms that are discussed below.

At low values of $\Delta K < 8$ MPa$\sqrt{m}$ Stage I crack [4] behaviour was observed. Rao et al.[5] reported that at low values of ΔK crack advance occurs along {100} or {110} planes. At the crack tip, localized slip bands are formed as Al_3Li precipitates are readily sheared by dislocations. In the peak aged 8090-T8771 material the slip bands are narrower and more widely spaced than for an under-aged material [6]. Consequently fatigue crack path tortuosity is increased as the crack moves along favourably oriented slip planes. Results obtained from a binary Al-Li alloy showed crack path tortuosity to be related to increasing lithium content [7]. An etch pit technique [8] could be used to determine 8090 slip band orientation at low ΔK values.

At $\Delta K > 8$-10 MPa$\sqrt{m}$, crack propagation frequently followed crystallographic planes to produce a faceted surface. It has been shown [9] that these facets are produced as the crack propagates locally along {111} crystallographic planes. The facets were covered with regular networks of shallow dimples, indicating that fracture on these planes occurred by a combination of shear and tensile rupture. Reductions in measured fatigue crack growth rates occurred with the initial formation of facets on the fracture surface.

Facets initially were limited to the grain size but exceeded this limiting dimension at higher stress intensities. It is believed that the crack tip plastic zone was not sufficiently large to overcome the hindering effect of the grain boundary until a critical stress intensity value was reached. However, additional research is required to support this belief. At $\Delta K > 10$-12 MPa$\sqrt{m}$, transgranular faces formed and increased the surface roughness and sites of contact between mating fracture surfaces. Higher levels of closure levels would then be expected to result. Closure analysis of the FCGR results are being carried out by DREP.

The initial occurrence of facets on the fracture surface varied with plate orientation and stress intensity. On L-(T+30) and L-(T+60) specimens intergranular facets were initially absent and then formed at the specimen sides before spreading toward the center thickness.

Grain boundaries played an increasing role during fatigue crack propagation as the crack plane of interest was rotated toward the rolling direction. Extensive grain boundary delamination on the T-L specimen allowed the crack front to change elevation in the through-the-thickness direction and allowed the largest amount of macroscopic crack deflection to occur. Grain boundaries segregation is believed to cause Li denuded precipitate free zones (PFZ) that are weaker than the grain matrix material [10]. The orientation of the PFZ is imposed by the 8090 grain shape which has its largest dimension lying in the L direction. Consequently the grain boundary would be the weakest when loading was applied in the T (or S) directions. Grain boundary fracture would enable crystallographic cracking to proceed in a direction governed by the material texture.

The effect of plate orientation on 8090-T8771 fatigue and fracture mechanisms has been studied fractographically. Changes in macroscopic surface roughness were caused as intergranular facets became transgranular in nature. Grain boundary fracture increased as the crack plane of interest was rotated from the T to the L direction. The importance of determining the expected fracture mechanisms for the intended material application has been emphasized.

ACKNOWLEDGMENT

The fractographic work was carried under contract W7708-2-0158/01 on behalf of the Defence Research Establishment Pacific.

REFERENCES

[1] Z. Zhai, J. Morrison, "Constant Amplitude Fatigue Crack Growth in Aluminum-Lithium," Research and Development on Aluminum-Lithium Alloys Proceedings, Defence Research Establishment Pacific Report 93-1, Victoria B.C., p. 306, April 1993

[2] D. A. Hull, "Fractographic Examination of Fatigue Crack Growth in Aluminum-Lithium Alloys", DREP Contractors Report 94-57, Victoria B.C., March 1994

[3] A.F. Smith, "Aluminum-Lithium Alloys for Helicopter Structures", **Metals and Materials**, p. 439, August 1987

[4] P.J.E. Forsyth, Proceedings of the Crack Propagation Symposium, Cranfield, U.K., 1961

[5] K.T. Venkateswara Rao, W. Yu, R.O. Ritchie, "Mechanisms of Fatigue Crack Growth Propagation in Commerical Aluminum-Lithium Alloys", Aluminum-Lithium Alloys Design, Development and Application Update, R.J. Kar, S.P. Agrawal, W.E. Quist, Eds., ASM International, Metals Park Ohio, p. 173, 1988

[6] K.V. Jata, E.A. Starke, "Fatigue Crack Growth and Fracture Toughness Behaviour of an Al-Li-Cu Alloy", **Met. Trans. A**, Vol 17A, p. 1011, June 1986

[7] A.K. Vasudevan, S. Suresh, "Lithium-Containing Aluminum Alloys: Cyclic Fracture", **Met. Trans. A**, Vol. 16A, p. 475, March 1985

[8] R.S. Piascik, R.P. Gangloff, "Environmental Fatigue of an Al-Li-Cu Alloy: Part II. Microscopic Hydrogen Cracking Processes", **Met. Trans A**, Vol 24A, p. 2751, Dec. 1993

[9] X.J. Wu, W. Wallace, M.D. Raizenne, A.K. Koul, "The Effects of Orientation on Fatigue Crack Growth and Fracture in 8090 Aluminum-Lithium Plate," Institute of Aerospace Research Report LTR ST 1912, National Research Council, 1993

[10] E.A. Starke, W.E. Quist, "The Microstructure and Properties of Aluminum-Lithium Alloys", AGARD Conference Proceedings, New Light Alloys, AGARD-CP-444, p. 4-1, 1989

THE PERFORMANCE OF THIN ANODIZED FILM OF 2024 ALUMINUM ALLOY UNDER LOW CYCLE FATIGUE

K. Habib [1] and R. Danials [2]

ABSTRACT

The aim of this study was to evaluate the performance of thin anodized film of the 2024 aluminum alloy under conditions of high-stress, low cycle fatigue. Of particular interest was the effect of second phase particles such as inclusions or precipitates on the process of short crack initiation in the developed oxide film, Al_2O_3, as a result of the anodizing process of the alloy. Comparisons were made of the short crack initiation for three heat treatments: The naturally aged T4 condition, and two artificially overaged conditions, (T71, 10 hours at 400°F, and T72, 10 hours at 500°F).
Both qualitative results based on the scanning electron microscopic SEM observations and quantitative results based on the number of cycle were obtained. The development of short crack initiation occurred at different numbers of accumulated cycles but at about the same percentage of fatigue life in all the samples. In the T4 condition, crack initiation occurred in the thin film preferentially at inclusion sites and subsequently, crack propagated from one inclusion to another on the

---

[1] Materials Application Department, Kuwait Institute for Scientific Research. P. O. Box 24885 Safat, 13109 Kuwait.

[2] School of Chemical Engineering and Materials Science, University of Oklahoma. 100 E. Boyd, Energy Center, Norman, O. K. 73019-0628, U.S.A.

metal surface. In the T71 condition, crack intitiation in the thin film took place preferrentially at precipitate clusters. Thereafter, the crack extended from one precipitate to another. In the T72 condition both precipitate clusters and grain boundaries were found preferential sites for crack initiation. Crack initiation at the grain boundaries in the samples overaged at 500°F (T72 condition) are explained by the large area of the surface depleted of precipitate clusters.

INTRODUCTION

In general documentation of fatigue crack nucleation has been accomplished through a number of direct experimental observations. Most of the data generated from such experimental observations have emphasized that there are three types of nucleation sites (1):

1. Fatigue slip bands
2. Grain boundaries
3. Second phase particles

It is worth noting that most of these observations were obtained from experiments conducted on pure metals or commercial alloys in tension-tension, tension-compression, and reverse bending fatigue conditions (2, 3, 4, 5). As far as we know, no experiment of fatigue crack nucleation on a commercial alloy such as 2024 aluminum alloy applying a rotating bending condition has been reported in literature. Therefore, the objective of the present work is to evaluate the performance of thin anodized film of the 2024 aluminum alloy at a high stress level, i.e, under condition described as low-cycle fatigue by applying the rotating bending-fatigue condition.

EXPERIMENTAL DETAILS

Fatigue specimens were fabricated from 2024-T4 aluminum bar stock according to the standard specifications of rotating bending fatigue samples. The chemical composition of 2024 aluminum alloy is 93.4% Al, 4.5% Cu, 0.6% Mn and 1.5% Mg.

Two sets of machined specimens were solution annealed at 515.5°C (960°F) for 60 minutes in an air furnace then water quenched. Thereafter, one set of the specimens was aged at 204.4°C (400°F) and the other set aged at 260°C (500°F) for 10 hours (these conditions will be referred to as T71 and T72, respectively) to produced overaged, stable microstructures. A third set of machined specimens was used without heat treatment. i.e.. in the as-received T4 (naturally aged) condition.

The surface of the fatigue specimens were hand ground longitudinally with SiC papers. The processes of progressive grinding were carried out until the finer grades of SiC papers were reached. Thereafter, the specimen were hand polished with emery cloth through 4/0 grade. Finally, the specimen anodized with a solution of 750 ml methanol and 250 ml nitric acid until bright scratch-free film was produced. The tensile properities of the three heat treated conditions of the alloy are given in Table 1. In addition the S-N curves (stress vs. number of cycle to failure) are plotted in Figure 1. The loading condition was completely reversed cyclic loading, i.e.. the fatigue stress ratio R --1.

Table 1. Tensile properties of 2024 aluminum alloy as a function treatment.

Materials and heat treatment conditions	2024-T4	2024-T71	2024-T72
Yield strength (0.2% of set strain) (Mpa)	410	338	292
Tensile strength (Mpa)	520	392	357
Percentage of reduction area	14.3	21	24

*Results are an average of two tests.

Fatigue crack nucleation was investigated by cyclic loading at a stress amplitude corresponding to 90% of yield strength. The test was periodically interrupted in order to examine the specimen surface. The stress levels used were 370 Mpa (T4), 304 Mpa (T71), and 207 Mpa (T72). The cyclic frequency was 5 Hz. Note that at 90% of yield strength, the fatigue lifes of the three conditions (T4,T71, and T72) are equivalent to 29,900 cycles, 16,200 cycles and 9,400 cycles, respectively (see Figure 1).

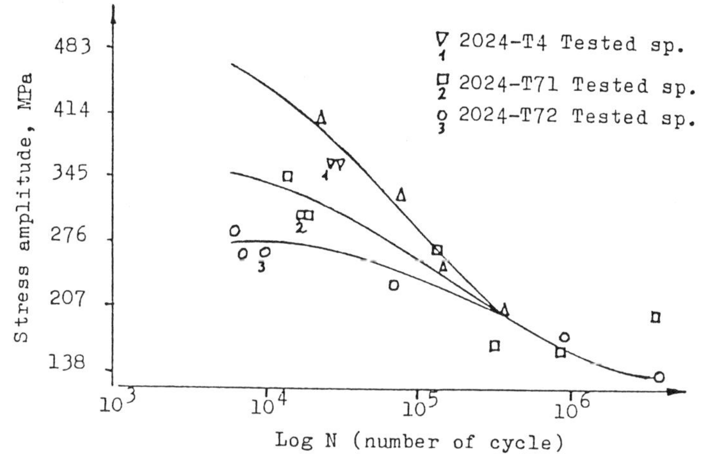

Figure 1. S-N fatigue curves for 2024 aluminum alloy in the three heat treatment conditions. T4, T71, and T72.

The specimen surface was examined in an ETEC scanning electron microscope. The aim of the SEM examinations was to record any microstructural irregularities that might form at the sample surface. The SEM examinations were performed at an intervals of cyclic loading. The intervals of cyclic loading were began with 820, 410 and 170 cycles for 2024-T4, 2024-T71, 2024-T72 specimens respectively. In each case, the interval of cyclic loading corresponds to approximately 2.75 (T4), 2.6 (T71), and 1.8 (T72) percent of fatigue life. Refer to Figure 1 for estimates of fatigue life percentage.

The SEM examinations were performed at intervals for all the heat treated conditions (T4, T71 and T72 specimens) until the first microcracks were detected at 22 (T4), 20.8 (T71) and 14 (T72) percent of accumulated fatigue life. Prior to the detection of the microcracks, some microstructural irregularities were observed at earlier stages of fatigue life on the anodized surface of all conditions of the alloy. Significant features of the microstructural irregularities were photographed.

After documenting the first microcracks, the SEM examinations were carried out periodically for larger increments of cyclic loading until final rupture occurred. The interval of cyclic loading were increased to 6,570; 3,340; 1,390 cycles for the T4, T71 and T72 specimens, respectively. These intervals of cyclic loading correspond to approximately 22 (T4), 20.8 (T71) and 14.7 (T72) percent of fatigue life. The purpose of SEM examinations at these larger intervals of cyclic loading was primarely to document the development of microcracks to macrocracks and eventually the processes leading to the final rupture. Examinations of the surface for fatigue crack nucleation was facilitated by the use of a special sample holder in SEM. A macrophotoghraph of the sample holder mounted in the SEM is shown in Figure 2.

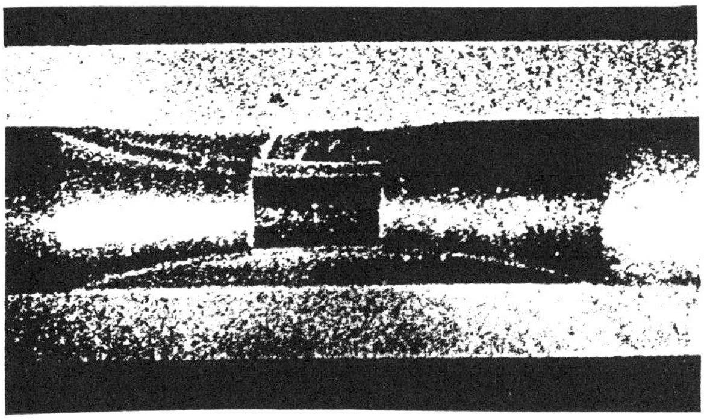

Figure 2. Depicts a macrophotograph of the sample holder in SEM. A bright oxide layer scractch-free film appears clearly in the middle of the specimen surface.

RESULTS AND DISCUSSION

Qualitative and quantitative results of the short crack initiation of all conditions of 2024 aluminum alloy, based on SEM examinations and number of cycles are presented in Table 2.

Illustrations 3-5 are some examples of the documentation of crazed oxide films which associated with the second phase particles in all the studied conditions. On the other hand, Figures 6-8 are examples of short crack initiation sites in all the studied conditions.

It is apparent from Figures 3-5 that the oxide layer was first shattered at sites preferentially related to the top of the second phase particles. This occurred because of the steep stress concentration at those sites due to the difference of the elastic properties of the metal base, second phase particles, and the oxide film. Consider the stress distribution in the rotation bending-fatigue sample, see Figure 9 , you find that the maximum stress distribution is maximum at the interface between the oxide layer and the metal base especially second phase particles present at the interface. As a result, the crazed oxide layer first because the layer, Al_2O_3, is more brittle than the second phase particles. Then, the short crack emenated from the fracture particles following a direction 45^0 with respect to the tensile axis.

Table 2. Comparison between results in all 2024 heat treatment conditions.

	HEAT TREATMENT CONDITIONS		
	T4	T71	T72
Fatigue life at 90% of yield strength (N cycles)	29,920	16,210	9,440
First cycle increment (N cycles)	820	410	170
% of fatigue life	2.75	2.6	1.84
First observation of surface damage (N cycles)	1,640	830	340
% of fatigue life	5.5	5.2	3.7
First crack initiation (N cycles)	6,570	3,340	1,390
% of fatigue life	22	20.8	14.7
Second cycle increment until failure (N cycles)	6,570	3,340	1,390
% of fatigue life	22	20.8	14.7
Sites of first surface damage	crazed oxide film at the surface of inclusions	crazed oxide film at the surface of precipitate clusters	crazed oxide film at the surface of precipitate clusters
Sites of fatigue microcrack nucleation	inclusions clusters	precipitate & grain boundaries	precipitate

CONCLUSIONS

A number of conclusions are drawn from this study:

1. Fatigue crack nucleation is strongly influenced by the strength of the alloy as determined by the heat treatment. The higher the strength the greater is the number of cycles for initiation of fatigue microcracks. However, in terms of percentage of fatigue life, i.e., cycles to failure, nucleation occurs at a constant percentage.
2. Crazed oxide layers at the surface of the second phase particles such as inclusions and precipitates are the first surface damages in all conditions of the alloy.
3. Second phase particles, such as inclusions and precipitates, are the preferential sites or short crack initiation in all the three conditions of the alloy.
4. Microcrack nucleation associated with grain boundaries was found in areas depleted from precipitates in 2024-T72 specimen.
5. The larger the size of the second phase particles the more susceptible they are to fracture.
6. In all the three heat treatment conditions, fatigue nucleated at a number of sites simultanuously, thereafter, microcracks extended in a direction approximately 45^0 with respect to the stress axis.

REFERENCES:

1) M. Klensil and P. Lukas, Fatigue of Metallic Materials, Elsevier Scientific Publisher Company, New York, (57, 66), 1980.
2) C. Laird and D.L. Duquette, "Initiation Processes". Corrosion Fatigue, NACE-2, National Association of Corrosion Fatigue, Houston, (88), 1970
3) N. Thompson, **phil Mag.**, 8th series, Vol. 26, (113), 1956.
4) W. Kim and C. Laird, "Crack Nucleation and Stage I Propagation of High Strain Fatigue". **Acta Metallurgica**, Vol. 26, (777), 1978.
5) D. Sigler, M. Montpetit and W. Haworth, Metallography of Fatigue Crack Initiation in an overaged High-Strength Aluminum Alloy. **Metallurgical Transaction A**, Vol. 14A., (931), 1983.

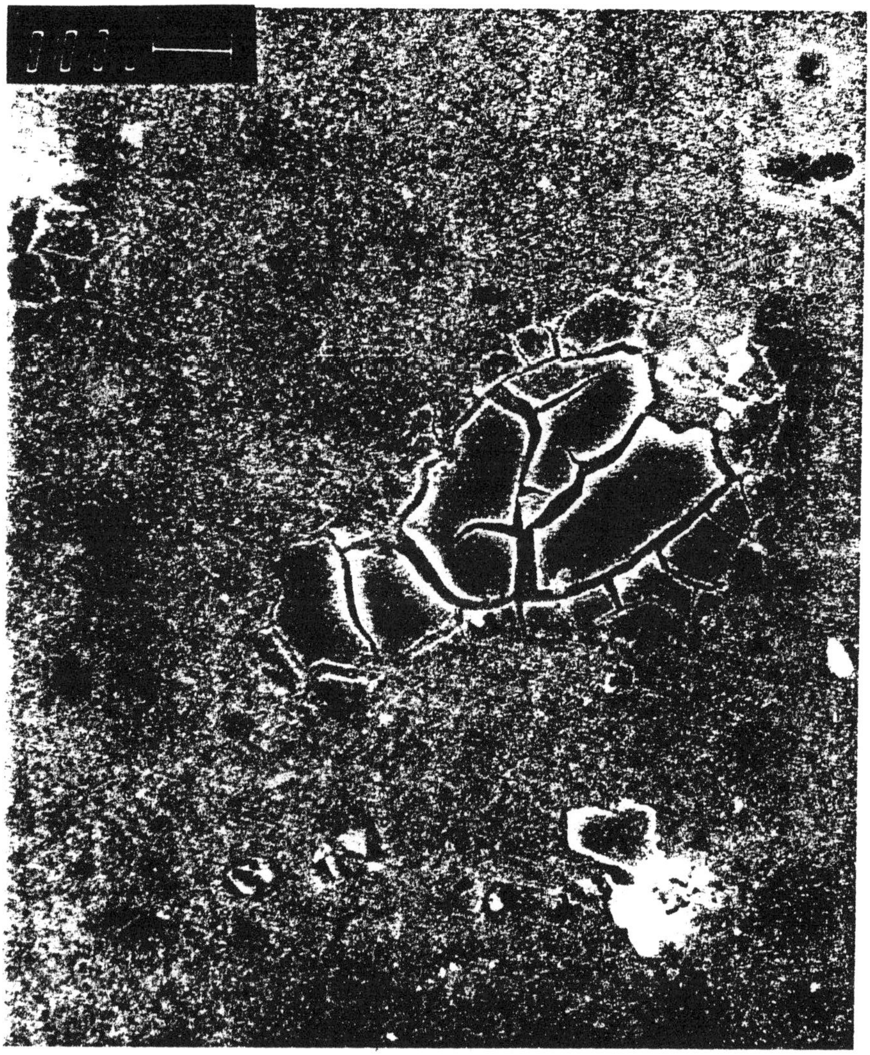

Figure 3. Shows a crazed oxide film at the surface of the inclusion in a 2024-T4 specimen. The specimen was cycled approximately 1,640 cycles (5.5% fatigue life) at stress amplitude = $\pm$53.4 Ksi (90% of yield strength).

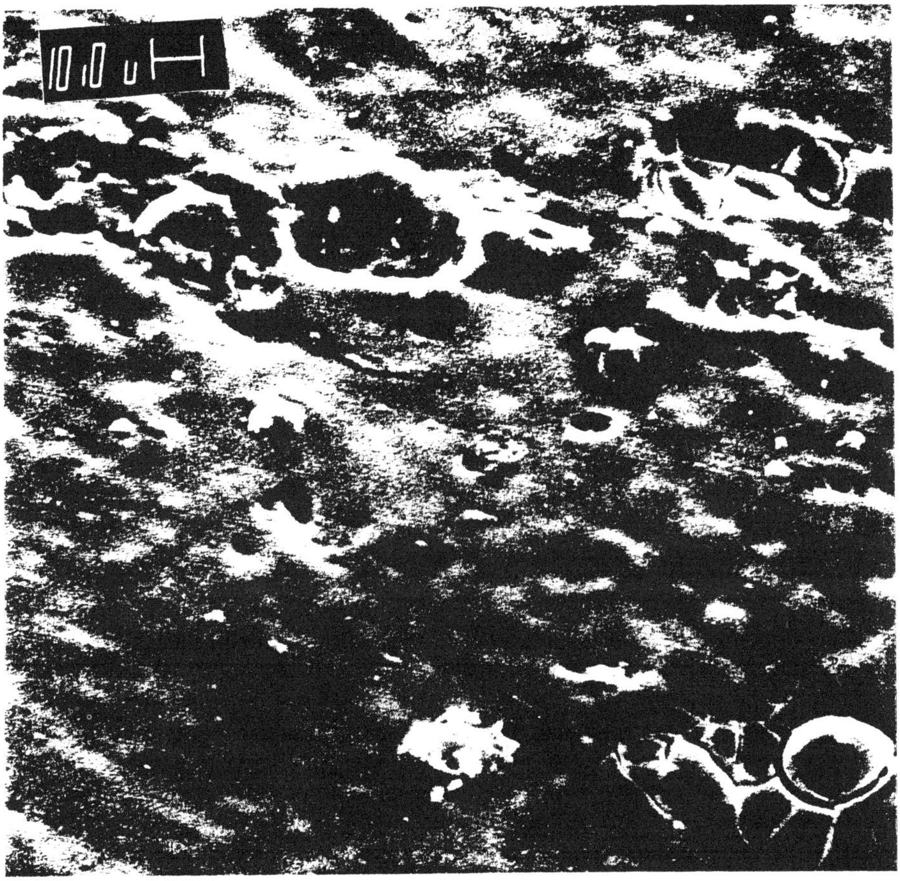

Figure 4. Shows a crazed oxide film at the surface of a precipitate cluster in 2024-T71 specimen. Furthermore, a localized disbonding of the fracture precipitate clusters appears on the surface of the specimen. The specimen was cycled 1,250 cycles (7.8% of fatigue life) at stress amplitude - ± 44.1 Ksi (90% of yield strength).

Figure 5. Depicts a localized disbonding of the fracture precipitate cluster in 2024-T72 specimen. The specimen was cycled 520 cycles (5.5% of fatigue life) at stress amplitude = ±38 Ksi (90% of yield strength).

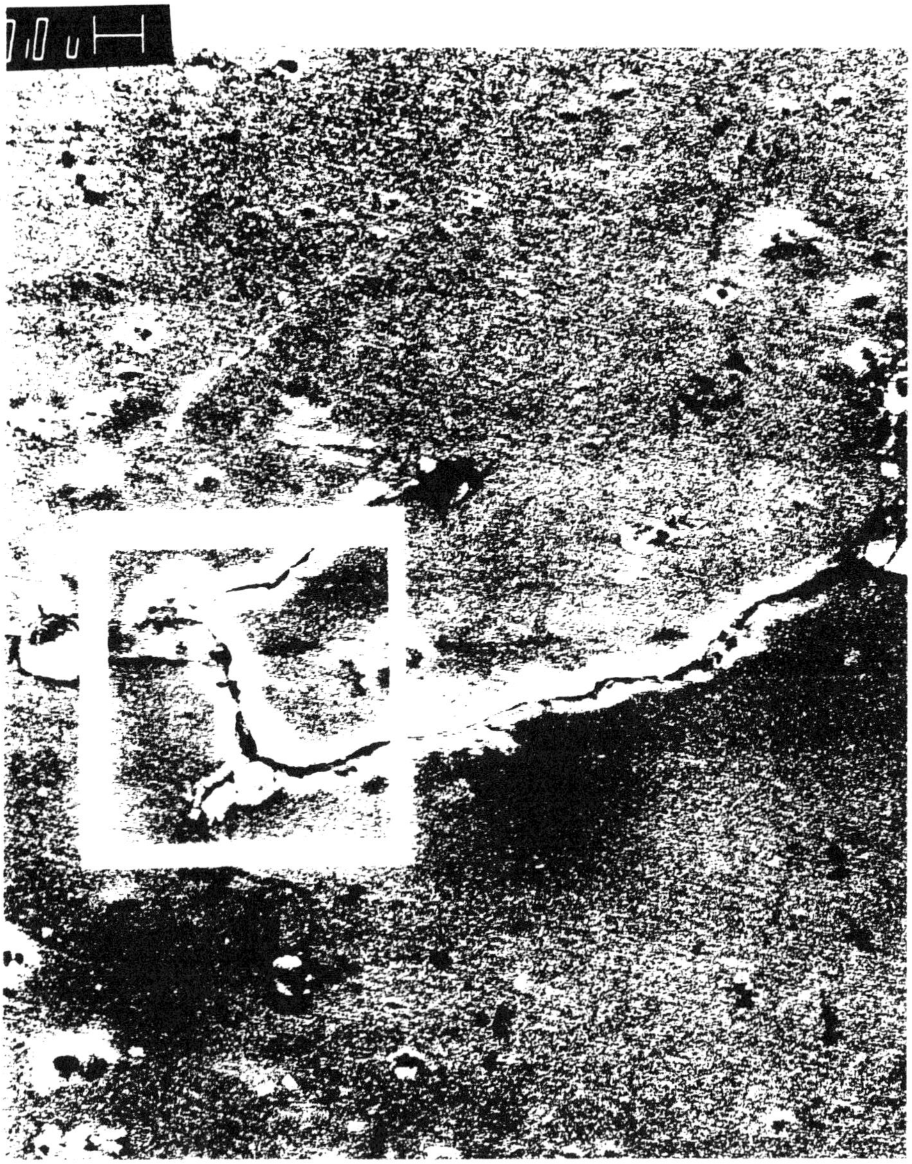

Figure 6. A photograph of short cracks emanated from inclusions in 2024-T4 specimen. The specimen was cycled approximately 6,580 cycles (22% of fatigue life) at stress amplitude = ± 53.4 Ksi (90% of yield strength).

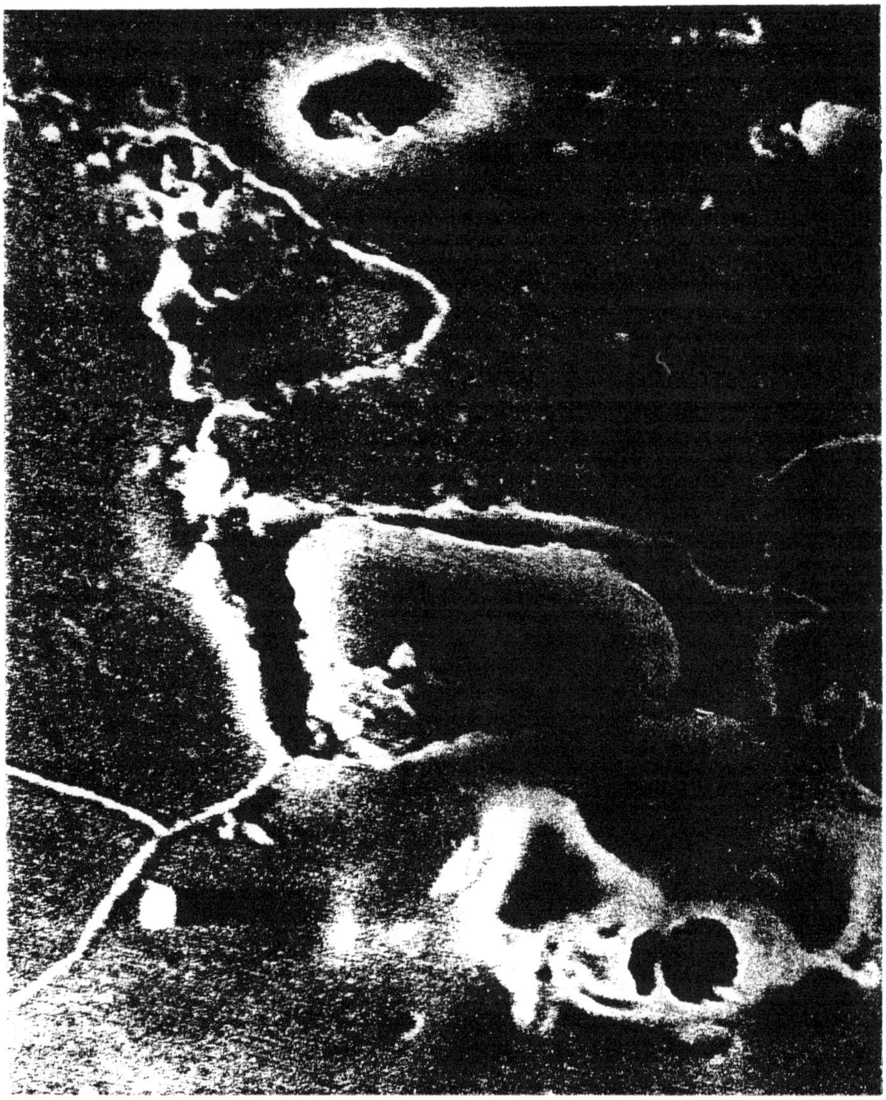

Figure 7. A view of short crack nucleation from a precipitate cluster in 2024-T71 specimen. The specimen was cycled 3,340 cycles (20.8% of fatigue life) at stress amplitude = $\pm$ 44.1 Ksi (90% of yield strength).

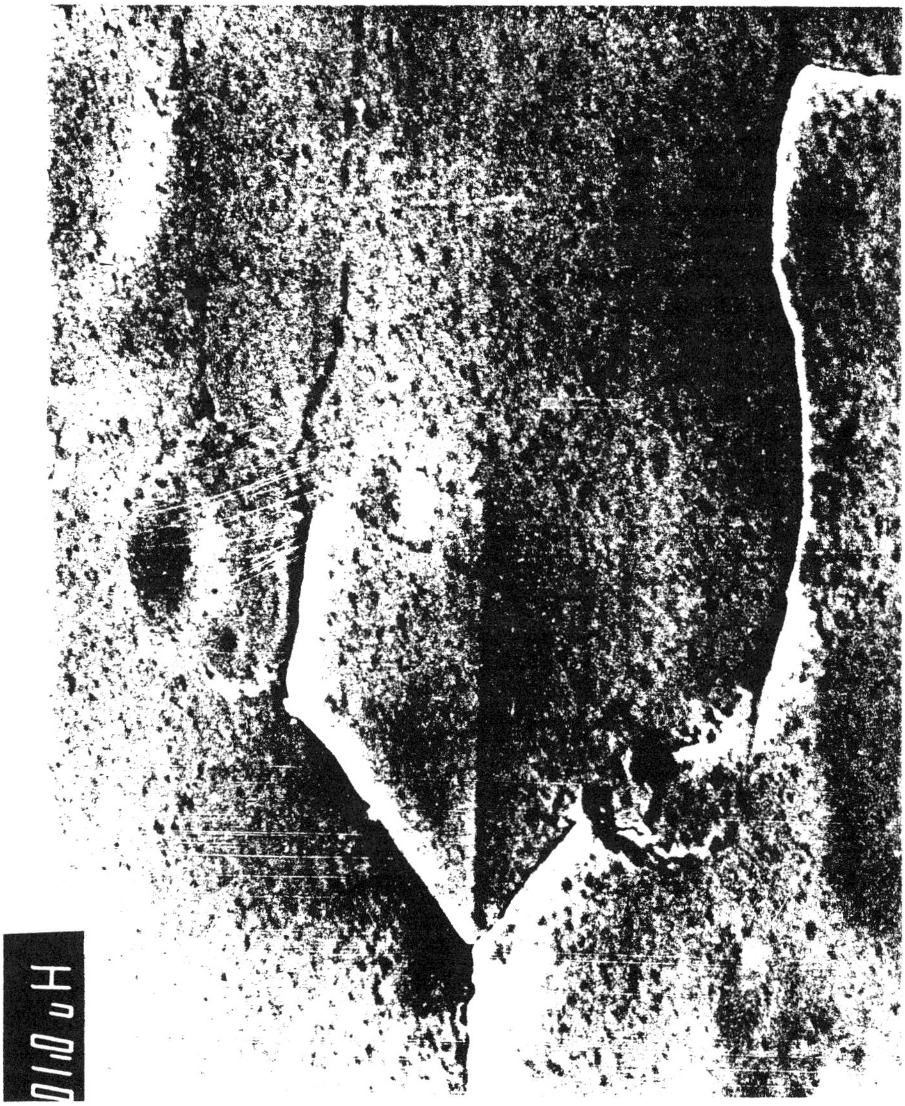

Figure 8. Depicts clearly that the triple grain boundaries intersect is one of the origin of the short crack nucleation in 2024-T72 specimen. The specimen was cycled 1,390 cycles (14.7% of fatigue life) at stress amplitude = $\pm$ 38 Ksi (90% of yield strength).

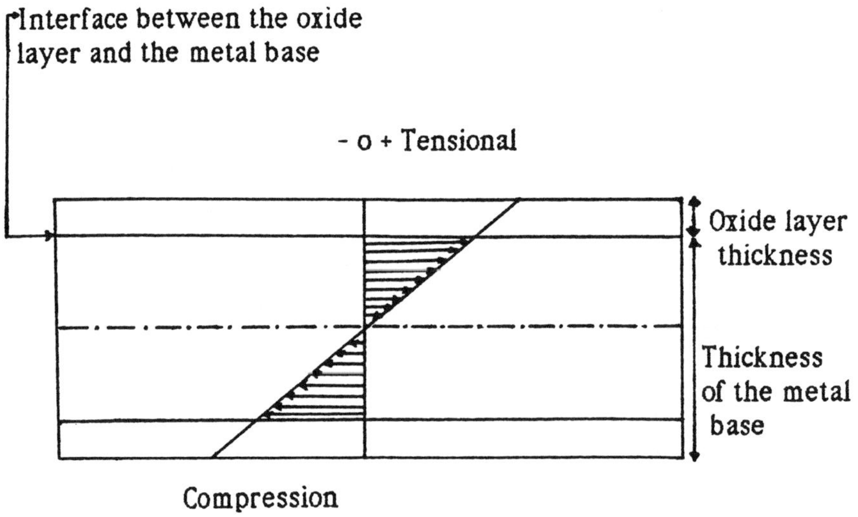

Figure 9. Stress distribution in rotating bending-fatigue sample

FAILURE ANALYSIS FOR A CARBON STEEL VAPORIZER COIL

Y. C. Lin and F. V. Ellis*

ABSTRACT

A root cause failure analysis was performed on a vaporizer coil removed from a horizontal forced circulation vaporizer. The carbon steel coil was wound in a right-hand helix with a coil centerline diameter of about 2 meters. The vaporizer was gas fired and used Dowtherm A as the heat transfer fluid. Design conditions are based on annular fluid flow to cool the coil wall. NDE, metallographic and fractographic examinations were performed. Numerous, circumferentially oriented, O.D. initiating cracks were found near the crown for two coils near the non-fired end of the vaporizer. The cracking was confined to the inner diameter of the vaporizer coil at positions from 4:00 to 7:00. The cracking was characterized as transgranular and the fracture surface had beach marks. The failure mechanism was thermal fatigue. The heat transfer calculation predicted that dryout of the coil would occur for coils at the non-fired end of the vaporizer during low flow transients. Dryout results in rapid increase in the tube wall temperature. Thermal cycling of the coil is completed by liquid quenching resulting from resumption of normal flow rates and the return to annular flow. The probable root cause of failure was low flow transient operation.

---

* Tordonato Energy Consultants, Inc.
4156 S. Creek Road, Chattanooga, TN 37406

INTRODUCTION

A metallurgical failure analysis was performed on a failed vaporizer coil removed from a horizontal forced circulation vaporizer. Three samples were removed for examination. The first section was an archived section of vaporizer coil adjacent to the leak site which was removed following a prior failure. The remaining two coil samples were from the leak position vaporizer coil 6 and an adjacent cracked vaporizer coil 5 following the second, most recent failure.

The root cause failure analysis method was used in the current study. This method was implemented successfully by the Electric Power Research Institute in their boiler tubes failure correction, prevention, and control research project [1]. Fundamental to an understanding of the method are several terms. First, root cause is defined as the circumstance that creates a suitable condition for a failure mechanism to exist. Second, the failure mechanism is defined as the process from which failure results. The inputs to the root cause failure analysis consist of failure location, failure appearance and prior events. The outputs from the root cause failure analysis include failure mechanism, probable root cause, corrective action, repair and non-destructive examination (NDE) recommendations. The current paper gives results of the metallurgical examination and the root cause failure analysis for the forced circulation vaporizer coil including identification of both the failure mechanism and the probable root cause of failure.

RESULTS AND DISCUSSION

The horizontal forced circulation vaporizer was designed to supply approximately seven million BTU per hour. The vaporizer was gas fired and used Dowtherm A as the heat transfer fluid. Dowtherm A is a mixture of diphenyl and diphenyl oxide and has superior heat transfer characteristics. The vaporizer coil material was specified as ASME SA-106 Grade B Schedule 80 pipe. The vaporizer coil was wound in a right-hand helix with a coil centerline diameter of approximately two meters. There were 24 3/4 total loops in the coil with the majority of the loops being tight wound and final three loops at the outlet end of the vaporizer being loose wound. The coil was design to ASME Code Section 1, Division 1, and the design conditions were an internal pressure of 1 MPa and temperature of 400 C.

Results are given for the failure location, failure appearance, and failure mechanism based on the metallurgical examination of the failed vaporizer and coil. The root cause failure analysis integrates these observations with manufacturing and operational variables in order to determine the probable root cause of failure.

Failure Location

Following the vaporizer coil failure, liquid penetrant testing (PT) was performed on the vaporizer coil. Results of this testing revealed several indications on the inner diameter of the coil near the crown of the pipe. The crack associated with the leak was in coil 6 (counting from the outlet end looking toward the inlet end) at about the 5:30 position. The indication was oriented in the circumferential direction of the pipe. Indications were also found in vaporizer coil 5 on the inner diameter of the vaporizer coil near the crown of the pipe. These indications were also oriented in the circumferential direction of the pipe and were confined to positions from approximately 4:00 to 7:00 based on the PT inspection. No indications were marked in vaporizer coil 4 or any other coil. Also there were no other indications found except at the position noted. Based on these results, sections of vaporizer coil 5 and 6 were removed for metallurgical examination.

Failure Appearance and Mechanism

Visual examination of the vaporizer coil material archived from the first vaporizer coil failure revealed a thick black scale on the inside surface at the inner diameter of the helical coil. The I.D. scale was apparently thicker at positions slightly off center of the crown of the pipe at the inner diameter of the helical coil but was symmetric around the crown position. The I.D. scale consists of at least two different layers. Dowtherm A heat transfer fluid decomposes at temperatures of approximately 480 C. It is believed that the scale found near the coil inner diameter was a result of high temperatures and the decomposition of the Dowtherm fluid. Because only the coil I.D. crown position had the scale, a circumferential temperature gradient around the pipe was inferred. This means that the fire side of the coil (inner diameter) reached metal temperature of at least 480 C while the metal temperature on the casing side of the coil (outer diameter) were significantly cooler. Temperature differences between the inner diameter and the outer diameter of the coil are important because they result in axial stresses being generated in the pipe.

Visual examination of the as-received vaporizer coil 6 and coil 5 samples revealed relatively uniformly spaced cracks. In order to better determine the distribution of the cracks prior to sectioning, wet fluorescent magnetic particle testing (WFMT) was preformed. This technique is more sensitive than liquid penetrant testing and allows for a more realistic determination of the representative crack distribution. The WFMT results revealed an abundance of crack indications with a much higher number than that revealed by either visual or PT inspection. Subsequent metallographic examination supported the WFMT results indicating numerous, nearly evenly spaced cracks. To document the wet MT results, a modified dry powder MT technique was used. Figure 1 documents the results of the dry powder MT inspection for coil 6 showing the numerous indications and the large indication at the leak site. The wet MT revealed more indications than the dry MT documentation has recorded. Low magnification examination of as ground or as polished surface is also effective means of crack detection. Figure 2 shows the pipe O.D. surface at the inner diameter of coil 6 following grinding through 320 grit. The number and axial spacing of the cracks are clearly visible.

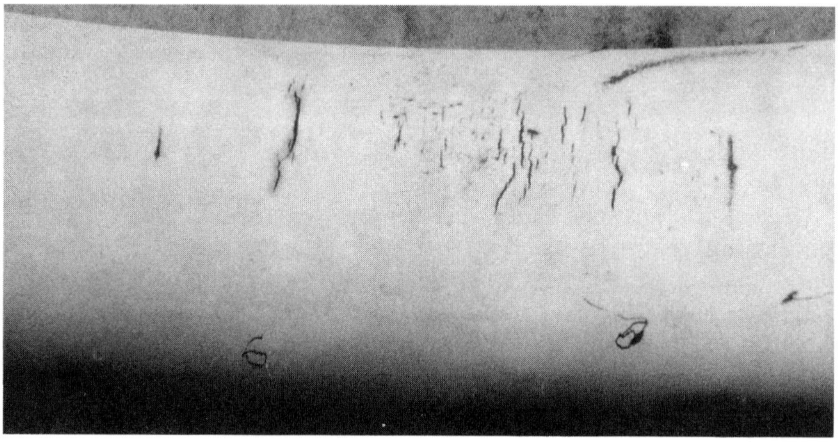

Figure 1. Documentation of dry MT powder inspection results for vaporizer coil 6.

To determine the pipe dimensions, a ring section was used. Prior to dimensional measurement, the ring sample was descaled in an inhibited acid solution. The I.D. and O.D. scale was removed by this procedure for all surfaces except the I.D. surface at the coil inner diameter where a thin adherent black scale persisted. This scale is in the same position as the cracking in the vaporizer coil and coincident with the position where the thick scale was found in the first sample. The thinnest wall at the extrados of the bend at the outer diameter of the coil was 11.8 mm. The outside diameter of the pipe was from 167 mm to 169 mm indicating some ovalness

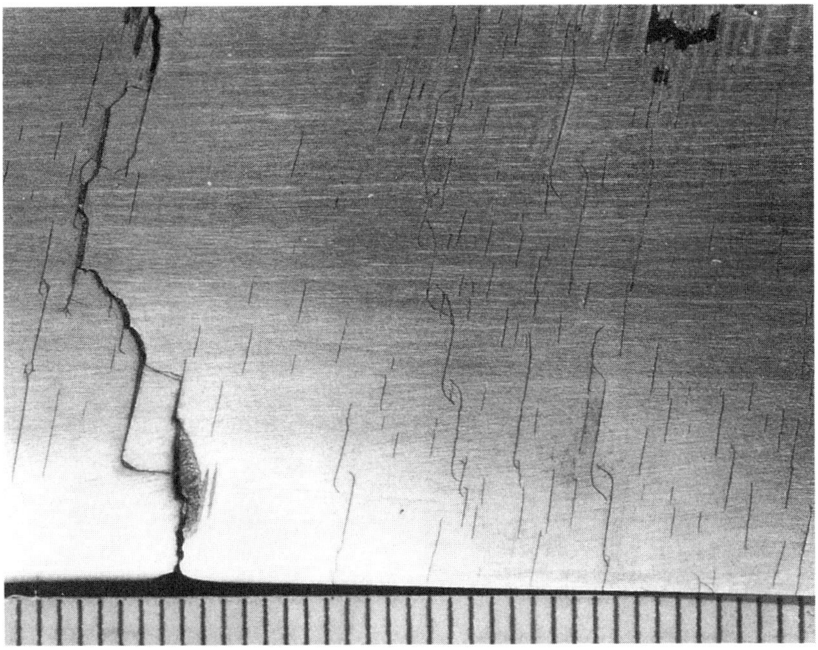

Figure 2. Photograph of circumferential cracking at pipe outside surface on inner diameter of vaporizer coil sample 6.

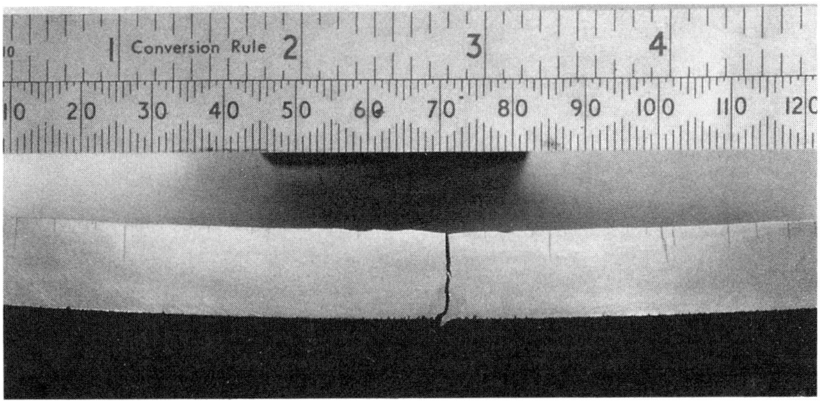

Figure 3. Photomacrograph of vaporizer coil 6 at leak site.

in the pipe. The pipe dimensions are consistent with those for a Schedule 80 pipe specification prior to the bending process.

Typical cracking found in vaporizer coil 6 at the leak position is shown in the cross sectional view of Figure 3. The cracking initiated at the pipe O.D. (inner diameter of header coil) and propagated in a radial direction through the pipe wall. The frequency distribution for crack depth shows that there are numerous, shallow, nearly uniformly spaced (axially) O.D. initiating cracks and a few deep cracks. This crack distribution is consistent with fatigue crack initiation and propagation mechanism. The shear lip at the pipe I.D. indicates the final ductile overload failure.

The microstructure of a typical crack for coil 5 is shown in Figure 4. The cracks are oxide filled and transgranular. Fracturing through the grain matrix is the typical cracking mode for fatigue. A portion of the coil 5 sample was fractured at liquid nitrogen temperature to expose the crack face. The sample was descaled in a commercial solution to reveal the fracture face. Figure 5 shows a photograph of one of the mating fracture surfaces for vaporizer coil 5. The fracture surface is crescent shaped. The crack length to crack depth is in the range of 3 to 4. On the fracture surface, there are "beach marks" which are typical for fatigue crack propagation.

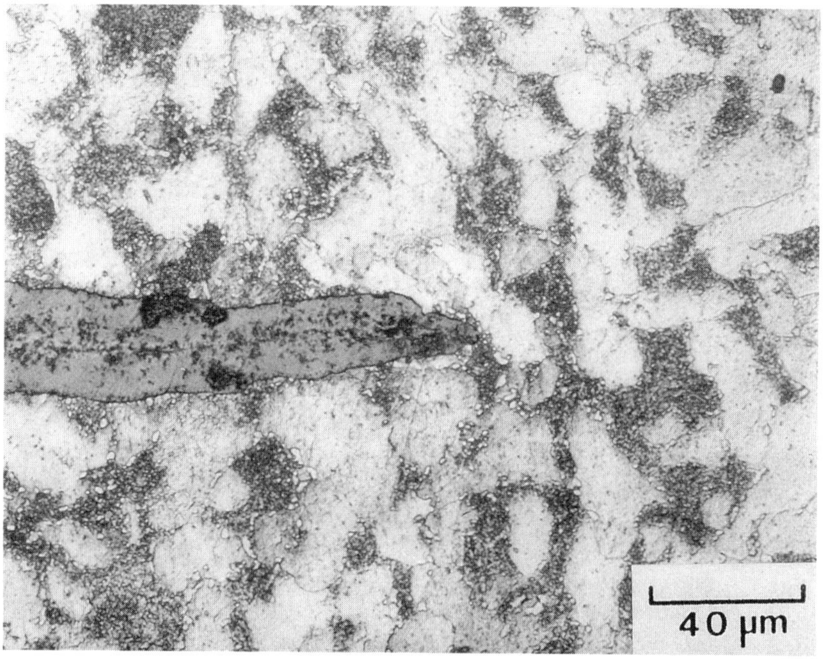

Figure 4. Transgranular thermal fatigue crack in vaporizer coil 5.

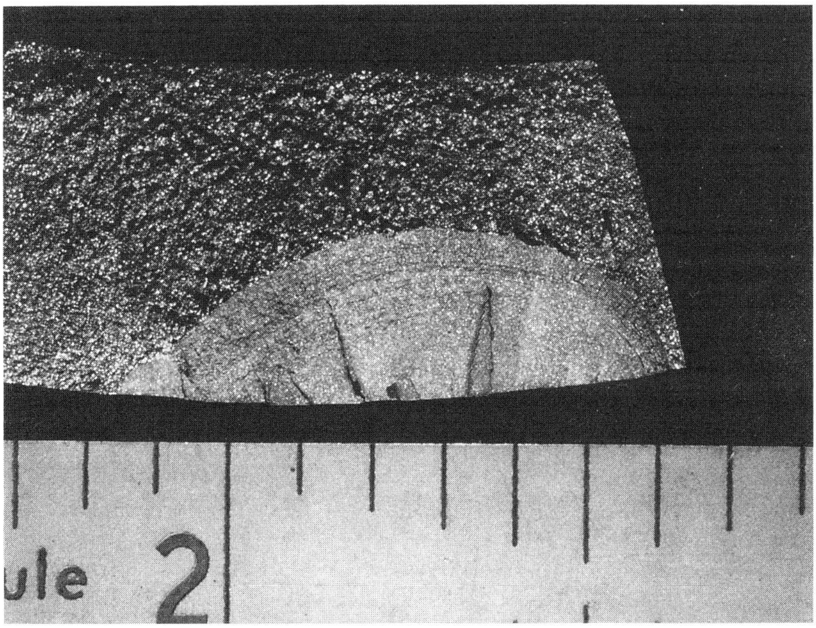

Figure 5. Photomacrograph of fracture surface for vaporizer coil 5.

Based on the results of metallurgical examination performed on the failed and cracked vaporizer coil material, the failure mechanism was determined to be thermal fatigue. The failure mechanism, thermal fatigue, was also reported for the first failure and by Becker [2] for the current failure.

Prior Events and Probable Root Cause

The possible root causes for the vaporizer coil failure evaluated in this study include the following: local flame impingement, piping system stresses, residual stresses from cold bending, thermal cycling from dryout/liquid quench operation. Based on the characteristics of the cracking (fracture appearance, location, and mechanism), heat transfer calculations by Ledford [3], and similar cracking in boiler waterwall tubing [4], it was concluded that the probable primary root cause of failure was thermal cycling from dryout/liquid quench operation. Based on the heat transfer calculations, the dryout was believed to be the direct result of deviations in flow rate from design values to low flow rate values.

Wright et al. [4] discussed the root cause study of circumferential cracking in boiler waterwall tubing. The nature of the thermal cycling and one sided heat transfer produce the high axial stresses that are necessary for the circumferential cracking [4]. This aspect of the vaporizer coil failure and waterwall cracking failure is similar. Basically, thermal and stress analysis has shown that high metal temperatures on the furnace side or fireside of the waterwall tube can cause high metal temperatures on the furnace side of the tube and lower or cooler metal temperatures for the casing or backside of the tubing for several boiler operating conditions. The temperature increases are transient and are followed by quench to complete the thermal cycle.

The evaporation of a liquid in a tube such as the vaporizer coil is a function of the flow pattern and the heat transfer region. The coil was designed to be in the annular flow regime with the liquid having good heat transfer properties in contact with the tube wall. Calculations [3] have shown that flow can enter the drop flow regime at low flow rates causing "dryout" of the tube wall. Because of the heat transfer properties of the vapor in contact with the tube wall, tube metal temperatures increase dramatically compared to the annular flow condition. At low flow conditions, a dryout zone was calculated to occur in the vaporizer coil only at those coils having cracking towards the outlet end of the vaporizer. To complete the thermal cycle, a quench or temperature decrease is required. The quench portion of the thermal cycle will occur after lower inlet liquid temperatures are achieved which return the coil to either single phase liquid, bubbly flow, slug flow or the annular flow regime causing liquid to be in contact with the tube wall. The tube metal temperature spike occurs on the fireside of the coil - i.e. coil inner diameter because of the higher heat flux on the fireside of the tube. High metal temperatures on the coil I.D. region were supported by the observation of decomposition product discussed earlier for the first failure. The thermal cycling from dryout/liquid quench operation is consistent with the observations of circumferential thermal fatigue cracking on the coil inner diameter near the outlet end of the forced circulation vaporizer. None of the other possible root causes can explain these failure features.

Any/all of the other possible root causes mentioned above could be contributing to the failure but are not believed to be the primary root cause of this failure. A complete root cause failure analysis would require confirmation of the probable root cause of failure including, for example, temperature monitoring during design and transient low flow operation. Because the fate of the equipment is uncertain at this time, the confirmation of the probable root cause and other corrective, repair and NDE actions were not pursued.

CONCLUSIONS

A root cause failure analysis was performed on a SA-106 Grade B Schedule 80 vaporizer coil in a horizontal forced circulation vaporizer. The failure location was on the inner diameter of the coil 6 (counting from the outlet end looking toward the inlet end) near the crown of the pipe at about the 5:30 position. The failure appearance is characterized by the following: transgranular cracking, numerous, circumferential oriented O.D. initiating cracks, radial crack propagation direction, crescent shaped crack - length/depth = 3 to 4, and beach marks in fracture surface. The failure mechanism was identified as thermal fatigue. The probable root cause of failure was thermal cycling from dryout/liquid quench operation. The dryout condition was believed to have resulted from flow rates which were below specified design minimum.

ACKNOWLEDGEMENTS

Financial support was provided by Heatec. The authors wish to acknowledge the major contribution of Mr. W. A. Ledford and Mr. M. D. Moon in the conduct of the heat transfer calculations and understanding of the operation and design of the horizontal forced circulation vaporizer.

REFERENCES

1. J. P. Dimmer, G. A. Lamping, and O. Jonas, "Boiler Tube Failure: Prevention, Correction, and Control," EPRI Report 65-6467, EPRI, Palo Alto, CA, 1989.

2. W. T. Becker, "Thermal Fatigue Failure in a Vaporizer," Handbook of Case Histories in Failure Analysis, Vol. 2, Ed. by K. A. Esaklul, ASM International, Materials Park, OH, 1993.

3. W. A. Ledford and M. D. Moon, Unpublished Research, Heatec, Chattanooga, TN, 1993.

4. I. G. Wright, D. Anson, H. J. Cialone, G. O. Davis, M. D. Kurre, and F.B. Stulen, "Root Cause of Circumferential Cracking in the Waterwalls of Supercritical Boilers," EPRI International Conference on Boiler Tube Failures in Fossil Plants, Electric Power Research Institute, Palo Alto, CA, Nov. 1991.

MICROSTRUCTURE AND TOUGHNESS RELATIONS IN WELDMETAL OF HIGH STRENGTH LOW ALLOY STEELS

M.R. Krishnadev[1], W.L. Zhang[1], Réjean Réne[2], Alain Gendron[3], Viwek Vaidya[3], and J.T. Bowker[4]

ABSTRACT

The study deals with the influence of alloying elements and welding practice on the toughness of the weldmetal of the new generation of copper containing 700 MPa HSLA steels. Instrumented Charpy has been used to characterize the toughness. Optical and transmission electron microscopy has been used to delineate the microstructural changes. Oblique section metallography has been used to enumerate the crack initiation and propagation resistance vis-à-vis the microstructural constituents. The toughness variations have been related to the influence of alloying elements on the weld metal transformation microstructures.

---

1 Department of Mining and Metallurgy, Laval University, Quebec, G1K 7P4
2 MIL Davie Inc., P.O.Box 130, 22, Geo D.Davie, Lévis, Quebec, G6V 6N7
3 Canadian Liquid Air Ltd., 5035, Rue Rouen, Montreal, P.Q., H1V 1J3
4 MTL, CANMET, 568, Booth St., Ottawa, Ont., K1A OG1

INTRODUCTION

In response to the need for 550 MPa (80 Ksi) to 700 MPa (100 Ksi) grade of steels with improved weldability and toughness for naval structural applications, a new generation of HSLA (High Strength Low Alloy) steels with extra low carbon levels and with fine tuned alloying additives (Cu, Ni, Cr, Mo, Nb), has been developed and these steels are usually referred to as HSLA-80 and HSLA-100 steels [1, 2]. There is considerable information regarding the base plate properties and the properties of the HAZ of these steels [3,4]. Efforts also have been made to improve their HAZ toughness through alloying modifications [5, 6]. However there is lack in the development of extra low carbon, extra clean welding electrodes that match the toughness and the strength levels of advanced HSLA steels. In addition, there is the need for a better understanding of the relations between welding parameters, compositional make up of the electrodes and weldment microstructure and properties. In this context, a research project has been initiated at Laval University to develop cleaner and stronger extra low carbon HSLA welding electrodes and to understand the influence of alloying elements and welding practice on the HAZ and weld metal of HSLA-100 steels. The present study deals with the development of covered electrodes for SMAW applications.

EXPERIMENTAL

Materials

Base Plate: Composition of the base plate is shown in Table 1. It contains extra low carbon content for good weldability. Copper provides precipitation strengthening. Nickel prevents hot shortness due to copper additions and improves toughness. Molybdenum and chromium retard premature precipitation of copper. Niobium acts as a grain refiner. The material was received in the form of 25 mm thick plates that had been subjected to a reaustenitization, quench and aging treatment to take full potential of the precipitation strengthening of copper. The plates were supplied by Lukens steel corporation. The steel is generally classified as a HSLA-100 (700 MPa) steel for naval structural applications [1].

Table 1 Chelmical Compostion of the Base Plate

steel	Element, wt%									
HSLA	C	Mn	Si	S	P	Ni	Cr	Mo	Cu	Nb
-100	0.05	1.00	0.34	0.003	0.009	1.77	0.61	0.51	1.23	0.037

Table 2 Targeted Weld Chemical Composition of Electrodes E1 And E2

Electrode	Element, wt %						
	C	Mn	Si	Cr	Ni	Mo	Cu
E1	0.06	1.50	0.3	0.3	3.0	0.5	- -
E2	0.06	1.50	0.3	0.3	3.0	0.5	0.5

Table 3 Chemical Composition of Commercial Electrode

Electrode	Element, wt %							
	C	S	P	Mn	Si	Cr	Ni	Mo
Commercial	.060	.010	.012	1.38	.26	.33	1.60	.38

Welding Electrodes: Specially formulated covered electrodes with the targeted weld metal chemistries as shown in Table 2 were prepared and supplied by Canadian Liquid Air Ltd. The chemistry is based on authors work which have shown that increased manganese and nickel levels confer good CGHAZ toughness even at high heat input levels and that increased strength levels coupled with excellent toughness can be realized by increased additions of nickel, manganese and copper [6, 7]. For the purpose of comparison, a commercial welding electrode (E10018) has also been used. Its typical chemical compositon is shown in Table 3.

Welding Procedure: Plates were welded at the research and development laboratories of MIL Davie Inc. Figure 1 shows the schematic of the weld joint. Manual SMAW process with a heat input of 1.3 KJ/mm was used. There was no preheating and no heating between the passes. All welds were subjected to radiography to ensure soundness.

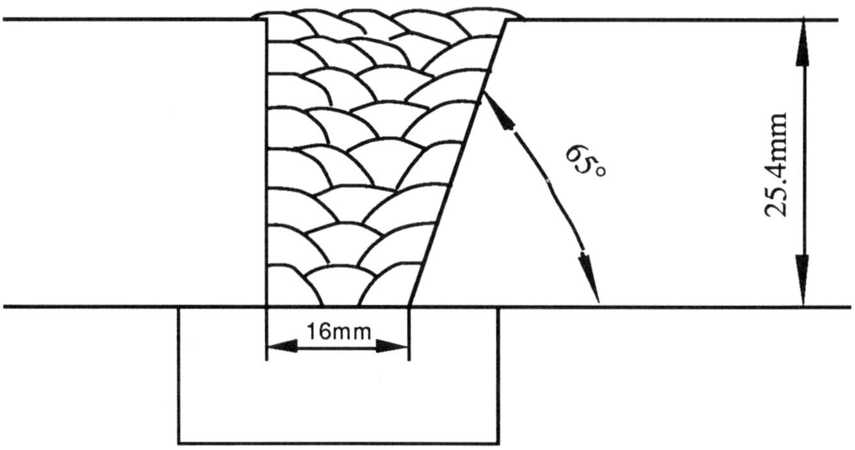

Figure 1 Scheme of weld joint.

Mechanical Testing

For assessing the toughness of weld metal, transverse Charpy V-notch specimens with their long axes perpendicular to the rolling direction and the notch through the center of weld metal have been used. The load-time and the energy-time traces from the instrumented impact test were recorded and analyzed via an ETI 830 data acquisition and analysis system. Testing has been conducted at -51 °C. For tensile testing, 12.7 mm (0.5 in.) dia round specimens with a gauge length of 50 mm (2 in.) were used. Their axis was parallel to the rolling direction. For hardness testing, a Vickers hardness testing machine with a 1 Kgf load was used.

Metallography

For optical microscopy, specimens were etched in 2% nital. For transmission electron microscopy, specimens have been prepared by an electrolytic jet polishing technique, and examined in a JEOL 1200-EX transmission electron microscope operating at 120 KV. Fracture surfaces have been examined in a JEOL JSM 840A scanning electron microscope operating at 25 KV. Oblique section metallography has been used to establish relations between microstructural features, crack initiation and propagation morphology, and fracture toughness.

Table 4 Actual Weld Metal Chemical Composition

No.	Element, wt %												
	C	S	P	Mn	Si	Cr	Ni	Mo	Al	V	Cu	Ti	Nb	B
E1	.065	.012	.016	1.69	.479	.236	3.24	.488	.0132	.011	.108	.007	.008	.0000
E2	.060	.013	.017	1.42	.317	.249	3.06	.529	.0144	.011	.513	.006	.008	.0001
COM	.052	.011	.015	.734	.202	.152	1.71	.388	.0099	.006	.046	.002	.003	.0004

RESULTS AND DISCUSSION

Microstructure

Weld metal: Experimental compositions exhibit essentially a fine grained bainite matrix (Fig. 2) whereas the commercial composition exhibits a coarse upper bainitic matrix (Fig. 3). This is due to the increased hardenability of the experimental compositions due to their increased nickel and molybdenum additions (see Table 4). Transmission electron microscopy revealed (see Fig.4) a highly dislocated matrix with laths and subgrains and, even in the copper containing composition, there was no evidence of any precipitation.

It is also worth observing that the actual weld chemistry of experimental compositions is quite close to that of the targets (Table 2 *vs* Table 4) attesting to the reliability and reproducibility of the covered electrode approach. The main difference between the two chemistries is the addition of around 0.5 percent copper in E2. The reason for this addition is the increased hardenability, corrosion resistance and mild solid solution strengthening provided by 0.5 percent copper [8]. Generally maximum copper content is restricted to 0.3 percent for most of the welding electrodes. However such restrictions, which probably arouse out of fear of hot shortness, are not necessary in the present case due to the presence of high levels of nickel.

Mechanical Properties of Welds

Hardness Profile: The hardness profiles across the weldment are shown in Fig. 5a and 5b. The hardness through the weld metal increases as the fusion line is approached. There is substantial increase in the hardness on crossing in to the HAZ region. For example, for weldment

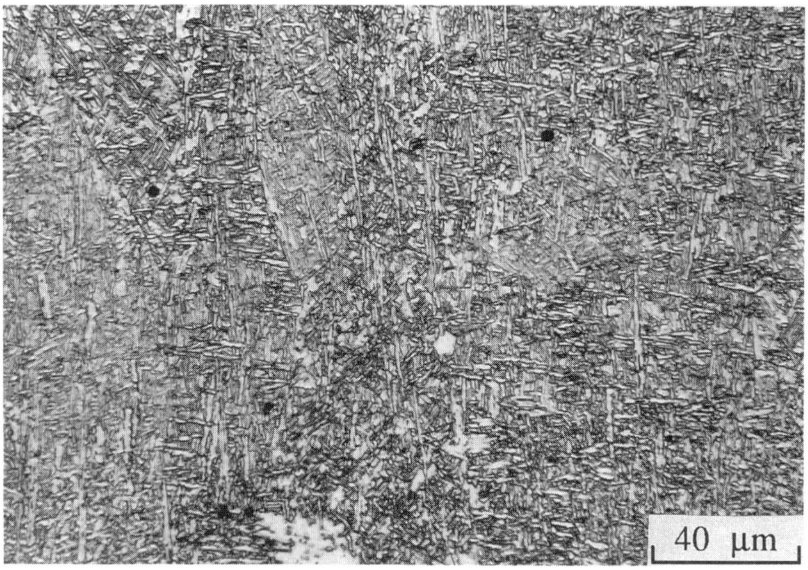

Figure 2 Optical microstructure of weld metal (Electrode E2).

El, the hardness increases from about 300 HV to 350 HV. The hardness then decreases gradually through the fine grained HAZ until it stabilizes around 240 HV as the base plate is reached. The value of peak hardness is slightly lower for the weldment E2. The average hardness of weld deposit is 300 HV for El and 270 HV for E2. This may be related to the slightly higher carbon and manganese level of El.

Tensile and Impact Properties: Table 5 shows the tensile property data and Table 6 shows the instrumented impact data. It is evident that, in the experimental compositions, strength levels in the neighborhood of 900 MPa are realized with decent toughness (around 55 J at - 51 °C). However in the commercial composition strength levels are quite low and ductility is also much reduced.

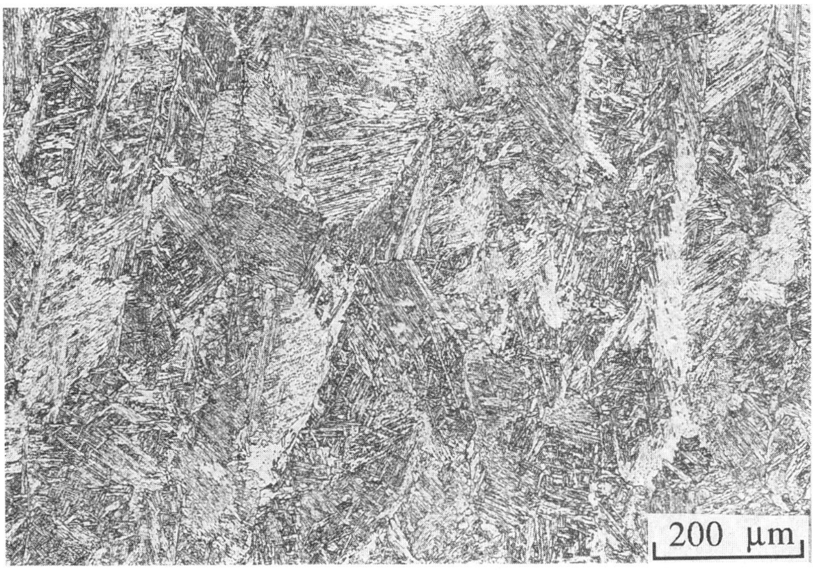

Figure 3 Optical microstructure of weld metal (commercial electrode).

Figure 4 TEM micrographs of weld metal (electrode E2).

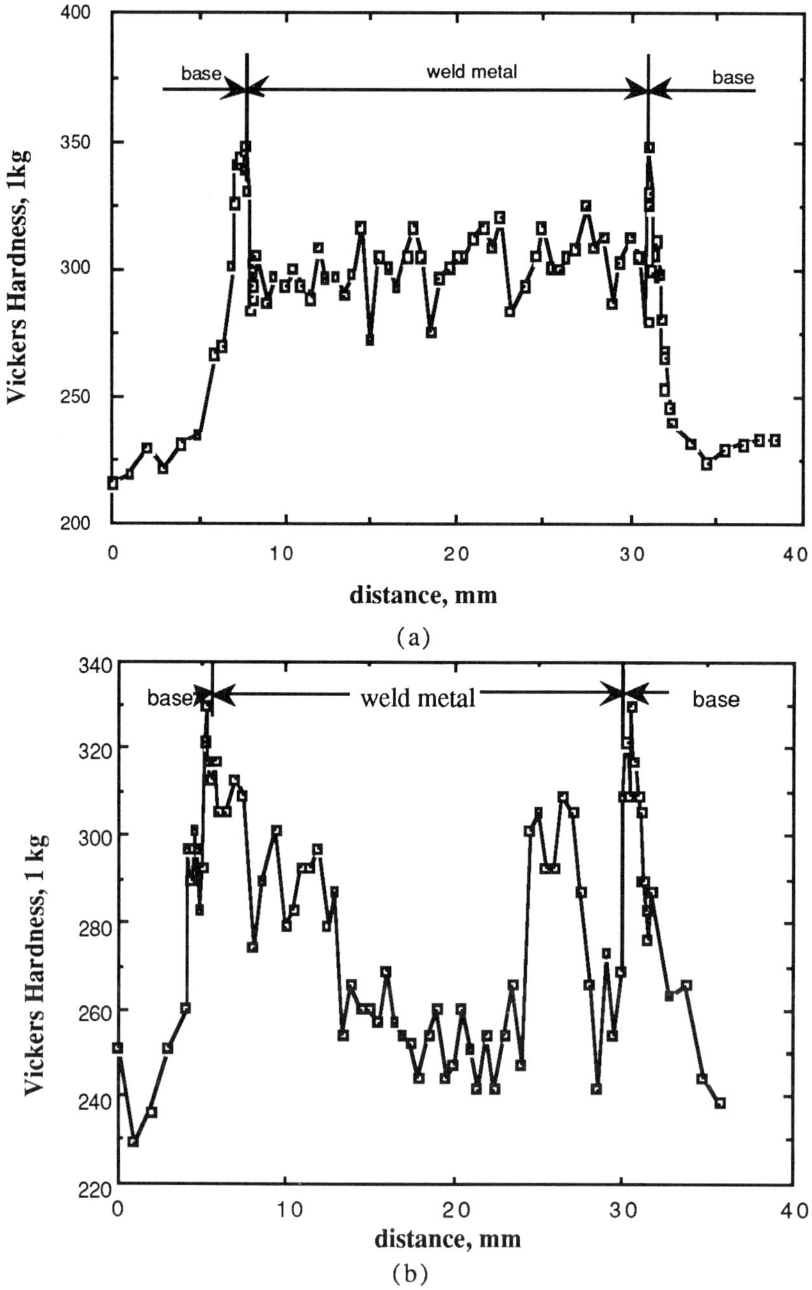

Figure 5 Microhardness traverse across weld deposit: (a) E1 and (b) E2.

Table 5 Tensile Properties of Weld

Electrode	Y.S. (MPa)	U.T.S. (MPa)	Elongation (%)	Reduction of Area (%)
E1	925	950	19	61
E2	870	900	18	61
Commercial	595	690	17	31

Table 6 Instrumented Charpy Impact Data at -51°C for Welds

Electrode	Et ft-lb (J)	Ei ft-lb (J)	Ep ft-lb (J)
E1	42 (57)	17 (23)	25 (34)
	40 (54)	15 (20)	25 (34)
E2	47 (63)	21 (28)	26 (35)
	37 (50)	12 (16)	25 (34)
Commercial	26 (35)	14 (19)	12 (16)
	19 (26)	8 (11)	11 (15)

Note: Et---total absorbed energy
Ei---initiation energy
Ep---propagation energy

Toughness-Microstructure-Fracture Relations

To relate the toughness variations to the microstructural characteristics, oblique section metallography was carried out. In the experimental compositions, microcracks are easily blunted (see arrows in Fig. 6) i.e. the matrix plastically deforms and accommodates the crack and prevents it from spreading and there is no linking up of the cracks. In contrast, in the commercial composition, micro cracks are quite sharp, long and link up easily (see arrows in Fig. 7).

Scanning electron fractography also confirmed the higher intrinsic toughness of the experimental compositions. The broken Charpy specimens exhibited a predominantly dimpled matrix interspersed with limited areas of quasi-cleavage (see Fig. 8). In the commercial composition, fracture surfaces were composed of large cleavage facets (see Fig.9) indicating very little resistance to crack propagation. These differences in crack propagation behaviour are also borne out by the marked differences in the crack propagation energies and the shape of

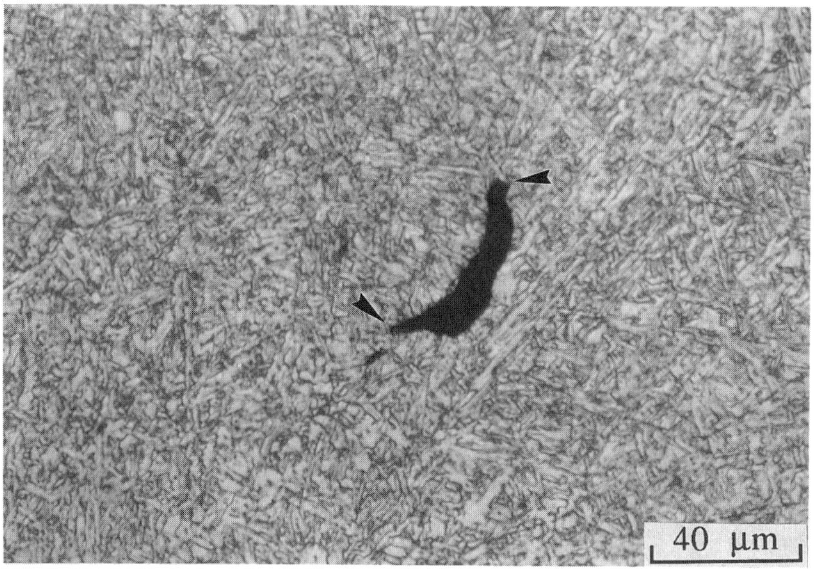

Figure 6 Oblique section metallographs from Charpy specimens of electrode E2 broken at -51°C.

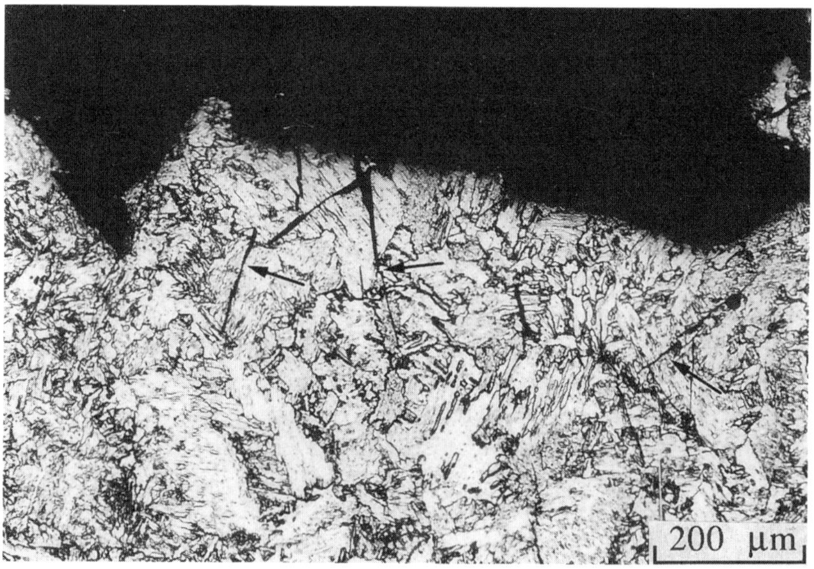

Figure 7 Oblique section metallograph from Charpy specimen of commercial electrode broken at -51°C.

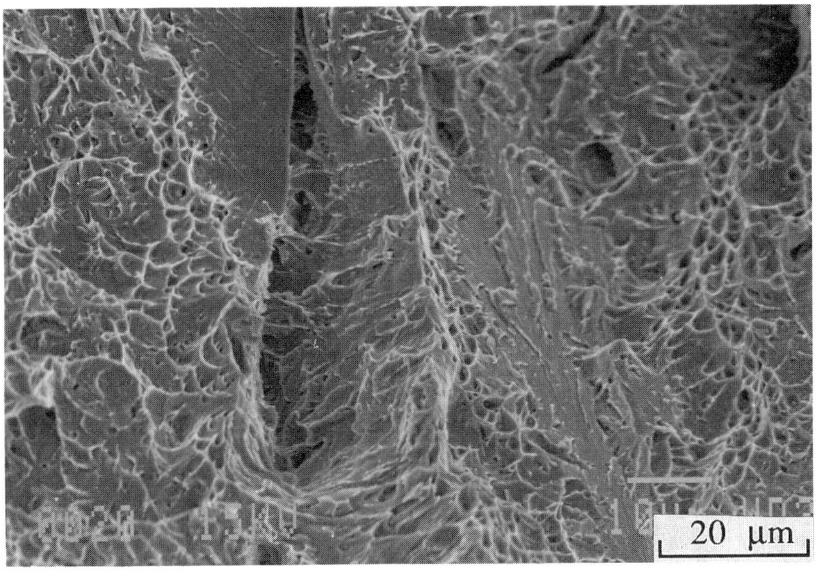

Figure 8 SEM fractograph of Charpy specimen of weld from experimental electrode, broken at -51°C.

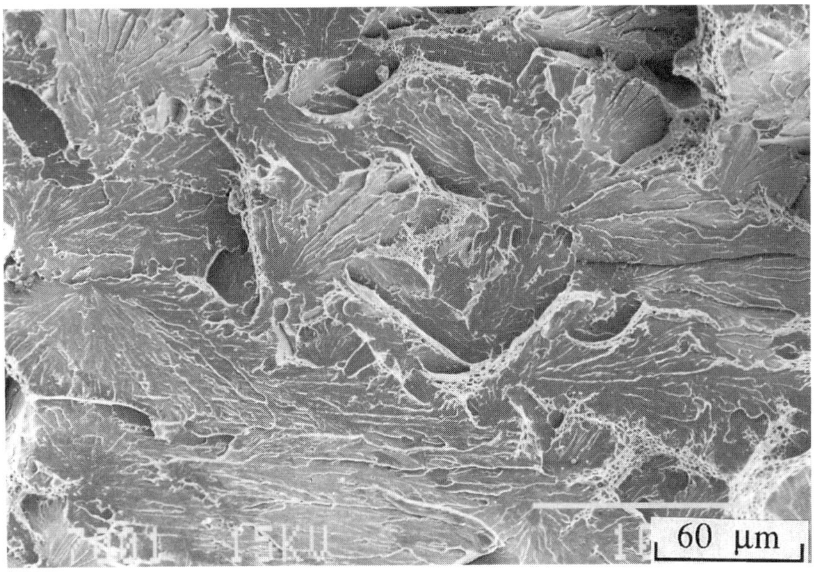

Figure 9 SEM fractograph of Charpy specimen of weld from commercial electrode, broken at -51°C.

Table 7 Calculated relative hardenabilities and transformation temperatures for weld metals

Electrode	CE(IIW)	Pcm	Ms(°C)	Bs(°C)
E1	0.737	0.255	405	484
E2	0.693	0.257	418	511
Commercial	0.539	0.198	446	576

note: Ms=561-474(C)-33(Mn)-17(Ni)-17(Cr)-21(Mo)
Bs=830-270(C)-90(Mn)-37(Ni)-70(Cr)-83(Mo)
CE(IIW)=C+Mn/6+(Cr+Mo+V)/5+(Ni+Cu)/15
Pcm+(Mn+Cu+Cr)/20+Ni/60+Mo/15+V/10+5B

the load-time traces in the instrumented impact testing (see Figures 10 and 11). In the experimental compositions, propagation energy in the neighborhood of 35 J (Fig. 10) whereas in the commercial composition, it is around 16 J (Fig. 11). Also, the load-time trace in the latter case shows that cleavage crack initiated propagates rapidly (as indicated by the sudden drop in the load immediately after the maximum load). In contrast, in the case of experimental compositions, there is considerable plastic deformation beyond the maximum load and the load decreases rather gradually.

A fine grained bainite offers greater resistance to crack propagation as prior austenite is divided in to finer units. In coarse bainite, once the crack is initiated, parallel laths of rather broad colonies offer little resistance. Both the experimental and commercial compositions have similar carbon levels. However the level of alloying elements used in the latter are not sufficient to provide a fine grained bainitic microstructure under the welding conditions used. It appears the around 3% Ni and 1.5 Mn are needed, in combination with normal levels of Cr and Mo, to provide the necessary hardenability. Calculation of the relative hardenabilities and transformation temperatures through relevant formulae [9, 10] indicate that experimental compositions have substantially lower Ms and Bs temperatures than the commercial composition (see Table 7). Although the experimental compositions have higher CE and Pcm values, it is worth while noting that they have better weldability and that they do not require preheating.

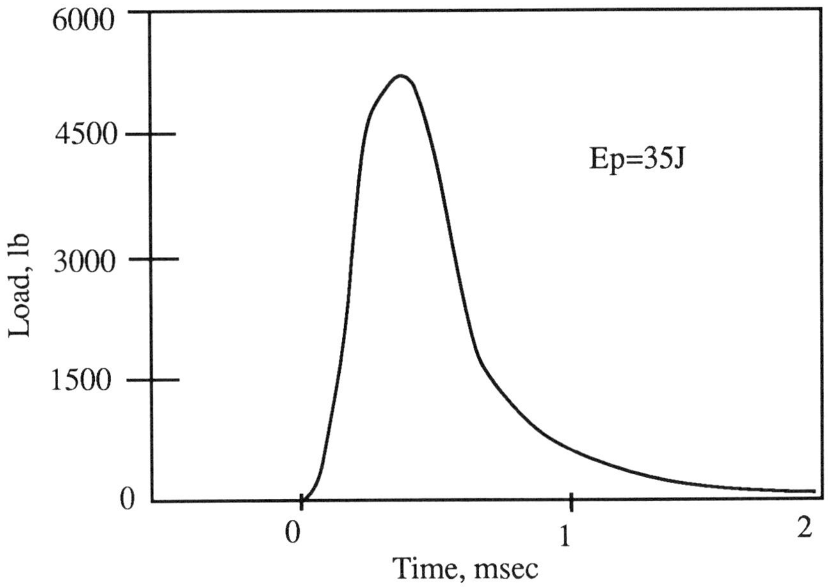

Figure 10 Instrumented impact load-time trace of Charpy specimen of experimental electrode (E2) broken at -51°C.

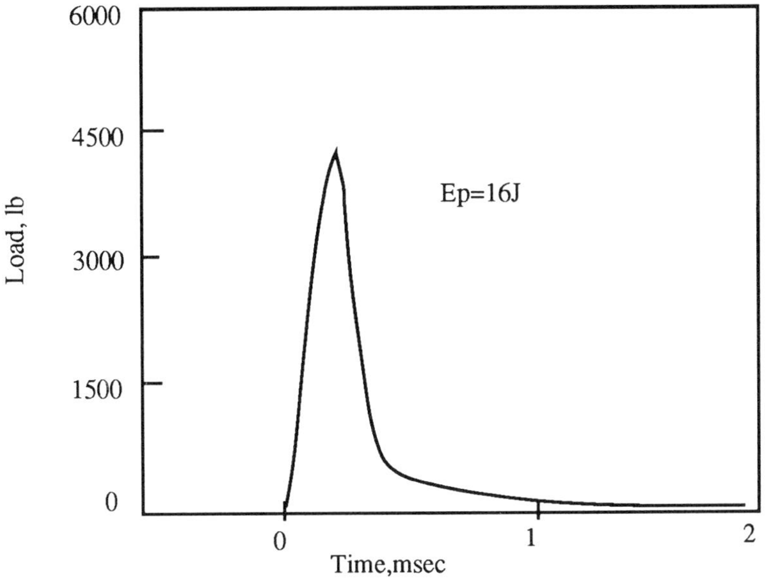

Figure 11 Instrumented impact load-time trace of Charpy specimen of commercial electrode broken at -51°C.

CONCLUSIONS

Following are the main conclusions from the study:

1. The higher nickel containing experimental electrodes provide higher strength levels and higher toughness levels than the commercial electrode.
2. The higher toughness realized with the experimental electrodes may be related to their finer weld microstructures with intrinsic ability to blunt cracking and with inherently superior resistance to cleavage.
3. The relatively lower toughness of the weld metal with commercial electrodes is due to its coarser upper bainitic matrix with inherently poor resistance to crack propagation.
4. The presence of 0.5 percent copper in the weld metal has no apparent deleterious effect on the toughness. Hence, at least for the case of HSLA-100 steels, the often imposed restriction of maximum of 0.3 percent copper for the electrodes, may be out dated

ACKNOWLEDGMENTS

Authors are grateful to NSERC (National Sciences and Engineering Research Council) of Canada, MIL Davie and Canadian Liquid Air for the financial provided. They also express their sincere thanks to Mr. Emil Hamburg, Lukens Steel corporation, for providing the base plates used in the study.

REFERENCES

1. T. W. Montemarano, B.P. Sack, J.P. Gudas, M.G. Vassilaros and H.H. Vandervelt, “High Strength Low Alloy Steels in Naval Construction”, **Journal of Ship Production**, Vol. 2, p.45, 1986.
2. L.G. Kvidahl, “An Improved High Yield Strength Steel for Ship Building”, Welding Journal, p.42, July 1985.
3. M.R. Krishnadev, S. Dionne, J.T.Bowker, L.E. Collins, V.K. Vasaudevan and K.D. Challenger, “Comparative Metallography of 690 MPa Grade High-Strength Steels” Microstructural Science, Vol. 18, p. 281, 1989.
4. M.R. Krishnadev, J.T. Bowker, J.T. McGrath, V.K. Vasudevan and K.D.Challenger, “Influence of Variability in Base Alloy Chemistry and Processing on HAZ Toughness and Microstructure in Advanced

Copper-HSLA Steels" in "**Recent Trends in Welding Science and Technology TWR'89**", ASM, p.799, 1990.
5. M.R. Krishnadev, J.T. McGrath, J.T. Bowker and S.Dionne, "Influence of Heat treatment on the HAZ Microstructure and Toughness in Precipitation Strengthened HSLA Steels", in **Welding Metallurgy of Structural Steels**, ed. J.Y. Koo, AIME, New York, p.137, 1987.
6. M.R. Krishnadev, W.L.Zhang and J.T. Bowker, "Copper Precipitation Strengthened HSLA Steels: Designing for Base Plate and HAZ Toughness", in **International Symposium on Low Carbon Steel for the 90's**, ed. R.Asafahani and G. Tither, The Mineral, Metals & Materials Society, Warrendale, Pa, p. 501,1993.
7. M.R.Krishnadev, W.L. Zhang and J.T. Bowker, "Influence of Alloying and Processing on the HAZ and Base Plate Properties of Experimental HSLA-80 Steels", in **International Trends in Welding Science and Technology**, ASM, p.599, 1992.
8. I. LeMay and M.R. Krishnadev, "Fundamental Metallurgical Considerations", **Copper in Iron and Steel**, ed. I. LeMay and Mcd. Schetky, John Wiley, New York, p.5, 1987.
9. N. Yourika, "Weldability of Modern High Strength Steels", in **Proceedings of the First United States-Japan Symposium on Advances in Welding Metallurgy**, AWS/JWS, p.51, 1990.
10. W.Steven, A.G. Haynes, "The Temperature of Formation of Martensite And Bainite In Low-Alloy Steel", **JISI**, Vol. 183, p. 349, 1956.

DEGRADATION OF STRESS RUPTURE PROPERTIES OF TYPE 304H STAINLESS STEEL AFTER PROLONGED EXPOSURE IN SERVICE

Susil K. Putatunda[1], Abhijit Sengupta,[2] and Jack Schaefer[3]

ABSTRACT

Type 304H stainless steel is used extensively in many high-temperature applications (e.g., superheater tubes in steam generators) which require long-term high-temperature service. The relationship between microstructure and mechanical properties has been investigated for type 304H stainless steel tube which had been exposed to a service temperature range of 565°-595°C for a period in excess of twenty years.

Detailed metallographic, X-ray, and transmission electron microscopic analysis were carried out on exposed material to determine the nature and type of precipitates formed in this material after prolong exposure in service.

----

1. Associate Professor, Department of Materials Science and Engineering, Wayne State University, Detroit, Michigan 48202, USA
2. Research Associate, Department of Materials Science and Engineering, Wayne State University, Detroit, Michigan 48202, USA
3. Metallurgical Research Engineer, Detroit Edison, Detroit, Michigan 48226, USA

The crack growth mechanism in creep in this exposed material was also determined.

INTRODUCTION

Type 304H stainless steel is used extensively in many high-temperature applications in the utility industries, e.g., superheater tubes in steam generators where it can undergo extensive creep. The creep properties of a material depend on grain boundary diffusion [1] and hence on the metallurgical microstructure. Prolonged exposure in service at elevated temperature can cause precipitation of carbides [2-4]. This carbide precipitation can influence the creep behavior, i.e., cavity growth [5-6], steady state creep rate and rupture time, and stress rupture properties of the material [7-9]. Investigation of the degradation of type 304H stainless steel after long term exposure in service will be useful for predicting the remaining life of such structural components.

The objective of the present investigation was to examine and evaluate changes in the metallurgical microstructure and stress rupture properties of type 304H stainless steel after prolonged exposure in service. Detailed metallographic, X-ray and transmission electron microscopic analysis was carried out on exposed material to determine the nature and the type of the precipitates formed after long term service exposure. The crack growth mechanism in creep in the exposed material has been documented through the use of fractographic analysis of the fracture surface.

EXPERIMENTAL PROCEDURE

The chemical composition of type 304H stainless steel is given in Table 1. This material was taken from a superheater tube which had been in use in Detroit Edison's Monroe power plant. The superheater tube had been exposed to 593° C (1100° F) for over twenty years. The metallographic techniques employed in this study were optical and electron microscopy and identification of phase by X-ray and electron diffraction analysis. The specimens were mechanically polished and etched with 2 percent nital. The microstructures of the exposed material is shown in Figure 1.

Table 1

Chemical Composition of the Material in wt. percent

C - 0.05	Si - 0.62
Mn - 0.50	P - 0.023
S - 0.0003	Ni - 8.50
Cr - 18.30	Cu - 0.14
Mo - 0.20	Fe - Balance

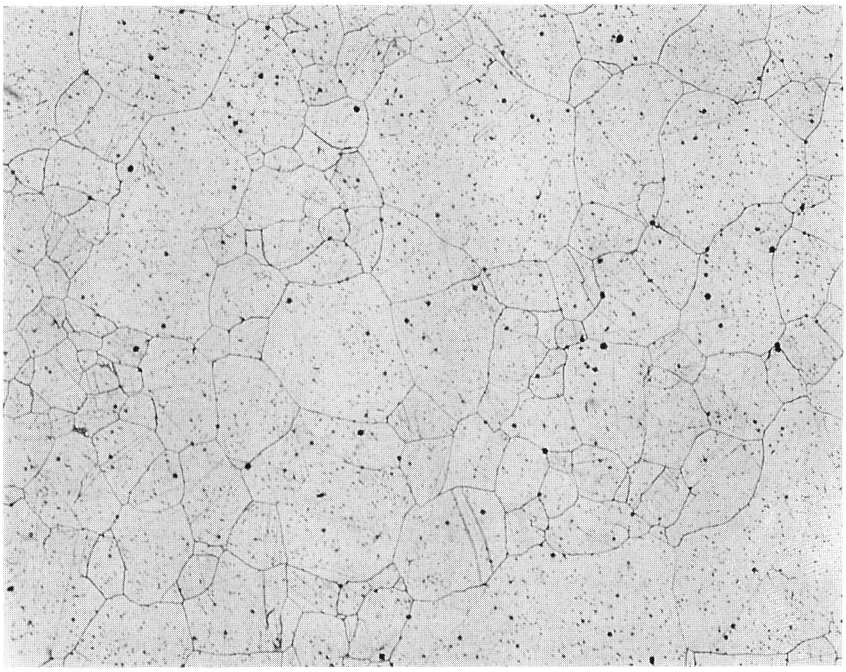

Figure 1 : Microstructure of the exposed material at a magnification of 50X

Phase extracts were prepared by anodic dissolution of the matrix in a 10 percent HCl-methanol electrolyte. One percent tartaric acid was added to avoid the precipitation of undesirable oxides. The current density for extraction was 10.1 Amp/cm^2. The extracts were separated from the electrolyte using a ceramic filter. The extracted precipitates were analyzed using an X-ray diffractometer, with a start angle of 10°, and stop angle of 150°. Voltage and current were 30 KeV and 30 mA. The Bragg angle θ was calculated from the peaks of intensity I versus 2θ plot. Interplanar spacing, d, was calculated and compared with the tables (in References 11 and 12) for various types of precipitates.

Tensile testing was performed at room temperature on three specimens from the exposed material. The average values from the tensile tests are reported in Table 2. Stress rupture specimens were prepared from the exposed material as per ASTM standard E-139 [13].

Table 2

Mechanical Properties of the Material

Material Condition	Yield Strength (KSi)	Ultimate Tensile Strength (Ksi)	%elongation	Hardness (HRB)
Unexposed	45	85	50	70
Exposed	60	90	12.5	98

These specimens were then tested to fracture at several stress levels and different temperatures. The details of these test results are reported in Table 3.

Fractographic samples were cut from these creep tested specimens. These were then examined under a 30 KeV scanning electron microscope.

The transmission electron microscopic studies were made with a 300 keV instrument. Specimens were cut parallel to the fracture plane using a diamond wafer blade to produce a slice approximately 0.5 mm thick. These specimens were then polished to a thickness of approximately 300 microns. The 300 micron thick specimens were then polished with emery papers, starting at 400 grit progressing to a final 2400 grit emery paper. A 3 mm disc was then produced by using

Table 3

Stress Rupture Properties of the Exposed Material

Stress (Ksi)	Temperature (°F)	Expected Time of Failure* (hours)	Actual Time of Failure (hours)
10	1344	1000	69.1
15	1270	100	63.5
20	1190	100	76.7
40	1040	100	92.5

*As per Larson-Miller plot of the unexposed material.

a specimen puncher. This disc was then thinned down to 100 microns by electropolishing with 6 percent perchloric acid, 35% butylcellusolve, and 59 percent methyl alcohol at a voltage of 35 volts and at a temperature maintained at 0°C. At least two specimens were examined to determine the dislocation substructure and the type of precipitates by the selected area diffraction technique.

RESULTS AND DISCUSSION

The microstructure of the service exposed material is shown in Figure 1. The microstructure clearly shows extensive precipitation of carbides, both within the grains and along the grain boundaries.

The stress rupture data of the exposed material is reported in Figure 2 in terms of the Larson-Miller plot. In Figure 3, the Larson-Miller plots of exposed and unexposed material (the unexposed material is the one which has not been used in service. The Larson-Miller plot of this material was obtained from ASTM data [14]) are compared. The steel exposed to service possessed lower rupture strength, especially in the high temperature range. However, at lower temperatures, the life has not been significantly affected. These test results indicate the degradation of the high temperature creep properties of the material.

The results of the tensile tests (Table 2) demonstrate that both

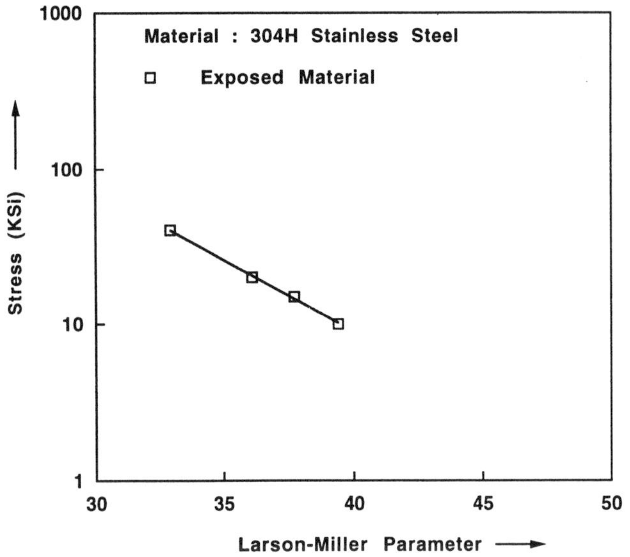

Figure 2. Larson-Miller plot for the exposed material

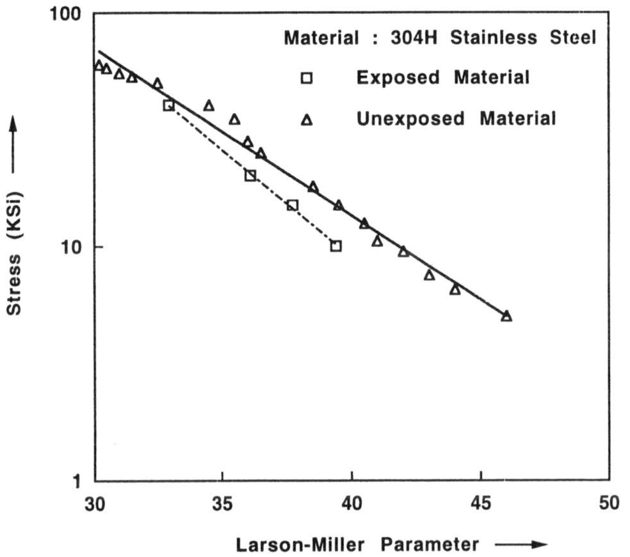

Figure 3. Larson-Miller plot for the exposed and unexposed material

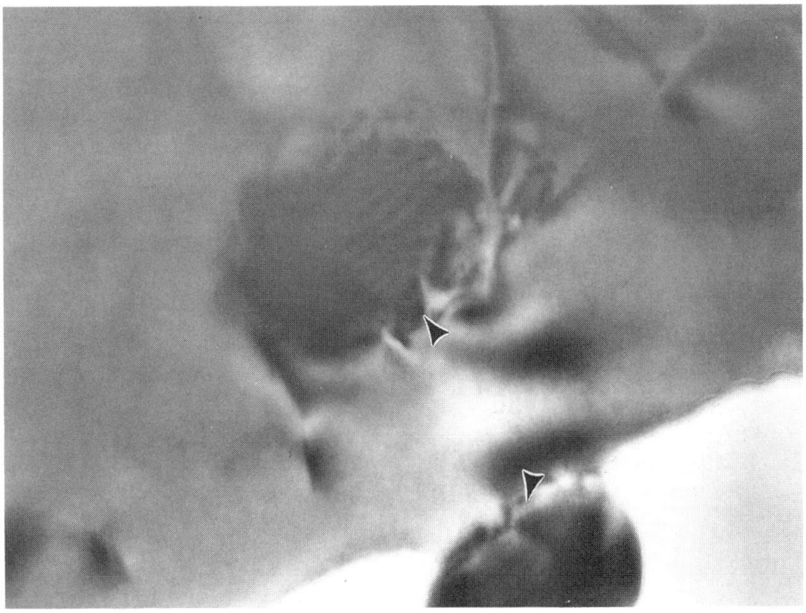

Figure 4. TEM picture of precipitates (Arrow indicates precipitates)
Magnification 26000X

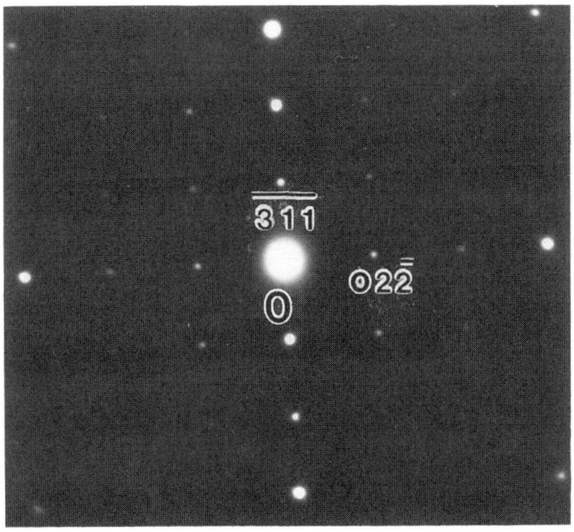

Figure 5. Selected Area Diffraction Pattern of the Precipitate shown above with zone axis [233]. O is the transmitted beam

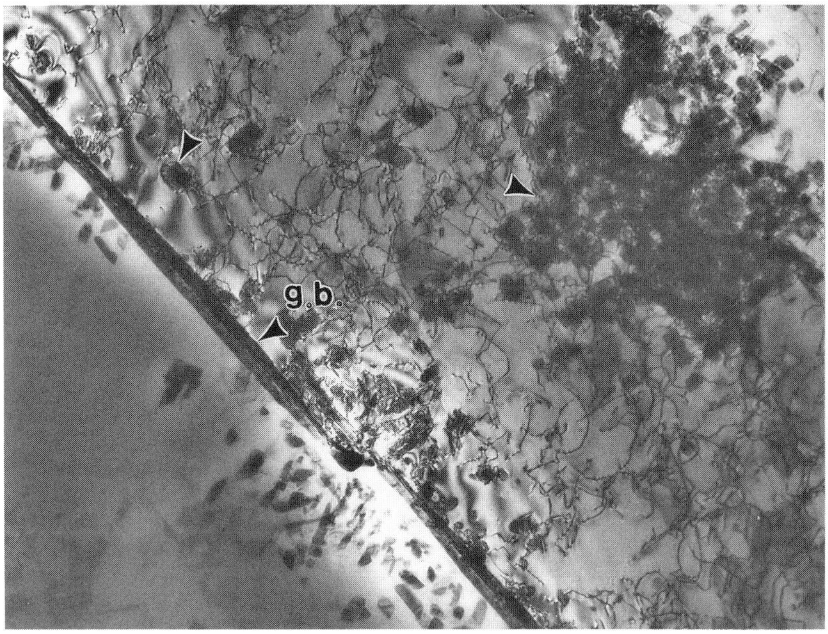

Figure 6. TEM picture showing dislocation-precipitate interaction (Arrow indicates grain boundary, precipitates and massive precipitation). Magnification 19400X

the yield and the ultimate tensile strength of the exposed material are higher than in the unexposed material. The ductility of the exposed material was drastically reduced.

Analysis of the precipitates indicates blocky $M_{23}C_6$ typeprecipitates in the exposed material. The $M_{23}C_6$ tends to appear in grain boundaries as isolated particles or continuous platelets [11-12]. The carbide phase also occasionally precipitates on dislocations and stacking faults. Figures 4, 5 and 6 show transmission electron microscopy of the precipitates and dislocation-precipitate interaction. An EDX analysis of the precipitates has shown that the iron and chromium level was much higher at the precipitates compared to the matrix. The nickel content of the precipitate was rather low. Selected area diffraction pattern (SADP) confirmed the presence of reflections from $Cr_{23}C_6$-type precipitates. X-ray studies also confirmed the presence of $M_{23}C_6$-type precipitate. Part of the intensity I versus 2θ plot is shown in Figure 7, and calculations are shown in Table 8. Though EDX analysis showed presence of iron in the precipitates, the

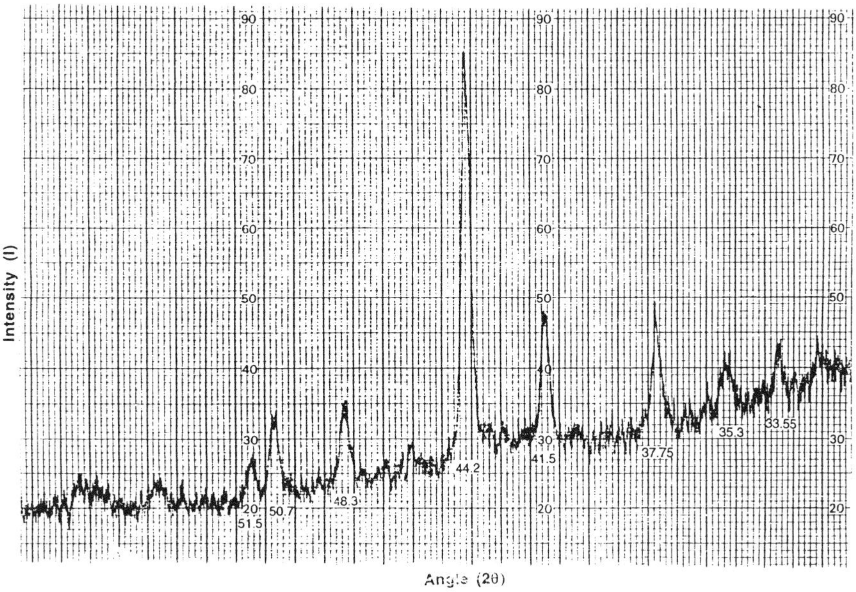

Figure 7. Part of the X-ray Spectrum

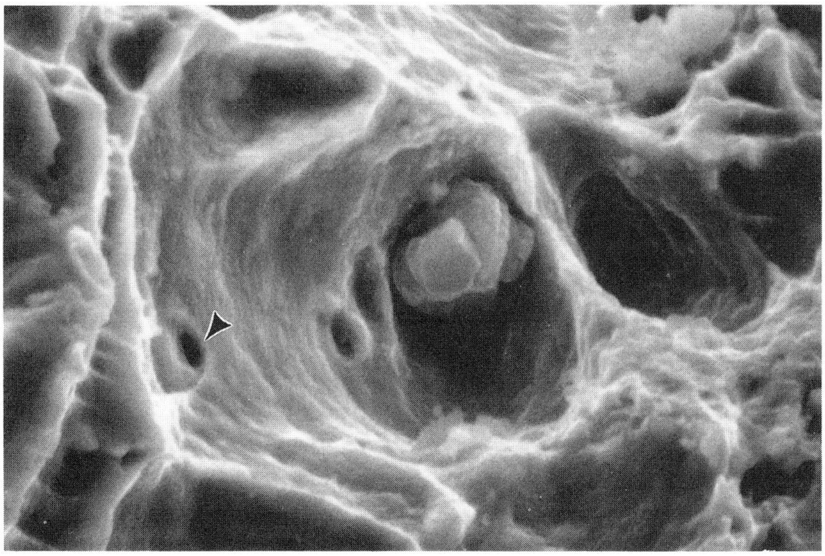

Figure 8. Fractograph of the Exposed Material at a magnification of 2720X (Arrow indicates voids)

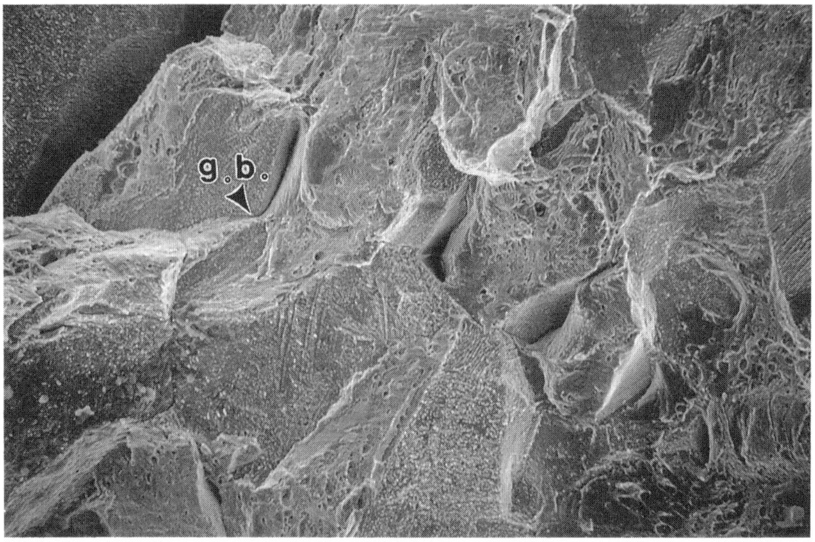

Figure 9. Fractograph of the Exposed Material at a magnification of 170X (Arrow indicates grain boundary)

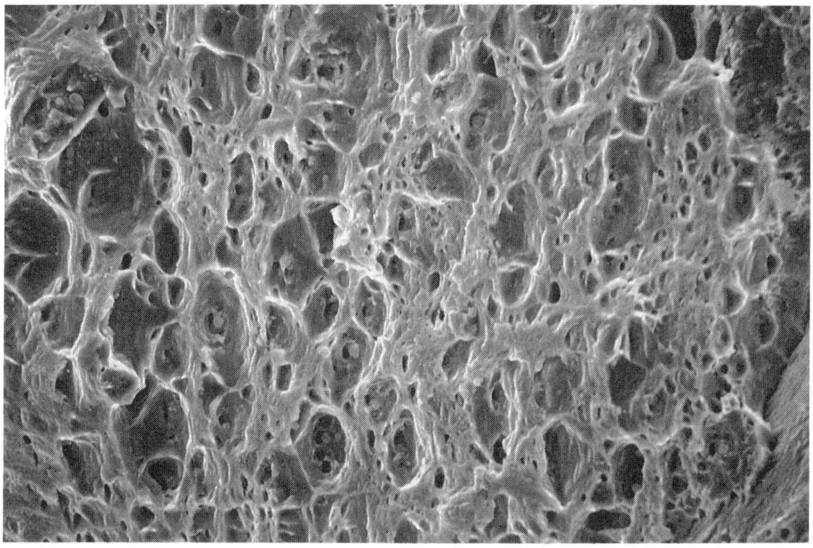

Figure 10. Fractograph of the Exposed Material at a magnification of 680X

Table 4.

X-ray Diffraction Analysis of Extracted Precipitate using Copper Kα Radiation

2θ	θ	Sinθ	2Sinθ	Bragg Spacing d (calculated)	Bragg Spacing d (from Ref. 10)	Type of lines
33.55	16.77	0.288	0.577	2.671006	2.66	Medium
35.40	17.70	0.304	0.608	2.535547	2.44	Weak
37.75	18.87	0.323	0.647	2.382934	2.38	Strong
41.50	20.75	0.354	0.708	2.175866	2.17	Strong
44.20	22.10	0.376	0.752	2.049017	2.07	V. Strong
48.30	24.15	0.409	0.818	1.884232	1.883	Medium
50.70	25.35	0.428	0.856	1.800528	1.800	Medium
51.50	25.75	0.434	0.868	1.774424	1.775	Weak
75.70	37.85	0.613	1.227	1.256347	1.292	Medium
77.65	38.82	0.626	1.253	1.229600	1.255	Weak
90.40	45.2	0.709	1.419	1.086417	1.065	Weak

* lattice parameter a = 10.65

SADP did not show any lines from iron.

The TEM micrographs show that dislocation-tangles acted as the preferential sites for nucleation of $M_{23}C_6$ precipitates within grains. The driving force for such massive precipitation has been proposed to be carbon supersaturation [15] and enhanced diffusion of chromium and carbon rejection from the austenitic matrix[16], with the increased vacancies concentrated during exposure at high temperature. High temperature causes the equilibrium concentration of vacancies to increase.

Fractographs of the exposed material are shown in Figures 8, 9 and 10. Fractograph in Figure 8 shows microvoid containing particles. Furthermore, the fractograph in Figure 9 shows the intergranular nature of crack growth in the material during stress rupture tests. These intergranular cracks form by the process of the grain boundary sliding. Fractograph in Figure 10 confirms the formation of microvoids or dimples preferentially along the carbides in the material.

CONCLUSIONS

1) After long term exposure in service in type 304H stainless steel, blocky $M_{23}C_6$ type precipitates were observed. These precipitates were mainly chromium carbides. Carbide coarsening was clearly visible in this steel.

2) This precipitation of carbides has caused reduction in stress-rupture life of the material, specifically at high temperature level.

3) The carbide precipitation was found both within the grains as well as along the grain boundaries in this material.

4) Prolong exposure in service has caused a significant reduction in ductility of the material.

5) The crack growth process appears to be intergranular in nature in exposed type 304H stainless steel.

ACKNOWLEDGEMENTS

The authors are thankful to Mr. Mike Balogh of General Motors Technical Center, Warren, Michigan for his help with transmission electron microscopy.

REFERENCES

1) G.E. Dieter, "**Mechanical Metallurgy**", 3rd ed., McGraw Hill, 1986, p. 395

2) B.Weiss and R.Stickler, "Phase Instabilities During High Temperature Exposure of 316 Austenitic Steel", **Metallurgical Transactions A**, vol.3, p. 851, 1972

3) A.A. Tavasoli, A. Bisson and R. Soulat, "Ferrite decomposition in austenitic stainless steel weld metals, **Metal Science**, vol. 18, p. 345, 1984

4) V. A. Biss and T. Wada, **Metallurgical Transactions A**, vol. 16, p. 109, 1985

5) R.A. Varin and J. Haftek, "Structural Changes in a Ferrite Heat-resistant Steel After Long-term Service", **Materials Science and Engineering**, vol. 62, p. 129, 1984

6) R.G. Baker and J. Nutting, "The tempering of 2.25Cr-1Mo Steel after quenching and normalizing", **Iron and Steel Institute Journal**, vol. 192, p. 257, 1959

7) T. Wada and V.A. Biss, "Restoration of Elevated Temperature Tensile Strength in 2.25Cr-1Mo Steel", **Metallurgical Transactions A**, vol. 16, p. 845, 1983

8) N.S. Cheruvu, "Degradation of Mechanical Properties of Cr-Mo-V and 2.25Cr-1Mo Steel Components After Long-term Service at Elevated Temperature, **Metallurgical Transactions A**, vol. 20, p. 87, 1989

9) W.B. Jones and J.A. Van Den Avyle : "Substructure and strengthening mechanisms in 2.25Cr-1Mo Steel at elevated temperature", **Metallugical Transactions A**, vol. 11, p. 1275, 1980

10) ASTM A213, "Standard specification for seamless ferritic and austenitic alloy steel boiler, superheater and heat exchanger tubes", **Annual book of ASTM stanards**, vol. 0.1.01, p. 109, 1994

11) M.J. Donachie, Jr. and O.H. Kriege, "Phase Extraction and Analysis in Superalloys-Summary of Investigations by ASTM Committee E-4 Task Group 1", **Journal of Materials**, vol. 7, no.3, p.269, 1972

12) M.J. Donachie, Jr., "Phase Extraction and Analysis in Superalloys-Second Summary of Investigations by ASTM Subcommittee E04.91"**Journal of Testing and Evaluation**, vol. 6, no. 3, p. 189, 1978

13) ASTM E-139, Standard practice for conducting creep, creep-rupture and stress-rupture tests of metallic materials", **Annual book of ASTM stanards**, vol. 03.01, p. 326, 1993

14) Metals Handbook, 8th Edition, American Society for Materials, ASM, vol. 1, p. 392, 1964

15) K.B.S. Rao, H. Schuster and G.R. Halford, "On Massive Carbide Precipitation during High Temperature Low Cycle Fatigue in Alloy 800H", **Scripta Metallurgica et Materialia**, vol. 31, no. 4, p. 381, 1994

16) M. Ahmad, K.A. Shoaib, M.A. Shaikh and J.A. Akhtar, " Identification of Surface Carbides and Spinels in Welded Austenitic Stainless Steels", **Journal of Materials Science**, vol. 29, p. 1169, 1994

TWO PHASE-FLOW INDUCED CORROSION IN NATURAL GAS WELLS

K. Habib*

ABSTRACT

A corrosion study of the influence of a mist flow (two phases) on materials commonly used in natural gas wells was conducted. The study was carried out in a way to simulate conditions normally found in natural gas wells in which a mist flow of natural gas and hydrocarbons are present. In other words, a new scheme of flowing rig has been built for such study. Different experimental parameters such as temperature, gas velocity, fluid flux rate, and pressure of the gas were controlled and regularly monitored during experiments. Furthermore, the corrosion rate of various materials was determined by the weight loss method. The results show that the L80 steel has higher corrosion rate than the 38Mn6 "J66" steel at pressures of 5 and 10 bar. In contrary, the AF22-100 steel and MW2832 steel show no sign of corrosion at all at both pressures of 5 and 10 bar in similar experimental conditions of the L80 steel and 38Mn6 "J66" steel.

----

*Materials Application Department, Kuwait Institute for Scientific Research. P.O. Box 24885 Safat, 13109 Kuwait.
This study conducted while the author was a post-doctoral fellow with Institut fur Technische Chemie und Petrolchemie, der RWTH Aachen 5100 Aachen, Germany.

INTRODUCTION

There is few conducted studies have been reported on evaluating the corrosion behavior of materials in extremely flow conditions , simulating environments normally found in natural gas wells in laboratories (1&2).Indeed it is a challenge to investigate the corrosion behavior of materials in dangerous environments especially in the present of toxic gases , i.e, H_2S, high flux rate, high temperature, and high pressure . However, the benefits of undertaking the challenge are essencial for many economical reasons in oil and gas industries.

The objective of the present investigation was to evaluate the corrosion behavior of the L80, 38Mn6 "J66", AF22-100, MW2832 steels under a mist-flow condition, simulating conditions normally exist in natural gas wells.

EXPERIMENTAL WORK

Corrosion specimens were fabricated from the L80, 38Mn6 "J66", AF22-100, MW2832 steels in a form of a semicircular tube with specifications of 60.3mm (diameter of the tube), 3.94mm (thickness of the tube), and 4omm (the width of the tube). The chemical compositions of the L80 , 38Mn6 "J66", AF22-100, MW2832 steels are respectively 0.28% C, 0.27% Si, 1.37% Mn, 0.009% P, 0.01% S, 0.2% Cr, and balanced Fe, 0.36% C, 0.21% Si, 1.30% Mn, 0.017% P, 0.013% S, 0.0074% Al, 0.03% Cu, 0.07% Cr, 0.05% Ni, 0.01% Mo, 0.002% As, and balanced Fe, 0.02% C, 0.44% Si, 1.66% Mn, 0.025% P, 0.0013% S, 0.006% Al, 0.11% Cu, 22.31% Cr, 5.49% Ni, 3.08% Mo, 0.07% V, 0.01% Ti, 0.14% N, and balanced Fe, 0.02% C, 0.27% Si, 1.58% Mn, 0.01% P, 0.004% S, 0.027% Al, 1.22% Cu, 26.50% Cr, 31.29% Ni, 3.50% Mo, 0.04% V, 0.01% Ti, 0.13% N, and balanced Fe. In addition, the yield strength of the L80, 38Mn6 "J66", AF22-100, MW2832 steels are respectively 435 N/mm2, 755 N/mm^2, 889.7 N/mm^2, and 841.4 N/mm^2.

Prior to corrosion tests, the specimens were cleaned in an ultrasonic cleaner in a petrolether solution for 5 minutes. Then, the specimens were ultrasonically cleaned in aceton for 10 minutes in order to remove any grease on the samples. After that the samples were immersed in water. Therafter, the samples were chemically etched in a10% HCl solution for 5 minutes. Then they immersed again in water. After that the samples were neutralized ultrasonically in a saturated solution of $Ca(OH)_2$. Then they immersed again in water. Eventually, the specimens were dried in methanol before mounting them in the flowing rig for testing.

In order to simulate the condition of the natural gas well, a chemical solution made of 90% water and 10% hydrocarbon was prepared. In this study, 27L of water were added to 1579.6 gm of NaCl and 396.9 gm of $CaCl_2$-$2H_2O$. In addition, the hydrocarbon made of 90% iso-octane (2430mL) and 10% xylene (270mL). All together, the solutions were prepared in an oxygen free environment.

Results of two different tests are reported in this article. The operational temperature of both tests was 100 C. On the other hand, the pressure of the CO_2 in the first test was P=5 bar, the gas velosity was Vg = 18.4m/sec., the flux rate of the fluid was V_l = 0.5L/min., and the calculated wall shear was Tw =10 N/m^2. However, in the second test, the pressure of the CO_2 was P = 10 bar, the gas velosity was Vg= 27.6m/sec., the flux rate of the fluid was V_l = 0.5L/min., and the calcutated wall shear was Tw=40 N/m^2. It is worth noting that both tests were terminated after 96 hours of testing. After the termination of the tests the corrosion rate was determined for all samples by the weight loss method (3). Detail information on the arrangement of the set up of the flowing rig are given elsewhere (2).

RESULTS AND DISCUSSION

Data of corrosion rate of the four different materials are given in Table 1. The Table shows that the L80 Steel has corroded the most among the investigated materials. This ocurred due to the low content of corrosion resistance-alloying elements in this steel. Also, the yield strength of the steel has the lowest value among the rest of the steel. In fact the corrosion rate was found slighltly independent on the pressure of the CO_2 gas. In other words, the corrosion rate of L80 steel was observed to decrease with increasing the pressure of the CO_2 gas. The opposite behavior was found in the case of the 38Mn6 "J66" steel in which the corrosion rate increased with increasing the pressure of the CO_2 gas. The tabulated results indcate that the 38Mn6 "J66" steel has better corrosion resistance than that of the L80 steel. This is probably resulted because of the addition of a small amount of Mo and Ni into the 38Mn6 "J66" steel which are further enhanced the strength of the steel compared to the L80 steel, see the yield strength of boh steels. In contrary, the AF22-100 and MW2832 steels were observed not susceptible to corrosion at all in the same testing environment. This observation is explained due to the present of a high content of elements,i.e., Cr and Ni, (3) which are known to improve the corrosion resistance of steels in aggressive environments .

Figure 1 shows a macrophotograph of the internal surface area of the four different samples after the first test. The photograph indicates clearly that both the AF22-100 and MW2832 steels have shiny surface area compared to the L80 and 38Mn6 "J66" steels. In other words, this result confirms the measurement of the corrosion rate by the weight loss method, which revealed that the L80 and 38Mn6 "J66" steels have suffered corrosion attack during the test relative to the AF22-100 and MW2832 steels which showed no susceptibility to corrosion. Furthermore, a magnified photograph of the the surface area of the L80 steel in figure 1 is shown in Figure 2. The surface area of the L80 steel appears to have a massive amount of pittings. The pits are oriented in a certain direction along the longtitudinal axis of the tube. The distribution of the pits is not uniform in the entire surface area. Some area are heavily pitted and some are not.

Table 1. Data of Corrosion Rate (mm/year)

	Materials			
	L80	38Mn6 "J66"	AF22-100	MW2832
Pressure (bar)				
5	2.62	0.54	_	_
10	2.14	1.47	_	_

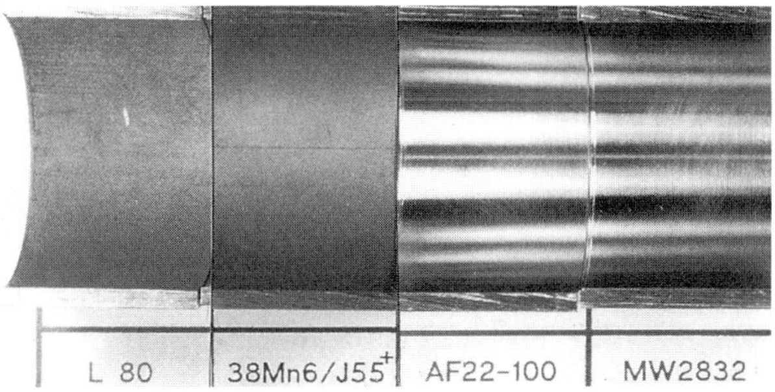

Figure 1. The enternal surface area of the four different samples after the first test, P = 5 bar, Vg=18.4 m/sec., V_l =0.5 L/min., and Tw = 10 N/m^2

Figure 2. A magnified photograph of the surface area of the L80 steel in Figure 1.

CONCLUSIONS

The following conclusions are drawn from the present study:

1- The L80 and 38Mn6 "J66" steels are found susceptible to corrosion in the mist flow condition , simulating environments normally exist in natural gas wells. In contrast, the AF22-100 and MW2832 steels are found to have an excellent corrosion resistance in the same conditions.

2- The corrosion rate of the 38Mn6 "J66" steel is found less than that of the L80 steel. This is due to the small addition of Mo and Ni into the 38Mn6 "J66" steel which enhanced the corrosion resistance of the alloy compared to the L80 steel.

3- The increase of the CO_2 pressure increased the corrosion rate of the 38Mn6 "J66" steel but not the L80 steel.

4 - The addition of a high content of corrosion resistance-alloying elements such as Cr and Ni into the AF22-100 and MW2832 steels has given these steels an excellent performnce in both conducted tests.

REFERENCES

1) G. Schmitt, W. Brockhoff, "Relevance of Laboratory Experiments for Investigation and Mitigation of Flow Induced Corrosion in Gas Prodution", CORROSION /88, Pape No.22 , NACE, Houston , TX (1988).

2) W. Bucken, Master Thesis, Technical University of Aachen, Aachen, Germany (1988).

3) H. Uhlig, Corrosion and Corrosion Control, John Wiley & Sons Inc., New York, PP.408, 292, 353 (1971).